MOLECULAR

ELECTRONICS

Commercial Insights, Chemistry, Devices,
Architecture and Programming

MOLECULAR

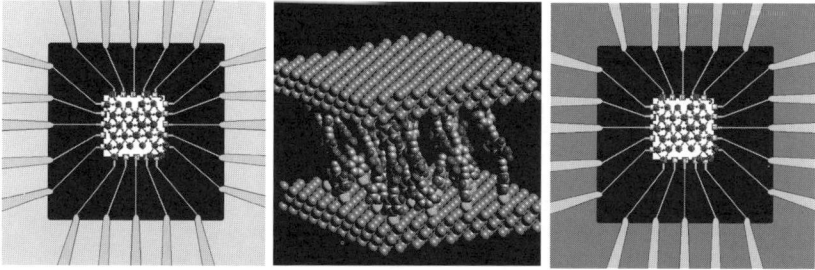

ELECTRONICS

Commercial Insights, Chemistry, Devices, Architecture and Programming

James M Tour

Rice University, USA

World Scientific
New Jersey • London • Singapore • Hong Kong

Published by

World Scientific Publishing Co. Pte. Ltd.

5 Toh Tuck Link, Singapore 596224

USA office: 27 Warren Street, Suite 401-402, Hackensack, NJ 07601

UK office: 57 Shelton Street, Covent Garden, London WC2H 9HE

British Library Cataloguing-in-Publication Data
A catalogue record for this book is available from the British Library.

First published 2003
Reprinted 2003, 2005

MOLECULAR ELECTRONICS: COMMERCIAL INSIGHTS, CHEMISTRY, DEVICES, ARCHITECTURE AND PROGRAMMING

ISBN 981-238-269-0
ISBN 981-238-341-7 (pbk)

Printed in Singapore by World Scientific Printers (S) Pte Ltd

110004531

Dedication

To Him Who demonstrated His love for me by giving

Himself for me…

Preface

I was asked by the editors to write a book on molecular electronics that focuses upon the work that has been conducted in my own research. A presentation of one's own work is far simpler to write than a compendium covering the work of the many superb researchers in the field. Although I will make mention of other's work in the context of the research that we have done, no science, at least not chemistry, is done in a perfect vacuum. I am deeply indebted to numerous colleagues around the world for their inspiring discussions and complements with work that far surpasses, in quantity and quality, that of my own. To them I say, thank you 1000 times.

The editors permitted me the liberty to insert some learned experiences regarding startup companies. Those experiences might prove insightful to the prospective founder, corporate officer, or potential investor in the molecular electronics business arena.

My entry into the field of molecular electronics occurred in 1988, at the start of my independent research career in synthetic organic chemistry at the University of South Carolina. I set out with my first graduate students, Jeffrey Schumm and Ruilian Wu, to synthesize the orthogonal switch as proposed by Ari Aviram, who was at IBM at the time. My students and I were blissfully unaware of the steep slopes of challenge that awaited us and the colorful comments that we would be obliged to entertain from concerned colleagues around the world. Their disgruntlement was fundamentally with the basic premise upon which molecular electronics is built, namely, molecules having to perform switching in an addressable array. In 1992 we teamed with Professor Mark Reed of Yale University. Mark's device testing ability, his "can-do" research spirit, his rich background in solid-state physics from his earlier corporate days at Texas Instruments, and his ability to address the sharpest attacks of the critics, became the sustaining grace of our molecular electronics effort. Though unbelievers remained for a time, and some even remain until today, the number of believers and true practitioners of molecular electronics has risen sharply.

With generous federal support, primarily from the Defense Advanced Research Projects Agency (DARPA), the Office of Naval Research (ONR), the Army Research Office (ARO), the National Science Foundation (NSF), the US Dept. of Commerce, National Institute of Standards and Testing (NIST), the

National Aeronautics and Space Administration (NASA), and the early risk-taking willingness by the program directors of those organizations (especially Jane Alexander, Bill Warren, Bruce Gnade, Christie Marrion, and Kwan Kwok of DARPA and John Pazik of ONR), the initial Reed/Tour research team could expand to add numerous complementary scientists who furthered the accomplishments. The team now includes David Allara (surface science) and Paul Weiss (probe microscopy) at Penn State University, Paul Franzon (device engineering) at North Carolina State University, Jorge Seminario (chemical theory) at the University of South Carolina, Patrick Lincoln (computer science) at the Stanford Research Institute, Herb Goronkin, Ray Tsui, and Islamshah Amlani (device physics and engineering) at Motorola Corp., Al Bard at the University of Texas (electronics measurements) and all of their respective research groups. There are many other groups with which we collaborate, a testament at the heart of molecular electronics, namely interdisciplinary research. Most importantly, many portions of this book were derived from collaboratively written works by and with my student and post doctoral associates, primarily: Darren Pearson, LeRoy Jones, II, Jeffrey Schumm, Ruilian Wu, Timothy Burgin, Adam Rawlett, Masatoshi Kozaki, Yuxing Yao, Raymond Jagessar, Shawn Dirk, David Price, Stephanie Chanteau, Dmitry Kosynkin, J. J. Hwang, Michael Stewart, Summer Husband, Christopher Husband, William Van Zandt, Lauren Wilson, Jonathan Daniels, Jay Henderson, and Dustin James. Dustin also helped enormously with the organizing of this book. I also thank Steve Currall, Professor of Business, for his insight and advice through the most trying business situations; he's never yet been incorrect in his counsel.

In summary, although the molecular electronics work I describe in this book is not comprehensive, it will be broad in its scope of covered topics. Specifically, the background, commercial landscape, synthetic chemistry, initial device construction, architecture and programming are covered in a manner to be of aid to the chemist, surface scientist, physicist, engineer, computer scientist, mathematician, investor and the casual reader alike.

James M. Tour, Rice University

Contents

Chapter 1

Commercialization of Molecular Electronics

1.1 Introduction

I have chosen to begin this book on molecular electronics with the business section. If that does not interest you, I understand. Simply skip on to Chapter 2 where a light introductory hors d'oeuvre to molecular electronics is served followed by ~350 pages of gut-filling scientific beef for the chemist, surface scientist, device physicist, computer scientist and applied mathematician, in that order.

Forget not molecular electronics' experimental historical progression. Many from the chemical and solid-state electronics establishments were pontiffs of skepticism. They capitalized on the easiest thing to do in science; namely, they offered a dozen scientific reasons why this new field will never break ground and thereby decreed its unworthiness of financial support. But, as always, the movers and shakers tried it anyway, with backing from risk-taking federal research program directors that were not bound by the traditional, and often good, peer-review evaluation processes. Once the scientific results were manifested, traditional grant-funding routes became more available, though it took over a decade to see general acceptance as a worthwhile exploratory science. As commonly occurs, many of the initially nay-saying pontiffs have converted to become avid researchers in the field, though never having confessed their past offenses. Therefore, I begin with the funding and commercialization aspects of molecular electronics since it suspends the canvas upon which the foundational scientific portions of molecular electronics were crafted.

This chapter is written in somewhat of a personal letter format, from an academic scientist to my academic scientist or engineering colleagues. Beginning with an outline on the commercialization landscape for molecular electronics and methods for insertion into such a well-developed electronics industry, it will cover some basic business concepts. These principles might seem pedestrian to the seasoned businessperson; nonetheless, they are foundational and critical for my non-businessperson colleagues to consider if they plan on launching a company of their own. Secondly, advice and forewarnings are provided to the academician interested in starting a company based upon their research. For example, all in the name of your company, you will subject your ear-sore cranium to hundreds of hours of phone conversations, fend off equity-ravenous attorneys, perform repetitive dog and pony shows for information-sucking venture capitalists (VCs), spend dozens of mealtimes smiling at the moneyed who want to press the flesh of the famous scientist, bear good-intentioned investors seeking updates on when their holdings will reach the 1999 e-commerce level, wrestle with financial spreadsheets that appear more confusing than chaos theory, and suffer repeated totalitarian episodes at the gloved-fingers of the cheeseburger-breath and orifice-examining airport security guards as they stroke their wands across your loins. Starting companies is not for cowards! Finally, a quick education is offered to the prospective Chief Executive Officer (CEO) so that he/she can become familiar with the far-too-typical idiosyncratic academic mindset before accepting a position with a university-founded startup.

1.2 Commercial Challenges of Molecular Electronics

The advances in the semiconductor industry came through Herculean efforts involving thousands of person-years of work and trillions of dollars of investment, hence any direct frontal assault by a new technology on the semiconductor industry will fail, period—end of discussion. Should we therefore retreat and leave the future to an industry that is struggling to meet the ever-increasing demand for electronics using a half-century old technology? Never! We must simply realize the facts for what they are, and devise a strategy that can overcome the hardened sites of the silicon fortress by focusing upon its weaknesses and leveraging its strengths. We must never lose sight of the fact that an insertion is possible if the fundamental and financial barriers facing the semiconductor industry are driving it to consider other options. And since molecular electronics has been demonstrating an impressive set of properties that have stunned even the greatest critics, the initial results could be a harbinger of the consumer electronics world to come.

A complementary technology for the electronics industry that is not solid state semiconductor-based will require the following.

Investment. More than just a few hundred million dollars of overall investment will be needed by a few laboratories or startup companies to compete with the trillions of dollars invested in the silicon market.

Insertion Strategy. A strategy, or business plan for insertion is needed that does not initially target the impenetrable high-end of logic and memory, i.e., there must be an assault on the unprotected or hitherto unforeseen product areas.[1]

These will be considered in succession.

1.2.1. *Investments in Molecular Electronics*

When considering investments in molecular electronics, there are noticeable similarities between molecular electronics and the biotechnology startup companies of the late 1970s and early 80s, which investigated areas that were outside the domain of the big pharmaceutical companies of their time. Angel investors and VCs filled the funding void even though the gestation periods (or time to market) for biotech companies' products were long, often on the order of 10-15 years. The lag in commercialization was due in part to Food and Drug Administration (FDA) approval guidelines needing to be fulfilled; nonetheless, much money could be raised and made. Once the startups showed prospects of insertion into markets by addressing fundamentally different biological mechanisms to bring health to the consumers, or simply the mega-companies feeling compelled to buy a "lottery ticket" to secure some hard-to-quantify future hopes, alliances and buy-outs became the order of the day. I believe that the funding trend for molecular electronics will be similar.

Angel investors are usually individuals or small consortia of investors, often the founder's friends, family or associates that have sufficient personal net worth to make high-risk investments. These investments generally follow trust in the founder of the company rather than a calculated assessment of the technology and the market risk vs. potential—it's "a bet on the jockey rather than the horse." Many characterize angels as unsophisticated investors and contrast them to "smart money", such as institutional or venture capital money which can bring domain experience and contacts to the table. Angel money can be simpler to secure and with lower dilution to the founder's interest. On the other hand, a network of angels rarely provides the asset base needed for further rounds of funding that are almost always required to bring a product to market. If a VC invests in a later financing round of a company that was originally angel-funded, regardless of the former terms of the deal with the angels, the VCs will generally be looking to give the angels a significant dilution to their money. In this manner, the VC's money buys more of the company per dollar than the angels' original money did, even though the risk could be considered lower at the later stage in the company's life. This is referred to as the angels "taking a haircut." It is not uncommon for the angels' investment to be worth only pennies on the dollar relative to the VC's money, and in this case it is referred to as a "cram down." Regardless of the angels' original terms, they have to accept the new terms for the ongoing of the company because their alternative is not rosy; namely, the company goes out of business. Hence the willingness of the angels to swallow the VC's terms, albeit with a lump in the angels' throats. So consider this before bringing your family and friends into the early stages of your corporate deal or else you might have nowhere to go for the holidays.

To prevent future funding trouble, it is advantageous to be associated with a lead VC firm that has the assets to further invest in the company through the entire series of funding phases (often four or five rounds of funding), to the point of commercialization of sellable product, public offering or other liquidity event. But if the company does not meet its milestones, the VC administers new haircuts, which give the VC a larger portion of the company per dollar invested.

VCs do not relish deals where large numbers of angels having invested in the company at earlier rounds. There are several reasons for this, but it gets back to the unsophistication of the angels and commonly their inability to heavily fund future rounds. There are often hard business decisions that are required by the VC as a company moves through its early phases of growth. The unsophisticated angel can be a thorn in the VC's flesh because it is difficult for the angel to comprehend the necessity of the corporate changes in direction and recapitalizations, particularly when the recapitalization is at the angel's expense. Moreover, a disgruntled angel can pose a litigation risk to the VC, especially if the angel is forced to take a dilution.

There is an old saying, "Never forsake the one who brought you to the dance." Don't be too eager to drive off with the VCs in their Corvette while leaving the angels to walk home. If the founder honors the angels and defends the angels' interests to the self-centered VCs, even at the expense of the founder's stock if need be, the angels will stand with the founder through the roughest times and the founder will never fear litigation from them. Recall, the angels bet on the founder when all the founder had was a vision, and they will be far less prone to forsake the founder than would an institutional investor. Never leave the angels at the dance; deal honorably with them and they'll stick closer than a brother.

Commercialization costs in excess of $100 million to VC consortia for a specific product line do not deter the larger firms from making investments as long as a clear path to product revenue can be identified and projected to undercut the cost of silicon in the commercialization timeframe. A single VC firm would rarely, by itself, write a check for $100 million even though their funds could be in excess of $1 billion. Although most VCs pride themselves on the ability to make assessments and insert top-notch board members and business teams to guarantee milestone deliverables, once it comes time to writing an investment check, the lead VC seeks other VC partners, even at the early stages. The partnership is not through benevolence to share the wealth, but to dilute the risk if failure is encountered and it expands the base of expertise available to make the company a success. Hence, VCs are often referred to as "lemmings" because they have a tendency to stand back and wait for another to take the lead, but when the lead is taken, especially by a larger and more well-known VC fund, there are dozens of smaller firms willing to throw their shoulder to the plow, praising the company's prospect for growth and accordingly writing supporting checks.

Angels are not alone in placing extraordinary value on the individuals. VCs do the same. If the company is led by a CEO that has a proven track record in the product domain of the company, i.e. has brought other such high tech companies from startup stage to become billion dollar market cap companies, the VCs are far more likely to fund that company. The VCs generally want to see (1) a first rate scientific team of founders with strong intellectual property (IP), (2) a first rate management team with a proven success record in the field, and (3) a business plan with a definable path to a manufactured product that has definite advantages over competitive mainline products at the time of insertion into the market. If a company lacks any one of these three fundamental building blocks, it is unlikely to be funded, unless the VC firm feels that it could build the business team itself or provide the skill set needed to define the product line. It is essential, however, that all three pieces be in place before a VC will fund the company.

Various other vermin-like names have been used to characterize VCs' modes of operation. But it is important to realize that the bottom line for a VC is their return on investment (ROI). In fact, they are mandated by law to bring the optimal return, within legal bounds, to the investors in their funds, who are referred to as "limited partners". Therefore, it is their primary objective to maximize ROI, regardless of the angels, even if the angels are friends and family of the founder. The reality is that only one or two in 10 VC investments, regardless of the field of investment, will ultimately be big winners. Therefore, the VC can justify, at least in their own minds, the dilution of others in each deal.

In general, if the VC sees the founder as a key component to the long-term survival and vitality of the company, the founder will be properly "incentivized" with stock or options to keep him/her properly engaged. Good VCs realize that people are their best assets. It is not equipment or their own experience. It is people—always has been and always will be. So incentives to the founders and management are needed to ensure the utmost ROI to the VC. Realize, however, if the founder becomes less important or even a liability to the ongoing progress of the company because of their waning interest or unneeded expertise, a re-capitalization can leave the founder with infinitesimal percentages of the company in comparison to the VCs and the essential business officers. There are examples where the VC and their officer buddies have well padded their stock portfolios while the scientific founder can't even afford a new suit.

The good VC can secure the expertise needed at the board and senior management level to propel the company to success in a manner that few scientists alone have the ability to do, regardless of the founder's reputation in the scientific arena. There are early-stage VC firms that will help the founding scientist through many of the initial planning phases as well as aiding in the insertion of an appropriate domain-experienced CEO and formulating a business plan. Some feel that the founder sacrifices too much of the company when approaching VCs at the very early stage before a business structure can be put in place. That might be the case, but I do not generally agree. I liken it to the experienced manager of a prizefighter. Sure, the manager receives a hefty cut of the champ's hard-earned winnings. But in return, the champ can focus on hitting and dodging and need not be overly concerned with bottom-dwelling lawyers, event promoters, scheduling, advertisers, maximization of prize negotiations etc. In the case of a startup, teaming with a good VC can result in a far more streamlined company where the VC manages the business plan, CEO selection, board constitution, real-estate decisions, legal concerns and the like. They do this, of course, after extracting their pound of flesh. But most VCs are quite good at establishing management, monitoring expenditures, minimizing legal expenses, etc. It is similar to university administration. The professor pays a significant part of their research funding to the university in indirect costs in return for laboratory space, utilities, maintenance, salary and benefits administration, insurance, tax statements, legal counsel etc. Therefore, there can be a significant upside to the founder for early VC partnership, not to mention some measure of stress relief, so that the academic scientist can focus on the technical side of product development. In return, the good VC provides sufficient incentive to the founder to keep them fully engaged in the technical aspects. As long as corporate founders and their angels realize the ultimate mission of the VCs, a healthy relationship can be enjoyed because "a small part of a big pie is much better than a large part of a small pie." If properly structured and financed, all can see substantial returns on the big winners.

Unlike the biotech industry, molecular electronics startup activities will not generally need FDA approval for their products to come to market. However, the products' time to market will often exceed the 2-4 year time frame that has become the main diet of the typical high tech VC. This is especially true in cases when the ultimate targets seek a market share dominated by advanced silicon manufacture, such as memory and logic. Therefore, another method to consider funding of the molecular electronics startup is through early alliances with major electronics companies that often (not always) can demonstrate a more patient time horizon, such as 10 years, until ROI is realized. The early alliance can provide the startup company, which is usually heavy on the academic side, with a healthy dose of commercial reality in understanding the enormous barriers to taking a blip on a laboratory oscilloscope into a scalable and marketable device. The alliance partner can use its domain-based infrastructure to carry out the needed marketing and product insertion through its established customer base.

One must also realize that even the mega-companies terminate far more exploratory projects than they ultimately convert to commercial product. Therefore, even with the deep pockets of the larger corporations, market factors drive them to make hard choices that must be in place to optimize return to their investors, which is often viewed in the short-termed quarterly bites. Additionally, depending on the structure of the alliance, the startup usually must negotiate terms on its technology that give that one mega-company a large portion of the rights to the IP. Therefore, the alliance has its restrictions.

In the context of a corporate alliance discussed so far, there is an understanding that the alliance is within a business sector of the large company. There are naturally other singular or hybridized relationships that can come through investments by the large company's VC wing into the startup. The VC units of the mega-companies have much the same ROI and focussed view that other VC firms maintain. When a mega-company's VC wing makes a small investment in a startup, in the $1-2 million range, this is more of a lottery ticket and or a placeholder to see the technology progress with a front-row seat via a board observer post. This level of investment might also give certain "rights to first refusal" to the mega-company wherein they have the rights to buy the technology before any other company could acquire the rights. Hence, there is an immediate limitation on the startup's future that must be balanced against their short-term cash needs and the prospects for development and commercialization aid that could ensue from the relationship. The large company will often prefer a board observer seat rather than a full board seat with voting rights. This results because a full board seat would cause the mega-company to have a fiduciary responsibility to the maximization of startup's shareholder value, which might conflict with the mission of the mega-company. An investment by the mega-company in an early stage startup in the $5-10 million range often comes with much deeper restrictions to the startup. Specifically, the startup becomes more of a subsidiary of the large company and gross limitations of the startup's interactions with other companies will be imposed. Again, such a relationship might be best for the startup once the long-term objectives are assessed with a desire to maximize return to its small pool of investors.

An alternative funding route available to molecular electronics startup companies during their early stages are through federal grants such as the NSF's Small Business Innovation Research (SBIR) grants, and other programs available through, for example, NIST and DARPA. The good thing about these grants is that they are exploratory in both the scientific and investment sense. Also, there is no formal payback needed to the government since it is an effort to spawn new industry with a very long-term vision of ROI. The ROI would come through job creation and an expanded tax base. Once a specific path to product for expectable ROI is realized, then private sector funding can ensue.

In summary, there are several funding mechanisms available to the startup molecular electronics companies including angels, VCs, alliances with larger corporations, and federal grants. They are not mutually exclusive and they each have their advantages and disadvantages. Maximization of shareholder value is the responsibility for any company, therefore several factors must be weighed accordingly.

1.2.2. *Molecular Electronics Market Insertion Strategy*

As described above, the semiconductor industry is so well entrenched that the near-term key to new technology insertion is integration through complementation, not substitution. Supplementation can occur, supplanting can not, at least in the next two decades. The silicon industry also moves at lightening speed with respect to other industries. Products are constantly being replaced by updated versions, which harries even the most emotionally stable process engineers and marketers.

The electronics industry is also like a well-balanced and integrated ecosystem. Insertion of a new technology that solves one problem will have to interface well with the other existing pieces of that overall product. There must be acceptance by the industry as a whole before insertion by even a complementary technology could ever take place. Therefore, although startup companies have the advantage of being able to turn on a dime and move swiftly through a new landscape like reconnaissance troops, alliances between the pocketed forces of the startup companies and the major electronics superpowers will be needed—and sooner rather than later. This can provide a win-win situation for all parties by capitalizing on the insertion experience, integration expertise and established markets of the major players. Moreover, the major electronics firms can absorb and justify the outlay of hundreds of millions of dollars for a single product line, while looking for avenues to configure this new technology with their existing manufacturing and marketing capabilities.

It is critical to realize that while molecular electronics holds the promise of smaller devices that can be assembled with greater ease, the initial adoption decisions by the industry will require cost advantages coupled with technological superiority. Most companies spend four times more on cost reduction programs (materials or manufacture processes) of present products than they do on developing new products. This makes good sense. Their markets and customer bases are established for these products, therefore any cost reduction in an existing product immediately impacts their bottom line. Most strategies would be welcomed that would use molecular electronics to reduce the cost of an existing product while maintaining or increasing the product's performance (i.e. speed, densification, resiliency). *Minimally*, a two-fold cost reduction in process or materials, or a two-fold technological enhancement at the same cost will be required before any mega-company would consider insertion of molecular electronics into its higher end systems. However, for simpler bulk consumer applications where products are sold on the millions of units scale, even a 10% cost reduction can be attractive because the margins are tighter.

Another way to view the insertion is by the development of products for niche markets where there is little interest by the mega-corporations and in areas where current products do not serve well. This is actually the method by which most "disruptive technologies" insert. Then, clever engineers see the technology and apply it to other applications. In that way, the disruptive technology gradually moves up the sophistication food chain taking larger and larger market shares of well established and deeply entrenched product lines, as described by Christensen.[1] Disruptive technologies are never successful when they immediately target the high-end applications. So initially keeping the target application "simple, cheap and stupid (unsophisticated)" is the key.

Initial products in molecular electronics will be silicon complements, for example hybrid silicon/molecular devices that are outside the mainstream targets for the traditional semiconductor industry. Using either small molecule or nanotube (nanotubes are small in diameter yet relatively large in length) systems, biological and chemical sensors could provide a venue for insertion of molecular electronic devices for medical diagnostics or chemical/biological weapons detection. It would involve the perfect union of a molecular-based technology with conduction, impedance, capacitance, or other simple electronic function responses resulting from molecule-molecule or molecule-cell interaction processes. A device response time in seconds is generally adequate, rather than having to cater to the state-of-the-art demands of the high-end computing and memory needs of billionths of a second per operation. Additionally, sensing functions on the order of once to a few hundred times can be adequate whereas traditional computing/memory functions often need to exceed trillions of operations for their resiliency specifications. The margins for sensors can be large for "first movers" (first to market), although the market sizes are generally smaller. These early applications would permit researchers in this embryonic field to hone their commercial skills while refining their talents to address higher-end products.

An even simpler commercial entry point for a molecular electronics market might be the use of molecule or nanotube-based systems for interconnects, namely molecular electronic entities serving as the conduit for electron flow between two solid state devices. The molecules could be passive, thereby merely serving as nano-scale wires. Or they could be active so as to pass current at certain voltages and retard electron transport at other voltages, for example. In this manner, a long-needed kick could be given to a technologically stagnant $40 billion interconnect industry that has seen little change in the past 30 years.

Following these advances might be electronics and optoelectronics-based products in places where such functionality rarely exists today, for example, in clothing fabric, plastics and related flexible substrates for bendable displays and imagers. These are host platforms where silicon devices do not generally function well due to silicon's restricted solid and brittle nature. Consider conformal electronics, where the added electronics are embedded within or on the surface of the plastic housing, thereby adding nothing to the size of the final electronics package, but adding form rather than functional enhancement. One could envision molecular electronic antenna arrays in wallpaper that would permit cellular technologies, such as cell phones and mobile modems, to function distortion-free even within the traditionally shielded central recesses of an office building. Therefore, one need not increase the density or speed of a system with a new technology provided it adds a novel function, form or cost advantage.

Memory or logic systems might be developed that add functionality to traditional transistors. Some have termed these "silicon with afterburners" to enhance the functionality of the cost- or function-limiting components in present devices. For example, imagine a memory system where the signal isolation is based upon conventional complementary metal oxide silicon (CMOS) transistor technology with conventional word and bit lines. But atop each CMOS transistor is embedded a layer of molecules that serve as a nonvolatile memory storage unit. Therefore, the molecules do not increase the densification of the array. But they add functionality by making the volatile CMOS memory (which normally needs to have its memory refreshed 1,000 times per second) retain its memory for hours, days or more, through the use of the molecules for the bit storage. This would then make the volatile CMOS memory flash-like in its qualities, and minimally SRAM-like in its capabilities. Present-day flash memory, although able to hold its memory for years, requires a deep trench capacitor; hence it can not be fabricated on a normal CMOS fab line. The CMOS/molecular hybrid would obviate a dedicated flash memory line. Likewise, SRAM is a footprint-intensive memory that can only hold its state when the system's power remains on. Therefore, there is a foreseeable complement to silicon through the marriage of molecules to that parent platform.

Finally, entirely new generations of products could be envisioned that use molecules for the central processing and memory components of top-of-the-line computers in configurations that bear little relation to present-day solid-state architectures. There are high prospects for using molecular electronics in this venue; however, commercially relevant delivery times for such systems are minimally a decade away. Specific possibilities will be considered in Chapter 5.

1.3 Molecular Electronics-Focused Companies

Listed below is the current corporate landscape for companies involved in molecular electronics. Although the list is not comprehensive, it is illustrative of the companies that presently exist. The information here was obtained in the public domain and in non-confidential discussions with the companies' associates. However, the accuracy of the data received could not be verified in many cases, therefore, it is provided with the disclaimer of non-substantiation.

The 2001-2002 downturn in the capital markets has produced a skittish investment environment that has made ongoing financing difficult for many of these companies. As a results, some may fail due to lack of investment rather than lack of technical prospects.

Molecular Electronics Corp. (MEC). Molecular Electronics Corp. (http://www.molecularelectronics.com) was founded in 1999 to capitalize on molecular electronics. In addition to other breakthroughs in molecular technology, members of this scientific team developed the first self-assembled reversible molecular switch and reversible memory that has functioned for nearly 1 billion cycles at room temperature with no degradation. MEC views molecular electronics as a technological platform, enabling the creation of numerous diverse products. For instance, MEC's research and development in nanometer-scale electronics creates the potential for constructing electronic devices that are less expensive, consume less power, and are thousands of times smaller per device (and hundreds of times smaller overall) than existing technologies. MEC is presently forming strategic alliances with industry partners for commercialization of its IP, and for manufacturing and marketing of specific products based upon MEC's IP. MEC is working on conventional/molecular electronic hybrid interconnects in an alliance with Amphenol Corp., hybrid CMOS/molecule arrays in an alliance with Motorola, and seeking partners to develop molecular electronic displays and imagers based upon their patented and exclusively licensed technologies which date back to 1992 patent filings. The 1992 filings disclose the use of molecular arrays between electrodes that are further part of an integrated circuit; therefore, they are quite encompassing. After a hiccup-laden start in identifying domain experienced leadership, MEC is led by CEO Chris Gintz. Gintz is one of the famed original employees to leave Texas Instruments to begin the then high-risk VC-backed (Sevin-Rosen) startup called Compaq Computers. Gintz is the inventor of the first laptop computer (Compaq's LTE Notebook, US Patent 317,442) and he served as Compaq's Director of Technology, Planning and Development. MEC was initially privately funded through individual angel investors, but licenses, alliances and federal research grants are being used to support further efforts. No venture capital funding has been utilized to date, however, MEC is open to their investment.

Nanosys, Inc. (http://www.nanosysinc.com/). Nanosys is focused on the development of nanotechnology-based systems. These systems incorporate nanometer-scale materials such as nanowires, nanotubes and quantum dots as their active elements. These exploit the fundamentally unique electronic, magnetic, optical and integration properties associated with the materials. Devices constructed with these systems will be directed toward chemical/biological sensors, nanoelectronics (electronic memory and logic) and optoelectronics. These devices are proposed to offer performance gains in speed, sensitivity, power consumption, device density and integration. This company is well VC-backed. Nanosys' CEO is Larry Bock, a highly successful business entrepreneur with multiple proven successes in the high tech startup space, hence he comes with high praises from the VC community. Bock is surrounded by a top-notch scientific team. Nanosys has raised $17 million, to date, in venture financing, and is funded by five leading venture capital funds: ARCH Ventures, CW Ventures, Polaris Ventures, Prospect Ventures Partners and Venrock Associates.

Nantero (http://www.nantero.com/tech.html). Nantero is focussed upon the formation of NRAM™, a type of memory potentially faster and denser than DRAM, with ambitiously proposed lower power consumption than DRAM or flash, as portable as flash memory and resistant to environmental perturbations. And as a nonvolatile chip, it might provide permanent data storage even without power. The NRAM™ design, invented by T. Rueckes, Nantero's Chief Scientific Officer, based, in part, on Professor C. Lieber's approach (Harvard), uses carbon nanotubes as the active memory elements. The design integrates nanotubes with traditional semiconductor technologies for claimed "immediate manufacturability." Nantero is funded by VCs, primarily Draper Fisher Jurvetson, Stata Venture Partners and Harris & Harris Group.

ZettaCore (http://www.zettacore.com/). ZettaCore is developing ultra-dense, low-power molecular memory chips based on technology from the University of California and North Carolina State University. The technical approach to molecular memory is based on specific properties of porphyrin molecules to store information. These molecules, called multiporphyrin nanostructures, can be oxidized and reduced (electrons removed or replaced) in a way that is stable, reproducible, and reversible. In addition, each molecule can store more than one bit of information and can maintain that information for relatively long periods of time before needing to be refreshed. The ability to integrate ZettaCore molecular technology with silicon semiconductor technology will hopefully accelerated development of hybrid chips that leverage both the advantages of molecular storage and the substantial capital investment in the silicon semiconductor manufacturing industry. With a development plan that leverages the existing semiconductor industry infrastructure, it is believed that products could be brought to market as early as 2005. ZettaCore's venture backers include Draper Fisher Jurvetson, Radius Venture Partners, Access Ventures and Stanford University.

Coatue Corp. (http://www.coatuecorp.com/). Coatue, a rather stealthy company with little public information being readily available, is developing microchips utilizing thin film organic polymers for non-volatile memory devices (possibly ferroelectric polymers) emanating in part from UCLA (not the Heath team). Coatue has received VC financing from Draper Fisher Jurvetson, New England and other investors.

Opticom ASA (http://www.opticomasa.com/). Opticom ASA is a Norwegian based research and development company. Opticom shares have been listed on the Oslo Stock Exchange since 1999. The subsidiary, Thin Film Electronics ASA (TFE), is a research and development company owned by Opticom ASA (87%) and Intel (13%) (Intel increased ownership in June 2001). TFE does research and development at its own facilities in Linkoping, Sweden and Albuquerque, New Mexico, and uses a network of institutions around the world, including academic labs in the US, for contract research. Opticom ASA technology is based on thin film electronics, and in some ways more of a molecular materials method for electronics rather than molecular electronics, per se, (vide infra). Opticom suggests that their layered technologies may be stacked, and there is potential for reel-to-reel production. Opticom ASA seeks early revenues through non-exclusive licenses of its technology.

Molecular Nanosystems, Inc.. Molecular Nanosystems (http://www.MolecularNanosystems.com) was founded in 2001 by Hongjie Dai, a Stanford University chemistry professor, and a group of other scientific and business professionals. Headquartered in Palo Alto, California, the company is engaged in research, development and production of carbon nanotube-based products. The core technology in controlled nanotube growth is slated to have a fundamental impact in electronic, biological and chemical industries. Early sales are hoped by marketing carbon nanotube-based scanning probe microscopy tips to researchers around the world through an alliance with NanoDevices Inc.

Interesting to note in this current landscape of molecular electronics companies is that most or all of them have derived their roots from academic research, especially federally funded research in the US. The funding by the US government is then built upon by private interests to transition the products into the market place.

Notice also a lead-funding role in several of these startups by the prominent Palo Alto-based VC firm, Draper Fisher Jurvetson, via the patriarch of nano-company ventures, Steve Jurvetson. A profitable exit strategy for Jurvetson might be to merge a subset of these venture-backed companies into a single entity that is then "flipped" (sold) to an Intel-like superpower semiconductor outfit. While that might not be the optimal financial outcome for the individual startups, it would combine the strengths of the various teams into a formidable force that downsizes the technical risk and capitalizes on the superpower's productization and marketing muscle.

Regardless of the hyperbole in the financial press, if one were to sum all the VC investments in molecular electronics and nanotechnology companies in general, it would be nano-sized relative to the federally funded and other private investments in the field. This is prudent to keep in mind when weighing fund-raising strategies and the time devoted to seeking support from each of the possible sources. It might even be better to initially label one's own company as a non-nanotech company and classify it as a "medical diagnostics" or a "sensors" company if seeking VC funding.

These companies have differing approaches to commercialization. The typical method for success of a small company is through focus, focus, focus. Specifically, choosing one particular target and driving toward that sole product target using every ounce of corporate energy that can be mustered. The reasons for the laser-focused business model are clear. Small companies have generally be unsuccessful when they strayed into trying to produce multiple products because of their inability to down-select one specific line; their resources were spread too thin. And any potential customer can cause an unfocused company to sway off course and thereby never hit any target. Like kids in a candy store, inability to choose candy fast enough can result in no candy at all.

Another strategy can be considered that exploits the platform technology—a base from which numerous product lines are spawned. Specifically, to have a parent company that generates technology based upon molecular electronics, and then spin-off companies are established for the focussed development. Each spin-off could have its own Board of Directors (BoD), CEO, business team, and even funding sources, which licenses the subset of IP needed from the parent company to build the particular focussed product. Personnel cross-fertilization, at the BoD level, would maintain the relationship, with some revenue flow from the spin-off back to the parent company. The advantage is that one can explore and exploit the newly discovered continent enabled by this platform technology rather than focussing exclusively on Plymouth Rock. Professors generally are good at identifying technologies but they have little experience in producing products to sell. The spin-off company can maintain the product focus.

Some VCs mandate the former model, namely the singular product focus. Others see value in capitalizing on the technology in its broader context while leaving the focus to independently funded spin-offs. Neither approach is all-correct, they each have value and dangers associated with them. Molecular Electronics Corp. is presently functioning in more of the spin-off mode, thereby leveraging its broad patent base for several applications. The danger, of course, is that the assets become too dispersed to remain effective. The upside is that a larger portion of the molecular electronics market can be secured. Conversely, ZettaCore appears, based on its publicly available information, to have the focussed approach to solely a memory-based product line based on porphyrin molecular arrays. This permits them to drive all their expertise and assets toward this singular desired target with overwhelming force.

As for the big companies, Hewlett-Packard and its outspoken proponent, Stan Williams, have a group of about 15 researchers working on developing memory arrays in conjunction with a first-rate academic team, headed by the talented Professor James Heath, formerly of UCLA and now the California Institute of Technology. The HP/Heath team has DARPA support for their joint research program. Motorola has a group of 12 scientist and engineers working on small molecule and nanotube molecular electronics systems headed by the productization expert, Herb Goronkin. In addition to Motorola working with MEC, they have had a nearly decade-long collaboration with Reed (Yale) and Tour (Rice) through the DARPA program, presently working toward the development of the NanoCell logic/memory system (see Chapters 5 and 6). Motorola has been a devoted believer in the prospects of this new technology with complementary programs in carbon nanotube research, small molecule devices and DNA-based diagnostic arrays. Hitachi has been involved in molecular electronics studies for nearly a decade, however, less is known about their recent developments. In the past they were working on multiple arrays of scanning probe tips and switching devices based on movement of single atoms—a steep hill to climb, indeed. More recent additions to the major players' list include the highly successful work by International Business Machines (IBM) with carbon nanotubes gates and nanoparticle memory work via teams headed by the eminent Phaedon Avouris and Chris Murray, respectively. Murray has the conviction that a small company will be the first entity to market a molecular electronic device, but never underestimate the prowess of IBM or the breadth of its IP portfolio. Many a bold CEO have sheepishly retreated when confronted with IBM's patent library—there are plenty of Rembrandts in their attic,[2] so an alliances might well rule in the end. Intel has shown investment interest in molecular electronics, as described above for Opticom, however, to date they have had little if any active internal program in the field, according to their officers. Intel is monitoring the technology's movement with an eagle eye—they'll not be caught off guard. Yet, in their conservative and possibly self-sustaining vein, they publicly point toward the challenges facing molecular electronics more than pointing toward the prospects of this new technology. Advanced Micro Devices (AMD) has invested in one of the above startup companies after seeing possible prospects for their successful implementation, while AMD itself has no formal research program in the area. At the highest levels within the corporate structure, AMD holds great hope for molecular electronics technology. Infineon has shown interest in molecular electronics and they have now assembled a research team. A large amount of recent work coming out of Lucent/Bell Labs has recently been disclosed on making organic transistors from self-assembled monolayers and even single organic molecules—initial presentations that have been stellar. However, some of the Bell Labs/Lucent work at molecular electronics and thin film electronics has been questioned for its scientific validity or accuracy. A competent external evaluation group organized by the commercial lab is presently assessing the work, and until that is defined, comment here will be withheld.[3]   Therefore, mega-company validation of the prospects for molecular electronics is quite apparent—better late than never.

In summary, the number of small and large research groups working in the area of molecular electronics is expanding rapidly. Although no commercial products have been realized, the hopes are high. The alliances between the startups and the major players are being established and the business strategies are becoming focused to pave the way for commercial success. The race has begun. Who will be the first to market and who will dominate in the long run? Bets are being placed. I have my favorites, as do most others.

1.4 Advice from the Trenches for the Wannabe Corporate Founder

Some lessons are provided here for the academician who is considering founding a company in molecular electronics. Naturally, these principles can also be applied beyond molecular electronics startups. At a minimum, reading the next few pages might spare you some sleepless nights. More importantly, it might save millions of wasted dollars, your embarrassment before your peers and litigation that could choke your scientific creativity for years to come. The examples are in generalities. There are plenty of exceptions, but knowing the terrain of possibilities is always an advantage. It is impossible to write without offending someone; therefore in an effort to minimize the affront, if anyone is offended by what is written, then consider that it is referring to your antithesis.

There can be great rewards for an academician in starting a business. For one, you can participate in taking your beloved research into the commercialization phase, for the good of humankind, in a timeframe that is shorter than if you had waited for another to recognize its potential and thereby capitalize upon it. The rewards further include fame, fortune, the envy of your peers, the meeting of people you would never normally have met if confined to the world of science, and an education in an area where few scientists ever delve. All these add significantly to life's experiences, so the upside can be profound and it should never be belittled.

Of course there are those faculty who feel that the hallowed halls of the academy are tarnished by professors who seek corporate gain, and such a view should be respected. But there was a time when the receipt of federal grants was considered defiling to the pristine and independent mission of the university—so the world is constantly changing. Likewise the spawning of startup companies is becoming an order of the day, and it will likely remain. Certainly, the neglect of one's basic academic duties of teaching, research and service is unacceptable. But provided one is willing to burn the midnight oil to make their corporate dreams a reality, there are significant tangibles and intangibles that can benefit the entire college. The tangibles include a share of market income into the university that could add significantly to the positive cash flow, albeit conceding that only one in ten small startups will ever truly be a success. The intangibles include notoriety for the university and the exposure of students to the rigors of corporate governance and management as they see it lived out in their professor. If properly shared by the faculty member, the students can benefit enormously in the breadth of their education.

There are downsides as well to academicians starting high tech companies, which I will dwell upon for some time. Unlike your National Science Foundation grants, in business, nobody will simply send you a check in the mail and leave you alone. If you plan to start a company, realize that grueling years of business challenges await you. You'll be busier than a Bombay traffic cop, and your teaching, research, and familial duties will take the commensurate hit unless time management and multitasking become your ever-present companions. Furthermore, most business and sales people are much higher maintenance than are graduate students. They dwell far too long on their compensation packages or perceived insufficiency thereof, and they have threatening quick-draw attorneys slung to both hips who are ready to rifle through the penny-bare pockets of your moribund academic lab coat. Therefore, be forewarned before flippantly becoming a wishful Porsche-driving corporate founder.

Throwing caution to the wind, how does one begin the process of commercializing their molecular electronics research? I am writing specifically to address some of the US-mandated approaches, while the protocols will differ in other nations. First, understand what is legally permitted. The Bayh-Dole Act permits federally funded research that is done in academic institutions to have its IP ownership reside with the academic institution. Accordingly, startup companies that were founded by the academic investigator(s) can then license much of the IP from the institutions. In that manner, the ultimate appreciation of the technology can often occur since the very professor and students that generated the IP are involved in the further development of the research. In many respects, this liberal approach of federally funded research being assigned to the university has generated an enormous growth of technology startups in the US and it is the envy of many other nations who are now following the lead of the Bayh-Dole Act. Private investment funds then stand ready to support the best of the technologies in an effort to parlay them into usable products for commerce through direct manufacture, license or sale to a larger asset company. It is capitalism at its finest for the ultimate public good of job creation and tax base growth.

Second, take a visit to your Technology or Patent Office within your university and work with them to file an invention disclosure on your discovery. If there is no such office at your university, the university's general counsel can often guide you through the process. Maintaining good laboratory notebook records will be essential and be sure to note every inventor, including student inventors, who contributed to the idea-generating process. An inventor is different than a co-author. An inventor is not necessarily even the individual who executed the key experiment. It is the person or persons who had the intellectual thought process that generated and/or enabled the idea to the extent that one normally skilled in the art would not have considered. For example, if the faculty member thought of an idea, and he/she directed their graduate student to set up an experiment to test the hypothesis or make the key compound, the graduate student would not be considered an inventor unless the student generated some unusual protocol to enable the process. Simply following standard techniques to test the professor's hypothesis does not constitute co-inventorship. Next, following the specific procedure of your institution, which usually involves a presentation to their equivalent of their Patent and Copyright Committee, and thereby request that they file a provisional patent on the invention. The provisional patent filing is less that a few hundred dollars, and it secures a filing date on the invention for one year. During that one year, the invention can be further refined and an assessment can be made as to whether the university should spend the $5,000-$10,000 (or more) needed for a US patent filing with the associated international coverage. While some legal counselors advise against provisional patents and suggest immediate US patent filings, most universities prefer the provisional patent route.

During the provisional patent filing period, or at any point along the process, you might wish to consider starting a company to exploit the technology. All universities differ on the level of involvement that their faculty can have in companies and they have policies that surround those interactions. Be as forthcoming as possible with the administration as they work with you to chart a mutually acceptable course which balances the strictures of federal and state grant support and student research labs with corporate commitments. There are always ways to structure deals and discussions with universities, regardless of their current policies. There is great liberality within the Bayh-Dole Act, therefore universities have substantial leeway on how they choose to function. In negotiations with the university, as difficult as they might become, do not let grace and truth leave you, and your university will be an ally to the end.

Next, seek some free counsel from the Technology or Patent Office and from business school faculty because they have few, if any, secret agendas. Have them guide you in the corporate governance process, and deeply consider their advice. You will appreciate their fellowship all the more as the corporate challenges begin to mount, and mount they will. So establish the university relationships early and if you ultimately choose to disregard their advice, at least do it graciously. (In every case, however, when I chose to disregard their counsel, I regretted it because they ultimately proved to be correct.)

After an investigation of the corporate landscape, discussions with your university's personnel, meeting with a lawyer, and a thorough psychiatric examination, if you still feel the need to start a company, begin to assemble a corporate BoD. A balanced initial BoD should be comprised of one founder, two individuals with experience in the specific field of commercialization, one with a legal/accounting background with startup expertise, and ultimately the senior company officer such as the CEO. The appointment of the CEO to the board will follow a BoD-directed search (see below). Generally, the corporate legal counsel should not be a member of the BoD. Keeping an odd number on the board can simplify tough votes. Never seek your academic scientist comrade to be a co-board member unless they have years of commercialization experience in the field of your product's insertion. I repeat, if you value your friendship and research relationship with that colleague, never invite them to join you on the board! Business governance hassles have destroyed many fine research relationships. If there is more than one founding scientist, establish a Technical Advisory Panel (TAP) on which they all will serve to advise the BoD. Selected individuals from the TAP can even fill the founder's seat on the BoD as they rotate in 12 month increments to serve on the BoD, for example. Undoubtedly, the senior investors will want strong board representation by selections of their choice, but that can await the first major funding round. Initially invite people onto the BoD that you trust and that have the array of expertise outlined above. Certainly, you might have to reach out to people you know in industries or other startup companies to fill these roles. But the building of the initial BoD is critical for the establishment of proper board governance and the selection of the CEO (see below). For early stage startups, compensation for board members should be limited to performance-based stock options or stock grants, never cash.

Use legal and accounting counsel for matters on which they have expertise. That does not generally include the establishment of a BoD or its governance practices, although all lawyers would willing chip in their 2¢ worth (or $2,000 worth). And be alerted that if your technology is attractive, most of the lawyers you engage will try to negotiate a decrease in their fees for a stock grant. Proceed with much caution because legal fee structures are difficult to assess. You will often end up paying the lawyers the same amount regardless of the stock you have given to them. Unless they are willing to carry out *all* their legal work for you for the first year, for example, in exchange for the stock grant, you would be advised to give them no stock at all.

Angels, due to their personal investment concerns, could have mixed motivations, especially when making painful decisions regarding dilutions. Therefore, an angel investor should only be considered as an early stage "stop-gap" board member before more permanent members can be secured, and only if they bring some level of expertise as outlined above. Typically, angels should have limited input into the governance of the firm.

Therefore, a balanced BoD is the first essential entity that should be established. Once that is in place, they can guide with the selection of proper legal counsel and the assessment of fair legal rates that should be paid for certain jobs. More importantly, the BoD can serve as the focal point for the selection of the CEO.

When a company based on molecular electronics is started, the prospects can be so exciting that there will be numerous people with business/management backgrounds who will suggest to you that they would be the "perfect" CEO for your company. The academic professor's credentials and passion for their research can make the would-be CEO see dollar signs and think that their *Forbes Magazine* cover-story fame is just around the corner. It is utterly amazing how many perfect CEOs, in their own eyes, immediately rise to the surface, especially if the founder should give a zealous presentation in a forum attended by under-employed "consultant-based" MBAs. The MBA's competency may be well-assured for certain types of companies, but high tech startups targeting insertion into a world dominated by silicon technologies, are unique businesses that demand inimitable expertise.

Why is CEO-based experience in the electronics world essential for a molecular electronics startup? First, because the electronics industry moves at a pace and intensity that is unlike any other industry. Outsiders, although they might initially concede the fact, really have no idea of the realities of the speed and intensity of the industry. The field-experienced CEO understands the barriers to product insertion in such an environment and they can help to refine or redirect the startup's target for a more acceptable application. The experienced officers are not overly impressed by the zealous pledges of the academician because they demand the specification chart comparisons. The inexperienced CEO either does not ask for precise projections until it is too late, or once it is provided, they have little way to assess it. One might argue that this is the VP of Engineering's job. Not in a small company. Small company CEOs must wear multiple hats; large staffs are not an option with the cash flow chokeholds that must be maintained in a startup. When the businessperson is solely dependent on the academician for technical assessment of the field, the CEO's failure is certain. The failed CEO might even pull the company down with them, while the founder and CEO are pointing fingers at each other en route to their respective legal counsels. Belaboring this point is needed because it is utterly essential: unless the CEO has technical experience in the domain where you hope to carry out your product insertion, they are not the CEO candidates for your company. Period. Good-hearted business-experienced candidates are ubiquitous and MBAs are a dime a dozen, while those with experience in the required field are much harder to find, but they are essential as you drill-down on specific product insertions.

Second, personal connections within the electronics industry are indispensable; outsiders garner no respect, and cold calls almost never lead to substantive long-term investments or alliances. The inexperienced will hunt down all prospects, from Persian rug salesmen to cattle ranchers, seeking bits of investment required to keep a comatose business on life-support. Electronics domain-experienced officers accomplish more with one phone call than a month of airport-hopping, door-knocking, ring-kissing CEOs with backgrounds in baking or candlestick making. It is not due to inferior intelligence or lack of hard work on anyone's part. Try as they might, the domain-lacking will spell corporate demise at the expense of time and money. Although many contacts will be initiated, no meaningful deal will close when the sphere-inexperienced is at the helm.

Third, without the domain experienced person, sophisticated investors are likely to pass on the deal, regardless of the science, because they must be assured of a first rate domain-experienced senior officer. A field-proven CEO for molecular electronics would be a person that has spent several years in the high tech solid state silicon world, for example, but holds promise for a new technology—you certainly do not want a pessimistic, nay-saying, silicon-stiff-necked CEO leading a molecular electronics company. The technically astute CEO will walk into any VC office or alliance presentation with confidence in their step that they have identified a specific product line with all the needed detail. Although open-minded, they are not storm-tossed by every potential investor's view on how the company should focus their productization.

The prospective CEO must have demonstrated successes in small company management, which is very different than having successfully functioned solely within big businesses. If the CEO candidate expects to have office space in a city's prime towers of real estate, fly first-class, rent stretch limos or stay at top-tier hotels, he/she is not the CEO for you. In the startup world, investors esteem cash flow conservation while they'll grind their teeth at other's lavish living at their expense. It is up to the CEO to manage and control cash flow by anyone associated with the company, including themselves, the founders and board members. To do otherwise is grossly irresponsible and the epitome of weak or misdirected management. The founding scientist is often not well versed in careful fund management, therefore the CEO must exercise restraint over the corporate funds even on behalf of the founder. Those who would argue to the contrary have been infected with the disease of big-business or big-ego. Either is fatal. Avoid such people lest you too catch that plague which initially attacks the mind of the unsuspecting to make them think that they are more than they really are. Remember that you are spending others' money, so respect it even more than you would your own, and hire a CEO that esteems those investments and the investors behind them.

There is a tendency for inexperienced founders to fall into a place termed "satisficing," namely accepting the immediately available option without searching the field for a more acceptable candidate.[4] You should be prepared to make phone calls and study the CEO candidate's past. The tendency to fall into the "satisficing" mode is enormous since there is hope on the part of the founder that he/she can be free of the business burden and thereby focus once again on the science or engineering. Be sure, however, if you make the wrong CEO selection, the business burdens will come right back to your doorstep and there might be litigation awaiting you from a disgruntled former CEO or frustrated investor.

So how does one locate the proper CEO? The BoD should have the contacts within the electronics domain that can identify key candidates and set performance-based compensation packages that are commensurate with the CEO's expertise and the job's challenges. If the BoD positions are properly filled, the CEO candidate will be far more likely to seriously consider the offer. A professional search firm can be used to locate a CEO but their fees can be quite substantial. However, search firms often make concessions for startup companies for stock rather than large amounts of up-front cash. But even using a search firm, there are no guarantees of suitability and the proper CEO choice is often one of the greatest growing pains of a startup company. Furthermore, the CEO choice for the initial phase may be different than the best person for the ongoing stages.

Although rewarding, the challenges in running a company can be substantial. Form an alliance with those in your university that can be a support to you at the early periods. Surround yourself with a skilled BoD that can help to absorb the burdens and provide the required corporate governance and funding advice. And select only a CEO that has domain experience. You'll sleep far better that way.

1.5 From a Front Row Observer to the Aspiring CEO of an Academically Founded Startup

Becoming the senior officer of a company that could be a world-changer is a once-in-a career opportunity, which few savvy businesspeople can resist. The potential for establishing fame and wealth are truly enormous if some of the proposals for molecular electronics technology insertion can be birthed. But before accepting an offer to become the senior officer of an academically founded startup, you would do well to read on. I am speaking stereotypically here, there are exceptions—some academics are wise and superb business partners. It will be up to you to make an informed evaluation.

The new CEO must be well aware of the academic scientist's mindset because it is often highly irrational from the business perspective. If there is more than one academic founder, the CEO's hemorrhaging is exponentially exacerbated. Many competent and experienced CEOs have met their match when dealing with teams of academic scientists because the scientists can vacillate in nanosecond increments. Likewise, founding scientists can find it grueling to function within constraints that are clear to any businessperson, namely, that if they start a company, they have a fiduciary responsibility to maximize shareholder value in the company. The academic-like investigations must often take a back seat to the business decisions. That is hard for the scientist to accept, although they often nod in agreement until the reality is thrust upon them.

It is not solely interactions with the founding professors that can be rough. University administrators have a propensity to believe their faculty's pontificating of the near-term commercial prospects of their over-inflated embryonic technology, even though their faculty have never even brought a single thimble to market, and there is nobody else on earth seeking to license the patents. Negotiating rights to the founder's patents can be nightmarish for the new CEO as he/she pleads with the university administration to accept some business realities of the risk at these early stages. Hence, the CEO can be faced with a university-based rationale that is utterly ridiculous from a business perspective.

In defense of the universities, often business people do not appreciate the concerns of the university and their requirements as a not-for-profit institution. Understanding the universities' constraints as an educational institution is therefore critical.

What is the solution for the prospective CEO? Realize that the CEO must have lots of energy left in him/her. It's not a semi-retirement job. Finding time for family will be like finding hens' teeth. Second, the CEO must establish clear lines of demarcation in initial discussions with the founding scientist, thus defining each other's respective roles and setting in stone the corporate objectives. Third, the CEO should ensure there is a competent BoD to whom they will be accountable and that will provide balance to the overall mission of the company. A board that is dominated by founding scientists or academicians will, in many cases, cause the CEO to become ineffective while ensuring that no institutional funding will be received. Support from a properly balanced board can be the key to keeping the CEO free from the cardiac-care unit as they slug through the waist-deep thicketed morass of the academic world. Therefore, definition of specific individual tasks and objectives with incisive and decisive board-level oversight by the domain-savvy business team can provide the stability to maximize the energetic CEO's likelihood of triumph.

Another thing to know is that university professors have trouble with focus because they are trained to think broadly. They are always dreaming up new ideas and tangential models. Every morning that they think of a new application, they defocus from the previous day's target. Even if the professor says that he/she will devote 90% of their corporate time to the product deliverables and 10% of their efforts to expanding the corporation's science base, don't believe them. When they wake up in the morning, they'll be dwelling on the 10% basic science part, which will then consume 90% of their brain's central processing capabilities. A VP of Engineering, on the other hand, will blow right past a Nobel Prize-winning idea if it will distract him/her from the product delivery date—they are laser-focussed, and rightly so. There is no other way to be when product delivery is expected. You do not sell technologies, you sell products that are generated from new technologies. Therefore, it makes enormous sense to keep the ethereal professors on the technology development efforts and the tobacco-chewing product engineers on the device delivery side.

In conclusion, although the benefits can be enormous, the prospective CEO should be aware of the academic mindset and motivations. Even with that understanding, however, be prepared for lifelong battle scars and lead-ridden limbs resulting from business-irrational dealings with hardheaded business-illiterate academicians who fear no hand-to-hand combat of opinions, even when their opinions are devoid of experience and fact.

Chapter 2

Molecular Electronics

2.1 Introduction

Molecular electronics, sometimes called moletronics, involves the use of single or small groups of molecules in device-based structures, that can be used as the fundamental units for electronic components such as wires, switches, memory and gain elements.[5] Molecular electronics is an area of research that is firing the imagination of scientists as few research topics have ever done in the past.[6] For instance, *Science* magazine labeled the hook-up of molecules into functional circuits as the breakthrough of the year for 2001,[7] and teams of chemists, engineers, materials scientists, physicists and computer scientists are learning each other's languages to hopefully turn this interdisciplinary new field into a worldwide product-bearing reality.

Molecular electronics is more than a single product line. For example, if one were to view the emergence of silicon in 1960 and suggest that silicon would be a replacement for the vacuum tube, that would have been correct, however, it would have also been myopic. Silicon was a technologic platform from which numerous products, most of them unfathomable at the time, were to be spawned. Likewise, molecular electronics is a platform technology, not a single product. From this new technological platform, many products will be generated that are presently unforeseeable.

A notable distinction needs to be made. Molecular materials for electronics[8] deals with films or crystals that contain many trillions of molecules per functional unit, the properties of which are measured on the macroscopic scale, while molecular scale electronics deals with one to a few thousand molecules per device. For example, thin film transistors (TFTs) and polymer-based light emitting diodes (LEDs) utilize molecular materials for electronics. The grain size of many of these crystalline features in TFTs and LEDs is in the

2 μm region, a size domain to which silicon electronics can already be reduced. Certainly, there are advantages to the molecular materials over silicon in that they can be fabricated on flexible substrates, for example, but they do not add the feature that will permit a dramatic reduction in size to the overall device dimensions.

The goal in molecular electronics is to use the molecules, designed from the "bottom-up" to have specific properties and behaviors, in place of present solid-state electronic devices that are constructed using lithographic technologies from the "top-down". Top-down is the approach that is currently used in the silicon industry wherein small features such as transistors are etched into silicon using resists (chemicals) and light, and the ever-increasing demand for densification is stressing the industry. This is analogous to making a table, for example, where a tree is cut down and fashioned into the table. Bottom-up, on the other hand, implies the construction of functionality, i.e. electron storage, into small features, such as molecules, with the opportunity to have the molecules further self-assemble into the higher ordered structural units such as transistors. Self-assembly is a thermodynamically favorable process, namely it is energetically favorable for the entities to interact to form some organized aggregate structure. Bottom-up self-assembling methodologies are quite natural in that all systems in nature are constructed bottom-up. Unlike the top-down making of a table from a tree, in nature, molecules with specific features assemble to form higher order structures such as lipid bilayers. Further assembly, some of it self-assembly and some of it enzyme-controlled assembly, albeit incomprehensibly complex, causes assembly into cells, and further into higher life forms. Using the tree example above, the tree was assembled from the bottom-up and not fashioned from a yet larger tree. While human beings are the only creatures to manufacture from the top-down, nature almost exclusively assembles its structures from the bottom-up. Therefore, bottom-up construction is bio-inspired, a natural method of manufacture, and a methodology which is certain to influence much of the use of nanotechnology, and molecular electronics in particular. Utilization of a diversity of self-assembly processes, from the bottom up, could lead to enormous advances in future manufacturing for electronics once scientists learn to further control specific molecular level interactions. Several self-assembled structures, in concert with traditional silicon platforms, form the basis of the first generation of molecular electronic test devices that have been made to date.

Ultimately, given advancements in our knowledge, it is thought by the proponents of molecular electronics that its purposeful bottom-up design will be more efficient than the top-down method, and that the incredible structure diversity available to the chemist will lead to more effective molecules that approach optimal functionality for each application. A single mole of molecular switches, weighing about 450 g and synthesized in small reactors (a 22 L flask might suffice for most steps of the synthesis), contains 6×10^{23} (that is the number six with 23 zeros after it) molecules, is more than the combined number of all transistors ever made in the history of the world. While we do not expect to be able to build a circuit in which each single molecule is addressable and is connected to a power supply (at least not in the first few generations), the extremely large numbers of switches available in a small mass illustrate one reason molecular electronics can be a powerful tool for future computing development.

Let us consider another point regarding memory, namely, the number of bits of information needed for particular applications;

- One color photo ~ 10^5 bytes
- Average book ~ 10^6 bytes
- Desktop computer ~ 10^8 bytes
- Genetic code ~ 10^{10} bytes
- Human brain ~ 10^{13} bytes
- Library of Congress ~ 10^{15} bytes
- Total human culture ~ 10^{20} bytes (which is doubling every three years)

Now imagine if we had a mole of bits, that is 10^{23} bytes! Certainly the big question remains how to access such a vast memory in any usable timeframe. Nonetheless, the shear numbers available to us is staggering when using molecules.

As a further comparison, there are 42 million transistors on a Pentium® 4 chip that occupy a few cm^2 of chip real estate. Using typical-sized organic molecules that have three branches, much like the source, drain and gate of a transistor, spanning 1-3 nm, one could fit approximately 10^{14} molecules in 1 cm^2. Thus, there is a density increase of about one million times that might be realized with molecules in comparison to present solid-state systems, and at least several hundred thousand times smaller than future projections for solid-state systems. To be fair to the transistor, there is no presently available method to hook more than one independent lead to each of those 10^{14} molecules, nonetheless, the size comparison is gripping.

There are other numerous significant problems to be solved before a molecular computer could be realized. Each one of those 42 million transistors in the Pentium® 4 is addressable and connected to a power supply. Molecules, though easier to synthesize in large quantities, can be difficult to arrange on a surface or in a 3-dimensional array such that each molecule is addressable. It is equally difficult to ensure that every molecule stays in place. If one must sacrifice the density advantages of molecules in order to individually address each one with a micron-sized lithographically formed address line or other macroscopic tip, then there is no size advantage to be gained from molecular-based systems. However, through the NanoCell approach, which is outlined in Chapters 5 and 6, we are devising methods to address the nanoscopic via the microscopic, but with no knowledge of the precise arrangement of the nanoscopic entities within.

Equally important is the consideration of heat dissipation in present-day systems, and the implications for further device size reductions. A Pentium® 4 chip, with 10^7-10^8 transistors operating at the present nanosecond rate can emit 40 watts of heat (100 watts in extreme cases)—similar in radiant heat to a home range-top cooking surface. Since molecular computing could take advantage of a million-fold increase in circuit density, an enormous cooling fan would be needed to prevent the ensuing meltdown! A way around this problem would be to devise molecular electronic systems that use, for example, 1-100 electrons per bit of information as opposed to the 15,000-20,000 electrons presently used. This would then result, fundamentally, in about 100-10,000 times less heat dissipation than present systems. Likewise, scenarios involving quantum cellular automata and electrostatic computing (see Chapter 5) could utilize as low as one millionth of an electron per bit of information; therefore, none of the hurdles proposed to date for molecular electronics is considered insurmountable. Merely stepping out of the silicon box, or off the silicon chip, is required to grapple with the fundamental challenges of molecular electronic computer design.

Chemistry alone is not the answer. While we concede that logic fundamentals predate electronic wiring and circuitry, insertion of molecular electronics into viable and competitive computation or memory architectures is difficult to envision without first interfacing them to present silicon input/output (I/O) platforms, namely the hybrid approach. Traditionally, chemists have proposed using "soups" of molecules moving randomly in beakers for "computing systems" without regard to realistic interface strategies. More specifically, designs in molecular computing should even have I/O signal homogeneity (e.g., voltage in and voltage out) and magnitudes within devices so that the second device can be driven with the same signal type and relative signal size that operated the first device; these are requirements in most any architecture that can be envisioned. Putting an electron into the system and getting a photon out may provide a fine laboratory experiment; however, construction of a device array based upon this typical chemical experiment is wholly impractical.

Though none of the current molecular electronic computer design challenges have been tackled with any degree of satisfaction, presented here are a few of the achievements and architectural scenarios being considered in order to construct molecular wires and devices, arrange large arrays of molecules, deal with the potential heat dissipation problems, and take advantage of the number of devices available from synthetic chemistry.

2.2 The DNA and Quantum Computing Distinctions

While solution-phase-based computing, including DNA computing,[9] is sometimes classified as molecular scale electronics, it is best to make the clear distinction. DNA computing is information storage and retrieval based on the formation and cleavage of many sets of chemical bonds. That is an inherently slow process, probably tenths of seconds at best. Since present-day computers operate in the nanosecond regime, DNA will only likely be useful for diagnostics, where analysis in the timeframe of seconds is usually sufficient. But for computation, DNA is impractical as a vehicle to compete with present or future non-DNA based system. Moreover, DNA it is wedded to the solution phases, a process that is hard to reliably integrate with electronic platforms, and the operational lifetime is generally inadequate for hard-core computation. Beyond diagnostics and sensors, it is difficult to see DNA computing as a commercially viable molecular electronics platform, therefore it is not covered here.

Quantum computing is a fascinating area of theoretical and laboratory study,[10] with several articles in the popular press concerning the technology.[11] However, since quantum computing is based on interacting quantum objects called qubits, and not molecular electronics, it is not covered here. Other interesting approaches to computing such as "spintronics"[12] and the use of light to activate switching[13] is also not covered.

2.3 Present Microelectronics Technology

An insatiable desire by the masses for consumer electronics, primarily for entertainment and communications, has fueled the semiconductor industry at unparalleled rates to produce smaller, faster, and more powerful logic, memory, display and imaging systems. Central to the advancements have been rapid scientific and technological developments across numerous fields that the semiconductor industry has capitalized upon to fulfill its customers' desires. The core component of the semiconductor industry's technology is the transistor, and the support electronics to drive the transistor. A transistor is the basic device for computational logic (using distinguishable on-off, or high-low switching states), electronic memory (by storing packets of 15,000-20,000 electrons per transistor for each of the smallest fundamental packets or "bits" of information), and gain (the boost in power so that the signals' information is not lost upon propagation through less-than-perfect wires; this boost in power comes at the expense of a power supply that is wired to every transistor). The trend of doubling the number of transistors per integrated circuit every 18-24 months (Figure 2.1), due to advancements in technology, is commonly referred to as "Moore's Law," after Intel founder Gordon Moore, who made the prediction in a 1965 paper with the prophetic title "Cramming more components onto integrated circuits."[14] It is a prediction and not a "Law" of physical science, in the sense of the Law of Thermodynamics, for example. Nonetheless, even though Moore did not believe that his prediction would hold far beyond 1975, the exponentially increasing rate of circuit densification has continued into the present. In 2000, when Intel introduced the Pentium® 4 containing 42 million transistors, that was an amazing engineering achievement indeed. The development of new fabrication techniques, materials, and processing technologies brought about these changes and they are a declaration of the dedication to the development of the industry by hundreds of thousands of scientists and engineers worldwide over the past half-century.

Moore's Law and the Densification of Logic Circuitry

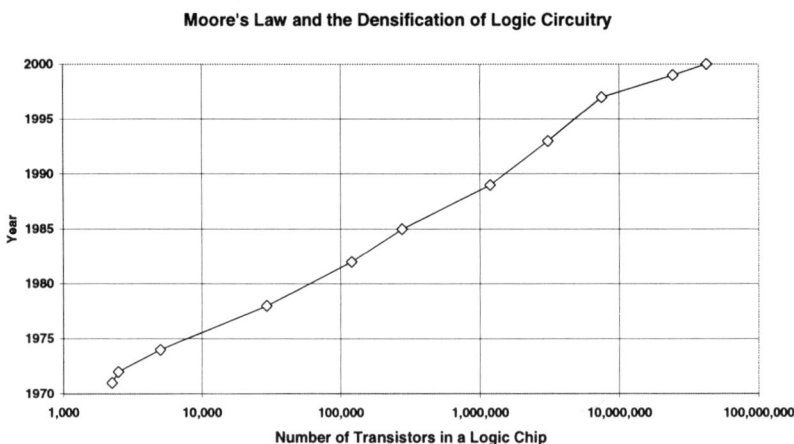

Figure 2.1 The number of transistors in a logic chip has increased   exponentially since 1972 (Intel Data).

2.4 Monetary and Fundamental Physical Limitations of Present Technology

Interestingly, theoreticians devised a trend of their own that tracks with Moore's Law. The theoreticians heralded the end to silicon's densification, namely, the now famous silicon brick wall, or the point at which silicon devices could become no smaller. However, their prediction of the theoretical demise of silicon had to be continually revised because it was based upon a paradigm of technological barriers, i.e. the inability to pattern ever smaller line widths in silicon. The brick wall was not predicated upon a physical science barrier. The engineers in the solid state industry thankfully paid little heed to the theorists' predicted day of doom for silicon, and the engineers proceeded to build smaller devices and denser integrated circuits.

And even now, as the lithography techniques used to create the circuitry on the wafers has neared its technological limits, extenuation technologies such as e-beam lithography, extreme ultraviolet lithography (EUV),[15] and x-ray lithography are being developed for commercial applications in the sub-100 nm regime.[16]

Altogether different, however, is the wall that is now being approached by silicon. The wall is not a technological wall but a fundamental physical wall related to the silicon material itself. That wall cannot be overcome by engineering. For instance, charge leakage becomes a problem when the insulating silicon oxide layers are thinned to about three silicon atoms thick, which will be reached commercially by 2004.[17] Moreover, silicon no longer possesses its band structure when it is restricted to very small sizes. Even the world's best engineers cannot overcome physical science barriers that are inherent in the materials properties. Therefore, the wall that silicon is now approaching is quite different than the wall that was previously overcome by clever technologists. Small materials modifications such as the use of silicon nitride might slow the collision, but industry insiders predict that silicon will hit that physical science barrier by 2008-2012, and the only way around this wall is a draconian change in the technology itself.

Secondly, financial roadblocks to continued increases in circuit density exist. Intel's Fab 22, a chip-fabrication facility (fab) which opened in Chandler, Arizona in October 2001, cost $2 billion to construct and equip, and is slated to produce logic chips using copper-based 0.13 μm technology on 200 mm wafers. The cost of building a fab is projected to rise to $15-30 billion by 2010 and could be as much as $200 billion by 2015. The staggering increase in cost is due to the extremely sophisticated tools that will be needed to form the increasingly small features of the devices. It is possible that manufacturers may be able to take advantage of infrastructure already in place in order to reduce the projected cost of the introduction of the new technologies, but much is uncertain since the methods for achieving further increases in circuit density are unknown or unproven.

Third, as devices increase in complexity, defect and contamination control become even more important. Since defect tolerance is very low in present systems, nearly every transistor must work perfectly, and when there are 42 million per chip, the challenge is colossal. For instance, cationic metallic impurities in the wet chemicals such as sulfuric acid used in the fabrication process are measured in the part per billion (ppb) range. With decreases in line width and feature size, the presence of a few ppb of metal contamination could lead to low chip yields. Therefore the industry has been driving suppliers to produce chemicals with part per trillion (ppt) contamination levels, further raising the cost of the chemicals used.

Fourth, depending on the complexity of the device, the number of individual processing steps used to make them can be in the thousands.[18] It can take 30-40 days for a single wafer to progress through the manufacturing process. Many of these steps are cleaning steps, requiring some fabs to use thousands of gallons of ultra-pure water per minute.[19] The reclaim of waste water is gaining importance in semiconductor fab operations.[20] The huge consumption of water and its subsequent disposal can lead to problems where aquifers are low and waste emission standards require expensive treatment technology. Interestingly, the environmental restrictions in the U.S. and western Europe are spawning the construction of chip manufacturing facilities in parts of east Asia where environmental concerns are, unfortunately, less stringent. However, a decade or so of ecological negligence will inevitably be followed by a continental or world outcry for prudent industrial practices because environmentally destructive effects are never confined to a single nation.

A subsitute being considered for chip cleaning is supercritical carbon dioxide.[21] This might reduce the volume of waste water produced. The use of carbon dioxide could also reduce the need for high energy drying processes while permitting smaller feature-sized lithography due to carbon dioxide's lower surface tension. But the technology's overall efficacy has yet to be fully assessed.

A new technology would be of interest to the semiconductor industry if it only addressed one of the problems in the chip manufacturing processes. But a new technology that produces faster and smaller logic and memory, reduces fabrication complexity, saves days to weeks of manufacturing time, and reduces the consumption of natural resources, would be revolutionary. Since silicon is approaching its fundamental materials science limits, a new platform would be the only solution. Can molecular electronics be the answer? Many think so.

Chapter 3

Chemical Synthesis

3.1 Iterative Approaches to oligo(2,5-thiophene ethynylene)s Molecular Wires, Properties and Experimental Details

3.1.1. *Introduction*

Though it is well-documented that bulk conjugated organic materials can be semiconducting or even conducting when doped,[22] we only recently determined how thiol-ended rigid rod conjugated molecules orient themselves on gold surfaces,[23] and how we could record electronic conduction through single undoped conjugated molecules that are end-bound onto a metal probe surface.[24] Here we describe the synthetic details for the formation of soluble oligo(3-ethyl-2,5-thiophene-ethynylene)s, potential molecular scale wires, by a rapid iterative divergent/convergent doubling approach.[25,26] Additionally, the syntheses and attachments of protected thiol moieties to one or both ends of the oligomers are presented. These thiols serve as molecular scale alligator clips for adhesion of the molecular scale wires to the gold probes.

There has been considerable recent effort to prepare large conjugated molecules of precise length and constitution.[27] Our approach to these compounds maintains several key features that make it well suited for the requisite large molecular architectures for molecular scale electronics studies. Specifically, the route involves (1) a rapid construction method that permits doubling molecular length at each coupling stage to afford an unbranched 100 Å oligomer, the approximate size of present nanopatterned probe gaps, (2) an iterative approach so that the same high yielding reactions can be used throughout the sequence, (3) the syntheses of conjugated compounds that are semiconducting in the bulk, (4) products that are stable to light and air so that subsequent engineering manipulations will not be impeded, (5) products that could easily permit independent functionalization of the ends to serve as molecular alligator clips that are required for surface contacts to metal probes, (6) products that are rigid in their frameworks so as to minimize conformational flexibility yet containing substituents for maintaining solubility and processability, (7) alkynyl units (cylindrically symmetric) separating the aryl units so that ground state contiguous π-overlap will be minimally affected by rotational variations, (8) molecular systems that do not have degenerate ground state resonance forms and are thus not subject to Peierls distortions, and finally, (9) products that serve as useful models for the understanding of bulk polymeric materials

The iterative divergent/convergent approach is outlined in Scheme 3.1.[28] A batch of monomer material M, with inactive end groups X and Y, is divided into two portions. In one portion, the end group X is activated by conversion to X'. In the second portion, Y is activated by conversion to Y'. The two portions are then brought back together to form the dimer XMMY with loss of X'Y'. Since the same end groups that were present in the monomer are now present in the dimer, the procedure can be repeated with a doubling of molecular length at each iteration. The advantages of this approach are that the molecular length grows rapidly, at a rate of 2^n were n = the number of iterations, and incomplete reactions yield unreacted material that is half the size of the desired compound. Thus, purification at each step is far simpler since separation involves, for example, an octamer from a 16-mer. This iterative divergent/convergent approach is therefore particularly attractive.

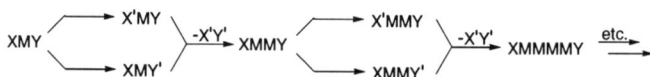

Scheme 3.1   Schematic presentation of the iterative divergent/convergent approach to molecular length doubling.

3.1.2. *Results and Discussion*

3.1.2.1. Monomer Syntheses

The syntheses of several monomers were conducted as shown in Schemes 3.2 and 3.3. Oligomers derived solely from **1** had minimal solubility. Oligomers derived from **9** were too difficult to purify since the butyl groups promoted excessively rapid migration on silica gel chromatography even with hexane as an eluent. Oligomers that were prepared from **10** or mixtures of **1** and **10** required protection of the hydroxyl moieties as t-butyldimethylsilyl ethers, and they suffered from silyl migration reactions. The use of monomer **8**, however, proved to be optimal; there were no protection/deprotection steps necessary and acceptable R_f values on silica gel could be maintained.

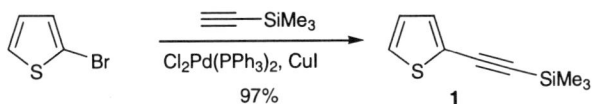

Scheme 3.2 Synthesis of monomers.

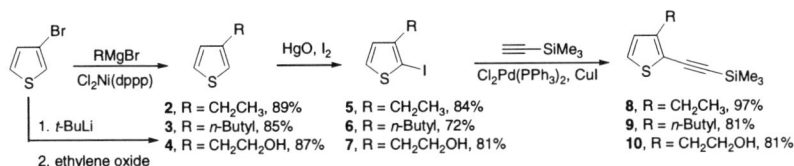

Scheme 3.3 Synthesis of functionalized monomers.

3.1.2.2. Controlled Oligomer Syntheses

The iterative divergent/convergent synthetic approach is outlined in Scheme 3.4. The sequence involves partitioning **8** into two portions; iodinating the 5-position in one of the portions to form **11** and protodesilylating the alkynyl end of the second portion to form **12**. Bringing the two portions back together in the presence of a soluble Pd/Cu catalyst mixture[29] couples the aryl iodide to the terminal alkyne, thus generating the dimer **13**. Iteration of this reaction sequence doubles the length of the dimer **13** to afford the tetramer **16**, and so on to the octamer **19**, and finally the 16-mer **22**. The silylated alkynes showed good oxidative stability, however, upon protodesilylation, the tetramer **18** and octamer **21** were air sensitive and immediate work-up and further coupling was necessary to minimize oxidative decomposition of these terminal alkyne intermediates. Similarly, the Pd/Cu coupling reactions to form **19** and **22** required the strict exclusion of oxygen; degassing and use of a dry box was required to attain reasonable yields.

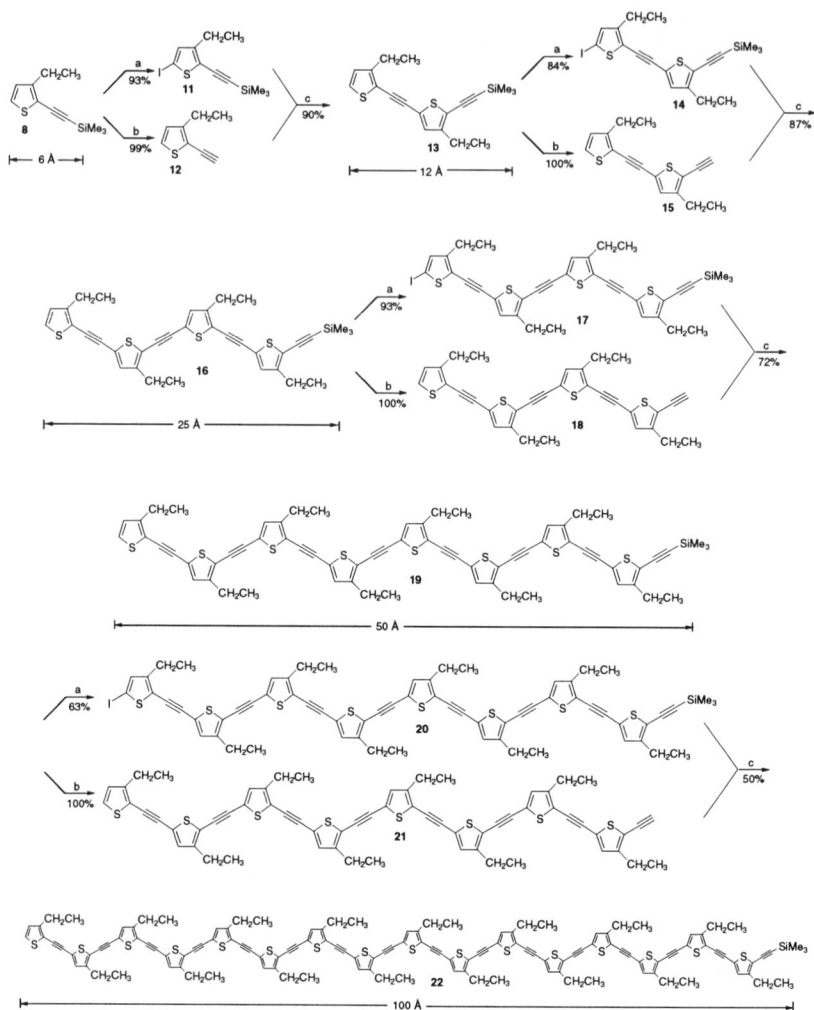

Reagents: (a) LDA, Et$_2$O, -78° to 0°C then I$_2$, -78°. (b) K$_2$CO$_3$, MeOH, 23°C. (c) Cl$_2$Pd(PPh$_3$)$_2$ (2 mol %), CuI (1.5 mol %), THF, *i*-Pr$_2$NH, 23 °C.

Scheme 3.4 Synthesis of 16-mer.

     The 16-mer, in its minimum-energy extended zigzag conformation, has a molecular length of approximately 100 Å. The zigzag conformation is significantly lower in energy, by molecular mechanics calculations,[30] than other non-extended forms. Once one end of the molecule binds to a gold probe, the other end could affix to the proximal probe with 2 eV stabilization[31] on each end from the Au-S bonds.

3.1.2.3. Oligomer Characterization

The monomer through 16-mer, **8**, **13**, **16**, **19**, and **22**, have been characterized spectroscopically. While the tetramer **16** and octamer **19** afforded molecular ions by direct exposure via electron impact mass spectrometry (MS), neither this method nor FAB or electrospray MS sufficed for obtaining a molecular ion of **22**. However, matrix assisted laser desorption MS (MALDI-MS) did afford an M+1 peak for **22** (sinapinic acid matrix, positive ion mode).

The optical spectra are interesting in that a near saturation of the systems appears to have occurred by the octamer stage so that doubling the conjugation length to the 16-mer caused little change in the absorbance maximum (Figure 3.1). We observed a similar near saturation for the third order (χ^3) nonlinear optical intensities.[32]

Optical Absorbance of Oligomers

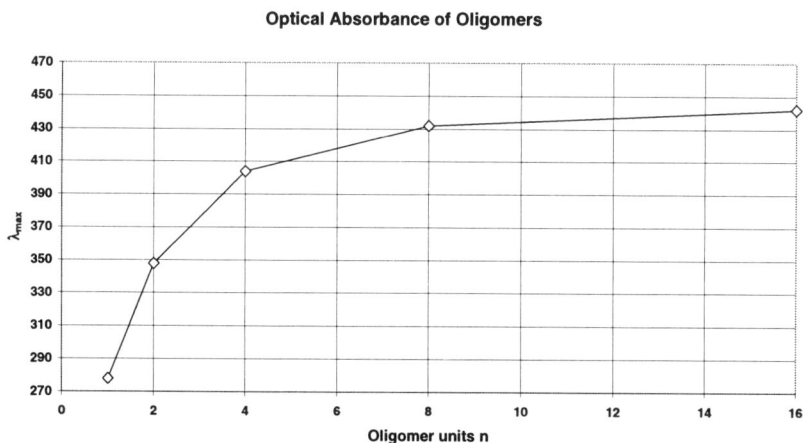

Figure 3.1 The optical absorbance maximum (λ_{max}) in CH_2Cl_2 versus the number of units in the oligomer (n) for 8, 13, 16, 19, and 22.

The results of the size exclusion chromatography (SEC) are also quite intriguing (Figure 3.2). SEC is not a direct measure of molecular weight but a measure of the hydrodynamic volume. Thus, by SEC using randomly coiled polystyrene standards, the number average molecular weights (M_n) of rigid rod polymers are usually greatly inflated relative the actual molecular weights (MW). Accordingly, the SEC recorded M_n values of the octamer **19** (M_n = 1610, actual MW= 1146) and 16-mer **22** (M_n = 3950, actual MW= 2218) were much greater than the actual MWs. Conversely, the monomer **8** through tetramer **16** had M_n values that were very close to the actual MWs (slope ~1.0

in Figure 3.2) because they are in the low MW region, prior to significant polystyrene coiling. Therefore Figure 3.2 could serve as a useful calibration chart for determining the MW of rigid rod polymers. In all cases, the SEC-determined values of M_w/M_n = 1.02-1.05 were within the detectable range limits.

The Values of M_n Determined by SEC

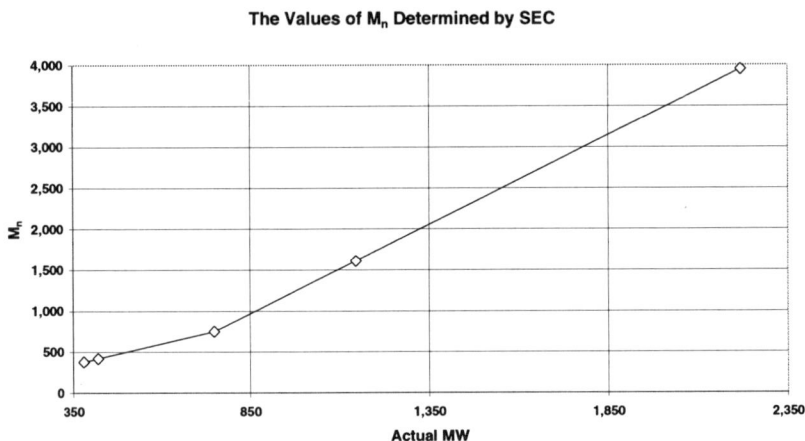

Figure 3.2 The values of M_n determined by SEC in THF (relative to polystyrene standards) versus the actual molecular weights of the monomer through 16-mer, 8, 13, 16, 19, and 22, respectively.

3.1.2.4. Attachment of Thiol End Groups

We then sought to affix protected thiol moieties to the ends of the oligomers. These thiols were used as molecular alligator clips for adhesion to gold probe surfaces. In some cases, we prepared monothiol-terminated systems for adhesion of these oligomers to single gold surfaces. In other cases, we prepared the α,ω-difunctionalized systems for adhesion between two gold probe surfaces. We found that the terminal aromatic thiols were most difficult to manipulate since they were unstable, undergoing disulfide formation in the presence of oxygen, often resulting in insoluble oligomers. However, the acetyl-protected thiols were resilient enough for manipulations in air, yet they could be readily hydrolyzed with NH4OH, in situ, once exposed to the gold probe surfaces. Initially, we attached phenylene bromides to the ends of the thiophene-ethynylene oligomers and then converted the halide moieties to protected thiols via a one pot lithiation-sulfide-acetylation protocol (Scheme 3.5).[33] This thioacetyl-generating protocol was low-yielding and it only worked for the monomeric thiophene case. For instance, we made the corresponding α,ω-dibromoaryl-16-mer **29** that we were unable to transform into the α,ω-dithioacetyl oligomer using the sequence of *t*-BuLi, S8, AcCl (Scheme 3.6).

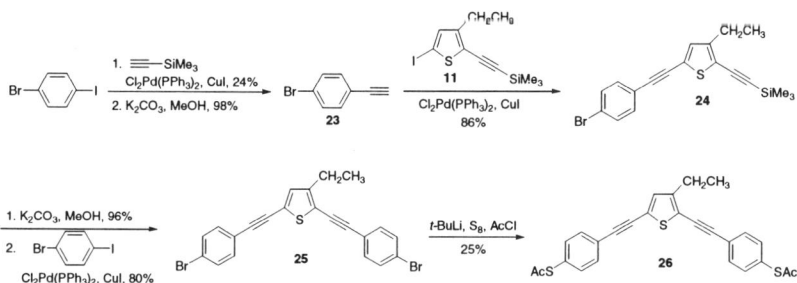

Scheme 3.5 Attachment of thiol end groups.

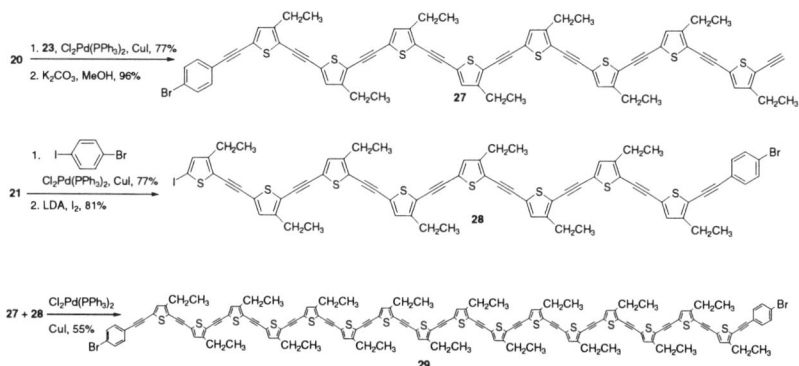

Scheme 3.6 Synthesis of α,ω-dibromoaryl-16-mer 29, potential precursor to α,ω-dithioacetyl oligomer.

We then decided to follow a more convergent approach by synthesizing the two complementary alligator clips, **30** and **31**, from a common intermediate, and then to affix the thioester-containing units directly to the ends of the oligomers (Scheme 3.7). During the final protodesilylation for the formation of **31**, if acidic conditions were not used, the fluoride was nucleophilic enough to cause deacylation as well as desilylation. Moreover, addition of acetic anhydride increased the yield of **31**, presumably by re-acylating any thiol intermediate that formed. The aryl iodide **30** can be coupled to the alkynyl end (head) of the oligomers while the alkynyl arene **31** can be coupled to the thienyl iodide end (tail) of the oligomers. A recent protocol by Wu, et. al. for the one-pot formation of **30** is very effective.[34]

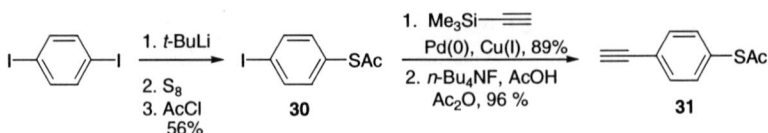

Scheme 3.7 Synthesis of alligator clip intermediates.

Scheme 3.8 shows the attachment of **30** to the heads of the monomer through octamer (**12**, **15**, **18**, and **21**, respectively). The yield for the attachment of **30** to the octamer **21** was low, and we were unable to obtain any desired material when we attempted the same reaction on the corresponding 16-mer, the protodesilylation product of **22**. Again, the sensitivity of the terminal alkynes increases with increased conjugation length.

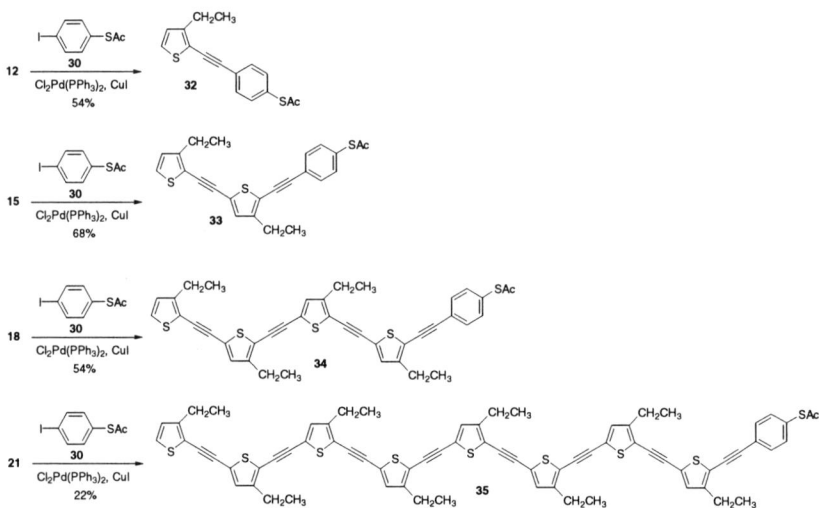

Scheme 3.8 Attachment of 30 to heads of monomer through octamer.

We then synthesized tail-functionalized dimer through 16-mer, **36-39**, respectively, as outlined in Scheme 3.9. As before, direct functionalization of the iodinated 16-mer was unsuccessful; however, coupling of the protodesilylated product of **38** did afford **39**, albeit in low yield.

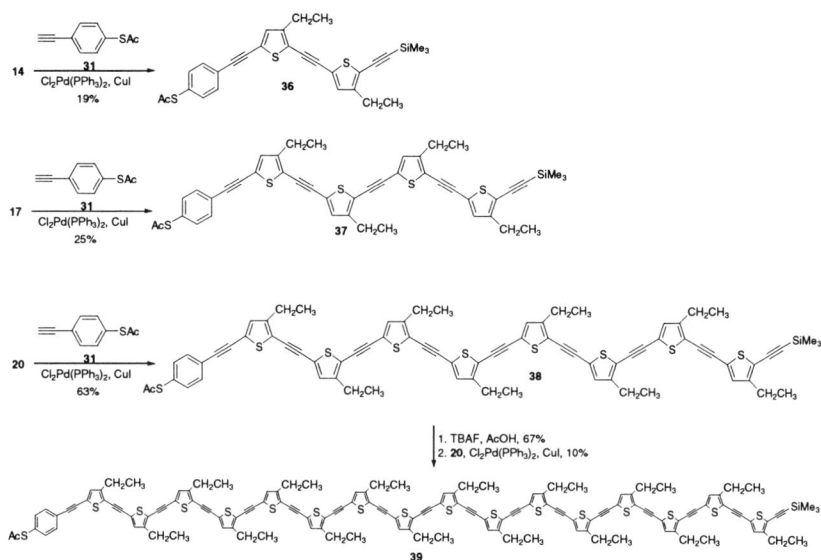

Scheme 3.9 Synthesis of tail-functionalized dimer through 16-mer.

We also prepared α,ω-difunctionalized oligomers based on some of our larger conjugated systems. For example, the α,ω-diphenylthioacetyl-octamer **40** has been prepared. Additionally, the phenylthioacetyl-containing octamer **37**, after protodesilylation, was coupled to the α,ω-diiodo-octamer **42** to afford the 128 Å long α,ω-diphenylthioacetyl-17-mer **43** (Scheme 3.10). These are the macromolecules that we are presently using in our attempts to bridge proximal gold probes.

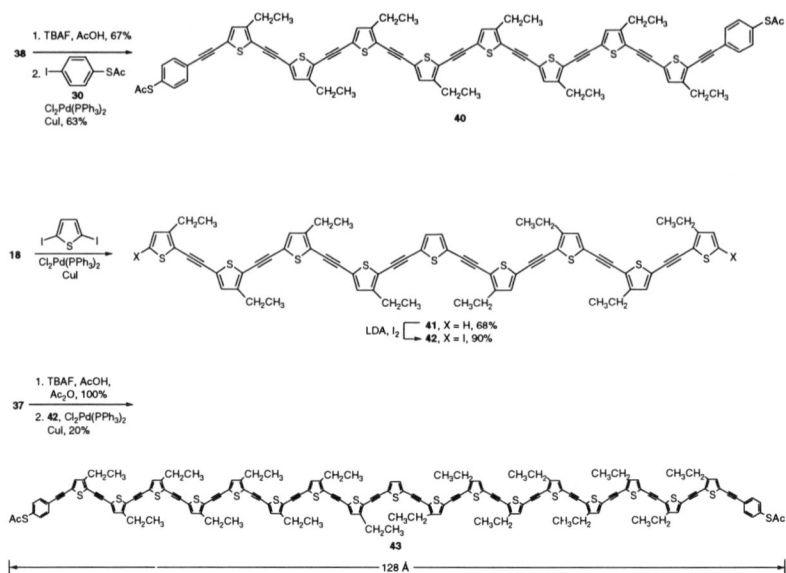

Scheme 3.10　Synthesis of α,ω-difunctionalized oligomers.

3.1.3. *Summary*

We outlined the rapid syntheses of oligo(3-ethyl-2,5-thiophene ethynylene)s via an iterative divergent/convergent approach starting from 3-ethyl-2-(trimethylsilylethynyl)thiophene (**8**). At each stage in the iteration, the length of the framework doubled. Only three sets of reaction conditions were needed for the entire iterative synthetic sequence: an iodination, a protodesilylation, and a Pd/Cu-catalyzed cross coupling. Convergent attachment of the complementary thiol end groups, protected as thioester moieties, was achieved. These serve as binding units for adhesion to gold surfaces. The rigid rod conjugated oligomers may act as molecular wires in molecular scale electronic devices. Moreover, they also serve as useful models for understanding analogous bulk polymers as judged by the optical spectra and SEC determined values.

3.1.4. *Experimental Procedures*

Unless otherwise noted, all operations were carried out under a dry, oxygen-free N_2 atmosphere. Molecular weight analyses were performed using two 30×75 cm GPC columns (10^5 Å 10 μ and 500 Å 5 μ) eluted with THF at 60 °C (flow rate 1.0 mL/min). Molecular weight results were based on eight polystyrene standards (M_W = 52000, 30300, 9200, 7000, 5050, 2950, 1060, and

580 with a correlation coefficient >0.9998) purchased from Polymer Laboratories Ltd. 3-Bromothiophene was purchased from Lancaster Synthesis Ltd. and used without purification. Alkyllithium reagents were obtained from Aldrich Chemical Company Inc. or FMC. Reagent grade diethyl ether and tetrahydrofuran (THF) were distilled under N_2 from sodium benzophenone ketyl. Reagent grade benzene and CH_2Cl_2 were distilled over calcium hydride. Bulk grade hexane was distilled prior to use. The 2-halogenations of 3-alkylthiophenes were performed according to the procedure of Uhlenbroek and Reinecke.[35] The terminal alkynes larger than the dimer stage were oxidatively unstable and they were used immediately after their preparation. Unless otherwise noted, all compounds were >95% pure as judged by NMR, GC, or combustion analyses. The oligomers' purities can be assessed by their polydispersity indexes M_w/M_n.

3-n-Butylthiophene (3). A procedure analogous to that of Kumada and co-workers was used as follows.[36] To magnesium turning (1.82 g, 75 mmol) in ether (20.0 mL) was added dropwise butyl bromide (10.3 g, 75 mmol) in ether (20.0 mL) at room temperature and an ice bath was used occasionally to maintain a mild reflux. The mixture was stirred at room temperature for 1 h and transferred via cannula to a solution of 3-bromothiophene (8.15 g, 50 mmol) and dichloro(diphenylphosphinopropane)nickel(II) (30 mg, 0.055 mmol) in ether (20.0 mL) at 0 °C. The mixture was then allowed to warm to room temperature and stir overnight before being poured into water with a few drops of 3 N HCl. The aqueous layer was extracted with ether and the organic extracts were washed with brine and dried over $MgSO_4$. The solvent was removed by rotary evaporation and the residue was purified by flash chromatography on silica gel (hexane) to provide 5.93 g (85%) of the title product as colorless liquid. IR (neat) 2929, 2859, 1466, 1079, 857, 834, 768 cm^{-1}. ^{1}H NMR (300 MHz, CDCl$_3$) δ 7.23 (m, 1 H), 6.91 (m, 2 H), 2.62 (t, J = 7.62 Hz, 2 H), 1.60 (pent, J = 7.58 Hz, 2 H), 1.32 (sext, J = 7.38 Hz, 2 H), 0.92 (t, J = 7.32 Hz, 3 H).

3-Hydroxyethylthiophene (4). To a one liter flame dried vessel was added ether (300 mL) and t-BuLi (212 mL, 467 mmol, 2.2 M in pentane) and the vessel cooled to -78 °C. To the solution was added 3-bromothiophene (19.5 mL, 208 mmol) dropwise and the reaction was stirred at -78 °C for 1 h at which time ethylene oxide (45.8 mL, 229 mmol, 5 M in ether) was added. The reaction was allowed to warm to room temperature and stir overnight. The mixture was poured into water and extracted with ether (3 × 20 mL) and the organic layer dried over $MgSO_4$ and filtered. The product was purified by flash chromatography (1:1 hexane:EtOAc) to afford 23 g (87 %) the title compound as a clear colorless liquid. IR (neat) 3356, 2933, 1410, 1047, 909, 835, 774 cm^{-1}. ^{1}H NMR (500 MHz, CDCl$_3$) δ 7.29-7.27 (m, 1 H), 7.04-7.03 (m, 1 H), 6.97-6.96 (m, 1 H), 3.83 (q, J = 6.1 Hz, 2 H), 2.89 (t, J = 6.4 Hz, 2 H), 1.43 (t, J = 5.9 Hz, 1 H).

3-n-Butyl-2-iodothiophene (6). To a solution of 3-butylthiophene (5.89 g, 42 mmol) in benzene (10.0 mL) was alternately added mercuric oxide (8.41 g, 38.9 mmol) and iodine (10.93 g, 43.1 mmol) in small portions at 0 °C. The mixture was stirred at room temperature overnight before filtration. The filtrate was poured into water and the aqueous layer was extracted with ether and the organic extracts were washed with brine and dried over $MgSO_4$. The solvent was removed by rotary evaporation and the residue was purified by a column chromatography on silica gel (hexane) to provide 8.06 g (72%) of the title product as colorless liquid. IR (neat) 2929, 2857, 2361, 1464, 1398, 966, 829, 715 cm^{-1}. ^{1}H NMR (300 MHz, CDCl$_3$) δ 7.36 (d, J = 5.51 Hz, 1 H), 6.73 (d, J = 5.46 Hz, 1 H), 2.48 (t, J = 7.63 Hz, 2 H), 1.54 (p, J = 7.70 Hz, 2 H), 1.35 (sext, J = 7.29 Hz, 2 H), 0.93 (t, J = 7.21 Hz, 2 H).

3-Hydroxyethyl-2-iodothiophene (7). See the preparation of **6** for the synthetic protocol. Used were 3-hydroxyethylthiophene (10.6 g, 82.8 mmol), benzene (100 mL), mercuric oxide (17.9 g, 82.9 mmol), iodine (21.0 g, 82.8 mmol), and flash chromatography on silica gel (3:2 hexane: EtOAc) to afford 17.11 g (81%) of the title product as a clear light yellow liquid. IR (neat) 2929, 2857, 1464, 1397, 965, 829 cm^{-1}. ^{1}H NMR (300 MHz, CDCl$_3$) δ 7.41 (d, J = 5.5 Hz, 1 H), 6.81 (d, J = 5.5 Hz, 1 H), 3.82 (t, J = 6.6 Hz, 2 H), 2.84 (t, J = 6.6 Hz, 2 H), 1.39 (br s, 1 H).

3-n-Butyl-2-(trimethylsilylethynyl)thiophene (9). To a 100 mL flame dried vessel containing 3-n-butyl-2-iodothiophene (20.93 g, 78.7 mmol), $((C_6H_5)_3P)_2PdCl_2$ (1.1 g, 1.57 mmol), CuI (0.44 g, 2.3 mmol), and THF (150 mL) was added diisopropylamine (11.9 mL, 85 mmol) at room temperature. The resulting clear brown solution was stirred for 5 min before trimethylsilylacetylene (TMSA), 12.0 mL, 85 mmol) was added. The N_2 outlet was removed and the septum capped. The reaction was stirred over night at room temperature and poured into water and the aqueous layer was extracted with ether (3 × 20 mL). The organic layer was dried over $MgSO_4$ and the crude product was concentrated in vacuo. The product was distilled at 100 °C /1 mm Hg to afford 15.0 g (81 %) of the title compound as a clear yellow liquid. IR (neat) 2958, 2144, 1458, 1250, 1084 cm^{-1}. ^{1}H NMR (500 MHz, CDCl$_3$) δ 7.10 (d, J = 5.2 Hz, 1 H), 6.81 (d, J = 5.1 Hz, 1 H), 2.67 (t, J = 7.6 Hz, 2 H), 1.58 (p, J = 7.5 Hz, 2 H), 1.32 (sext, J = 7.3 Hz, 2 H), 0.91 (t, J = 7.3 Hz, 3 H), 0.22 (s, 9 H).

2-Hydroxyethyl-1-(trimethylsilylethynyl)thiophene (10). See the preparation of **9** for the synthetic protocol. Used were 3-hydroxyethyl-2-iodothiophene (34.6 g, 136 mmol), $((C_6H_5)_3P)_2PdCl_2$ (1.9 g, 2.7 mmol), CuI (1.3 g, 6.8 mmol), THF (200 mL), diisopropylamine (19.0 mL, 136 mmol), TMSA (19.2 mL, 136 mmol), and flash chromatography (3:1 hexane:EtOAc) to afford 16.37 g (81 %) of the title compound as a clear yellow liquid. IR (neat) 3356, 2958, 2898, 2144, 1418, 1250, 1047, 843 cm^{-1}. ^{1}H NMR (300 MHz,

CDCl$_3$) δ 7.15 (d, J = 5.1 Hz, 1 H), 6.88 (d, J = 5.1 Hz, 1 H), 3.86 (q, J = 6.4 Hz, 2 H), 2.96 (t, J = 6.5 Hz, 2 H), 1.41 (t, J = 5.9 Hz, 1 H), 0.23 (s, 9 H).

Iodinated Tetramer 17. To a solution of diisopropylamine (7.89 g, 11 mL, 78 mmol) in ether (40 mL) at -78 °C was added dropwise n-BuLi (44 mL, 65 mmol, 1.49 M in hexanes). The solution was warmed to 0 °C for 30 min and then recooled to -78°C. **16** (4.35 g, 7.1 mmol) in ether (15 mL) at room temperature was then added to the -78 °C lithium diisopropylamide solution. This solution was warmed from -78 °C to 0 °C for 10 min then recooled to -78 °C. While at -78 °C, iodine (19.98 g, 78 mmol) in ether (140 mL) was added via cannula, and the solution was allowed to warm to room temperature overnight. The reaction was quenched with water and the aqueous layer was extracted with CH$_2$Cl$_2$ and the organic extracts were washed with brine and aqueous sodium thiosulfate. The organic layers were dried over MgSO$_4$. The solvent was removed by rotary evaporation and the residue was purified by flash chromatography (silica gel, hexane) to provide 4.86 g (93 %) of the title product as a fluorescent yellow liquid that darkened upon standing. IR (neat) 2966, 2140, 1459, 1249, 1020, 844 cm^{-1}. ^{1}H NMR (300 MHz, CDCl$_3$) δ 7.05 (s, 1 H), 7.03 (s, 1 H), 7.01 (s, 1 H), 2.69 (11 line m, 8 H), 1.25 (t, J = 7.60 Hz, 3 H), 1.24 (t, J = 7.56 Hz, 6 H), 1.22 (t, J = 7.63 Hz, 3 H), 0.25 (s, 1 H). ^{13}C NMR (100 MHz, CDCl$_3$) δ 151.24, 149.88, 149.59, 149.57, 137.78, 132.58, 132.51, 132.39, 123.43, 123.38, 123.34, 122.54, 120.07, 119.46, 119.35, 102.69, 96.65, 89.98, 89.76, 89.67, 86.61, 86.01, 85.75, 75.03, 23.02, 23.02, 22.86, 22.78, 22.71, 14.62, 14.55, 14.54, 14.32, -0.04. HRMS calcd for C$_{35}$H$_{33}$IS$_4$Si: 736.0279. Found: 736.0294.

Octamer 19. To a solution of **17** (0.23 g, 0.32 mmol) in THF (5 mL) was added **18** (0.36 g, 0.67 mmol) and diisopropylamine (0.10 g, 0.10 mL, 0.71 mmol). This solution was allowed to stir under N$_2$ in a dry box for 1 h. The catalysts ((C$_6$H$_5$)$_3$P)$_2$PdCl$_2$ (0.01 g, 0.02 mmol) and CuI (0.002 g, 0.01 mmol) (degassed in vacuo for 1 h) were added and the reaction was allowed to stir for 3 days. The reaction was quenched with water and the aqueous layer was extracted with ether and the organic extracts were washed with brine. The ether layers were dried over MgSO$_4$. The solvent was removed by rotary evaporation and the residue was purified by silica gel flash chromatography by first using hexane then slowly increasing to 9:1 hexane:CH$_2$Cl$_2$ to provide 0.26 g (72 %) of the title product as yellow-orange solid. IR (neat) 2967, 1511, 1460, 1321, 1264, 1187, 1061, 900, 844, 739 cm^{-1}. ^{1}H NMR (300 MHz, CDCl$_3$) δ 7.21 (d, J = 5.15 Hz, 1 H), 7.06 (s, 4 H), 7.04 (s, 2 H), 7.02 (s, 1 H), 6.90 (d, J = 5.14 Hz, 1 H), 2.72 (overlapping q, J = 7.6 Hz, 16 H), 1.26 (overlapping t, J = 7.6 Hz, 24 H), 0.26 (s, 9 H). ^{13}C NMR (125 MHz, CDCl$_3$) δ 150.30, 150.24, 150.07, 150.03, 150.00, 133.00, 132.94, 132.85, 132.80, 132.63, 128.27, 127.40, 124.44, 124.40, 124.17, 132.95, 123.87, 123.85, 123.83, 123.73, 122.94, 120.47, 119.88, 119.83, 119.82, 119.80, 119.71, 119.29, 119.24, 119.06, 117.67, 117.66, 103.10, 97.04, 90.17, 90.13, 90.08,

89.83, 89.78, 88.69, 88.68, 87.62, 97.60, 87.19, 87.13, 87.10, 87.03, 86.40, 23.57, 23.44, 23.42, 23.26, 15.10, 15.03, 14.94, 14.71, 0.34. UV (CH2Cl2) λ_{max} 432 nm, ε_{max} 2.5 × 10^5. LRMS calcd for $C_{67}H_{58}S_8Si$ at statistical isotopic maximum with one ^{13}C: 1147. Found: 1147. SEC: M_n = 1610, M_W = 1660.

Iodinated octamer 20. See the preparation of **17** for the synthetic protocol. Used were diisopropylamine (3.18 g, 4.4 mL, 31.4 mmol), ether (20 mL), n-BuLi (16 mL, 23.8 mmol, 1.49 M in hexanes), **19** (1.00 g, 0.87 mmol) in ether (10 mL), iodine (6.48 g, 25.5 mmol) in ether (45 mL), flash chromatography (silica gel, 9:1 hexane:CH2Cl2) to provide 0.70 g (63 %) of the title product as a fluorescent yellow liquid that darkened upon standing. IR (film) 2967, 2930, 2872, 2180, 2140, 1698, 1514, 1460, 1384, 1319, 1262, 1249, 1062, 902, 842, 740 cm^{-1}. ^1H NMR (300 MHz, CDCl3) δ 7.062 (s, 3 H), 7.058 (s, 2 H), 7.036 (s, 2 H), 7.014 (s, 1 H), 2.71 (overlapping q, J = 7.6 Hz, 16 H), 1.26-1.25 (overlapping t, J = 7.6 Hz, 24 H), 0.25 (s, 9 H). ^{13}C NMR (100 MHz, CDCl3) δ 151.22, 150.78, 150.33, 149.90, 149.64, 149.59, 137.79, 132.93, 132.59, 132.50, 132.40, 123.97, 123.47, 123.40, 122.56, 120.08, 119.50, 119.45, 119.36, 102.74, 96.71, 92.61, 90.49, 90.07, 89.84, 89.77, 86.80, 86.72, 86.09, 85.85, 75.18, 23.28,, 23.08, 22.90, 22.82, 14.82, 14.77, 14.68, 14.60, 14.37, 14.26, 14.01, 0.00. LRMS calcd for $C_{67}H_{57}IS_8Si$: 1272. Found: 1272.

Deprotected octamer 21. To a solution of **19** (0.77 g, 0.67 mmol) in MeOH (5 mL) and CH2Cl2 (5 mL) was added K2CO3 (0.69 g, 5.0 mmol). The solution was allowed to stir for 6 h before being poured into water. The aqueous layer was extracted with CH2Cl2 and the organic extracts were washed with brine. The combined organic layers were dried over MgSO4. The solvent was removed by rotary evaporation. No further purification was necessary to afford 0.72 g (100 %) of the title compound as yellow-orange solid. IR (film) 3298, 2967, 2932, 2872, 2180, 2097, 1460, 1321, 1262, 1187, 1061, 901, 843, 739, 661 cm^{-1}. ^1H NMR (300 MHz, CDCl3) δ 7.21 (d, J = 5.17 Hz, 1 H), 7.064 (s, 3 H), 7.059 (s, 2 H), 7.039 (s, 1 H), 7.030 (s, 1 H), 6.90 (d, J = 5.15 Hz, 1 H), 3.489 (s, 1 H), 2.71 (overlapping q, 16 H), 1.25 (overlapping t, J = 7.6 Hz, 24 H). ^{13}C NMR (100 MHz, CDCl3) δ 152.40, 150.31, 149.90, 149.84, 149.67, 149.63, 132.87, 132.61, 132.55, 132.45, 132.35, 132.23, 127.89, 127.03, 124.52, 124.05, 124.01, 123.77, 123.56, 123.46, 123.01, 119.42, 119.37, 119.33, 119.11, 118.90, 118.78, 118.68, 117.28, 89.78, 89.55, 89.48, 89.44, 88.33, 87.25, 87.02, 86.84, 86.76, 86.71, 86.08, 84.65, 81.34, 76.08, 23.19, 23.06, 23.04, 22.82, 14.73, 14.65, 14.57, 14.52.

16-mer 22. See the preparation of **19** for the synthetic protocol. Used were **20** (0.68 g, 0.53 mmol), THF (3 mL), **21** (0.70 g, 0.65 mmol), diisopropylamine (0.30 g, 0.42 mL, 3.0 mmol), ((C6H5)3P)2PdCl2 (0.02 g, 0.02 mmol), CuI (0.01 g, 0.03 mmol) and silica gel flash chromatography by first using hexane then slowly increasing to 9:1 hexane:CH2Cl2 and finally 7:3 hexane:CH2Cl2 to provide 0.60 g (50 %) of the title product as red-orange solid.

IR (neat) 2968, 2933, 2180, 2140, 1460, 1321, 1249, 1187, 1062, 906, 844, 734 cm^{-1}. ^{1}H NMR (300 MHz, CDCl$_3$) δ 7.21 (d, J = 5.15 Hz, 1 H), 7.064 (s, 7 H), 7.057 (s, 3 H), 7.045 (s, 2 H), 7.035 (s, 3 H), 6.90 (d, J = 5.17 Hz, 1 H), 2.71 (overlapping q, J = 7.6 Hz, 32 H), 1.25-1.24 (overlapping t, J = 7.6 Hz, 48 H), 0.26 (s, 9 H). ^{13}C NMR (125 MHz, CDCl$_3$) δ 150.30, 150.24, 150.08, 150.03, 133.02, 132.95, 132.87, 132.71, 132.63, 128.96, 128.86, 128.27, 127.39, 124.86, 124.37, 124.04, 123.92, 123.83, 119.79, 119.68, 119.56, 119.25, 119.05, 117.64, 90.08, 90.02, 89.79, 88.65, 87.57, 87.09, 23.62, 23.54, 23.41, 23.23, 15.13, 15.08, 15.01, 14.92, 14.68, 14.31, 0.33. UV (CH$_2$Cl$_2$) λ$_{max}$ 442 nm, ε$_{max}$ 2.1 × 10^5. Laser desorption MS (sinapic acid matrix) M^{+1} calcd for C$_{131}$H$_{106}$S$_{16}$Si^{+1} at statistical isotopic maximum with two ^{13}C: 2221.37. Found 2219.98 ± 1.20. SEC: M$_n$ = 3950, M$_w$ = 4160.

1-Bromo-4-trimethylsilylethynylbenzene. See the preparation of **19** for the synthetic protocol. Used were 1-bromo-4-iodobenzene (2.83 g, 10 mmol) in THF (15 mL), TMSA (1.47 g, 2.10 mL, 15 mmol), ((C$_6$H$_5$)$_3$P)$_2$PdCl$_2$ (0.21 g, 0.30 mmol), CuI (0.06 g, 0.30 mmol), diisopropylamine (2.16 g, 3.00 mL, 21.42 mmol) for 1 day. The reaction was passed through a silica gel plug to remove the catalyst using a 1:1 hexane:CH$_2$Cl$_2$ eluent. The solvent was removed by rotary evaporation and the residue was purified by gravity chromatography (silica gel) using hexane as the eluent to provide 0.62 g (24 %) of the title product as a light yellow-brown solid. IR (KBr) 2960, 2157, 1896, 1581, 1484, 1393, 1247, 1070, 1009, 846 cm-1. 1H NMR (300 MHz, CDCl$_3$) δ 7.41 (d, J = 8.39 Hz, 2 H), 7.31 (d, J = 8.37 Hz, 2 H), 0.27 (s, 9 H). ^{13}C NMR (100 MHz, CDCl$_3$) δ 133.38, 131.49, 122.79, 122.15, 103.97, 95.56, -0.03. HRMS calcd for C$_{11}$H$_{13}$BrSi: 251.9970. Found: 251.9966.

1-Bromo-4-ethynylbenzene (23). See the preparation of **21** for the synthetic protocol. Used were 1-bromo-4-trimethylsilylethynylbenzene (1.42 g, 5.61 mmol), MeOH (10 mL), and K$_2$CO$_3$ (2.77 g, 20.04 mmol) for 5 h. No purification was necessary to afford 1.00 g (98 %) of the title compound as brown liquid. IR (film) 3267, 2976, 1902, 1584, 1484, 1069, 821 cm^{-1}. ^{1}H NMR (400 MHz, CDCl$_3$) δ 7.44 (d, J = 8.45 Hz, 2 H), 7.33 (d, J = 8.49 Hz, 2 H), 3.11 (s, 1 H). ^{13}C NMR (100 MHz, CDCl$_3$) δ 133.55, 131.60, 123.15, 121.06, 82.59, 78.37.

Monofunctionalized Monomer 24. See the preparation of **19** for the synthetic protocol. Used were **11** (1.02 g, 3.04 mmol), THF (5 mL), **23** (0.64 g, 3.54 mmol), ((C$_6$H$_5$)$_3$P)$_2$PdCl$_2$ (0.063 g, 0.09 mmol), CuI (0.017 g, 0.09 mmol) and N,N-diisopropylethylamine (0.74 g, 1.0 mL, 5.74 mmol) for 1 day. Gravity chromatography (silica gel) using hexane as the eluent provided 1.01 g (86 %) of the title product as yellow liquid. IR (neat) 2962, 2931, 2872, 2198, 2146, 1489, 1248, 1068, 1007, 842 cm-1. 1H NMR (400 MHz, CDCl$_3$) δ 7.46 (d, J = 8.60 Hz, 2 H), 7.33 (d, J = 8.61 Hz, 2 H), 7.01 (s, 1 H), 2.66 (q, J = 7.60 Hz, 2 H), 1.21 (t, J = 7.65 Hz, 3 H), 0.25 (s, 9 H). ^{13}C NMR (100 NMR, CDCl$_3$) δ 149.84, 132.81, 132.46, 131.67, 122.84, 122.60, 121.66, 119.85, 102.44, 96.58,

92.41, 83.81, 22.79, 14.24, -0.11. HRMS calcd for $C_{19}H_{19}BrSSi$: 386.0160. Found: 386.0150. Anal. calcd for $C_{19}H_{19}BrSSi$: C, 59.07; H, 4.92. Found: C, 58.82; H, 5.01.

2-(4'-Bromophenylethynyl)-4-ethyl-5-ethynylthiophene. See the preparation of **21** for the synthetic protocol. Used were **24** (1.01 g, 2.61 mmol), MeOH (5 mL), CH_2Cl_2 (5 mL), and K_2CO_3 (1.60 g, 11.58 mmol) for 5 h. No purification was necessary to afford 0.79 g (96 %) of the title compound as brown liquid. IR (neat) 3298, 2968, 2933, 2874, 2204, 2098, 1899, 1487, 1393, 1085, 1068, 1010, 846 cm^{-1}. ^{1}H NMR (400 MHz, CDCl$_3$) δ 7.46 (d, J = 8.48 Hz, 2 H), 7.33 (d, J = 8.47, 2 H), 7.03 (s, 1 H), 3.47 (s, 1 H), 2.68 (q, J = 7.60 Hz, 2 H), 1.21 (t, J = 7.60 Hz, 3 H). ^{13}C NMR (100 MHz, CDCl$_3$) δ 150.25, 132.83, 132.41, 131.70, 123.10, 122.94, 121.57, 118.53, 92.49, 84.37, 83.58, 76.03, 22.75, 14.43.

Dibromo Difunctionalized Monomer 25. See the preparation of **19** for the synthetic protocol. Used were 2-(4'-bromophenylethynyl)-4-ethyl-5-ethynylthiophene (0.79 g, 2.51 mmol), THF (5 mL), 1-bromo-4-iodobenzene (1.45 g, 5.13 mmol), $((C_6H_5)_3P)_2PdCl_2$ (0.056 g, 0.08 mmol), CuI (0.015 g, 0.08 mmol), and diisopropylamine (1.08 g, 1.50 mL, 10.71 mmol) for 1 day. Gravity chromatography (silica gel) using hexane as the eluent to provided 0.94 g (80 %) of the title product as yellow solid. IR (film) 2960, 2202, 1631, 1530, 1481, 1391, 1064, 1003, 818 cm^{-1}. ^{1}H NMR (400 MHz, CDCl$_3$) δ 7.47 (d, J = 8.53 Hz, 2 H), 7.46 (d, J = 8.53 Hz, 2 H), 7.35 (d, J = 8.55 Hz, 2 H), 7.34 (d, J = 8.55 Hz, 2 H), 7.07 (s, 1 H), 2.73 (q, J = 7.60 Hz, 2 H), 1.26 (t, J = 7.60 Hz, 3 H). ^{13}C NMR (100 MHz, CDCl$_3$) δ 149.30, 132.82, 132.74, 132.72, 131.71, 123.14, 122.92, 122.78, 121.84, 121.62, 119.49, 95.37, 92.92, 83.81, 83.00, 22.95, 14.53. HRMS calcd for $C_{22}H_{14}Br_2S$: 467.9183. Found: 467.9186.

Dithioester Functionalized Monomer 26. To a solution of **25** (0.04 g, 0.085 mmol) in ether (2 mL) at -78 °C was added dropwise t-BuLi (1.0 mL, 2.1 mmol, 2.1 M in hexanes). The solution was stirred at -78 °C for 15 min. Sulfur powder (0.08 g, 2.5 mmol) was added as a solid, and the reaction was warmed to 0 °C for 30 min.[16] The solution was recooled to -78 °C and acetyl chloride (3.31 g, 3.0 mL, 4.22 mmol) was added in one portion. The solution was allowed to warm to room temperature overnight. The mixture was extracted with CH_2Cl_2 and dried over $MgSO_4$. The solvent was removed by rotary evaporation and the residue was purified by silica gel flash chromatography using 9:1 hexane:CH_2Cl_2, and finally a 7:3 hexane:CH_2Cl_2 eluent to provide 0.01 g (25 %) of the title product as yellow-brown solid. IR (film) 2925, 2860, 1712, 1591, 1198, 1116, 1022, 958, 836 cm^{-1}. ^{1}H NMR (400 MHz, CDCl$_3$) δ 7.52 (d, J = 8.35 Hz, 2 H), 7.51 (d, J = 8.42 Hz, 2 H), 7.38 (d, J = 8.37 Hz, 2 H), 7.37 (d, J = 8.41 Hz, 2 H), 7.09 (s, 1 H), 2.73 (q, J = 7.60 Hz, 2 H), 2.42 (s, 6 H), 1.25 (t, J = 7.59 Hz, 3 H). ^{13}C NMR (100 MHz, CDCl$_3$) δ 193.40, 149.51, 134.27, 134.24, 132.87, 131.95, 131.89, 128.29, 124.11,

123.87, 123.18, 95.69, 93.80, 84.00, 83.51, 30.31, 22.94, 14.51. HRMS calcd for $C_{26}H_{20}O_2S_3$: 460.0625. Found: 460.0631.

Coupling of 20 with 23. See the preparation of **19** for the synthetic protocol. Used were **20** (0.33 g, 0.26 mmol), THF (5 mL), **23** (0.18 g, 0.99 mmol), $((C_6H_5)_3P)_2PdCl_2$ (0.022 g, 0.03 mmol), CuI (0.006 g, 0.03 mmol), and N,N-diisopropylethylamine (0.37 g, 0.50 mL, 2.86 mmol) for 1 day. The residue was purified by silica gel flash chromatography by first using hexane then slowly increasing to a 9:1 hexane:CH_2Cl_2 to provide 0.26 g (77 %) of the title product as yellow-orange solid. IR (film) 2966, 2933, 2873, 2180, 2140, 1460, 843 cm^{-1}. ^{1}H NMR (400 MHz, CDCl$_3$) δ 7.46 (d, J = 8.45 Hz, 2 H), 7.34 (d, J = 8.41 Hz, 2 H), 7.05 - 7.01 (4 line m, 8 H), 2.72 (overlapping q, J = 7.6 Hz, 16 H), 1.26 (overlapping t, J = 7.6 Hz, 24 H), 0.25 (s, 9 H).

Compound 27. See the preparation of **21** for the synthetic protocol. Used were the **20/23** coupled product (0.22 g, 0.17 mmol), MeOH (5 mL), CH$_2$Cl$_2$ (5 mL), and K$_2$CO$_3$ (1.42 g, 10.3 mmol) for 5 h. No purification was necessary to afford 0.20 g (96 %) of the title compound as yellow-orange solid that, due to its oxidative instability, was carried on immediately for the preparation of **29**.

Coupling of 21 with 1-bromo-4-iodobenzene. See the preparation of **19** for the synthetic protocol. Used were **21** (0.18 g, 0.17 mmol), THF (5 mL), 1-bromo-4-iodobenzene (0.04 g, 0.14 mmol), $((C_6H_5)_3P)_2PdCl_2$ (0.022 g, 0.03 mmol), CuI (0.006 g, 0.03 mmol), and N,N-diisopropylethylamine (0.37 g, 0.50 mL, 2.86 mmol) for 1 day. The residue was purified by silica gel flash chromatography by first using hexane then slowly increasing to a 9:1 hexane:CH$_2$Cl$_2$ to provide 0.16 g (77 %) of the title product as yellow-orange solid. IR (film) 3074, 2960, 2932, 2872, 2181, 1460, 1069, 843 cm-1. 1H NMR (400 MHz, CDCl$_3$) δ 7.47 (d, J = 8.43 Hz, 2 H), 7.35 (d, J = 8.45 Hz, 2 H), 7.21 (d, J = 5.15 Hz, 1 H), 7.06 (s, 5 H), 7.04 (s, 2 H), 6.90 (d, J = 5.14 Hz, 1 H), 2.72 (overlapping q, J = 7.60 Hz, 16 H), 1.26 (overlapping t, J = 7.60 Hz, 24 H). ^{13}C NMR (100 MHz, CDCl$_3$) δ 152.39, 149.89, 149.84, 149.67, 149.63, 149.34, 132.80, 132.72, 132.61, 132.54, 132.45, 132.23, 131.70, 127.87, 126.99, 123.44, 123.04, 122.78, 121.85, 119.70, 119.39, 117.25, 89.70, 89.40, 88.27, 87.17, 86.69, 23.15, 23.02, 22.97, 22.68, 14.68, 14.61, 14.52, 14.15.

Compound 28. See the preparation of **17** for the synthetic protocol. Used were diisopropylamine (1.59 g, 2.2 mL, 15.71 mmol) in THF (2 mL), n-BuLi (8.5 mL, 12.83 mmol, 1.51 M in hexanes), the **21/1-bromo-4-iodobenzene** coupled product (0.28 g, 0.23 mmol) in ether (10 mL), and iodine (3.50 g, 13.79 mmol) in ether (40 mL). The residue was purified by silica gel flash chromatography by first using hexane and then slowly increasing to a 9:1 hexane:CH$_2$Cl$_2$ eluent to provide 0.25 g (81 %) of the title product as an reddish-orange solid. 1H NMR (400 MHz, CDCl$_3$) δ 7.47 (d, J = 8.48 Hz, 2 H), 7.35 (d, J = 8.55 Hz, 2 H), 7.06 (s, 4 H), 7.03 (s, 4 H), 2.71 (overlapping q, J = 7.6 Hz, 16 H), 1.25 (overlapping t, J = 7.6 Hz, 24 H).

Compound 29. See the preparation of **19** for the synthetic protocol. Used were **28** (0.23 g, 0.18 mmol), THF (1.0 mL), **27** (0.08 g, 0.06 mmol), ((C_6H_5)$_3$P)$_2$PdCl$_2$ (0.004 g, 0.006 mmol), CuI (0.002 g, 0.01 mmol), and N,N-diisopropylethylamine (0.370 g, 0.40 mL, 2.30 mmol) for 1 day. The residue was purified by silica gel flash chromatography by first using hexane then slowly increasing to a 9:1 hexane:CH$_2$Cl$_2$ eluent then a 7:3 hexane:CH$_2$Cl$_2$ eluent to provide 0.01 g (55 %) of the title product as yellow-orange solid. ^{1}H NMR (400 MHz, CDCl$_3$) δ 7.46 (d, J = 8.41 Hz, 4 H), 7.34 (d, J = 8.42 Hz, 4 H), 7.07 (s, 4 H), 7.06 (s, 10 H), 7.04 (s, 2 H), 2.72 (overlapping q, J = 7.6 Hz, 32 H), 1.26 (overlapping t, J = 7.6 Hz, 48 H). ^{13}C NMR (100 MHz, CDCl$_3$) δ 149.67, 132.95, 132.80, 132.69, 132.61, 131.70, 123.41, 119.39, 89.69, 86.66, 23.20, 23.00, 22.66, 14.72, 14.51, 14.13, 13.90.

1-Iodo-4-thioacetylbenzene (30). To a solution of 1,4-diiodobenzene (4.95 g, 17.5 mmol) in ether (2 mL) at -78 °C was added dropwise t-BuLi (16 mL, 33 mmol, 2.10 M in hexanes). The solution was stirred at -78 °C for 5 min. Sulfur powder (0.08 g, 2.5 mmol) in THF (75 mL) at 0 °C was added and the reaction was warmed to 0 °C for 30 min.[16] The solution was recooled to -78 °C and acetyl chloride (1.77 g, 1.6 mL, 22.5 mmol) was added in one portion. The solution was allowed to warm to room temperature overnight. The mixture was extracted with CH$_2$Cl$_2$ and dried over MgSO$_4$. The solvent was removed by rotary evaporation and the residue was purified by silica gel flash chromatography using a 25:1 hexane:ether eluent to provide 2.34 g (56 %) of the title product as yellowish white solid. IR (film) 3080, 1907, 1694, 1466, 1382, 1122, 1005, 811 cm^{-1}. ^{1}H NMR (400 MHz, CDCl$_3$) δ 7.72 (d, J = 8.35 Hz, 2H), 7.11 (d, J = 8.35 Hz, 2 H), 2.40 (s, 3 H). ^{13}C NMR (100 MHz, CDCl$_3$) δ 193.10, 138.36, 135.94, 127.79, 95.93, 30.25. HRMS calcd for C$_8$H$_7$IOS: 277.9262. Found: 277.9272. Anal. calcd for C$_8$H$_7$IOS: C, 34.53; H, 2.52. Found: C, 34.67; H: 2.42.

1-Thioacetyl-4-(trimethylsilylethynyl)benzene. See the preparation of **19** for the synthetic protocol. Used were **30** (1.11 g, 3.99 mmol), THF (5.0 mL), TMSA (0.58 g, 0.84 mL, 5.90 mmol), ((C_6H_5)$_3$P)$_2$PdCl$_2$ (0.14 g, 0.2 mmol), CuI (0.04 g, 0.02 mmol), and N,N-diisopropylethylamine (0.83 g, 1.12 mL, 6.42 mmol) for 1 day. The residue was purified by silica gel flash chromatography by first using hexane then slowly increasing to 9:1 hexane:CH$_2$Cl$_2$ to provide 0.88 g (89 %) of the title product as an off white solid. IR (neat) 2960, 2159, 1713, 1484, 1250, 864 cm^{-1}. ^{1}H NMR (300 MHz, CDCl$_3$) δ 7.46 (d, J = 8.10 Hz, 2 H), 7.33 (d, J = 8.07 Hz, 2 H), 2.40 (s, 3 H), 0.23 (s, 9 H). ^{13}C NMR (100 MHz, CDCl$_3$) δ 193.28, 134.04, 132.49, 128.32, 124.36, 104.15, 96.20, 30.24, -0.13. HRMS calcd for C$_{13}$H$_{16}$OSSi: 248.0691. Found: 248.0694. Anal. calcd for C$_{13}$H$_{16}$OSSi: C, 62.90; H, 6.45. Found: C, 62.66; H, 6.19.

4-Ethynyl-1-thioacetylbenzene (31). To a solution of 1-thioacetyl-4-(trimethylsilylethynyl)benzene (0.17 g, 0.69 mmol) in THF (2 mL) at 0 °C was added acetic acid (0.01g, 0.01 mL, 0.167 mmol) and acetic anhydride (0.01 g, 0.01 mL, 0.10 mmol) followed by the dropwise addition of tetrabutylammonium fluoride (2.88 g, 3.20 mL, 11.03 mmol). The solution was allowed to warm to room temperature for 5 min. The reaction was run through a silica gel plug to remove the solid impurities. The solvent was removed by rotary evaporation. No further purification was necessary to afford 0.20 g (96 %) of the title compound as yellow-orange solid. IR (film) 3288, 2924, 1708, 1483, 1398, 1353, 1126, 951, 829 cm^{-1}. ^{1}H NMR (300 MHz, CDCl$_3$) δ 7.50 (d, J = 8.46 Hz, 2 H), 7.35 (d, J = 8.55 Hz, 2 H), 3.13 (s, 1 H), 2.41 (s, 3 H). ^{13}C NMR (75 MHz, CDCl$_3$) δ 193.26, 134.18, 132.73, 128.80, 123.35, 82.84, 78.85, 30.31. LRMS calcd for C$_{10}$H$_8$OS: 176. Found: 176.

Compound 32. See the preparation of **19** for the synthetic protocol. Used were **12** (0.22 g, 1.61 mmol), THF (2.0 mL), **30** (0.29 g, 1.04 mmol), ((C$_6$H$_5$)$_3$P)$_2$PdCl$_2$ (0.041g, 0.06 mmol), CuI (0.013 g, 0.07 mmol), and N,N-diisopropylethylamine (0.374 g, 1.0 mL, 5.74 mmol) for 1 day. The residue was purified by silica gel flash chromatography by first using hexane then slowly increasing to 9:1 hexane:CH$_2$Cl$_2$ and finally 7:3 hexane:CH$_2$Cl$_2$ to provide 0.16 g (54 %) of the title product as yellow-orange solid. IR (neat) 3103, 2967, 2932, 2873, 2201, 1709, 1592, 1486, 1459, 1396, 1258, 1117, 1015, 950, 827 cm^{-1}. ^{1}H NMR (300 MHz, CDCl$_3$) δ 7.51 (d, J = 8.54 Hz, 2 H), 7.37 (d, J = 8.51 Hz, 2 H), 7.20 (d, J = 5.17 Hz, 1 H), 6.90 (d, J = 5.17 Hz, 1 H), 2.77 (q, J = 7.60 Hz, 2 H), 2.42 (s, 3 H), 1.25 (t, J = 7.60 Hz, 3 H). ^{13}C NMR (100 MHz, CDCl$_3$) δ 193.35, 149.63, 134.25, 131.82, 127.99, 127.88, 126.65, 124.57, 117.51, 94.63, 84.14, 30.27, 22.98, 14.70. HRMS calcd for C$_{16}$H$_{14}$OS$_2$: 286.0486. Found: 286.0496. Anal. calcd for C$_{16}$H$_{14}$OS$_2$: C, 67.13; H, 4.89. Found: C, 66.87; H: 4.91.

Compound 33. See the preparation of **19** for the synthetic protocol. Used were **15** (0.26 g, 0.196 mmol), THF (0.5 mL), **30** (0.19 g, 0.70 mmol), ((C$_6$H$_5$)$_3$P)$_2$PdCl$_2$ (0.014 g, 0.02 mmol), CuI (0.004 g, 0.02 mmol), and N,N-diisopropylethylamine (0.21 g, 0.28 mL, 1.62 mmol) for 1 day. The residue was purified by silica gel flash chromatography by first using hexane then slowly increasing to 9:1 hexane:CH$_2$Cl$_2$ and finally 7:3 hexane:CH$_2$Cl$_2$ to provide 0.20 g (68 %) of the title product as a yellowish brown liquid. IR (neat) 3103, 2967, 2932, 2193, 1710, 1591, 1120, 827 cm^{-1}. ^{1}H NMR (300 MHz, CDCl$_3$) δ 7.52 (d, J = 8.34 Hz, 2 H), 7.38 (d, J = 8.29 Hz, 2 H), 7.21 (d, J = 5.16 Hz, 1 H), 7.05 (s, 1 H), 6.90 (d, J = 5.16 Hz, 1 H), 2.76 (q, J = 7.57 Hz, 2 H), 2.73 (q, J = 7.53 Hz, 2 H), 1.26 (t, J = 7.56 Hz, 3 H), 1.25 (t, J = 7.61 Hz, 3 H). ^{13}C NMR (100 MHz, CDCl$_3$) δ 193.31, 149.81, 149.47, 134.24, 132.24, 131.85, 128.27, 127.85, 126.92, 124.18, 123.64, 119.13, 117.25, 95.59, 88.89, 83.61, 30.27, 22.96, 14.64, 14.50. HRMS calcd for C$_{24}$H$_{20}$OS$_3$: 420.0676. Found: 420.0674.

Compound 34. See the preparation of **19** for the synthetic protocol. Used were **18** (0.85 g, 1.58 mmol), THF (1.0 mL), **30** (0.28 g, 1.01 mmol),

$((C_6H_5)_3P)_2PdCl_2$ (0.01 g, 0.02 mmol), CuI (0.002 g, 0.01 mmol), and N,N-diisopropylethylamine (0.37 g, 0.50 mL, 2.87 mmol) for 1 day. The residue was purified by silica gel flash chromatography by first using hexane then slowly increasing to 9:1 hexane:CH_2Cl_2 and finally 7:3 hexane:CH_2Cl_2 to provide 0.37 g (54 %) of the title product as reddish brown liquid. IR (neat) 2968, 2933, 2874, 2249, 2194, 1709, 1460, 1120, 907, 827 cm^{-1}. ^{1}H NMR (400 MHz, CDCl$_3$) δ 7.52 (d, J = 8.21 Hz, 2 H), 7.38 (d, J = 8.19 Hz, 2 H), 7.20 (d, J = 5.12 Hz, 1 H), 7.07 (s, 1 H), 7.06 (s, 1 H), 7.04 (s, 1 H), 6.90 (d, J = 5.13 Hz, 1 H), 2.74 (overlapping q, J = 7.6 Hz, 8 H), 2.42 (s, 3 H), 1.25 (overlapping t, J = 7.6 Hz, 12 H). ^{13}C NMR (100 MHz, CDCl$_3$) δ 193.33, 149.85, 149.64, 149.63, 149.54, 134.26, 132.62, 132.55, 132.23, 131.87, 128.33, 127.86, 126.97, 124.10, 123.94, 123.45, 123.12, 119.29, 118.83, 117.21, 95.86, 89.59, 89.33, 88.21, 87.10, 86.68, 86.42, 83.52, 30.28, 22.98, 22.95, 14.65, 14.50. HRMS calcd for $C_{40}H_{32}OS_5$: 688.1057. Found: 688.1046.

Compound 35. See the preparation of **19** for the synthetic protocol. Used were **21** (0.20 g, 0.19 mmol), THF (0.5 mL), **30** (0.03 g, 0.11 mmol), $((C_6H_5)_3P)_2PdCl_2$ (0.01 g, 0.02 mmol), CuI (0.002 g, 0.01 mmol), and N,N-diisopropylethylamine (0.37 g, 0.50 mL, 2.87 mmol) for 1 day. The residue was purified by silica gel flash chromatography by first using hexane then slowly increasing to 9:1 hexane:CH_2Cl_2 and finally 7:3 hexane:CH_2Cl_2 to provide 0.03 g (22 %) of the title product as yellow-orange film. IR (film) 2966, 2931, 2872, 2181, 1709, 1459, 1186, 1118, 903, 842 cm^{-1}. ^{1}H NMR (400 MHz, CDCl$_3$) δ 7.52 (d, J = 8.50 Hz, 2 H), 7.38 (d, J = 8.50 Hz, 2 H), 7.21 (d, J = 5.19 Hz, 1 H), 7.07 (s, 1 H), 7.065 (s, 3 H), 7.058 (s, 1 H), 7.04 (s, 1 H), 6.90 (d, J = 5.16 Hz, 1 H), 2.71 (overlapping q, J = 7.6 Hz, 16 H), 2.42 (s, 3 H), 1.25 (overlapping t, J = 7.6 Hz, 24 H).

Compound 36. See the preparation of **19** for the synthetic protocol. Used were **14** (0.14 g, 0.30 mmol), THF (0.5 mL), **31** (0.06 g, 0.34 mmol), $((C_6H_5)_3P)_2PdCl_2$ (0.01 g, 0.02 mmol), CuI (0.003 g, 0.02 mmol), and N,N-diisopropylethylamine (0.37 g, 0.50 mL, 2.87 mmol) for 1 day. The residue was purified by silica gel flash chromatography by first using hexane then slowly increasing to 9:1 hexane:CH_2Cl_2 and finally 7:3 hexane:CH_2Cl_2 to provide 0.03 g (19 %) of the title product as yellow liquid. IR (neat) 2966, 2140, 1712, 1488, 1457, 1250, 1121, 1016, 949, 844 cm^{-1}. ^{1}H NMR (300 MHz, CDCl$_3$) δ 7.51 (d, J = 8.45 Hz, 2 H), 7.38 (d, J = 8.45 Hz, 2 H), 7.07 (s, 1 H), 7.01 (s, 1 H), 2.70 (q, J = 7.58 Hz, 2 H), 2.66 (q, J = 7.56 Hz, 2 H), 2.42 (s, 3 H), 1.21 (t, J = 7.65 Hz, 3 H), 0.24 (s, 9 H). ^{13}C NMR (100 MHz, CDCl$_3$) δ 193.20, 149.89, 149.55, 134.21, 132.84, 132.41, 131.94, 128.53, 123.84, 123.41, 122.51, 120.03, 119.37, 102.65, 96.60, 93.41, 89.56, 85.94, 84.30, 30.28, 22.97, 22.83, 14.49, 14.28, -0.08. HRMS calcd for $C_{29}H_{28}OS_3Si$: 516.1072. Found: 516.1066.

Compound 37. See the preparation of **19** for the synthetic protocol. Used were **17** (1.83 g, 2.49 mmol), THF (3 mL), **31** (0.47 g, 2.67 mmol),

$((C_6H_5)_3P)_2PdCl_2$ (0.07 g, 0.1 mmol), CuI (0.02 g, 0.01 mmol), and N,N-diisopropylethylamine (0.74 g, 1.00 mL, 5.72 mmol) for 2 days. The residue was purified by silica gel flash chromatography by first using hexane then slowly increasing to a 9:1 hexane:CH_2Cl_2, then a 7:3 hexane:CH_2Cl_2 eluent to provide 0.48 g (25 %) of the title product as yellow-brown liquid. IR (film) 2967, 2933, 2874, 2184, 2140, 1712, 1460, 1249, 1121, 845 cm^{-1}. ^{1}H NMR (300 MHz, CDCl$_3$) δ 7.51 (d, J = 8.32 Hz, 2 H), 7.38 (d, J = 8.29 Hz, 2 H), 7.082 (s, 1 H), 7.062 (s, 1 H), 7.056 (s, 1 H), 7.010 (s, 1 H), 2.71 (overlapping q, J = 7.6 Hz, 8 H), 2.42 (s, 3 H), 1.25 (overlapping t, J = 7.6 Hz, 12 H), 0.24 (s, 9 H). ^{13}C NMR (100 MHz, CDCl$_3$) δ 193.26, 149.90, 149.64, 149.60, 134.22, 132.86, 132.63, 132.40, 131.94, 128.52, 123.83, 123.52, 123.40, 123.29, 122.50, 120.03, 119.44, 119.37, 119.30, 102.67, 96.60, 93.48, 89.73, 89.67, 89.51, 86.65, 86.59, 85.96, 84.30, 30.30, 23.00, 22.83, 22.68, 14.53, 14.30, 14.17, -0.07. HRMS calcd for $C_{45}H_{40}OS_5Si$: 784.1452. Found: 784.1440.

Compound 38. See the preparation of **19** for the synthetic protocol. Used were **20** (0.23 g, 0.18 mmol), THF (0.5 mL), **31** (0.09 g, 0.50 mmol), $((C_6H_5)_3P)_2PdCl_2$ (0.01 g, 0.02 mmol), CuI (0.002 g, 0.01 mmol), and N,N-diisopropylethylamine (0.37 g, 0.50 mL, 2.87 mmol) for 1 day. The residue was purified by silica gel flash chromatography by first using hexane then slowly increasing to 9:1 hexane:CH_2Cl_2 and finally 7:3 hexane:CH_2Cl_2 to provide 0.15 g (63 %) of the title product as yellow-orange solid. IR (neat) 2967, 2933, 2140, 1711, 1459, 1249, 904, 843 cm^{-1}. ^{1}H NMR (300 MHz, CDCl$_3$) δ 7.51 (d, J = 8.50 Hz, 2 H), 7.38 (d, J = 8.54 Hz, 2 H), 7.08 (s, 1 H), 7.06 (s, 5 H), 7.05 (s, 1 H), 7.01 (s, 1 H), 2.69 (overlapping q, J = 7.6 Hz, 16 H), 2.42 (s, 3 H), 1.23 (overlapping t, J = 7.6 Hz, 24 H), 0.25 (s, 9 H). ^{13}C NMR (100 MHz, CDCl$_3$) δ 193.22, 149.89, 149.67, 134.20, 132.84, 132.61, 132.39, 131.94, 128.52, 123.83, 123.41, 119.37, 89.66, 86.64, 30.28, 23.19, 22.99, 22.81, 14.71, 14.49, 14.26, 13.89, -0.10.

Desilylation of 38. To a solution of **38** (0.11 g, 0.08 mmol) in THF (1 mL) at 0 °C was added acetic acid (0.01g, 0.01 mL, 0.18 mmol) followed by tetra-n-butylammonium fluoride (0.36 g, 0.40 mL, 1.38 mmol). The solution was allowed to warm to 23 °C for 5 min. The reaction mixture was run through a silica gel plug using 1:1 hexane:CH_2Cl_2 eluent. The solvent was removed by rotary evaporation. No further purification was necessary to afford 0.07 g (67 %) of the title compound as yellow-orange solid. IR (film) 3298, 2966, 2932, 2872, 2183, 2096, 1711, 1186, 903, 843 cm^{-1}. ^{1}H NMR (300 MHz, CDCl$_3$) δ 7.51 (d, J = 8.53 Hz, 2 H), 7.38 (d, J = 8.55 Hz, 2 H), 7.08 (s, 1 H), 7.06 (s, 6 H), 7.03 (s, 1 H), 3.48 (s, 1 H), 2.71 (overlapping q, J = 7.6 Hz, 16 H), 2.42 (s, 3 H), 1.23 (overlapping t, J = 7.6 Hz, 24 H).

Compound 39. See the preparation of **19** for the synthetic protocol. Used were **20** (0.18 g, 0.14 mmol), THF (1.0 mL), desilylated **38** (0.08 g, 0.06 mmol), $((C_6H_5)_3P)_2PdCl_2$ (0.004 g, 0.006 mmol), CuI (0.002 g, 0.01 mmol), and N,N-diisopropylethylamine (0.30 g, 0.40 mL, 2.32 mmol) for 3 days. The residue was purified by silica gel flash chromatography by first using hexane then slowly increasing to 9:1 hexane:CH$_2$Cl$_2$ and finally 7:3 hexane:CH$_2$Cl$_2$ to provide 0.01 g (10 %) of the title product as yellow-orange solid. IR (neat) 2967, 2930, 2181, 2140, 1712, 1634, 1461, 1320, 1250, 1186, 1092, 1015, 948, 902, 844 cm^{-1}. ^{1}H NMR (400 MHz, CDCl$_3$) δ 7.51 (d, J = 8.42 Hz, 2 H), 7.38 (d, J = 8.46 Hz, 2 H), 7.08 (s, 3 H), 7.06 (s, 11 H), 7.01 (s, 2 H), 2.71 (overlapping q, J = 7.6 Hz, 32 H), 2.42 (s, 3 H), 1.25-1.23 (overlapping t, J = 7.6 Hz, 48 H), 0.24 (s, 9 H).

Compound 40. See the preparation of **19** for the synthetic protocol. Used were desilylated **38** (0.07 g, 0.06 mmol), THF (0.3 mL), **30** (0.03 g, 0.11 mmol), ((C$_6$H$_5$)$_3$P)$_2$PdCl$_2$ (0.01 g, 0.02 mmol), CuI (0.002 g, 0.01 mmol), and N,N-diisopropylethylamine (0.07 g, 0.10 mL, 0.57 mmol) for 3 days. The residue was purified by silica gel flash chromatography by first using hexane then slowly increasing to 9:1 hexane:CH$_2$Cl$_2$, followed by 7:3 hexane:CH$_2$Cl$_2$ and finally 6:4 hexane:CH$_2$Cl$_2$ to provide 0.05 g (63 %) of the title product as yellow-orange solid. IR (neat) 2966, 2930, 2180, 1712, 1459, 1119, 843 cm^{-1}. ^{1}H NMR (400 MHz, CDCl$_3$) δ 7.518 (d, J = 8.41 Hz, 2 H), 7.512 (d, J = 8.43 Hz, 2 H), 7.384 (d, J = 8.36 Hz, 2 H), 7.380 (d, J = 8.34 Hz, 2 H), 7.08 (s, 1 H), 7.07 (s, 1 H), 7.06 (s, 6 H), 2.71 (overlapping q, J = 7.6 Hz, 16 H), 2.422 (s, 3 H), 2.420 (s, 3 H), 1.25 (overlapping t, J = 7.6 Hz, 24 H). ^{13}C NMR (125 MHz, CDCl$_3$) δ 193.68, 193.63, 150.08, 150.04, 149.93, 134.64, 134.60, 133.24, 133.02, 132.34, 132.26, 128.93, 128.76, 124.50, 124.24, 123.92, 123.81, 119.81, 119.77, 119.75, 119.69, 93.83, 90.04, 87.01, 83.91, 30.67, 30.09, 23.38, 23.35, 14.88.

Compound 41. See the preparation of **19** for the synthetic protocol. Used were **18** (1.42, 2.64 mmol), THF (2.5 mL), 2,5-diiodothiophene18 (0.38 g, 1.1 mmol), ((C$_6$H$_5$)$_3$P)$_2$PdCl$_2$ (0.07 g, 0.1 mmol), CuI (0.008 g, 0.04 mmol), diisopropylamine (0.74 g, 1.0 mL, 5.74 mmol) for 1 day. The residue was purified by gravity chromatography (silica gel) using a 9:1 hexane:CH$_2$Cl$_2$ eluent to provide 0.89 g (68 %) of the title product as yellow-orange solid. IR (film) 3101, 2966, 2930, 2872, 2183, 1459, 1188, 1060, 842 cm^{-1}. ^{1}H NMR (400 MHz, CDCl$_3$) δ 7.21 (d, J = 5.14 Hz, 2 H), 7.15 (s, 2 H), 7.07 (s, 2 H), 7.06 (s, 2 H), 7.05 (s, 2 H), 6.91 (d, J = 5.14 Hz, 2 H), 2.73 (overlapping q, J = 7.6 Hz, 16 H), 1.26 (overlapping t, J = 7.6 Hz, 24 H). ^{13}C NMR (100 MHz, CDCl$_3$) δ 149.86, 149.82, 149.71, 149.65, 132.62, 132.56, 132.24, 132.09, 127.88, 127.01, 124.61, 123.98, 123.56, 119.26, 119.23, 118.86, 117.24, 89.61, 89.40, 89.27, 88.27, 87.18, 86.98, 86.77, 23.03, 23.01, 22.70, 14.70, 14.55, 14.18. LRMS calcd for C$_{68}$H$_{52}$S$_9$: 1156. Found: 1156.

Compound 42. See the preparation of **17** for the synthetic protocol. Used were n-BuLi (16 mL, 25.0 mmol, 1.56 M in hexanes), diisopropylamine (5.06 g, 7.0 mL, 50.0 mmol) in ether (5.0 mL), **41** (0.90 g, 0.78 mmol) in ether (30 mL), and iodine (12.94 g, 51.0 mmol) in ether (80 mL). The residue was purified by silica gel flash chromatography using a 9:1 hexane:CH$_2$Cl$_2$ eluent to provide (0.99 g) of the title product as a red-orange solid that was 90 % pure by spectral analysis. IR (film) 2965, 2929, 2871, 2184, 1459, 1262, 1185, 1061, 902, 841, 801, 739 cm^{-1}. ^{1}H NMR (400 MHz, CDCl$_3$) δ 7.08 (s, 2 H), 7.06 (s, 3 H), 7.04 (s, 5 H), 2.71 (overlapping q, J = 7.6 Hz, 16 H), 1.25 (overlapping t, J = 7.6 Hz, 24 H).

Desilylation of 37. See the desilylation of **38** for the synthetic protocol. Used were acetic acid (0.01g, 0.01 mL, 0.18 mmol), **37** (0.11 g, 0.14 mmol), THF (5.0 mL), and TBAF (2.71 g, 3.0 mL, 10.36 mmol). The solution was allowed to warm to 23°C over 8 min. The reaction mixture was passed through a silica gel plug to remove the solid impurities using a 1:1 hexane:CH$_2$Cl$_2$ eluent. The solvent was removed by rotary evaporation to afford (0.10 g, 100 %) of the title compound as brown liquid. IR (neat) 3298, 2967, 2932, 2874, 2184, 2097, 1710, 1591, 1460, 1396, 1121, 845, 738 cm^{-1}. ^{1}H NMR (400 MHz, CDCl$_3$) δ 7.51 (d, J = 8.46 Hz, 2 H), 7.38 (d, J = 8.52 Hz, 2 H), 7.08 (s, 1 H), 7.06 (s, 1 H), 7.05 (s, 1 H), 7.03 (s, 1 H), 3.49 (s, 1 H), 2.70 (overlapping q, J = 7.60 Hz, 8 H), 2.42 (s, 3 H), 1.25 (overlapping t, J = 7.60 Hz, 12 H). ^{13}C NMR (100 MHz, CDCl$_3$) δ 193.32, 150.31, 149.68, 149.64, 134.22 (2C), 132.86, 132.63, 132.61, 132.36, 131.94 (2C), 128.51, 123.82, 123.52, 123.41, 123.38, 122.96, 119.34, 119.30, 118.71, 93.48, 89.63, 89.50, 89.46, 86.65, 86.62, 85.99, 84.60, 84.30, 76.02, 30.30, 23.00, 22.77, 14.52, 14.48, 14.16.

Compound 43. See the preparation of **19** for the synthetic protocol. Used were desilylated **37** (0.10 g, 0.14 mmol), **42** (0.09 g, 0.06 mmol), THF (0.5 mL), ((C$_6$H$_5$)$_3$P)$_2$PdCl$_2$ (0.084 g, 0.12 mmol), CuI (0.089 g, 0.47 mmol), and N,N-diisopropylethylamine (0.37 g, 0.5 mL, 2.87 mmol) for 2 days. The residue was purified by silica gel flash chromatography. The eluent was initially hexane and then increased to 9:1 hexane:CH$_2$Cl$_2$ to provide (0.03 g, 20 %) of the title product as red-orange solid. IR (film) 2966, 2936, 2880, 2188, 1712, 1466, 1192, 1128, 850 cm^{-1}. ^{1}H NMR (400 MHz, CDCl$_3$) δ 7.51 (d, J = 8.24 Hz, 4 H), 7.38 (d, J = 8.27 Hz, 4H), 7.08 (s, 5 H), 7.07 (s, 13 H), 2.71 (overlapping q, J = 7.6 Hz, 32 H), 2.42 (s, 6 H), 1.25 (overlapping t, J = 7.6 Hz, 48 H). ^{13}C NMR (300 MHz, CDCl$_3$) δ 193.67, 152.80, 150.31, 150.12, 150.09, 150.05, 134.61, 133.24, 133.02, 132.87, 132.35, 128.91, 124.85, 124.24, 124.03, 123.92, 123.86, 123.81, 119.75, 119.67, 119.54, 119.04, 93.82, 90.02, 89.95, 89.83, 87.24, 87.11, 87.01, 84.65, 30.68, 23.57, 23.51, 23.38, 15.08, 14.97, 14.87, 14.50, 14.27.

3.2 Iterative Approaches to oligo(1,4-phenylene ethynylene)s Molecular Wires, Properties and Experimental Details

3.2.1. *Introduction*

The expectation and potential of molecular-sized computational devices has attracted the interest of the synthetic chemist, and rapid synthetic architectural approaches to these and related structures have been dramatically increasing.[25] Here we describe the synthetic details for the formation of soluble oligo(2-alkyl-1,4-phenylene-ethynylene)s, potential molecular scale wires, by a rapid iterative divergent/convergent doubling approach. These 1,4-phenylene-ethynylene compounds exhibit even further rigidity (no main-chain conformers that alter the molecular length) over the 2,5-thiophene-ethynylene compounds described in Section 3.1. The syntheses and attachments of protected thiol moieties to one or both ends of the phenylene-ethynylenes are presented. These thiols serve as molecular scale alligator clips for adhesion of the molecular scale wires to the gold probes.[23] Additionally, synthesis of the dodecyl-containing 16-mer was also achieved on Merrifield's resin (chloromethyl polystyrene) using the iterative divergent/convergent approach; a method which significantly streamlined the preparation.[37]

3.2.2. *Results and Discussion*

3.2.2.1. Monomer Syntheses for Solution-Based Oligomerizations

The syntheses of the monomers are described in Schemes 3.11, 3.12, and 3.13. Initially, we thought an ethyl substituent would provide enough solubility for the conjugated oligomers up to the 16-mer, the size necessary for bridging patterned proximal gold probes. We therefore started with 3-nitroacetophenone and easily formed the ethyl side chain by hydrogenation under acidic conditions. N-Acetyl protection and Friedel-Crafts acylation afforded a 4:1 mixture of the methyl ketones in which the desired 4-acylation product predominated. However, tert-butoxycarbonyl (BOC) protection of the amine and Friedel-Crafts acylation afforded only the desired 4-acylation product **2**. The BOC protecting group was conveniently lost upon aqueous work-up. Conversion of the amine to the diethyltriazene **3** followed by one-pot conversion of the methyl ketone to the trimethylsilylalkyne[38] afforded the desired monomer **4** for the iterative divergent/convergent oligomerization (Scheme 3.1). Notice that the triazyl moiety was quite robust and able to survive the LDA treatment.

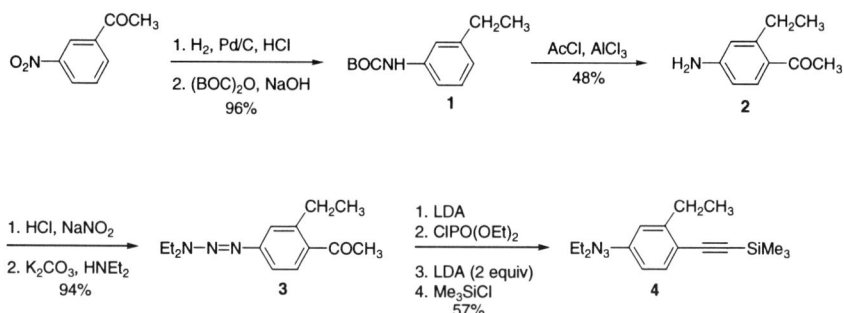

Scheme 3.11 Synthesis of monomers.

We subsequently discovered that the ethyl moiety did not provide sufficient solubility (vide infra), thus we prepared the analogous monomer bearing a 3-ethylheptyl substituent (Scheme 3.12). It was surmised that a monomer bearing a stereogenic center would afford diastereomers on successive dimerizations, thereby retarding crystallization and increasing solubility. We started with 3-nitrobenzoic acid and treated the corresponding acid chloride **5** with the cyanocuprate to afford the desired ketone **6**. We tried numerous methods to concomitantly reduce both the ketone and the nitro moieties of **6**; however, all methods failed, including the Pd-C/H$_2$/HCl reduction that worked in the synthesis of **1**. Therefore, the multi-step reaction sequence was used which involved reduction of the ketone to the alcohol, acetoxy formation, and finally reduction under dissolving metal conditions for both reductive removal of the acetoxy moiety and amine formation. Additionally, the standard NaNO$_2$/HCl conditions that were used in eq 1 did not permit diazonium formation on this more lipophilic substrate; however, tert-BuONO permitted formation of the intermediate diazonium species which was followed by capture with diethylamine to form the triazene **8**.[39] Preparation of **9** then proceeded under the standard alkynylation conditions (Scheme 3.12).

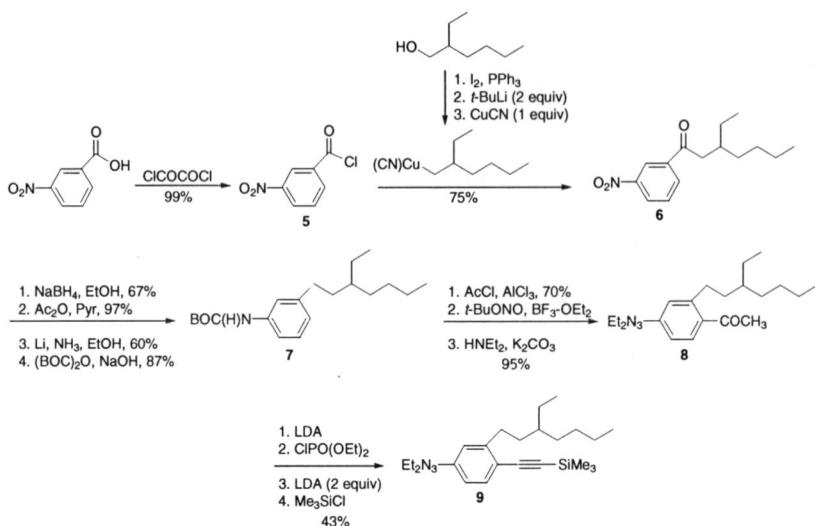

Scheme 3.12 Synthesis of monomer **9** with 3-ethylheptyl side chain.

After discovering that the 3-ethylheptyl substituent in **9** permitted solubility through to the 16-mer (vide infra), we desired to prepare an analogous version that would permit formation of soluble homogeneous material. Hexadecyl chains often cause side chain crystallization in polymers while dodecyl chain can impart solubility without the crystallization difficulties.[29,40] Thus a dodecyne moiety was affixed to 3-nitroiodobenzene, via a Pd/Cu-catalyzed coupling,[41] followed by reduction and conversion to the desired **13**, again via exclusive 4-acylation of the BOC-protected amine (Scheme 3.13).

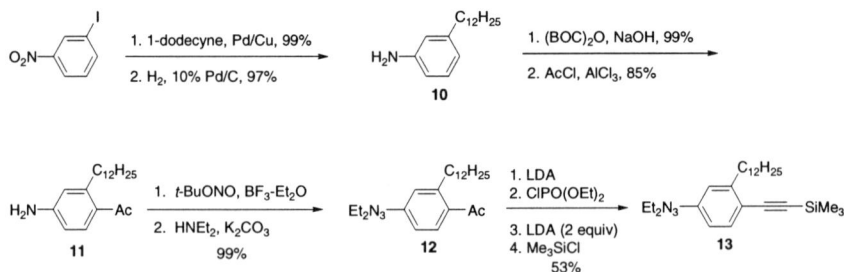

Scheme 3.13 Synthesis of monomer **13** with dodecyl side chain.

3.2.2.2. Oligomer Syntheses in Solution

The solution phase iterative divergent/convergent synthetic approach is outlined in Scheme 3.14. The sequence involved, for example, partitioning **4** into two portions; conversion of the triazene in one portion to the aryl iodide **14** with CH₃I, and protodesilylation of the alkynyl end of the second portion to form **17**. Bringing the two portions back together in the presence of a soluble Pd/Cu catalyst mixture coupled the aryl iodide to the terminal alkyne, thus generating the dimer **20**. Since the dimer had the same end groups as the monomer, the same process could be repeated. Iteration of this reaction sequence doubled the length of the dimer **20** to afford the tetramer **29**, and so on to the octamer **38**. At that point, the octamer was nearly insoluble and we were only able to obtain a UV-visible spectrum and a direct exposure mass spectrum (MS). However, carrying out the same sequence starting from 3-ethylhexyl or dodecyl bearing monomers **9** and **13**, respectively, we were able to achieve the syntheses of the 16-mers **45** and **46**, respectively. The silylated alkynes showed good oxidative stability, however, upon protodesilylation, the tetramers and octamers were air sensitive.

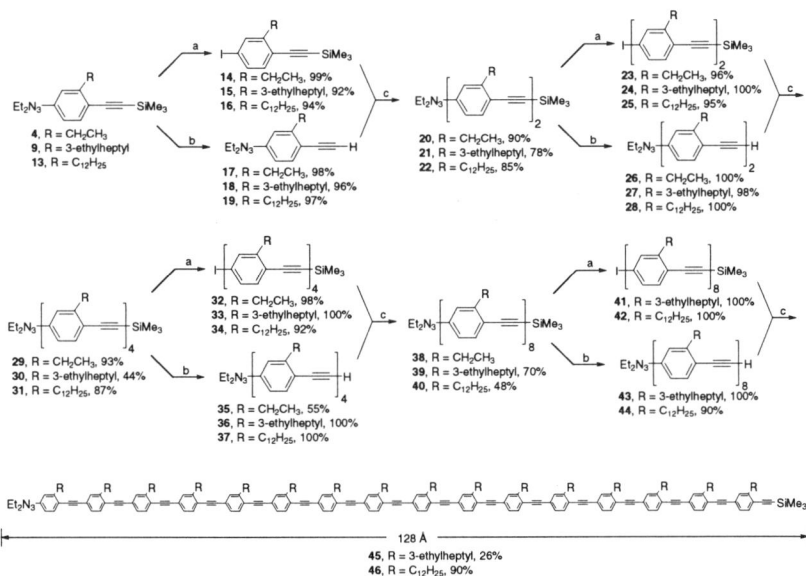

Reagents: a. MeI. b. K₂CO₃, MeOH, or n-Bu₄NF, THF. c. Pd(dba)₂ (5 mol %), CuI (10 mol %), PPh₃ (20 mol %), i-Pr₂NH/THF (1:5).

Scheme 3.14 Coupling of monomers to produce the 16-mers **45** and **46**.

3.2.2.3. Monomer Syntheses for the Polymer Supported Approach

The monomers needed for the polymer support synthesis were prepared as depicted in Scheme 3.15. Compound **11** was converted to the diazonium intermediate which was captured with 2-(ethylamino)ethanol, to form the hydroxytriazene according to Moore's protocol. Silylation to form **47** followed by acetyl conversion to the silylalkyne afforded **48**. Compound **48** was divided into two portions; the first portion was desilylated to form **49**, the anchor unit to be attached to the polymer support. The second portion was iodinated[42] to form the iodoarene **16**.

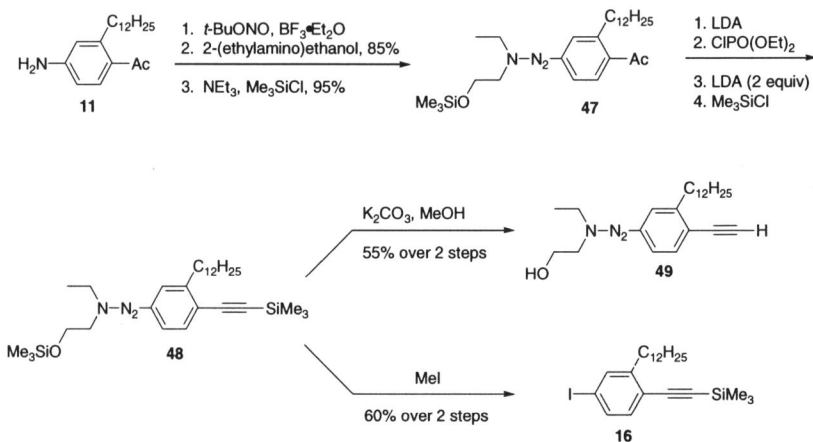

Scheme 3.15 Synthesis of the monomers for the polymer supported approach.

3.2.2.4. Oligomer Syntheses on the Polymer Support

Attachment of **49** to the polymer support resin, and the oligomer syntheses on the polymer support are depicted in Scheme 3.16. Etherification of Merrifield's resin with an excess of anchor **49** provided the polymer-supported monomer **50**. Unfortunately, it was difficult to determine the etherification reaction completion point by IR analysis (*vide infra*). Even though the reaction was allowed to stir over several days at elevated temperature with a 3-fold excess of anchor **49**, gel-phase ^{13}C NMR[43] on the supported material revealed a small resonance at δ 46.77 corresponding to benzylic chloride residues. This was further confirmed by elemental analysis. Fortunately, it was found that the unreacted benzylic chloride sites on the resin did not appear to hinder the oligomer synthesis on the polymer support.

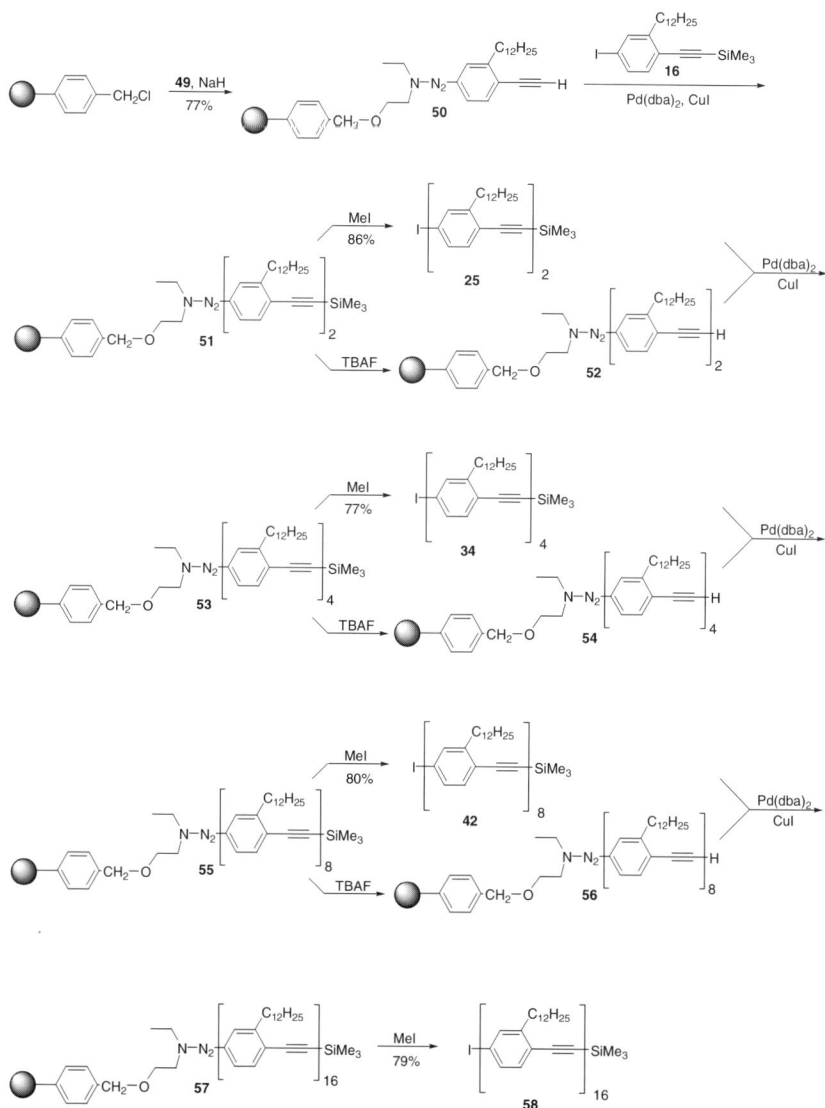

Scheme 3.16 Synthesis of the 16-mer on the polymer.

The successive dimerization sequence was pursued as illustrated in Scheme 3.16. **50** was subjected to a Pd/Cu cross-coupling with monomer **16** to afford the polymer-supported dimer **51**. Two-thirds of **51** was then treated with CH$_3$I affording liberated dimer **25** in 71% yield over two steps. The remaining one-third of the polymer-supported dimer **51** underwent desilylation with tetra-n-butylammonium fluoride (TBAF) to afford the polymer-supported dimer **52**. **52** was then treated with all of the liberated iodide **25** under Pd/Cu cross coupling conditions to afford the polymer-supported tetramer **53**. The sequence was repeated to generate the tetramer **34**, octamer **42**, and 16-mer **58** in the yields indicated over three steps each. Remarkably, we found that the liberated oligomers were contaminated with only small amounts of impurities as assessed by chromatographic and spectroscopic analysis. These impurities were easily removed after passing the compounds through silica gel.

3.2.2.5. Spectroscopic Reaction Monitoring Techniques on the Resin

Since the aforementioned reactions were conducted on an insoluble support, conventional techniques such as TLC could not be used to monitor the progress of the reactions. Consequently, completion of each reaction was estimated by infrared analysis of the polymer bound substrate.[36,44] Polymer supported material was placed on a NaCl plate. After the beads were swollen with 2-3 drops of CCl4, a second NaCl plate was pressed onto the beads, and an FTIR spectrum was recorded. Absorptions at 3311 cm^{-1} (strong) and 2109 cm^{-1} (weak) are characteristic of the terminal alkynyl carbon-hydrogen and carbon-carbon stretches, respectively, and an absorption at 2156 cm^{-1} (strong) is characteristic of the carbon-carbon stretch of the trimethylsilyl-terminated acetylene.[36,45] As expected, we observed that the coupling reaction was accompanied by the disappearance of the 3311 cm^{-1} band and the appearance of the 2156 cm^{-1} absorption. In the reverse manner, the trimethylsilyl removal step was accompanied by the disappearance of the 2156 cm^{-1} band and the appearance of the 3311 cm^{-1} band. Suitability of this monitoring technique was confirmed by removal and characterization of the trimethylsilyl-protected product.

Gel-phase ^{13}C NMR spectra were also recorded on the polymer-supported material at each step in the reaction sequence. Again, the loss or gain of the trimethylsilyl resonance at δ 0.3--0.5 ppm was a diagnostic feature.

3.2.2.6. Assessing Solid Phase Reaction Yields

Though the reaction progress was monitored by both IR and gel-phase NMR, these spectroscopic tools are of relatively low sensitivity when used for resin analysis. Yield estimation by these methods was difficult, especially for the longer oligomers. For these reasons, the approaches used for the calculation of yield are described.

The yield can be determined based on the weight difference of the resin.[46] The weight increase caused by chemical modification of the resin (ΔW) is related to the yield (Y) by the following relation:

$$Y = \Delta W / N_{si}\, \Delta M \qquad\qquad (3.1)$$

where N_{si} is the moles of starting functionality in 1 g of initial resin, and ΔM is the change in molecular weight upon reaction. If mass is lost over the course of the reaction, both ΔW and ΔM should be negative. The results of these calculations for the formation of compounds **50 58** are given in Table 3.1. Unfortunately, the yields determined by this method were found to be unreliable. Small mechanical mass losses naturally occurred during resin manipulations. Small mass increases due to residual solvent or impurities trapped in the polymer matrix were also observed. These small mass changes had an enormous effect on the calculated product yields since the weight changes due to the coupling or deprotection reactions were small relative to the weight of the initial resin. These errors were exacerbated when the weight change due to the reaction was small (for example, in a protodesilylation step), or when performing the reactions on a small scale. As a result, weight changes were only used as a rough guide to the success of a reaction.

Table 3.1 Data for Polymeric Support Reactions Derived using Gravimetric Analysis.

Cmpd	Theor. Wt. Change, ΔW_{theor} (g)	Actual Wt. Change, ΔW (g)	% Yield by Wt. Diff., Y	Theor. Prod. Conc., N_{pf} (theor.) (mmol/g)	Exp. Prod. Conc., N_{pf} (mmol/g)
50	17.41	17.95	103	0.61	0.61
51	12.25	9.98	81	0.51	0.42
52	-0.84	-0.78	103	0.52	0.45
53	5.47	4.43	93	0.40	0.33
54	-0.20	-0.08	48	0.41	0.34*b*
55	2.56	1.71	168	0.28	0.21*b*
56	-0.04	0.03	----------	0.29	0.21*b*
57	0.98	0.23	37*a*	0.17	0.07*b*
50	17.41	17.95	103	0.61	0.61

a Yield was calculated over 2 steps. *b* Based on quantitative yield for the deprotection step and the average two-step yield for the coupled product.

The molar concentrations of the products in the final resins were also monitored for compounds **50-58**. The molar concentration decreases as the resin's mass increases due to modification of the reactive moiety. The actual concentration (mol/g) of the product in the final resin (N_{pf}) is given by

$$N_{pf} = Y N_{si} / W_f = Nsi \, Y / (1 + \Delta W) = Y \, Nsi / (1 + Y \, Nsi \, \Delta M) \qquad (3.2)$$

where W_f is the weight of the final resin. The theoretical concentration (mol /g) of product in the final resin is

$$N_{pf}(theor.) = N_{si} / 1 + N_{si} \Delta M \qquad (3.3)$$

The theoretical and experimentally determined values for N_{pf} are listed in Table 3.1. In both cases the concentration of **50** in the final resin was determined via elemental analysis using a method that will be described later. This initial value of 0.61 mmol/g was then used as the basis for the remainder of the calculations. The errors of the weight measurement caused significant discrepancies in the calculated concentrations. As a result, the moles of reagent needed in the polymer-supported reactions were determined assuming quantitative yields for all previous steps other than the formation of **50**.

The yield of these reactions was also derived from the elemental analysis data. The calculations assumed that no side reactions were taking place and that unreacted functionality did not participate in any subsequent reactions. The element used for the calculations was unique either to the moiety being added to the resin or to the starting functionalized resin. Three distinct situations were encountered which required different methods of yield determination: (1) The tag element was added to the resin over the course of the reaction. (2) The tag element was removed from the resin. (3) The tag element was unique to the resin and present before and after the reaction. The derivations of the equations used for each case are outlined here.

Case 1: When the tag element is added to the resin during the course of the reaction the yield (Y) is given by

$$Y = (N_{pi} / N_{si}) \qquad (3.4)$$

where N_{pi} is the moles of product relative to 1 g of the initial resin and N_{si} is again the moles of starting functionality in 1 g of the initial resin. Elemental analysis yields the weight fraction of the tag element (E). E is the weight in g of an element (W_e) relative to the weight in g of the final resin (W_f). Since the element is unique to the product, this can then be related to the moles of product in 1 g of the final resin (N_{pf}) by

$$E = W_e / W_f = N_{pf} M_e (N_e / N_p) \qquad (3.5)$$

where M_e is the molecular weight of the tag element in g/mol, N_e is the moles of the tag element, and N_p is the moles of product. N_{pf} is related to N_{pi} by

$$N_{pf} = N_p / W_p = N_p / (W_i + \Delta W) = N_{pi} / (1 + \Delta W / W_i) \qquad (3.6)$$

where W_f is the weight in g of the final resin, W_p is the weight in g of the product, and W_i is the weight in g of the initial resin. With this relationship in hand, E can be expressed in terms of N_{pi}.

$$E = N_{pi} M_e (N_e / N_p) / (1 + \Delta W / W_i) \qquad (3.7)$$

Rearrangement of this expression and substitution into Eq. 3.1 gives

$$Y = (N_p / N_e) (E / M_e N_{si}) (1 + \Delta W / W_i) \qquad (3.8)$$

Substituting the expression for ΔW (Eq. 3.1.2.6.1) into Eq. 3.5 gives the derivation of the yield from the elemental analysis.

$$Y = (N_p / N_e) (E / M_e N_{si}) (1 + Y N_{si} \Delta M / W_i)$$
$$= 1 / [(N_e / N_p) (M_e N_{si} / E) - N_{si} \Delta M] \qquad (3.9)$$

The yield for the formation of **50** based on the elemental analysis for N was Y = 1 / [3 (14.0 g/mol) (0.001 mol/g) / 0.0256] - (0.001 mol/g)(349 g/mol) = 0.77 = 77%

The mole/g of Merrifield's resin has a reported error of ± 0.05 mmoles. The error in the nitrogen determination is estimated to be ± 0.2%. Therefore, an approximate expression for the error (e) in Y based on elemental analysis for nitrogen is

$$e = (5 \times 10^{-5} \text{ mol} / 1.00 \times 10^{-3} \text{ mol}) + (0.002 / \text{wt. fraction. N}) = \pm 0.13$$

This relatively high error is largely due to the low weight percent of nitrogen present in the final resin. Nevertheless, this method proved to be the most accurate for calculating the molar concentration of **50** on the polymer support. Calculation of the chloride concentration of Merrifield's resin by elemental analysis has a higher error than the titration method used by the manufacturer.

Case 2: When the tag element is removed during the course of the reaction the yield is given by

$$Y = 1 - (N^*_{si} / N_{si}) \qquad (3.10)$$

where N^*_{si} is the moles of unreacted starting functionality in the final resin relative to 1 g of starting resin. Because the tag element is unique to the unreacted resin, the weight fraction is now

$$E = W_e / W_f = N_{sf} M_e (N_e / N_p) \qquad (3.11)$$

where N_{sf} is the moles of starting functionality present in 1 g of final resin. N^*_{si} is related to N_{sf} by

$$N_{sf} = N^*_{si} / (1 + \Delta W / W_i) \qquad (3.12)$$

E can then be expressed in terms of N^*_{si}.

$$E = N^*_{si} \, M_e \, (N_e / N_s) / (1 + \Delta W / W_i) \qquad (3.13)$$

where N_s is the moles of starting functionality.   Rearrangement of this expression and substitution into Eq. 3.10 gives

$$Y = 1 - [(N_s / N_e) (E / M_e N_{si}) (1 + \Delta W / W_i)] \qquad (3.14)$$

Substituting the expression for ΔW (Eq. 3.1) into Eq. 3.11 gives the calculation of the yield from element loss based on elemental analysis. Taking advantage of the ratio of tag element to product, $(N_p / N_e) = (N_s / N_e) = 1$ (for monosubstituted tag element on resin), the expression for calculating the yield becomes

$$\begin{aligned} Y &= 1 - (N_p / N_e) (E / M_e N_{si}) (1 + Y N_{si} \Delta M / W_i) \\ &= [1 - [(N_p / N_e) (E / M_e N_{si})]] / [[\Delta M (N_p / N_e) E / M_e N_e] + 1] \end{aligned} \qquad (3.15)$$

The yield for the formation of **50** based on loss of Cl is

$$Y = [1-[1 \ (0.0061) / (35.45 \text{ g / mol}) (0.001 \text{ mol/g})]]/[1 \ (349 \text{ g/mol}) (0.0061) / (35.45 \text{ g/mol}) + 1] = 0.78 = 78\%$$

This yield is in good agreement with that based on analysis for nitrogen. The error for this measurement is estimated at $\pm$ 0.29.

Case 3: If the tag element is unique to the resin but is not lost on reaction, the moles of the element are no longer directly related to the amount of product formed. The change in the elements' weight fraction is entirely due to the weight change of the resin. The weight change of the resin can be related to the change in the weight fraction of the tag element by

$$\Delta E = E_f - E_i = (W_{ef} / W_f) - (W_{ei} / W_i) = W_{ef} / (W_i + \Delta W) - W_{ei} / W_i \qquad (3.16)$$

Recognizing that $W_{ei} = W_{ef}$, and expressing in terms of 1 g of initial resin, therefore $E_i = W_{ei}$, and Eq. 3.16 simplifies to

$$\Delta E = \left[\left(\frac{1}{1 + \Delta W} \right) - 1 \right] W_{ei} \qquad (3.17)$$

Substituting Eq. 3.1 for ΔW and rearrangement gives the yield expressed in terms of weight fraction of the tag element before and after the reaction.

$$Y = \frac{\left(\dfrac{1}{1 + (\Delta E / E_i)} \right) - 1}{N_{si} \, \Delta M} \qquad (3.18)$$

The yield of the formation of **51** is then

$$Y = \{\{1 / [1 + (-0.0025 / 0.0256)]\} - 1\}/(0.00061 \text{ mol/g}) (341 \text{ g / mol}) = 0.52 = 52\%$$

The estimated error for this measurement is ± 0.92. The error was higher than the calculated value for all the compounds prepared. Clearly, the accuracy of this method is unacceptable for yield determination at the low nitrogen levels present in these systems.

Due to the inherent inaccuracies of yield determination based on elemental and gravimetric analysis, yields were also calculated based on the amount of cleaved product (by CH_3I treatment) recovered from the resin after each coupling step. The formula used for these determinations was

$$Y = N_p / N_s = N_p \, W_i / N_{si} \qquad (3.19)$$

The value of 0.61 mmol/g, determined from elemental analysis, was used for the concentration of **50** in the polymer matrix. From that point onward, the yields were calculated over three steps. The moles of products were determined directly from purified compounds after cleavage from the resin. Subsequent to **50**, N_{si} for the initial polymer supported compound of the 3 step sequence was determined using Eq. 3.3. Using this method, the yields of the cleaved dimer **25**, tetramer **34**, octamer **42**, and 16-mer **58** were 86%, 77%, 80% and 79% respectively. This corresponds to an average yield of 92-93% over the deprotection, coupling and cleavage steps. The amount of material that remained uncleaved by the CH_3I treatment is not known, but the error in the nitrogen analysis excludes the possibility of this aiding in the determination.

In summary, the determination of the coupling efficiency on the polymer support has severe difficulties due to the fact that the resin comprises the vast majority of the materials' mass. Though the absolute error remains constant, the percent error increases for each iteration along the sequence. Therefore, yield determinations based on elemental composition changes along a series of steps eventually have such significant errors that the assessment becomes completely unreliable. However, if the initial linker's coupling efficiency is reasonably established, the yield determinations over the three-step sequence can be sufficiently utilized.

3.2.2.7. Characterization of the Oligomers

As mentioned above, while the ethyl-containing octamer **38** was nearly insoluble, the longer alkyl bearing octamers **39** and **40**, as well as the corresponding 16-mers **45** and **46**, were quite soluble, and they could be adequately spectroscopically characterized. While many of the lower molecular weight oligomers could be characterized by direct exposure via electron impact MS, in order to obtain MS data on the soluble octamer **39** and 16-mer **45**, it was necessary to use matrix assisted laser desorption/ionization mass spectrometry (MALDI-MS) using a sinapinic acid matrix, positive ion mode. The dodecyl-containing octamer **40** and 16-mer **46** were simpler to characterize spectroscopically than the 3-ethylheptyl-containing oligomers since the former are not mixtures of diastereomers. However, **40** and **46**, being more lipophilic than **39** and **45**, were too insoluble in the polar MALDI-MS matrices, thus no MS data could be obtained. FAB-MS did not work on these samples, presumably due to charging difficulties. The 16-mers are 128 Å long as determined by a conformational minimization using MM2 with extended π-Hückel and multiconformational parameters.[30] This length is long enough to span the gaps between proximal patterned electronic probes.

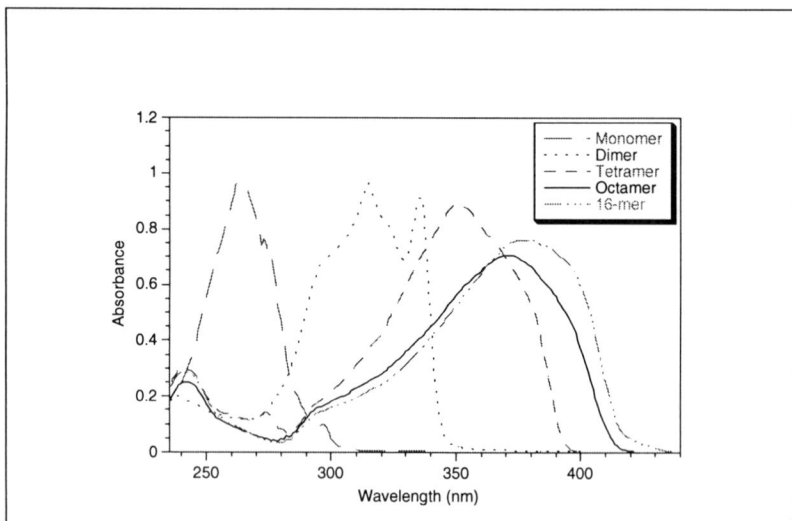

Figure 3.3 The UV-Vis absorbance spectra (THF) of the monomer through 16-mer iodides, **16**, **25**, **34**, **42**, and **58**, respectively.

Absorption (Figure 3.3) and emission studies (Figure 3.4) were conducted on the oligomers. As expected, we observed an increase in the absorption maxima from shorter wavelength (263 nm) for the monomer **16** to longer wavelength (377 nm) for the 16-mer **58**. Likewise, an increase in the emission intensity from shorter wavelength (359 nm) for the dimer **25** to longer wavelength (414 nm) for the 16-mer **58** was observed (Figure 3.4). The monomer **16** did not fluoresce. Note that a near saturation of the systems appears to have occurred by the octamer stage so that doubling the conjugation length from the octamer to the 16-mer caused only a small bathochromic shift in absorption and emission maxima. As illustrated in Figure 3.3, we also noticed that the π-extended triazene substituents shifted the absorption maxima bathochromically, relative to the corresponding iodides, for the monomers through octamers. However, the saturation effects are apparent at the 16-mer stage since the two absorptions are nearly identical (Figure 3.3).

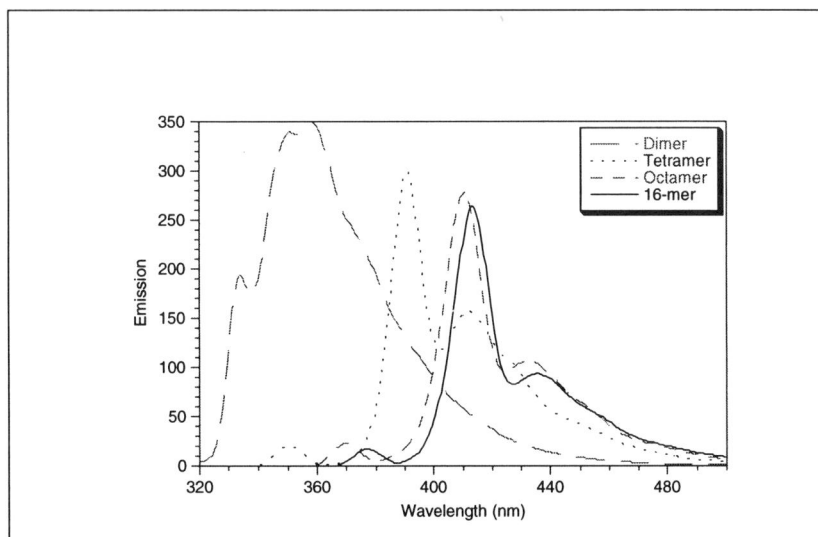

Figure 3.4 The fluorescence emission spectra (THF) of the dimer through 16-mer iodides, **25**, **34**, **42**, and **58**, respectively.

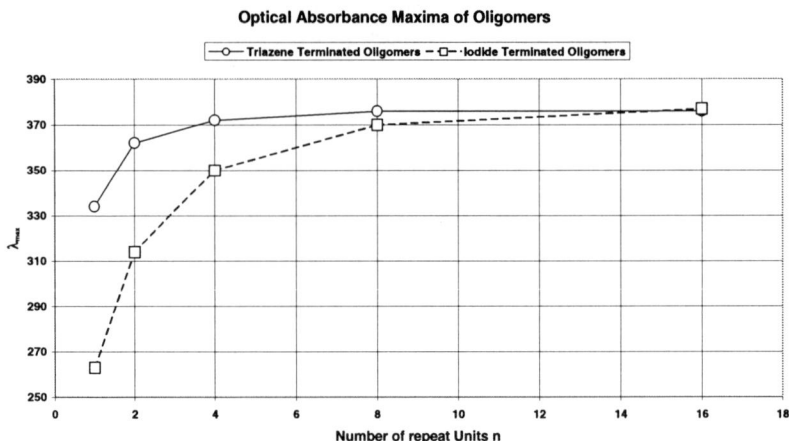

Optical Absorbance Maxima of Oligomers

Figure 3.5 The optical absorbance maxima (λ_{max}) versus the number of repeat units (n) in the triazene-terminated oligomers **9, 21, 30, 39**, and **45** (circles) and the iodide terminated oligomers **16, 25, 34, 42** and **58** (squares).

The results of the size exclusion chromatography (SEC), which is commonly used in the separation and identification of oligomers, are shown in Figure 3.6.   SEC is not a direct measure of MW, but a measure of the hydrodynamic volume.   Thus, the number average molecular weight (Mn) values of the rigid rod oligomers should be significantly inflated relative to the randomly coiled polystyrene standards.   As expected, the M_n values of the monomer **9** (M_n = 390, actual MW = 400), dimer **21** (M_n = 715, actual MW = 626), dimer **25** (M_n = 900, actual MW= 736), tetramer **30** (M_n = 1335, actual MW = 1079), and tetramer **34** (M_n = 1705, actual MW= 1,274) were reasonably close to the actual MWs (initial slope $\cong$ 1.0 in Figure 3.5), prior to significant polystyrene coiling.   However, the M_n values of the octamer **39** (M_n = 2,790, actual MW= 1,981), octamer **42** (M_n = 3,625, actual MW= 2,348), 16-mer **45** (M_n = 6,650, actual MW= 3,789), and 16-mer **58** (M_n = 8,856, actual MW= 4,495) were much greater than the actual MWs.   As expected, these curves in Figure 3.6 are quite similar in their appearance to the curves generated from related thiophene-ethynylene oligomers of similar molecular weights.

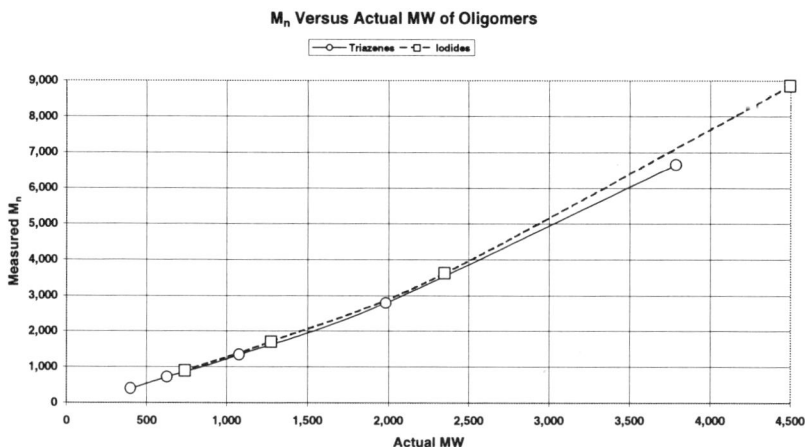

Figure 3.6 The values of Mn determined by SEC (relative to polystyrene standards in THF) versus the actual molecular weights for the triazene-terminated oligomers **9, 21, 30, 39**, and **45** (circles) and the iodide terminated oligomers **25, 34, 42** and **58** (squares).

3.2.2.8. Attachment of Thiol End Groups

With the requisite precisely defined oligo(phenylene-ethynylene)s in hand, we then addressed methods to affix protected thiols to the ends of the oligomers. The thiols serve as molecular alligator clips for adhesion to gold probe surfaces. We recently used these thiol-terminated oligo(phenylene-ethynylene)s to record electronic conduction through single undoped conjugated molecules that were end-bound onto a metal probe surface. These nearly linear molecules worked quite well for single molecule insertions into self-assembled monolayers of dodecane thiolate on gold, thereby providing a means of molecular isolation. Monothiol-terminated systems were prepared for adhesion of these oligomers to single gold surfaces, and in other cases, α,ω-difunctionalized systems were prepared for adhesion between proximal gold surfaces. Though the aromatic thiols could be synthesized under reductive base conditions, due to their oxidative instabilities, they proved difficult to use in actual surface deposition experiments. However, the acetyl-protected thiols were resilient enough for manipulations in air, yet they could be readily hydrolyzed with NH4OH in the presence of the gold probe surfaces.

Using a convergent method of molecular alligator clip formation and attachment, the two thioesters **59** and **60** were synthesized. The aryl iodide **59** can be coupled to the alkynyl end (head) of the oligomers while the alkynyl arene **60** can be coupled to the aryl iodide end (tail) of the oligomers. Of course, our oligomers would permit orthogonal end group functionalization, for example, one end containing a thiol moiety for adhesion to gold and the other end a carboxy moiety for adhesion to an oxide. However, for all of our initial studies on difunctionalized systems, proximal gold connects were considered, thus identical end groups sufficed.

Scheme 3.17 shows the efficient attachment of **59** to the heads of both phenylacetylene and the dimeric system **62**. Scheme 3.8 shows an approach to a concomitant head and tail functionalization of the ethyl-containing monomer, dimer, and tetramer. Deprotection of the thiols was accomplished with NaOH under reductive conditions to minimize disulfide formation. Though this deprotection method proved effective for laboratory isolation and characterization of the thiols, it could not be used for molecular scale electronics study since exogenous metals such as zinc or Na+ would interfere with electronic measurements.[23, 24] Thus we developed the complementary in situ technique using NH$_4$OH.

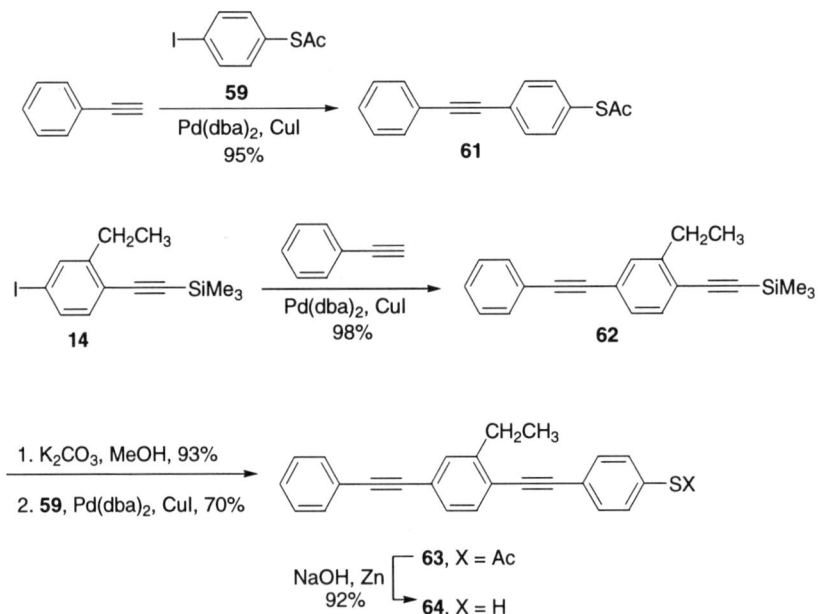

Scheme 3.17 Attachment of thiol end groups.

Both ends of the oligomers could be functionalized as shown in Scheme 3.18. However, the α,ω-bis-terminal alkynes **68-70** showed increasingly low yields in the couplings with **59**. Therefore, we used a stepwise procedure (Scheme 3.19) of tail functionalization, followed by protodesilylation, to afford **80-82**. The use of TBAF alone for the desilylation reactions also caused deacylation of sulfur; however, the buffered TBAF/HF conditions worked well. Finally, coupling with **59** permitted formation of the α,ω-dithioesters **83** and **84**.

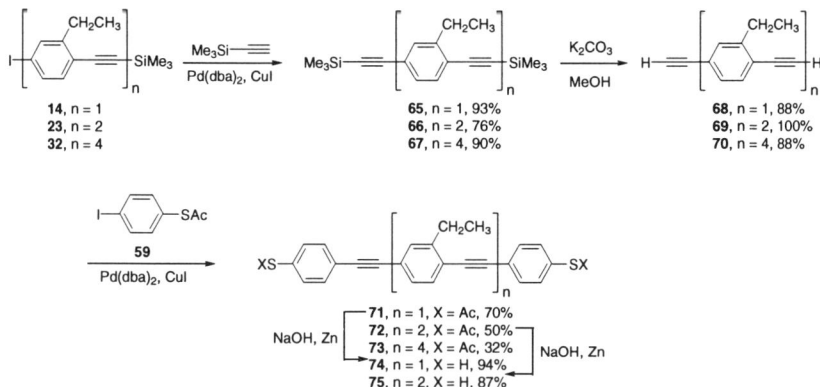

Scheme 3.18 Functionalization of both ends of the oligomers through the intermediacy of the α,ω-bis-terminal alkynes.

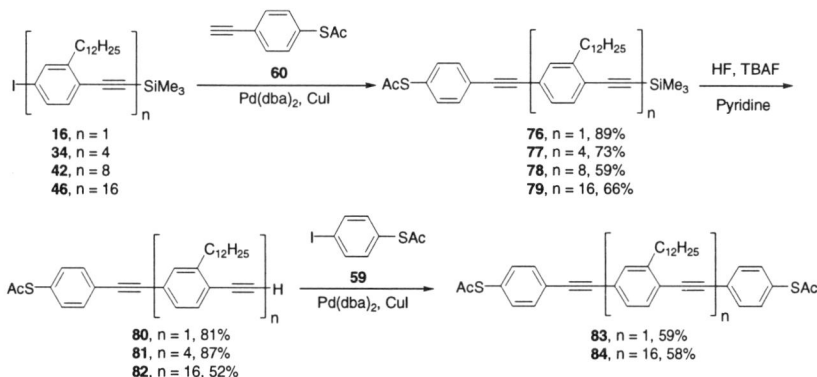

Scheme 3.19 Functionalization of both ends of the oligomers through the stepwise procedure.

3.2.3. Summary

We outlined the rapid syntheses of oligo(2-alkyl-1,4-phenylene-ethynylene)s via an iterative divergent/convergent approach that needed only three sets of reaction conditions for the entire iterative synthetic sequence: an iodination, a protodesilylation, and a Pd/Cu-catalyzed cross coupling. The syntheses were conducted in solution and on a polymer support. At each stage in the iteration, the length of the framework doubled. Equations were derived for assessing the efficiency of the polymer-supported reactions based on resin weight differences, molar concentration differences, and elemental analysis data. Each of these methods had severe limitations. Attachment of the complementary thiol end groups, protected as thioester moieties, was achieved. These serve as binding units for adhesion to gold surfaces, and these rigid rod conjugated oligomers have been shown to exhibit single molecule conductivity.[23,24] This represents a step in the development of organic architectures for molecular scale electronic instruments.

3.2.4. *Experimental Procedures*

Merrifield's resin was purchased from Aldrich. The tert-butyl nitrite diazonium formations,[38] methyl ketone to trimethylsilylalkyne conversions,[37] Pd/Cu-catalyzed aryl halide-terminal alkyne couplings,[29,40] triazene formation and iodinations,[38,41] and the protodesilylation reactions,[47] were conducted according to reported procedures. Terminal alkynes beyond the dimer stage were often oxidatively unstable, and they were used immediately after preparation; thus spectroscopic characterization was minimized. The 3-ethylheptyl-containing oligomers could not be characterized by ^{13}C NMR due to the existence of multiple diastereomers. FTIR characterization of the polymer-supported reactions was carried out by placing ca. 10 mg of the polymer-supported material on a NaCl plate. After the beads were swollen with 2-3 drops of CCl4, a second NaCl plate was pressed onto the beads, and an FTIR spectrum was recorded. Gel-phase ^{13}C NMR spectra on the polymer-supported material were recorded at 125 MHz on a Brüker AM-500 spectrometer. The NMR gel samples were prepared as follows:[48] polymers were dried under vacuum at 60 °C until constant weight was obtained. 300 mg were weighed into a 10 mm NMR tube and CDCl₃ was added. Complete swelling of the polymers was normally achieved by allowing them to equilibrate with the solvent for 15-30 min. Sometimes, ultrasonic agitation (Bransonic 5200) was necessary to obtain well-dispersed samples. The syntheses of **59** and **60** have been described previously. Unless otherwise noted, all compounds were >95% pure as judged by NMR, GC, or combustion analyses. The oligomers' puritites can be assessed by their polydispersity indexes M_w/M_n.

Calculation of elemental percentages for Merrifield's resin. Merrifield's resin is a copolymer of styrene (x), 4-vinylbenzyl chloride (y), and divinylbenzene (z). The commercial resin was listed to contain 1% w/w divinylbenzene and 1.00± 0.05 meq/g 4-vinylbenzyl chloride. Thus, the weight fraction of each unit can be determined as follows:

wt. fraction styrene (x) + wt. fraction 4-vinylbenzyl chloride (y) + wt. fraction divinylbenzene (z) = 1.00

wt. fraction z = 0.0100

wt. fraction y = (0.001 mol y × 152.63 g/mol) / 1 g resin = 0.1526

wt. fraction x = 1.000 - (0.1526 + 0.0100) = 0.8374

The percentages of the elements present in each unit were:

x: C: 92.24%, H: 7.76%

y: C: 70.82%, H: 5.96%, Cl: 23.23%

z: C: 90.83%, 9.17%

The percentage of a given element in the resin was then the sum of the weight fractions of each unit multiplied by the percentage of the element in each unit. Therefore, the percentage of each element in Merrifield's resin was:

%C = (%C in x) (wt. Frac. x) + (%C in y) (wt. Frac. y) + (%C in z) (wt. Frac. z)
= (92.24%) (0.8374) + (70.82%) (0.1526) + (90.83%) (0.0100) = 88.96%

%H = (%H in x) (wt. Frac. x) + (%H in y) (wt. Frac. y) + (%H in z) (wt. Frac. z)
= (7.76%) (0.8374) + (5.96%) (0.1526) + (9.17%) (0.0100) = 7.50%

% Cl = (% Cl in y) (wt. Frac. y) = (23.23%) (0.1526) = 3.54%

Calculation of elemental percentages for the functionalized resins. The percentage of each element in the functionalized resins was determined in a manner analogous to that used for Merrifield's resin. Calculations were performed based on a 77% yield for the formation of **50** on the polymer support. For compounds **50-57**, the resin was treated as a mixture of x, y, z and the functionalized unit (y'). The relative percentages of y and y' were assigned as 23% and 77% respectively, based on the yield derived from analysis for nitrogen. The formation of **51-57** was assumed to be quantitative. Sequential modification of the initial y unit caused a change in the relative percentages of the monomer units comprising the resin. The weight increase for 1 g of Merrifield's resin caused by the modification was:

wt. increase for 1 g of Merrifield's resin = [(molecular wt. of y') - (molecular wt. of y)]   (mol y in 1 g of Merrifield's resin)

Using **50** as an example:

= [(501.8 g/mol) - (152.6 g/mol)] (0.001 mol × 0.77) = 0.269 g

The weight fraction of each unit was then:

wt. fraction styrene (x) + wt. fraction 4-vinylbenzyl chloride (y) + wt. fraction functionalized 4-vinylbenzyl chloride (y') + wt. fraction divinylbenzene (z) = 1.00

Again using **50** as an example:

wt. fraction z = 0.0100 g z / 1.269 g resin = 0.0079
wt. fraction y = (0.001 mol y × 0.23 × 152.6 g/mol) = 0.0351
wt. fraction y' = (0.001 mol y' × 0.77 × 501.8 g/mol) / 1.269 g resin = 0.3045
wt. fraction x = 0.8374 g x / 1.269 g resin = 0.6599

The percentage of each element could then be determined as before.

%C = (%C in x) (wt. fraction x) + (%C in y) (wt. fraction y) + (%C in y')
(wt.fraction y') + (%C in z) (wt. fraction z) = (92.24%) (0.6599) +
(70.82%)(0.0351)+ (78.98%) (0.3045) + (90.83%) (0.0079) = 88.12%

%H = (%H in x) (wt. fraction x) + (%H in y) (wt. fraction y) + (%H in y') (wt.
fraction y') + (%H in z) (wt. fraction z) = (7.76%) (0.6599) + (5.96%) (0.0351)
+ (6.64%) (0.3045) + (9.17%) (0.0079) = 7.42%

%N = (%N in y') (wt. fraction y') = (8.37%) (0.3045) = 2.55%

% Cl = (% Cl in y) (wt. fraction y)= (23.23%) (0.0351) = 0.82%

3-Ethylaniline.[25] Anhydrous ethanol (100 mL), 3-nitroacetophenone (16.52 g, 100 mmol), concentrated HCl (20.0 mL), and 10% palladium-on-carbon (2 g) were placed in a Parr flask. The flask was purged with hydrogen before pressurizing to 60 psi and heating to 70 °C. The pressure was maintained overnight. The palladium on carbon was then filtered and the reaction mixture poured into 3 M NaOH (200 mL). The solution was extracted with ether (3×) and the combined organic layers washed with brine and dried over $MgSO_4$. The $MgSO_4$ was filtered and the solvent removed in vacuo to afford 12.29 g of the title compound as a brown oil which was taken on to the next step with no further purification. R_f = 0.14 (9:1 hexane:ether). ^{1}H NMR (500 MHz, $CDCl_3$) δ 7.08 (t, J = 7.7 Hz , 1 H), 6.62 (d, J = 7.1 Hz, 1 H), 6.54 (s, 1 H), 6.53-6.51 (m, 1 H), 3.55 (br s, 2 H), 2.56 (q, J = 7.6 Hz, 2 H), 1.21 (t, J = 7.6 Hz, 3 H). HRMS calcd for $C_8H_{11}N$: 121.0891. Found: 121.0888.

3-Ethyl-N-(tert-butoxycarbonyl)aniline (1). To a solution of crude 3-ethylaniline (**7**) (100 mmol) and di-t-butyl dicarbonate (34.50 g, 150 mmol) in THF (300 mL) was added 1 M NaOH (300 mL) in 3 portions. The reaction was heated to reflux for 2 h and cooled to room temperature. The reaction was poured into water and the organic layer extracted with ether (3×). The organic layer was dried over $MgSO_4$, filtered and concentrated in vacuo. The crude mixture was purified by flash chromatography on silica gel (9:1 hexane:ether) to afford 21.12 g (96%) of the title compound as a yellow oil. R_f = 0.39 (9:1

hexane:ether). IR (neat) 3332, 2971, 2932, 2875, 1704, 1610, 1534, 1492, 1441, 1368, 1309, 1247, 1162, 1054, 986, 871, 774, 738, 721, 697 cm^{-1}. ^{1}H NMR (300 MHz, CDCl$_3$) δ 7.19 (d, J = 7.8 Hz, 1 H), 7.15 (s, 1 H), 7.10 (d, J = 8.1 Hz), 6.86 (d, J = 7.4 1 H), 6.42 (br s, 1 H), 2.62 (q, J = 7.6 Hz, 2 H), 1.50 (s, 9 H), 1.21 (t, J = 7.6 Hz, 3 H). ^{13}C NMR (75 MHz, CDCl$_3$) δ 153.0, 145.3, 138.4, 128.9, 122.6, 118.1, 116.0, 80.3, 29.0, 28.4, 15.6. HRMS calcd for C$_{13}$H$_{19}$NO$_2$: 221.1416. Found: 221.1420.

4-Amino-2-ethylacetophenone (2). To a 250 mL 3-neck round bottom flask equipped with a reflux condenser and drying tube containing AlCl$_3$ (4.93 g, 37 mmol) and CS$_2$ (50 mL) at 0 °C were added **1** (2.32 g, 10.5 mmol) in CS$_2$ (50 mL) and acetyl chloride (1.49 mL, 21 mmol). The reaction mixture was heated to reflux at 70 °C overnight. After cooling to 60 °C the reflux condenser was removed to allow the CS$_2$ layer to boil off before pouring it into ice water (10 mL). The solution was then poured into 3 M KOH (100 mL) and extracted with ether (5×). The combined organic extracts were washed with brine and dried over MgSO$_4$. The solution was filtered and the solvent removed in vacuo to give a brown oil that was purified by flash chromatography on silica gel (2:1, hexane:ether) to afford 0.82 g (48%) of the title compound as a dark brown liquid which solidified upon standing. R$_f$ = 0.27 (2:1, hexane:ether). MP 53-55 °C. IR (KBr) 3456.5, 3347.7, 3218.7, 2966.0, 1635.2, 1601.8, 1553.6, 1455.7, 1358.4, 1336.8, 1293.8, 1256.0, 1186.6, 1145.7, 1059.7, 1017.8, 954.0, 874.6, 857.6, 821.8, 585.7, 583.8, 547.8 cm^{-1}. ^{1}H NMR (300 MHz, CDCl$_3$) δ 7.59 (d, J = 7.9 Hz, 1 H), 6.46 (s, 1 H), 6.44-6.42 (m, 1 H), 4.15 (s, 2 H), 2.89 (q, J = 7.4 Hz, 2 H), 2.47 (s, 3 H), 1.15 (t, J = 7.4 Hz, 3 H). ^{13}C NMR (75 MHz, CDCl$_3$) δ 199.0, 150.3, 148.5, 133.4, 126.5, 116.1, 111.0, 29.0, 28.0, 15.6. HRMS calcd for C$_{10}$H$_{13}$NO: 163.0997. Found: 163.0992.

4-(Diethyltriazenyl)-2-ethylacetophenone (3). To **2** (12.60 g, 77.3 mmol) in water (316 mL) and HCl (53 mL) was added sodium nitrite (5.48 g, 79.42 mmol) in water (10 mL) at 0 °C. The reaction was stirred at 0 °C for 30 min then poured into water (530 mL), K$_2$CO$_3$ (84 g) and diethylamine (63 mL) at 0 °C. The reaction was stirred for 30 min at 0 °C then poured into water. The aqueous layer was extracted with diethyl ether (3 × 25 mL) and the organic layer was dried over MgSO$_4$, and the product concentrated in vacuo to afford 17.90 g (94%) of the title compound as dark brown liquid. IR (neat) 2975, 2935, 2873, 1680, 1597, 1556, 1243, 1109, 952 cm^{-1}. ^{1}H NMR (500 MHz, CDCl$_3$) δ 7.68 (d, J = 8.2 Hz, 1 H), 7.29-7.25 (m, 2 H), 3.77 (q, J = 7.2 Hz, 4 H), 2.93 (q, J = 7.4 Hz, 2 H), 1.27 (br t, 6 H), 1.22 (t, J = 7.5 Hz, 3 H). ^{13}C NMR δ 201.1, 154.1, 146.9, 133.8, 131.6, 123.1, 117.4, 30.0, 28.1, 16.2. HRMS calcd for C$_{14}$H$_{21}$N$_3$O: 247.1685. Found: 247.1687.

1-(Diethyltriazenyl)-3-ethyl-4-(trimethylsilylethynyl)benzene (4). To a stirred solution of diisopropylamine (9.4 mL, 67 mmol) in THF (70 mL) was added at 0 °C under N$_2$ n-BuLi (40.6 mL, 65 mmol, 1.6 M in hexanes). The reaction mixture was stirred at 0 °C for 5 min then cooled to -78 °C. To the LDA solution was added **3** (15.5 g, 62.8 mmol) in THF (40 mL). The reaction mixture was stirred at -78 °C for 1 h before adding, at the same temperature, diethylchlorophosphate (9.1 mL, 63 mmol). The reaction mixture was allowed to warm to room temperature and stirred for 3 h. The reaction solution was then transferred via cannula into 2.25 equivalents of LDA [made as above with diisopropylamine (19.6 mL, 140 mmol) and n-BuLi (88.3 mL, 141.3 mmol) in THF (150 mL)] at -78 °C. The ensuing reaction mixture was stirred at -78 °C for 30 min and then allowed to warm to room temperature overnight at which time the reaction was cooled to 0 °C and chlorotrimethylsilane (8.2 mL, 65 mmol) was added. The solution was poured into water (100 mL) and extracted with ether. The combined ether extracts were washed with brine and dried over MgSO$_4$. Filtration and removal of the solvent in vacuo left a brown oil which, upon purification by flash chromatography on silica gel (5:1 hexane:EtOAc), afforded 10.81 g (57 %) of the title compound as clear, colorless oil.   IR (neat) 3290, 2105, 1585, 1482, 1390, 1090, 1065, 1005, 820, 620 cm^{-1}. ^{1}H NMR (300 MHz, CDCl$_3$) δ 7.36 (d, J = 8.2 Hz, 1 H), 7.23 (d, J = 1.7 Hz, 1 H), 7.16 (dd, J = 8.2, 2.1 Hz, 1 H), 3.74 (q, J = 7.2 Hz, 4 H), 2.78 (q, J = 7.5 Hz, 2 H), 1.24 (t, J = 7.2 Hz, 9 H), 0.23 (s, 9 H). ^{13}C NMR (75 MHz, CDCl$_3$) δ 151.4, 147.6, 133.1, 120.3, 118.6, 117.5, 104.7, 97.2, 28.0, 14.7, 0.18. HRMS calcd for C$_{17}$H$_{27}$N$_3$Si: 301.1974. Found: 301.1983.

2-Ethyl-1-iodohexane. To CH$_2$Cl$_2$ (150 mL) in a 500 mL oven dried vessel equipped with a condenser was added sequentially (C$_6$H$_5$)$_3$P (15.7 g, 60 mmol), imidazole (4.1 g, 60 mmol) and I$_2$ (15.2 g, 60 mmol). The mixture was stirred until the I$_2$ had dissolved and a reddish-yellow solid appeared. To the mixture was added dropwise 2-ethyl-1-hexanol (7.8 mL, 50 mmol) keeping the temperature of the reaction just below reflux. After addition of the alcohol was complete, the reaction was stirred for 2 h and monitored by GC. The completed reaction was concentrated in vacuo, and silica gel (10 g) was added to the resulting crude mixture in hexane (100 mL). The mixture was stirred for 15 min and filtered through a plug of silica gel. The resulting clear solution was concentrated in vacuo to afford the title compound in nearly quantitative yield. IR (neat) 2960, 2928, 2858, 1458, 1187 cm^{-1}. ^{1}H NMR (500 MHz, CDCl$_3$) δ 3.25 (2 sets of overlapping dd, J = 14.1, 9.0 Hz, 2 H), 1.4-1.1.0 (m, 9 H), 0.88 (t, J = 6.5 Hz, 3 H), 0.84 (t, J = 7.4 Hz, 3 H). ^{13}C NMR (125 MHz, CDCl$_3$) δ 40.7, 34.1, 29.2, 27.4, 23.2, 16.5, 14.5, 11.3.

3-Nitrobenzoyl chloride (5). To an oven dried vessel was added 3-nitrobenzoic acid (83.56 g, 500 mmol) and benzene (1 L). The vessel was fitted with a reflux condenser and drying tube. To the reaction was added oxalyl chloride (65.4 mL, 750 mmol) at room temperature. After seeing no reaction after a period of 10 min, 1 drop of DMF was added. Stirring was continued until evolution of HCl and CO was complete. The reaction was concentrated in vacuo and CaH_2 (2 g) and benzene (200 mL) were added. Stirring was continued for 1 h at which time the reaction was filtered through celite and concentrated in vacuo to afford 92.3 g (99 %) of the title compound as brown solid. ^1H NMR (300 MHz, $CDCl_3$) δ 8.94, (t, J = 2.0 Hz, 1 H), 8.53 (ddd, J = 8.2, 2.3, 1.0 Hz, 1 H), 8.42 (ddd, J = 7.9, 1.8, 1.2 Hz, 1 H), 7.75 (t, J = 8.0 Hz, 1 H).

1-(3'-Nitrophenyl)-3-ethylheptane-1-one (6). To t-BuLi (37.4 mL, 80 mmol, 2.14 M in hexane) in ether (20 mL) was added 2-ethyl-1-iodohexane (9.6 g, 40 mmol) at -78 °C. The reaction mixture was stirred for 30 min, then added to flame dried $ZnCl_2$ (7.6 g, 56 mmol) in THF (30 mL) at room temperature. The resulting cloudy solution was stirred for 30 min. In a separate vessel was combined tetrakis((C_6H_5)$_3$P)palladium(0) (0.7 g, 0.6 mmol) and 3-nitrobenzoyl chloride (5.57 g, 30 mmol) in THF (10 mL), and the vessel was flushed with N_2. To the resulting Pd(0) solution was then added the Zn reagent at room temperature. The reaction was stirred at room temperature overnight and then poured into water. The aqueous layer was extracted with EtOAc and dried over $MgSO_4$. The organic layer was filtered and concentrated in vacuo. The crude product was filtered and purified by flash chromatography on silica gel (5:1, hexane:EtOAc) to afford 1.8 g (39 %) of the title compound as clear yellow oil. IR (neat) 3086, 2929, 2873, 1694, 1614, 1538, 1351 cm^{-1}. ^1H NMR (300 MHz, $CDCl_3$) δ 8.75 (t, J = 0.8 Hz, 1 H), 8.40 (dt, J = 7.7, 0.8 Hz, 1 H), 8.26 (dt, J = 7.7, 0.8 Hz, 1 H), 7.66 (t, J = 7.7 Hz, 1 H), 2.90 (d, J = 6.6 Hz, 2 H), 2.03 (m, 1 H), 1.41-1.23 (m, 8 H), 0.88 (t, J = 7.2 Hz, 3 H), 0.85 (t, J = 7.4 Hz, 3 H). ^{13}C NMR (75 MHz, $CDCl_3$) δ 198.3, 138.7, 133.6, 129.9, 127.1, 124.5, 123.0, 43.2, 35.6, 33.2, 28.9, 26.5, 22.9, 14.0, 10.9.

1-(3'-Nitrophenyl)-3-ethylheptane-1-ol. To a suspension of $NaBH_4$ (1.89 g, 50 mmol) in EtOH (100 mL) at 0 °C was added **6** (26.3 g, 100 mmol) in EtOH (25 mL), and the reaction was stirred at 0 °C for 1 h. The reaction was quenched by the slow addition of 3 N HCl at 0 °C, and the organic layer was extracted with EtOAc. The crude product was purified by flash chromatography on silica gel (7:1, hexane:EtOAc) to afford 17.88 g (67 %) of the title compound as brown oil. IR (neat) 3418, 2923, 1538, 1348, 1065, 807, 739, 691 cm^{-1}. ^1H NMR (300 MHz, $CDCl_3$) δ 8.21 (t, J = 2.0 Hz, 1 H), 8.10 (dd, J = 7.9, 2.1 Hz, 1 H), 7.67 (dd, J = 8.4, 0.9 Hz, 1 H), 7.50 (t, J = 7.9 Hz, 1 H), 4.87 (td, J = 8.4, 4.2 Hz, 1 H), 1.90 (d, J = 3.6 Hz, 1 H), 1.78-1.69 (m, 1 H), 1.61-1.22 (m, 9 H), 0.90-0.81 (m, 6 H).

1-(3'-Nitrophenyl)-1-acetoxy-3-ethylheptane. To a solution of 4-N,N-dimethylaminopyridine (0.16 g, 1.3 mmol) and 1-(3'-nitrophenyl)-3-ethylheptane-1-ol (36.02 g, 135.9 mmol) in pyridine (150 mL) at 0 °C was added Ac_2O (18.9 mL, 200 mmol) dropwise over 10 min. The reaction was stirred for 30 min and monitored by TLC. The reaction was poured into cold 3 N HCl, and the organic layer extracted with EtOAc. The organic layer was dried over MgSO$_4$, filtered and concentrated in vacuo to afford 40.49 g (97 %) of the title compound as brown oil. No purification was necessary. IR (neat) 2926, 1746, 1538, 1463, 1351, 1232, 1023 cm^{-1}. ^{1}H NMR (300 MHz, CDCl$_3$) δ 8.19 (s, 1 H), 8.13 (d, J = 7.1 Hz, 1 H). 7.62 (d, J = 7.7 Hz, 1 H), 7.50 (t, J = 8.0 Hz, 1 H), 5.83 (2 overlapping d, J = 8.9, 8.9 Hz, 1 H), 2.07 (s, 3 H), 1.97-1.82 (m, 1 H), 1.65-1.57 (m, 2 H), 1.44-1.16 (m, 8 H), 0.89-0.79 (m, 6 H).

1-(3'-Aminophenyl)-3-ethylheptane. To a 1 L oven dried round bottom flask equipped with a dry ice condenser was condensed liquid NH$_3$ (400 mL), and then Li metal (13.9 g, 2 meq) was added in small portions over 30 min. The resulting blue solution was stirred at -33 °C for 30 min and 1-(3'-nitrophenyl)-1-acetoxy-3-ethylheptane (61.4 g, 200 mmol) was added dropwise in THF (400 mL) at -78 °C over 30 min. The reaction was stirred at -33 °C for 1 h and quenched by the slow addition of EtOH. The crude mixture was subjected to aqueous work up and the organic layer extracted with EtOAc (4 × 50 mL). The organic layer was dried over MgSO$_4$ and purified by flash chromatography on silica gel (5:1, hexane:EtOAc) to afford 26.1 g (60 % yield) of the title compound as light brown oil. IR (neat) 3450, 3370, 2926, 1621, 1462, 1392, 1166 cm^{-1}. ^{1}H NMR (300 MHz, CDCl$_3$) δ 7.05 (t, J = 8.0 Hz, 1 H), 6.57 (d, J = 8.0 Hz, 1 H), 6.51 (s, 1 H), 6.49 (d, J = 7.7 Hz, 1 H), 3.58 (br s, 2 H), 2.47 (2 sets of overlapping ddd, J = 16.6, 5.7, 5.5 Hz, 2 H), 1.57-1.43 (m, 2 H), 1.33-1.22 (m, 9 H), 0.9-0.82 (m, 6 H). ^{13}C NMR (75 MHz, CDCl$_3$) δ 146.4, 144.7, 129.2, 118.8, 115.3, 112.5, 38.7, 35.1, 33.2, 32.7, 28.9, 25.8, 23.2, 14.2, 10.8.

1-(3'-Amino-N-(tert-butoxycarbonyl)phenyl)-3-ethylheptane   (7). To a solution of 1-(3'-aminophenyl)-3-ethylheptane (1.1 g, 5 mmol) and di-t-butyl dicarbonate (1.5 mL, 6.5 mmol) in THF (15 mL) was added 1 N NaOH (15 mL) in 3 portions. The reaction was heated to reflux for 2 h and cooled to room temperature. The reaction was poured into water and the organic layer extracted with EtOAc. The organic layer was dried over MgSO$_4$, filtered and concentrated in vacuo. The crude mixture was purified by flash chromatography on silica gel (6:1, hexane:EtOAc) to afford 1.39 g (87 %) of the title compound as brown oil. IR (neat) 3336, 2928, 1704, 1594, 1531, 1442, 1367, 1237, 1164, 1055 cm^{-1}. ^{1}H NMR (300 MHz, CDCl$_3$) δ 7.19 (s, 1 H), 7.15 (d, J = 7.3 Hz, 1 H), 7.12 (2 line m, 1 H), 6.84 (d, J = 7.2 Hz, 1 H), 6.40 (br s, 1 H), 2.53 (2 sets of overlapping ddd, J = 17.1, 6.0, 5.1 Hz, 2 H), 1.53-1.49 (m, 1 H), 1.50 (s, 9 H), 1.32-1.22 (m, 10 H), 0.88 (t, J = 6.3 Hz, 3 H), 0.84 (t, J = 7.7 Hz, 3 H). ^{13}C NMR (75 MHz, CDCl$_3$) δ 152.8, 144.4, 138.3, 128.8, 123.1, 118.5, 115.8, 80.3, 39.3, 38.7, 35.2, 33.3, 32.7, 28.9, 28.4, 25.7, 23.2, 14.2, 10.8.

4-Amino-2-(3'-ethylheptyl)acetophenone. To a 25 mL 3-neck round bottom flask equipped with a reflux condenser and drying tube containing AlCl$_3$ (0.93 g, 7.0 mmol) and CS$_2$ (10 mL) at 0 °C was added **7** (0.64 g, 2.0

mmol) in CS_2 (10 mL) and acetyl chloride (0.28 mL, 4.0 mmol). The reaction mixture was heated to reflux at 70 °C for 3 h. After cooling to 60 °C, the reflux condenser was removed to allow the CS_2 layer to boil off before pouring it into ice water (20 mL). The solution was then poured into 3 N NaOH (20 mL) and extracted with ether. The combined organic extracts were washed with brine and dried over $MgSO_4$. The solution was filtered and the solvent removed in vacuo to give brown oil that was purified by flash chromatography on silica gel (3:1, hexane:EtOAc) to afford 0.36 g (70%) of the title compound as dark liquid. IR (neat) 3500, 3362, 2926, 1601, 1455, 1252, 1141, 1054 cm^{-1}. ^{1}H NMR (300 MHz, CDCl$_3$) δ 7.59 (d, J = 9.1 Hz, 1 H), 6.47 (s, 1 H), 6.45 (d, J = 9.9 Hz, 1 H), 3.96 (br s, 2 H), 2.86-2.80 (4 line m, 2 H), 2.49 (s, 3 H), 1.47-1.43 (m, 2 H), 1.39-1.22 (m, 9 H), 0.88 (t, J = 7.5 Hz, 3 H), 0.86 (t, J = 7.5 Hz, 3 H). ^{13}C NMR (75 MHz, CDCl$_3$) δ 199.23, 150.05, 148.11, 133.55, 127.58, 117.25, 111.39. 39.75, 35.69, 33.15, 32.76, 29.41, 29.39, 26.15, 23.53, 14.58, 11.24.

4-(Diethyltriazenyl)-2-(3'-ethylheptyl)acetophenone (8). To an oven dried vessel containing boron trifluoride etherate (0.37 mL, 3.0 mmol) at 0 °C was added 4-amino-2-(3'-ethylheptyl)acetophenone (0.52 g, 2.0 mmol) in CH_2Cl_2 (2 mL) followed by t-butyl nitrite (0.3 mL, 2.5 mmol). The reaction was stirred for 30 min at 0 °C and K_2CO_3 (1 g) and diethylamine (1 mL) were added sequentially. The reaction was stirred for 1 h and poured into water. The organic layer was extracted with EtOAc and dried over $MgSO_4$. The sample was concentrated in vacuo to afford 0.65 g (95 %) of the title compound as dark liquid. IR (neat) 2929, 1677, 1352, 1240, 1107 cm^{-1}. ^{1}H NMR (300 MHz, CDCl$_3$) δ 7.66 (d, J = 8.8 Hz, 1 H), 7.25 (2 line m, 2 H), 3.77 (q, J = 7.1 Hz, 4 H), 3.02-3.00 (4 line m, 1 H), 2.88-2.82 (4 line m, 1 H), 2.55 (s, 3 H), 1.55 (br s, 1 H), 1.46-1.15 (m, 16 H), 0.95-0.82 (m, 6 H).

1-(Diethyltriazenyl)-3-(3'-ethylheptyl)-4-(trimethylsilylethynyl)benzene (9). To a stirred solution of diisopropylamine (1.8 mL, 13 mmol) in THF (15 mL) was added at 0 °C under N_2 n-BuLi (8.1 mL, 12 mmol, 1.49 M in hexanes). The reaction mixture was stirred at 0 °C for 5 min then cooled to -78 °C. To the LDA solution was added **8** (3.71 g, 10.8 mmol) in THF (2 mL). The reaction mixture was stirred at -78 °C for 1 h before adding, at the same temperature, diethylchlorophosphate (1.7 mL, 12 mmol). The reaction mixture was allowed to warm to room temperature and then stirred for 3 h. The reaction solution was then transferred via cannula into 2.25 equivalents of LDA [made as above with diisopropylamine (3.1 mL, 22 mmol) and n-BuLi (16.3 mL, 24.3 mmol)] in THF (25 mL) at -78 °C. The ensuing reaction mixture was stirred at -78 °C for 30 min and then allowed to warm to room temperature for 1 h, at which time the reaction was cooled to 0 °C and chlorotrimethylsilane (1.4 mL, 11 mmol) was added. The reaction was stirred for 2 h at room temperature and the solution was poured into water (20 mL) and extracted with ether. The combined ether extracts were washed with brine and dried over $MgSO_4$. Filtration and removal of the solvent in vacuo left a brown oil which upon purification by flash chromatography on silica gel (9:1, hexane:EtOAc) afforded 1.9 g (43 %) of the title compound as yellow oil. IR (neat) 2959, 2148, 1598, 1402, 1248, 1103, 843, 759 cm^{-1}. ^{1}H NMR (500 MHz, CDCl$_3$) δ 7.36 (d, J = 8.2 Hz, 1 H), 7.21 (s, 1 H), 7.16 (d, J = 8.2 Hz, 1 H), 3.74 (q, J = 7.1 Hz, 4 H), 2.78 (m, 2 H), 1.63 (m, 1 H), 1.30-1.17 (m, 16 H), 0.95-0.81 (m, 6 H), 0.23 (s, 9 H). LRMS calcd for C$_{23}$H$_{41}$N$_3$Si: 399. Found: 399. λ$_{max}$ (CH$_2$Cl$_2$) = 334 nm. UV 10% edge value = 388 nm. M$_n$ = 390, M$_w$ = 417, M$_w$/M$_n$ = 1.07.

General procedure for the coupling of a terminal alkyne with an aryl halide. Before beginning the reaction, all of the starting materials were thoroughly flushed with argon and/or N$_2$. To a stirring solution of the aryl halide, bis(dibenzylideneacetone)palladium(0) (2 mol %), (C$_6$H$_5$)$_3$P (2.5 equivalents based on Pd(0)) and CuI (2 mol %) in THF was added the terminal alkyne followed by the amine (4 equivalents based on aryl halide) at room temperature (unless otherwise stated) under argon and/or N$_2$ in a screw cap tube. The vessel was flushed with argon and/or N$_2$, sealed and allowed to stir overnight. The reaction mixture was then subjected to an aqueous work-up and the aqueous layer extracted with ether, CH$_2$Cl$_2$ or EtOAc (3×). After drying the combined organic layers over MgSO$_4$, the solvent was removed in vacuo to afford a crude product that was purified by column chromatography (silica gel). Eluents and other slight modifications are described below for each reaction.

3-(1'-Dodecynyl)nitrobenzene. 1-Iodo-3-nitrobenzene (12.45 g, 50 mmol), 1-dodecyne (11.76 mL, 55 mmol), bis(dibenzylideneacetone)palladium(0) (0.57 g, 1.0 mmol), (C$_6$H$_5$)$_3$P (0.66 g, 2.50 mmol), CuI (0.19 g, 1.0 mmol), N,N-diisopropylethylamine (35 mL, 200 mmol), and THF (50 mL) afforded 14.21 g (99%) of the title compound as light brown oil after gravity liquid chromatography (9:1 hexane:ether). R$_f$ = 0.76 (9:1 hexane:ether). IR (neat) 3084, 2921, 2855, 2229, 1959, 1902, 1825, 1795, 1728, 1572, 1534, 1467, 1351, 1301, 1165, 1096, 900, 805, 736, 675 cm^{-1}. ^{1}H NMR (400 MHz, CDCl$_3$) δ 8.21 (s, 1 H), 8.10 (ddd, J = 8.3, 2.2, 1.0 Hz, 1 H), 7.65 (d, J = 7.7 Hz, 1 H), 7.43 (t, J = 8.1 Hz, 1 H), 7.24 (s, 1 H), 2.40 (t, J = 7.1 Hz, 2 H), 1.60 (p, J = 7.0 Hz, 2 H), 1.43 (p, J = 7.0 Hz, 2 H), 1.28-1.25 (m, 12 H), 0.86 (t, J = 7.1 Hz, 3 H). ^{13}C NMR (100 MHz, CDCl$_3$) δ 148.01, 137.22, 129.04, 126.26, 125.94, 122.10, 93.58, 78.46, 31.90, 29.59, 29.52, 29.32, 29.14, 28.93, 28.47, 22.68, 19.33, 14.06. HRMS calcd for C$_{18}$H$_{25}$NO$_2$: 287.1885. Found: 287.1891.

3-Dodecylaniline (10). Anhydrous ethanol (50 mL), 3-(1'-dodecynyl)nitrobenzene (14.21 g, 49.51 mmol) and 10% Pd/C (2.50 g) were placed in a Parr flask. The flask was purged with H$_2$ before pressurizing to 65 psi and heating to 65 °C. The pressure was maintained at 65 psi for 5 h. The Pd/C was then filtered and the solvent removed in vacuo to afford 12.54 g (97%) of the title compound as a brown liquid that solidified upon standing into a tan solid. $R_f = 0.63$ (4:1 hexane:EtOAc). IR (neat) 3464, 3379, 3036, 2925, 2854, 1620, 1590, 1494, 1461, 1265, 1167, 909, 738 cm^{-1}. ^{1}H NMR (400 MHz, CDCl$_3$) δ 7.04 (dt, J = 7.4, 0.90 Hz, 1 H), 6.57 (d, J = 7.7 Hz, 1 H), 6.49 (d, J = 7.5 Hz, 1 H), 6.48 (d, J = 7.3 Hz, 1 H), 3.57 (s, 2 H), 2.49 (t, J = 7.9 Hz, 2 H), 1.56 (p, J = 7.3 Hz, 2 H) 1.28-1.24 (m, 18 H), 0.87 (t, J – 7.0 Hz, 3 H).). ^{13}C NMR (100 MHz, CDCl$_3$) δ 146.51, 144.22, 129.18, 118.86, 115.40, 112.67, 36.19, 32.14, 31.60, 29.91, 29.87, 29.84, 29.78, 29.62, 29.59, 22.89, 14.28. HRMS calcd for C$_{18}$H$_{31}$N: 261.2456. Found: 261.2448.

1-Amino(N-tert-butoxycarbonyl)-3-dodecylbenzene. To a solution of 3-dodecylaniline (**10**) (119.84 g, 459 mmol) and di-t-butyl dicarbonate (116 mL, 505 mmol) in THF (500 mL) was added 1 M NaOH (500 mL) in 3 portions. The reaction mixture was heated to reflux overnight and then cooled to room temperature. The reaction mixture was poured into water and the organic layer extracted with ether (3×). The organic layer was dried over MgSO$_4$, filtered and concentrated in vacuo. The crude mixture was purified by flash liquid chromatography on silica gel (24:1 hexane:ether) to afford 163.86 g (99%) of the title compound as a light brown oil which solidified upon standing. $R_f = 0.41$ (24:1 hexane:ether). IR (neat) 3320, 3145, 2920, 2853, 2356, 1693, 1597, 1539, 1440, 1366, 1251, 1162, 1054, 873 cm^{-1}. ^{1}H NMR (500 MHz, CDCl$_3$) δ 7.21 (br s, 1 H), 7.16 (t, J = 7.9 Hz, 1 H), 7.10 (d, J = 8.1 Hz, 1 H), 6.83 (d, J = 7.5 Hz, 1 H), 6.41 (br s, 1 H), 3.55 (t, J = 7.6 Hz, 2 H), 1.58 (p, J = 7.3 Hz, 2 H), 1.50 (s, 9 H), 1.28-1.23 (m, 18 H), 0.89 (t, J = 6.9 Hz, 3 H). ^{13}C NMR (125 MHz, CDCl$_3$) δ 153.20, 144.42, 138.64, 129.13, 123.57, 118.90, 116.21, 80.73, 36.44, 32.31, 31.84, 30.06, 30.05, 30.02, 29.98, 29.90, 29.77, 29.74, 28.75, 23.07, 14.49. HRMS calcd for C$_{23}$H$_{39}$NO$_2$: 361.2981. Found: 361.2993.

4-Amino-2-dodecylacetophenone (11). To a 2-neck round bottom flask equipped with a reflux condenser containing aluminum chloride (23.48 g, 176.10 mmol) and chlorobenzene (50 mL) at 0 °C was added 1-amino(N-tert-butoxycarbonyl)-3-dodecylbenzene (18.15 g, 50.30 mmol) in chlorobenzene (50 mL) and acetyl chloride (3.93 mL, 55.3 mmol). The reaction mixture was heated to reflux overnight and then allowed to cool to room temperature. The solution was then slowly poured into 3 M KOH (100 mL) and extracted with ether (3×). The combined organic extracts were washed with brine and dried over MgSO$_4$. The solution was filtered and the solvent removed in vacuo to give a brown oil that was purified by gravity liquid chromatography on silica gel (1:1 hexane:ether) to afford 12.90 g (85%) of the title compound as a brown liquid which solidified upon standing. $R_f = 0.37$ (1:1 hexane:ether). IR (neat) 3484, 3369, 3228, 3051, 2957, 2926, 2855, 1660, 1622, 1602, 1563, 1460, 1356, 1252, 1143, 909, 814, 738 cm^{-1}. ^{1}H NMR (400 MHz, CDCl$_3$) δ 7.61 (d, J = 9.1 Hz, 1 H), 6.48-6.44 (m, 2 H), 3.96 (br s, 2 H), 2.85 (t, J = 7.7 Hz, 2 H), 2.49 (s, 3 H), 1.50 (p, J = 7.7 Hz, 2 H), 1.40-1.23 (m, 18 H), 0.86 (t, J = 7.0 Hz, 3 H). ^{13}C NMR (100 MHz, CDCl$_3$) δ 198.78, 150.46, 147.23, 133.44, 126.30,

116.76, 110.96, 35.17, 31.98, 31.62, 31.61, 30.03, 29.79 (2 C), 29.73, 29.70, 29.43, 28.87, 22.74, 22.68, 14.15. HRMS calcd for $C_{20}H_{33}NO$: 303.2562. Found: 303.2568.

4-(Diethyltriazenyl)-2-dodecyl-1-acetophenone (12). To an oven dried vessel containing boron trifluoride etherate (0.21 mL, 1.68 mmol) at 0 °C was added **11** (0.26 g, 0.86 mmol) in CH_2Cl_2 (2 mL) followed by t-butyl nitrite (0.17 mL, 1.4 mmol). The mixture was stirred for 30 min at 0 °C and K_2CO_3 (0.56 g) and diethylamine (0.56 mL) were sequentially added. The reaction was stirred for 2 h and poured into water. The organic layer was extracted with EtOAc and dried over $MgSO_4$. The sample was concentrated in vacuo and the crude oil that was purified by flash liquid chromatography (9:1, hexane:ether) to afford 0.33 g (99%) of the title compound as a light brown oil. R_f = 0.36 (9:1, hexane:ether). IR (neat) 2927, 2855, 1677, 1597, 1559, 1449, 1431, 1395, 1348, 1239, 1105, 952, 893, 824 cm^{-1}. ^{1}H NMR (400 MHz, CDCl$_3$) δ 7.62 (d, J = 8.3 Hz, 1 H), 7.26 (d, J = 1.9 Hz, 1 H), 7.23 (dd, J = 8.3, 2.0 Hz, 1 H), 3.72 (q, J = 7.2 Hz, 4 H), 2.88 (t, J = 7.8 Hz, 2 H), 2.50 (s, 3 H), 1.59 (p, J = 8.3 Hz, 2 H), 1.40-1.22 (m, 24 H), 0.86 (t, J = 7.0 Hz, 3 H). ^{13}C NMR (125 MHz, CDCl$_3$) δ 200.76, 153.83, 145.51, 133.98, 131.47, 123.85, 117.38, 35.05, 32.31, 32.17, 30.29, 30.09, 30.07, 30.06, 30.04, 29.95, 29.88, 29.74, 23.07, 14.49. HRMS calcd for $C_{24}H_{41}N_3O$: 387.3250. Found: 387.3254.

1-(Diethyltriazenyl)-3-dodecyl-4-(trimethylsilylethynyl)benzene (13). To a stirred solution of diisopropylamine (0.79 mL, 6.05 mmol) in THF (5 mL) was added at -78 °C under N_2 n-BuLi (3.59 mL, 5.50 mmol, 1.53 M in hexanes). The reaction mixture was allowed to warm to 0 °C for 10 min then cooled to -78 °C. To the LDA solution was added **12** (1.94 g, 5.00 mmol) in THF (5 mL). The reaction mixture was stirred at -78 °C for 1 h before adding, at the same temperature, diethylchlorophosphate (0.73 mL, 5.05 mmol). The reaction mixture was allowed to warm to room temperature over 30 min before being recooled to -78 °C. The reaction solution was then transferred via cannula into 2.25 equivalents of LDA [made as above with diisopropylamine (1.62 mL, 12.38 mmol) and n-BuLi (7.35 mL, 11.25 mmol, 1.53 M in hexane)] in THF (15 mL) at -78 °C. The ensuing reaction mixture was allowed to warm to 0 °C over 30 min and then recooled to -78 °C before chlorotrimethylsilane (1.59 mL, 12.5 mmol) was added. The solution was allowed to stir overnight at room temperature before being poured into water and extracted with ether (3×). The combined ether extracts were washed with brine and dried over $MgSO_4$. Filtration and removal of the solvent in vacuo left a brown oil which, upon purification by gravity liquid chromatography on silica gel (100:1 hexane:ether), afforded 1.17 g (53%) of the title compound as a light brown oil. R_f = 0.64 (9:1 hexane:ether). IR (neat) 2957, 2925, 2854, 2149, 1598, 1453, 1434, 1402, 1341, 1238, 1203, 1104, 867, 843, 759, 654 cm^{-1}. ^{1}H NMR (400 MHz, CDCl$_3$) δ 7.36 (d, J = 8.2 Hz, 1 H), 7.21 (d, J = 1.9 Hz, 1 H), 7.16 (dd, J = 8.3, 2.0 Hz, 1 H), 3.74 (q, J = 7.2 Hz, 4 H), 2.74 (t, J = 7.8 Hz, 2 H), 1.68 (p,

J = 7.7 Hz, 2 H), 1.40-1.24 (m, 24 H), 0.86 (t, J = 7.0 Hz, 3 H), 0.24 (s, 9 H). ^{13}C NMR (100 MHz, CDCl$_3$) δ 151.12, 146.48, 133.01, 120.92, 118.76, 117.33, 104.79, 96.96, 34.99, 31.92, 30.75, 29.74, 29.69, 29.65, 29.57, 29.36, 22.69, 14.12, 0.10. HRMS calcd for C$_{27}$H$_{47}$N$_3$Si: 441.3539. Found: 441.3549.

General procedure for iodide formation from triazenes. To a thick-walled oven dried screw cap tube was added the corresponding triazene and CH$_3$I (8 eq) as the solvent. The tube was flushed with N$_2$ or argon, sealed and heated to 120 °C overnight. The reaction was cooled to room temperature and diluted with hexane. The smaller oligomers were then filtered through a plug of silica gel with hexane, and the larger oligomers were filtered through a plug of silica gel with EtOAc.

General procedures for the desilylation of alkynes. (a) The silylated alkyne was dissolved in MeOH, or in some cases, MeOH and CH$_2$Cl$_2$ to aid in the solubility of the substrate. K$_2$CO$_3$ (2 equivalents) was added and the reaction was stirred overnight. The reaction mixture was then subjected to an aqueous work-up and the aqueous layer extracted with ether, CH$_2$Cl$_2$ or EtOAc (3×). After drying the combined organic layers over MgSO$_4$, the solvent was removed in vacuo. The crude product was then filtered through a plug of silica gel yielding the free alkyne that was used without further purification. Eluents and other slight modifications are described below for each material. (b) The silylated alkynes was dissolved in THF and cooled to 0 °C. TBAF (1.5 equivalents, 1.0 M in THF) was added and the reaction was stirred overnight. The reaction mixture was then subjected to an aqueous work-up and the aqueous layer extracted with ether, CH$_2$Cl$_2$ or EtOAc (3×). After drying the combined organic layers over MgSO$_4$, the solvent was removed in vacuo. The crude product was then filtered through a plug of silica gel yielding the free alkyne that was used without further purification. Eluents and other slight modifications are described below for each material. (c) The silylated alkyne was dissolved in pyridine (0.5 M) in a plastic vessel. A pre-formed solution of concentrated HF (1.1 eq) in TBAF (2.2 eq, 1.0 M in THF) was added at room temperature and allowed to stir for 15 to 30 min before being quenched with silica gel. The reaction mixture was then filtered, subjected to an aqueous work-up, and the aqueous layer extracted with ether, CH$_2$Cl$_2$ or EtOAc (3×). After drying the combined organic layers over MgSO$_4$, the solvent was removed in vacuo. The crude product was then purified by column chromatography (silica gel) yielding the free alkyne that was immediately taken onto the next step. Eluents and other slight modifications are described below for each material.

1-Iodo-3-ethyl-4-(trimethylsilylethynyl)benzene (14). 4 (3.05 g, 10.0 mmol) and CH$_3$I (5 mL) afforded 3.25 g (99%) of the title compound as a dark brown liquid. R$_f$ = 0.55 (hexane). IR (neat) 2964, 2156, 1579, 1472, 1393, 1250, 1183, 1125, 1081, 841, 760, 699, 657 cm^{-1}. ^{1}H NMR (300 MHz, CDCl$_3$)

δ 7.53 (d, J = 1.5 Hz, 1 H), 7.44 (dd, J = 8.1, 1.8 Hz, 1 H), 7.11 (d, J = 8.1 Hz, 1 H), 2.71 (q, J = 7.6 Hz, 2 H), 1.20 (t, J = 7.5 Hz, 3 H), 0.23 (s, 9 H). HRMS calcd for $C_{13}H_{17}ISi$: 328.0144. Found: 328.0135. λ_{max} (CH_2Cl_2) = 266 nm. UV 10% edge value = 300 nm.

1-(Diethyltriazenyl)-3-ethyl-4-(ethynyl)benzene (17). 4 (2.98 g, 9.9 mmol), MeOH (50 mL), and K_2CO_3 (4.1 g, 30 mmol) afforded 2.22 g (98 %) of the title compound as clear yellow oil. IR (neat) 3294, 2970, 2099, 1600, 1402, 1236, 1105, 891, 829 cm^{-1}. ^{1}H NMR (300 MHz, CDCl$_3$) δ 7.41 (d, J = 8.2 Hz, 1 H), 7.25 (d, J = 1.9 Hz, 1 H), 7.18 (dd, J = 8.2, 2.1 Hz, 1 H), 3.75 (q, J = 7.2 Hz, 4 H), 3.22 (s, 1 H), 2.80 (q, J = 7.6 Hz, 2 H), 1.25 (t, J = 7.5 Hz, 9 H). HRMS calcd for $C_{14}H_{19}N_3$: 229.1579. Found: 229.1574.

Ethyl-containing dimer 20. 14 (2.95 g, 9 mmol), 17 (2.24 g, 9.78 mmol), THF (20 mL), bis(dibenzylideneacetone)palladium(0) (0.26 g, 0.45 mmol), $(C_6H_5)_3P$ (0.59 g, 2.25 mmol), CuI (0.01 g, 0.90 mmol), and N,N-diisopropylethylamine (6.23 mL, 36 mmol), for 1 d afforded 3.48 g (90%) of the title compound as yellow oil after gravity liquid chromatography (24:1 hexane:ether). R_f = 0.56 (24:1 hexane:ether). IR (neat) 2971, 2200, 2153, 1595, 840 cm^{-1}. ^{1}H NMR (300 MHz, CDCl$_3$) δ 7.43 (d, J = 8.2 Hz, 1 H), 7.37 (d, J = 7.9 Hz, 1 H), 7.33 (d, J = 1.2 Hz, 1 H), 7.29 (d, J = 1.7 Hz, 1 H), 7.25 (dd, J = 7.9, 1.6 Hz, 1 H), 7.22 (dd, J = 8.3, 2.0 Hz, 1 H), 3.76 (q, J = 7.2 Hz, 4 H), 2.87 (q, J = 7.6 Hz, 2 H), 2.78 (q, J = 7.6 Hz, 2 H), 1.30 (t, J = 7.5 Hz, 6 H), 1.25 (t, J = 7.4 Hz, 6 H), 0.25 (s, 9 H). ^{13}C NMR (75 MHz, CDCl$_3$) δ 151.4, 147.1, 146.7, 133.0, 132.4, 130.8, 128.6, 124.2, 121.8, 120.4, 118.5, 117.7, 103.7, 99.5, 92.7, 90.5, 28.0, 27.7, 14.9, 14.5, 0.04. HRMS calcd for $C_{27}H_{35}N_3Si$: 429.2600. Found: 429.2592.

Ethyl-containing iodide dimer 23. 20 (1.93 g, 4.50 mmol) and CH_3I (2.26 mL) afforded 1.98 g (96%) of the title compound as brown oil. IR (neat) 2964, 2152, 1579, 1490, 1393, 1250, 1183, 1125, 1081, 841, 760, 699, 657 cm^{-1}. ^{1}H NMR (500 MHz, CDCl$_3$) δ 7.58 (d, J = 1.6 Hz, 1 H), 7.49 (dd, J = 8.1, 1.8 Hz, 1 H), 7.39 (d, J = 7.9 Hz, 1 H), 7.34 (d, J = 1.1 Hz, 1 H), 7.27 (dd, J = 7.9, 1.6 Hz, 1 H), 7.18 (d, J = 8.1 Hz, 1 H), 2.81 (q, J = 7.5 Hz, 2 H), 2.80 (q, J = 7.5 Hz, 2 H), 1.28 (t, J = 7.6 Hz, 3 H), 1.26 (t, J = 7.5 Hz, 3 H), 0.26 (s, 9 H). ^{13}C NMR (125 MHz, CDCl$_3$) δ 148.5, 147.2, 137.5, 135.3, 133.9, 132.8, 131.2, 129.0, 123.7, 122.9, 122.3, 103.8, 100.3, 95.1, 94.6, 89.1, 28.0, 27.9, 15.7, 15.0, 14.8, 0.4. HRMS calcd for $C_{23}H_{25}ISi$: 456.0770. Found: 456.0769. λ_{max} (CH_2Cl_2) = 316 nm. UV 10% edge value = 350 nm.

Ethyl-containing terminal alkynyl dimer 26. 20 (1.07 g, 2.50 mmol), MeOH (10 mL) and K_2CO_3 (1.04 g, 7.50 mmol) afforded 0.89 g (100%) of the title compound as brown oil. ^{1}H NMR (500 MHz, CDCl$_3$) δ 7.43 (dd, J = 8.0, 0.3 Hz, 1 H), 7.41 (d, J = 7.9 Hz, 1 H), 7.35 (d, J = 1.3 Hz, 1 H), 7.28 (d, J = 1.8 Hz, 1 H), 7.27 (ddd, J = 8.1, 1.8, 0.3 Hz, 1 H), 7.22 (dd, J = 8.2, 1.9 Hz, 1 H), 3.76 (q, J = 7.2 Hz, 4 H), 3.31 (s, 1 H), 2.86 (q, J = 7.5 Hz, 3

H), 2.80 (q, J = 7.6 Hz, 3 H), 1.30 (t, J = 7.6 Hz, 3 H), 1.25 (t, J = 7.5 Hz, 6 H), 1.19 (t, J = 7.0 Hz, 3 H). ^{13}C NMR (125 MHz, CDCl$_3$) δ 151.9, 147.5, 147.3, 133.4, 133.2, 131.2, 129.0, 124.9, 121.2, 120.8, 118.9, 118.1, 92.9, 91.0, 82.6, 82.5, 28.4, 27.9, 15.3, 15.1. HRMS calcd for C$_{24}$H$_{27}$N$_3$: 357.2205. Found: 357.2202.

Ethyl-containing tetramer 29. 23 (0.91 g, 2.0 mmol), **26** (0.73 g, 2.06 mmol), THF (4 mL), bis(dibenzylideneacetone)palladium(0) (57.40 mg, 0.10 mmol), (C$_6$H$_5$)$_3$P (105 mg, 0.40 mmol), CuI (38.1 mg, 0.20 mmol), N,N-diisopropylethylamine (1.38 mL, 8 mmol), for 5 d afforded 1.17 g (93%) of the title compound as a yellow solid after gravity liquid chromatography (24:1 hexane:ether). R$_f$ = 0.26 (24:1 hexane:ether). IR (KBr) 2965, 2872, 2151, 1590, 1503, 1401, 1246, 841 cm^{-1}. ^{1}H NMR (300 MHz, CDCl$_3$) δ 7.46 (d, J = 7.9 Hz, 1 H), 7.45 (d, J = 8.0 Hz, 1 H), 7.45 (d, J = 8.2 Hz, 1 H), 7.40 (d, J = 7.8 Hz, 1 H), 7.39 (br m, 2 H), 7.35 (d, J = 1.3 Hz, 1 H), 7.32 (dd, J = 7.9, 1.7 Hz, 1 H), 7.32 (dd, J = 7.9, 1.7 Hz, 1 H), 7.29 (d, J = 1.9, 1 H), 7.27 (dd, J = 8.2, 1.6 Hz, 1 H), 7.23 (dd, J = 8.2, 2.1 Hz, 1 H), 3.76 (q, J = 7.2 Hz, 4 H), 2.885 (q, J = 7.5 Hz, 2 H), 2.884 (q, J = 8.2 Hz, 2 H), 2.87 (q, J = 7.5 Hz, 2 H), 2.79 (q, J = 7.6 Hz, 2 H), 1.32 (t, J = 7.6 Hz, 6 H), 1.31 (t, J = 7.6 Hz, 3 H), 1.26 (t, J = 7.6 Hz, 3 H), 0.25 (s, 9 H). ^{13}C NMR (75 MHz, CDCl$_3$) δ 151.75, 147.50, 147.22, 146.69, 146.62, 133.29, 132.75, 132.54, 132.50, 131.33, 131.26, 131.21, 129.18, 129.06, 129.03, 124.45, 123.85, 123.79, 122.78, 122.68, 122.16, 120.72, 118.88, 117.96, 103.82, 100.29, 94.99, 94.80, 93.08, 90.93, 89.98, 89.73, 28.33, 28.05, 28.03, 27.99, 15.21, 15.05, 15.01, 14.80, 0.33. HRMS calcd for C$_{47}$H$_{51}$N$_3$Si: 685.3852. Found: 685.3843.

Ethyl-containing iodide tetramer 32. 29 (136.70 mg, 0.20 mmol) and CH$_3$I (4 mL) afforded 142.40 mg (98%) of the title compound as light yellow solid. IR (KBr) 2965, 2930, 2151, 1500, 1459, 1249, 888, 841 cm^{-1}. ^{1}H NMR (500 MHz, CDCl$_3$) δ 7.60 (d, J = 1.6 Hz, 1 H), 7.51 (dd, J = 8.1, 1.7 Hz, 1 H), 7.47 (d, J = 7.7 Hz, 2 H), 7.41-7.38 (4 line m, 3 H), 7.37-7.28 (m, 4 H), 7.20 (d, J = 8.1 Hz, 1 H), 2.89 (q, J = 7.5 Hz, 2 H), 2.88 (q, J = 7.5 Hz, 2 H), 2.83 (q, J = 7.6 Hz, 2 H), 2.81 (q, J = 7.6 Hz, 2 H), 1.33 (t, J = 7.6 Hz, 3 H), 1.29 (t, J = 7.6 Hz, 6 H), 1.27 (t, J = 7.6 Hz, 3 H), 0.27 (s, 9 H). ^{13}C NMR (75 MHz, CDCl$_3$) δ 148.57, 147.23, 146.73, 146.72, 137.47, 135.26, 133.88, 132.77, 132.72, 132.57, 131.35, 131.33, 131.27, 129.22, 129.19, 129.05, 128.38, 126.07, 123.77, 123.72, 123.63, 122.87, 122.83, 122.31, 103.83, 100.34, 95.16, 95.11, 94.66, 89.76, 89.70, 89.16, 28.06, 28.01, 27.93, 15.03, 14.96, 14.82, 0.35. HRMS calcd for C$_{43}$H$_{41}$ISi: 712.2022. Found: 712.2028. λ$_{max}$ (CH$_2$Cl$_2$) = 352 nm. UV 10% edge value = 385 nm.

Ethyl-containing terminal alkynyl tetramer 35. 29 (0.29 g, 0.42 mmol), THF (5 mL), and TBAF (0.6 mL, 0.6 mmol, 1.0 M in THF) afforded 0.14 g (55 %) of the title compound as a light yellow solid. A ^{1}H NMR was run to examine the loss of the trimethylsilyl peak, and the material was used directly in the next reaction.

Ethyl-containing octamer 38. Bis(dibenzylideneacetone)palladium(0) (3.4 mg, 0.006 mmol), $(C_6H_5)_3P$ (7.9 mg, 0.03 mmol), CuI (2.3 mg, 0.012 mmol), **32** (41.9 mg g, 0.06 mmol), **35** (42.9 mg, 0.07 mmol), THF (2 mL), and diisopropylamine (0.5 mL, 3.5 mmol) at 60 °C overnight afforded a solid that was filtered and rinsed with water and CH_2Cl_2. Due to its insolubility, the solid was only characterized by direct exposure mass spectrometry and UV spectroscopy. Direct exposure MS statistical isotopic range calcd for $C_{87}H_{83}N_3Si$: 1198 (95%), 1199 (100%), 1200 (55%). Found: 1098.6 (signifying loss of the triazene moiety at 100 amu). λ_{max} (CH_2Cl_2) = 364 nm. UV 10% edge value = 500 nm.

3-Ethylheptyl-containing iodide monomer 15. **9** (0.79 g, 2.0 mmol) and CH_3I (2 mL) afforded 0.78 g (92 %) of the title compound as a yellow liquid. IR (neat) 2957, 2156, 1579, 1472, 1249, 1184, 844, 760 cm^{-1}. ^{1}H NMR (300 MHz, CDCl$_3$) δ 7.52 (s, 1 H), 7.43 (d, J = 8.1 Hz, 1 H), 7.11 (d, J = 8.1 Hz, 1 H), 2.72-2.62 (m, 2 H), 1.60 (br s, 1 H), 1.32-1.21 (m, 8 H), 0.95-0.84 (m, 6 H), 0.23 (s, 9 H).

3-Ethylheptyl-containing terminal alkynyl monomer 18. **9** (0.93 g, 2.3 mmol), THF (6 mL), and TBAF (2.6 mL, 2.6 mmol, 1.0 M in THF) afforded 0.72 g (96 %) of the title compound as yellow liquid. IR (neat) 3311, 2931, 2099, 1599, 1402, 1239, 891 cm^{-1}. ^{1}H NMR (300 MHz, CDCl$_3$) δ 7.39 (d, J = 8.2 Hz, 1 H), 7.22 (s, 1 H), 7.18 (d, J = 8.2 Hz, 1 H), 3.74 (q, J = 7.2 Hz, 4 H), 3.21 (s, 1 H), 2.79-2.74 (4 line m, 2 H), 1.64-1.55 (m, 1 H), 1.50-1.19 (m, 16 H), 0.99-0.83 (m, 6 H).

3-Ethylheptyl-containing   dimer   21. Bis(dibenzylideneacetone)palladium(0) (57.4 mg, 0.1 mmol), $(C_6H_5)_3P$ (130 mg, 0.5 mmol), CuI (38.1 mg, 0.2 mmol), **15** (0.78 g, 1.8 mmol), **18** (0.61 g, 1.8 mmol) THF (5 mL), diisopropylamine (1.0 mL) and purification by flash chromatography on silica gel (19:1, hexane:EtOAc) afforded 0.88 g (78 %) of the title compound as yellow oil. IR (neat) 2957, 2151, 1464, 1402, 1236, 1103, 843 cm^{-1}. ^{1}H NMR (300 MHz, CDCl$_3$) δ 7.45-7.27 (m, 6 H), 3.76 (q, J = 7.2 Hz, 4 H), 2.83-2.62 (m, 4 H), 1.80-1.60 (m, 2 H), 1.40-1.11 (m, 26 H), 0.99-0.80 (m, 12 H), 0.25 (s, 9 H). LRMS calcd for $C_{41}H_{63}N_3Si$: 625. Found: 625. λ_{max} (CH_2Cl_2) = 362 nm. UV 10% edge value = 410 nm. M_n = 715, M_w = 744, M_w/M_n = 1.04.

3-Ethylheptyl-containing iodide dimer 24. **21** (0.22 g, 0.35 mmol) and CH_3I (2 mL) afforded 0.23 g (100 %) of the title compound as yellow liquid. IR (neat) 2956, 2152, 1464, 1249, 843, 760 cm^{-1}. ^{1}H NMR (300 MHz, CDCl$_3$) δ 7.57 (s, 1 H), 7.48 (d, J = 8.1 Hz, 1 H), 7.38 (d, J = 8.0 Hz, 1 H), 7.31 (s, 1 H), 7.23 (d, J = 7.7 Hz, 1 H), 7.17 (d, J = 8.1 Hz, 1 H), 2.74 (m, 4 H), 1.64-1.58 (m, 2 H), 1.43-1.20 (m, 20 H), 0.96-0.82 (m, 12 H), 0.25 (s, 9 H).

3-Ethylheptyl-containing terminal alkynyl dimer 27. **21** (0.23 g, 0.37 mmol), THF (2 mL), and TBAF (0.5 mL, 0.5 mmol, 1.0 M in THF) afforded 0.2 g (98 %) of the title compound as yellow oil. IR (neat) 3300, 2956, 2200, 2102, 1597, 1464, 1237, 1104 cm^{-1}. ^{1}H NMR (500 MHz, CDCl$_3$) δ 7.44-7.40 (m, 2 H), 7.33 (s, 1 H), 7.26-7.21 (m, 3 H), 3.75 (q, J = 7.1 Hz, 4 H), 3.30 (s, 1 H), 2.85-2.63 (m, 4 H), 1.63-1.57 (m, 2 H), 1.44-1.22 (m, 26 H), 0.98-0.81 (m, 12 H).

3-Ethylheptyl-containing **tetramer** **30**. Bis(dibenzylideneacetone)palladium(0) (13.8 mg, 0.024 mmol), (C$_6$H$_5$)$_3$P (31.5 mg, 0.12 mmol), CuI (9.1 mg, 0.048 mmol), **24** (0.29 g, 0.48 mmol), **27** (0.27 mg, 0.48 mmol), THF (5 mL), diisopropylamine (0.5 mL) and purification by flash chromatography on silica gel (3:1::hexane:CH$_2$Cl$_2$) afforded 0.23 g (44 %) of the title compound as a yellow oil. IR (neat) 2956, 2151, 1595, 1464, 1402, 1236, 1103 cm^{-1}. ^{1}H NMR (300 MHz, CDCl$_3$) δ 7.48-7.27 (m, 12 H), 3.76 (q, J = 7.1 Hz, 4 H), 2.98-2.70 (m, 8 H), 1.82-1.63 (m, 4 H), 1.59-1.05 (m, 46 H), 1.05-0.78 (m, 24 H), 0.27 (s, 9 H). LRMS calcd for C$_{75}$H$_{107}$N$_3$Si: 1077.8. Found: 1077.9. λ_{max} (CH$_2$Cl$_2$) = 372 nm. UV 10% edge value = 421 nm. M$_n$ = 1335, M$_w$ = 1388, M$_w$/M$_n$ = 1.06.

3-Ethylheptyl-containing iodide tetramer 33. **30** (0.13 g, 0.12 mmol) and CH$_3$I (3 mL) afforded 0.12 g (100 %) of the title compound as a yellow oil. IR (neat) 2955, 2151, 1464, 1378, 1249, 859 cm^{-1}. ^{1}H NMR (500 MHz, CDCl$_3$) δ 7.59 (s, 1 H), 7.50-7.24 (m, 10 H), 7.19 (d, J = 8.5 Hz, 1 H), 2.96-2.65 (br m, 8 H), 1.80-1.59 (m, 4 H), 1.59-1.10 (m, 40 H), 1.10-0.89 (m, 24 H), 0.28 (s, 9 H). LRMS calcd for C$_{73}$H$_{97}$ISi: 1104.6. Found: 1104.7.

3-Ethylheptyl-containing terminal alkynyl tetramer 36. **30** (0.11 g, 0.1 mmol), THF (2 mL), and TBAF (0.15 mL, 0.15 mmol, 1.0 M in THF) afforded 0.1 g (100 %) of the title compound as a yellow oil. IR (neat) 2928, 2360, 2344, 1464, 1401, 1236, 1103, 827 cm^{-1}. ^{1}H NMR (500 MHz, CDCl$_3$) δ 7.42-7.21 (m, 12 H), 3.76 (q, J = 7.2 Hz, 4 H), 3.36 (s, 1 H), 2.95-2.76 (m, 8 H), 1.80-1.60 (m, 4 H), 1.60-1.10 (m, 46 H), 1.10-0.75 (m, 24 H).

Octomer 39. Bis(dibenzylideneacetone)palladium(0) (6.9 mg, 0.012 mmol), (C$_6$H$_5$)$_3$P (15.7 mg, 0.06 mmol), CuI (2.3 mg, 0.012 mmol), **33** (0.13 g, 0.12 mmol) **36** (0.1 mg, 0.1 mmol), THF (2 mL), and diisopropylamine (0.5 mL) at 60 °C for 12 h, and flash chromatography on silica gel (3:1::hexane:CH$_2$Cl$_2$) afforded 0.17 g (70 %) of the title compound as a yellow solid. IR (neat) 2952, 2198, 2147, 1464, 827 cm^{-1}. ^{1}H NMR (300 MHz, CDCl$_3$) δ 7.48-7.27 (m, 24 H), 3.77 (q, J = 7.1 Hz, 4 H), 2.98-2.63 (m, 16 H), 1.84-1.58 (m, 8 H), 1.58-1.10 (m, 86 H), 1.10-0.78 (m, 48 H), 0.27 (9 H). λ_{max} (CH$_2$Cl$_2$) = 376. UV 10% edge value = 422 nm. M$_n$ = 2800, M$_w$ = 2960, M$_w$/M$_n$ = 1.05. MALDI MS (sinapinic acid matrix, positive ion mode) average molecular weight calculated for C$_{143}$H$_{195}$N$_3$Si: 1984. Found peak maximum (M + 1): 1791± 9 (large error due to broad signal), (loss of -SiMe$_3$ at

73 amu, -N3Et2 at 100 amu, -C≡C at 24 amu which is a common phenyl-alkynyl cleavage route).[49]

3-Ethylheptyl-containing iodide octamer 41. **39** (83.7 mg, 0.04 mmol) and CH_3I (3 mL) afforded 80 mg (100 %) of the title compound as a yellow solid. IR (neat) 2955, 2926, 2858, 2160, 1592, 1502, 1461, 826 cm⁻¹. ¹H NMR (500 MHz, $CDCl_3$) δ 7.58-7.18 (m, 24 H), 2.98-2.62 (m, 16 H), 1.82-1.55 (m, 8 H), 1.55-1.10 (m, 80), 1.10-0.70 (m, 48 H), 0.26 (s, 9 H).

3-Ethylheptyl-containing terminal alkynyl octamer 43. **39** (28.6 g, 0.01 mmol), THF (2 mL) and TBAF (0.1 mL, 0.1 mmol, 1.0 M in THF) afforded 19.1 mg (100 %) of the title compound as a yellow solid. IR (neat) 2927, 2361, 2343, 1458 cm⁻¹. ¹H NMR (500 MHz, $CDCl_3$) δ 7.47-7.23 (m, 24 H), 3.74 (q, J = 7.9 Hz, 4 H), 3.22 (s, 1 H), 2.98-2.65 (m, 16 H), 1.80-1.60 (m, 8 H), 1.60-1.05 (m, 86 H), 1.05-0.70 (m, 48 H).

3-Ethylheptyl-containing   16-mer   45. Bis(dibenzylideneacetone)palladium(0) (1.6 mg, 0.0028 mmol), $(C_6H_5)_3P$ (3.7 mg, 0.014 mmol), CuI (0.5 mg, 0.0028 mmol), **41** (37.0 g, 0.018 mmol), **43** (0.26.3 mg, 0.014 mmol), THF (2 mL), and diisopropylamine (0.25 mL) at 60 °C for 12 h, and flash chromatography on silica gel (3:1::hexane:CH_2Cl_2) afforded 30.8 mg of a 1:1 mixture of the title compound and unreacted iodide **41** (approximately 15.4 mg (26 %) of the title compound). The two compounds could not be separated by flash chromatography but they could be detected by SEC using UV detection (λ_{max} = 376). One of the two peaks indeed corresponded to the iodooctamer **41** and the second peak had an M_n value more than double that of the **41**. An analytically pure sample of the 16-mer **45** was obtained by adding diethyl ether and centrifuging the mixture. The liquid was decanted away from the remaining solid, and the procedure was repeated. After four such washing cycles the remaining solid was examined by SEC to reveal a single product, that was subsequently used for analytical analysis. IR (neat) 2956, 2927, 2360, 2340, 1506, 1458 cm⁻¹. ¹H NMR (300 MHz, $CDCl_3$) δ 7.47-7.24 (m, 48 H), 3.77 (q, J = 7.1 Hz, 4 H), 2.95-2.65 (m, 32 H), 1.82-1.58 (m, 16 H), 1.58-1.10 (m, 166 H), 1.10-0.70 (m, 96 H), 0.26 (s, 9 H). λ_{max} (CH_2Cl_2) = 376 nm. UV 10% edge value = 422 nm. M_n = 6649, M_w =7114, M_w/M_n = 1.07. MALDI-MS (sinapinic acid matrix, positive ion mode) average molecular weight calculated for **23** with $C_{279}H_{371}N_3Si$: 3795. Found (M + 1): 3486±14 (large error due to broad signal), (signifying loss in the MS of the -SiMe3 at 73 amu, -N3Et2 at 100 amu, -C≡C at 24 amu which a common phenyl-alkynyl cleavage route, and a -C8H17 fragment at 113 amu which represents a typical benzylic methylene-ethylene cleavage site.)[49]

1-Iodo-3-dodecyl-4-(trimethylsilylethynyl)benzene (16) prepared from 13. **13** (4.70 g, 10.64 mmol) and CH_3I (5.30 mL) afforded 4.70 g (94%)

of the title compound as light brown oil. R_f = 0.88 (hexane). IR (neat) 2956, 2924, 2854, 2156, 1579, 1543, 1472, 1392, 1250, 1183, 1128, 844, 816, 760, 658 cm^{-1}. ^{1}H NMR (400 MHz, CDCl$_3$) δ 7.51 (d, J = 1.7 Hz, 1 H), 7.43 (dd, J = 8.1, 1.8 Hz, 1 H), 7.11 (d, J = 8.1 Hz, 1 H), 2.66 (t, J = 7.7 Hz, 2 H), 1.58 (p, J = 7.2 Hz, 2 H), 1.30-124 (m, 18 H), 0.86 (t, J = 7.1 Hz, 3 H), 0.23 (s, 9 H). ^{13}C NMR (100 MHz, CDCl$_3$) δ 147.59, 137.65, 134.61, 133.63, 122.10, 103.04, 99.24, 94.69, 34.53, 31.95, 30.53, 29.70, 29.68 (2 C), 29.65, 29.61, 29.49, 29.39, 22.72, 14.16, -0.050.   Anal. calcd for C$_{23}$H$_{37}$ISi: C, 58.96; H, 7.96; I, 27.09.   Found: C, 58.70; H, 7.91; I, 27.27.   HRMS calcd for C$_{23}$H$_{37}$ISi: 468.1709. Found: 468.1700. λ (THF) = <u>263</u> (ε 2.50 × 10^4 M^{-1} cm^{-1}), 273, 275, 296 nm.

1-Diethyltriazyl-3-dodecyl-4-(ethynyl)benzene (19).   **13** (4.70 g, 10.64 mmol), MeOH (10 mL), CH$_2$Cl$_2$ (10 mL) and K$_2$CO$_3$ (4.41 g, 31.92 mmol) afforded 3.79 g (97%) of the title compound as yellow oil. R_f = 0.20 (hexane).   IR (neat) 3310, 2926, 2855, 2099, 1677, 1600, 1458, 1400, 1340, 1237, 1101, 829 cm^{-1}. ^{1}H NMR (500 MHz, CDCl$_3$) δ 7.45 (d, J = 8.2 Hz, 1 H), 7.30 (d, J = 2.0 Hz, 1 H), 7.25 (dd, J = 8.2, 2.0 Hz, 1 H), 3.76 (q, J = 7.1 Hz, 4 H), 3.24 (s, 1 H), 2.33 (t, J = 7.7 Hz, 2 H), 1.7i (p, J = 7.3 Hz, 2 H), 1.43-1.24 (m, 24 H), 0.92 (t, J = 7.1 Hz, 3 H). ^{13}C NMR (125 MHz, CDCl$_3$) δ 151.81, 146.76, 133.89, 121.46, 118.25, 117.86, 83.54, 80.40, 35.11, 32.40, 31.12, 30.18 (2 C), 30.13, 30.09, 30.01, 29.98, 29.84. 23.16, 14.57.   HRMS calcd for C$_{24}$H$_{39}$N$_3$: 369.3144. Found: 369.3144.

Dodecyl-containing dimer 22.   **16** (0.38 g, 0.82 mmol), **19** (0.42 g, 1.15 mmol), THF (4 mL), bis(dibenzylideneacetone)palladium(0) (0.024 g, 0.041 mmol), (C$_6$H$_5$)$_3$P (0.055 g, 0.21 mmol), CuI (0.016 g, 0.082 mmol), and N,N-diisopropylethylamine (0.57 mL, 3.28 mmol), for 3 d afforded 0.50 g (85%) of the title compound as yellow oil after gravity liquid chromatography (24:1 hexane:ether). R_f = 0.44 (24:1 hexane:ether).   IR (neat) 2922, 2853, 2202, 2151; 1596, i495, 1465, 1404, 1378, 1332, 1249, 1210, 1103, 862, 760, 633 cm^{-1}. ^{1}H NMR (400 MHz, CDCl$_3$) δ 7.42 (d, J = 8.2 Hz, 1 H), 7.37 (d, J = 7.9 Hz, 1 H), 7.30 (s, 1 H), 7.26-7.25 (m, 1 H), 7.22 (d, J = 8.2 Hz, 1 H), 7.21 (d, J = 8.3 Hz, 1 H), 3.76 (q, J = 7.2 Hz, 4 H), 2.82 (t, J = 7.7 Hz, 2 H), 2.73 (t, J = 7.7 Hz, 2 H), 1.69 (p, J = 7.2 Hz, 2 H), 1.63 (p, J = 7.2 Hz, 2 H), 1.32-1.22 (m, 42 H), 0.86 (t, J = 7.0 Hz, 6 H), 0.24 (s, 9 H). ^{13}C NMR (100 MHz, CDCl$_3$) δ 151.19, 145.97, 145.59, 132.84, 132.30, 131.49, 128.42, 124.03, 121.99, 121.15, 118.78, 117.59, 103.85, 99.24, 92.58, 90.71, 35.03, 34.73, 31.99, 31.65, 30.91, 30.64, 29.78, 29.76, 29.73, 29.64, 29.61, 29.43, 22.75, 22.71, 14.17, 0.01 HRMS calcd for C$_{47}$H$_{75}$N$_3$Si: 709.5730. Found: 709.5697.

Dodecyl-containing iodide dimer 25 from 22.   **22** (0.25 g, 0.35 mmol) and CH$_3$I (3 mL) afforded 0.24 g (95%) of the title compound as yellow oil. R_f = 0.63 (hexane).   IR (neat) 2923, 2854, 2152, 1945, 1892, 1775, 1600, 1579, 1541, 1490, 1465, 1406, 1393, 1378, 1351, 1302, 1249, 1225, 1194, 1144, 1117, 1092, 1015, 947, 843, 760, 722, 700, 687, 633 cm^{-1}. ^{1}H NMR (400

MHz, CDCl$_3$) δ 7.56 (d, J = 1.5 Hz, 1 H), 7.48 (dd, J = 8.1, 1.7 Hz, 1 H), 7.37 (d, J = 7.9 Hz, 1 H), 7.30 (d, J = 1.0 Hz, 1 H), 7.23 (dd, J = 8.0, 1.4 Hz, 1 H), 7.17 (d, J = 8.1 Hz, 1 H), 2.76-2.71 (m, 4 H), 1.64-1.62 (m, 4 H), 1.32-1.22 (m, 36 H), 0.86 (t, J = 7.0 Hz, 6 H), 0.24 (s, 9 H). ^{13}C NMR (100 MHz, CDCl$_3$) δ 147.10, 145.70, 137.76, 134.80, 133.43, 132.34, 131.58, 128.52, 123.14, 122.64, 122.01, 103.57, 99.70, 94.55, 94.08, 88.88, 34.68, 34.50, 31.95, 30.60, 30.64, 29.70, 29.68, 29.65, 29.56, 29.50, 29.39, 22.71, 14.15, 14.14, -0.031. Anal. calcd for C$_{43}$H$_{65}$ISi: C, 70.08; H, 8.89; I, 17.22. Found: C, 69.84; H, 8.77; I, 17.42. HRMS calcd for C$_{43}$H$_{65}$ISi: 736.3900. Found: 736.3904. M_n = 900, M_w = 929, M_w/M_n = 1.03. λ (THF) = <u>314</u> (ε 5.94 × 10^4 M^{-1} cm^{-1}), 335 nm. Emiss (THF) = 334, 351, <u>386</u> nm.

Dodecyl-containing terminal alkynyl dimer 28. **22** (2.84 g, 4 mmol), MeOH (4 mL), CH$_2$Cl$_2$ (4 mL), and K$_2$CO$_3$ (1.66 g, 12 mmol) afforded 2.55 g (100%) of the title compound as a dark yellow oil. IR (neat) 3302, 2925, 2853, 2201, 2102, 1596, 1496, 1465, 1401, 1332, 1237, 1210, 1103, 891, 828, 721, 644 cm^{-1}. ^{1}H NMR (300 MHz, CDCl$_3$) δ 7.48 (d, J = 8.2 Hz, 1 H), 7.45 (d, J = 8.0 Hz, 1 H), 7.38 (d, J = 1.2 Hz, 1 H), 7.33 (dd, J = 4.3, 1.8 Hz, 1 H), 7.29 (d, J = 8.2 Hz, 1 H), 7.28 (d, J = 7.9 Hz, 1 H), 3.77 (q, J = 7.2 Hz, 4 H), 3.33 (s, 1 H), 2.89 (t, J = 7.5 Hz, 2 H), 2.81 (t, J = 7.5 Hz, 2 H), 1.77 (p, J = 7.3 Hz, 2 H) 1.69 (p, J = 7.3 Hz, 2 H), 1.43-1.20 (m, 42 H), 0.91 (t, J = 6.9 Hz, 6 H). ^{13}C NMR (100 MHz, CDCl$_3$) δ 151.24, 145.98, 145.61, 132.86, 132.81, 131.48, 128.47, 127.54, 124.39, 121.18, 121.03, 118.73, 117.62, 92.43, 90.78, 82.31, 81.86, 35.06, 34.39, 32.01, 30.93, 30.54, 29.81, 29.78, 29.75, 29.69, 29.67, 29.56, 29.54, 29.47, 29.46, 22.77, 14.19. HRMS calcd for C$_{44}$H$_{67}$N$_3$: 637.5335. Found: 637.5325.

Dodecyl-containing tetramer 31. **25** (2.69 g, 3.65 mmol), **28** (2.55 g, 4.01 mmol), THF (8 mL), bis(dibenzylideneacetone)palladium(0) (0.10 g, 0.18 mmol), (C$_6$H$_5$)$_3$P (0.24 g, 0.91 mmol), CuI (0.070 g, 0.37 mmol), and N,N-diisopropylethylamine (2.54 mL, 14.60 mmol), for 2 d to afforded 3.97 g (87%) of the title compound as a yellow oil which solidified upon standing after gravity liquid chromatography (24:1 hexane:ether). R$_f$ = 0.54 (24:1 hexane:ether). IR (neat) 2921, 2851, 2898, 2151, 1592, 1502, 1467, 1406, 1378, 1352, 1247, 1235, 1212, 1095, 884, 862, 843, 828, 759, 721, 634 cm^{-1}. ^{1}H NMR (400 MHz, CDCl$_3$) δ 7.50 (d, J = 8.2 Hz, 1 H), 7.49 (d, J = 7.9 Hz, 2 H), 7.44 (d, J = 8.0 Hz, 1 H), 7.42 (s, 1 H), 7.38-7.34 (m, 4 H), 7.32 -7.29 (m, 2 H), 3.78 (q, J = 7.1 Hz, 4 H), 2.90 (m, 6 H), 2.82 (t, J = 7.9 Hz, 2 H), 1.90-1.70 (m, 8 H), 1.40-1.28 (m, 78 H), 0.94-0.89 (m, 12 H), 0.31 (s, 9 H). ^{13}C NMR (100 MHz, CDCl$_3$) δ 151.21, 145.98, 145.71, 145.19, 145.11, 132.88, 132.38, 132.16, 132.12, 131.73, 131.64, 131.61, 128.76, 128.63, 128.61, 124.02, 123.42, 123.38, 122.62, 122.53, 122.03, 121.20, 118.84, 117.63, 103.69, 99.65, 94.58, 94.40. 92.70, 90.86, 89.90, 89.65, 35.08, 34.76, 32.03, 30.94, 30.74, 30.65, 29.83, 29.82, 29.78, 29.69, 29.67, 29.65, 29.63, 29.49, 29.46, 22.78,

22.73, 14.20, 0.01. LRMS calcd for $C_{87}H_{131}N_3Si$: 1246 (96%), 1247 (100%), 1248 (54%). Found: 1246 (100%), 1247 (91%), 1248 (55%).

Dodecyl-containing iodide tetramer 34 from 31. **31** (1.99 g, 1.60 mmol) and CH_3I (2 mL) afforded 1.88 g (92%) of the title compound as yellow oil that solidified upon standing. IR (KBr) 2924, 2853, 2203, 2151, 1697, 1657, 1598, 1543, 1500, 1463, 1384, 1252, 1183, 1005, 656, 761, 723, 634 cm^{-1}. ^{1}H NMR (400 MHz, CDCl$_3$) δ 7.57 (d, J =1.7 Hz, 1 H), 7.49 (dd, J = 8.0, 1.7 Hz, 1 H), 7.45 (d, J = 7.9 Hz, 2 H), 7.44 (d, J = 7.8 Hz, 1 H), 7.39 (d, J = 1.0 Hz, 1 H), 7.35 (d, J = 1.3 Hz, 1 H), 7.31-7.26 (m, 4 H), 7.18 (d, J = 8.2 Hz, 1 H), 2.85-2.75 (m, 8 H), 1.67 (m, 8 H), 1.40-1.20 (m, 72 H), 0.86-0.84 (m, 12 H), 0.25 (s, 9 H). ^{13}C NMR (100 MHz, CDCl$_3$) δ 147.11, 145.70, 145.21, 137.78, 134.82, 133.44, 132.87, 132.36, 132.15, 131.70, 131.62, 128.72, 128.57, 123.28, 123.21, 123.11, 122.68, 122.62, 122.12, 103.63, 99.70, 94.68, 94.61, 94.57, 94.17, 89.60, 89.55, 89.00, 34.69, 34.53, 34.36, 31.97, 30.67, 30.62, 30.49, 29.75, 29.71, 29.66, 29.59, 29.53, 29.42, 28.00, 27.15, 22.73, 19.70, 14.15, 11.42, -0.018. Anal. calcd for $C_{83}H_{121}ISi$: C, 78.26; H, 9.57. Found: C, 77.30; H, 9.24. LRMS calcd for $C_{83}H_{121}ISi$: 1273 (99%), 1274 (100%), 1275 (53%). Found: 1273 (100%), 1274 (98%), 1275 (50%). M_n = 1706, M_w = 1777, M_w/M_n = 1.04. λ (THF) = 241, 273, <u>350</u> (ε 1.10 × 10^5 M^{-1} cm^{-1}) nm. Emiss (THF) = <u>391</u>, 412 nm.

Dodecyl-containing terminal alkynyl tetramer 37. **31** (0.14 g, 0.11 mmol), MeOH (2 mL), CH$_2$Cl$_2$ (4 mL) and K$_2$CO$_3$ (0.14 g, 1.0 mmol) afforded 0.13 g (100%) of the title compound as yellow oil which solidified upon standing. IR (neat) 3314, 3029, 2921, 2851, 2198, 2100, 1594, 1503, 1465, 1403, 1350, 1234, 1095, 885, 828, 722, 648 cm^{-1}. ^{1}H NMR (400 MHz, CDCl$_3$) δ 7.49-7.44 (m, 4 H), 7.40-7.31 (m, 7 H), 7.29-7.25 (m, 1 H), 3.77 (q, J = 7.1 Hz, 4 H), 3.33 (s, 1 H), 2.90 (m, 6 H), 2.80 (t, J = 7.1 Hz, 2 H), 1.75-1.67 (m, 8 H), 1.43 -1.26 (m, 78 H), 0.90-0.87 (m, 12 H). ^{13}C NMR (100 MHz, CDCl$_3$) δ 151.20, 146.00, 145.74, 145.21, 145.11, 132.86, 132.15, 132.10, 131.73, 131.62, 128.74, 128.61, 123.99, 123.70, 123.44, 122.43, 121.98, 121.60, 121.15, 118.79, 117.60, 94.34, 92.66, 90.83, 89.89, 89.66, 82.15, 35.04, 34.72, 34.37, 32.00, 30.91, 30.71, 30.51, 29.80, 29.78, 29.76, 29.74, 29.69, 29.66, 29.64, 29.52, 29.45, 29.43, 22.75, 14.17. LRMS calcd for $C_{84}H_{123}N_3$: 1174 (100%), 1175 (98%), 1176 (47%). Found: 1174 (100%), 1175 (50%), 1176 (33%).

Dodecyl-containing octamer 40. **34** (1.88 g, 1.47 mmol), **37** (1.73 g, 1.47 mmol), THF (8 mL), bis(dibenzylideneacetone)palladium(0) (0.042 g, 0.074 mmol), (C$_6$H$_5$)$_3$P (0.097 g, 0.37 mmol), CuI (0.029 g, 0.15 mmol), and N,N-diisopropylethylamine (1.02 mL, 5.88 mmol), for 3 d to afforded 1.62 g (48%) of the title compound as a yellow solid after flash liquid chromatography (8:1 hexane:CH$_2$Cl$_2$). R$_f$ = 0.12 (8:1 hexane:CH$_2$Cl$_2$). IR (KBr) 2922, 2851, 2199, 2149, 1594, 1503, 1460, 1402, 1348, 1235, 1096, 835, 722, 643 cm^{-1}. ^{1}H NMR (500 MHz, CDCl$_3$) δ 7.49-7.42 (m, 8 H), 7.40-7.38 (m, 4 H), 7.36-7.25

(m, 12 H), 3.77 (q, J = 7.1 Hz, 4 H), 2.87-2.75 (m, 16 H), 1.73-1.64 (m, 16 H), 1.54-1.25 (m, 150 H), 0.89-0.85 (m, 24 H), 0.27 (s, 9 H). ^{13}C NMR (125 MHz, CDCl$_3$) δ 151.60, 146.39, 146.14, 146.11, 145.61, 145.51, 145.31, 133.25, 132.76, 132.54, 132.49, 132.12, 132.02, 132.00, 131.91, 131.77, 129.15, 129.00, 128.96, 124.37, 124.09, 123.83, 123.79, 123.68, 123.64, 123.04, 123.02, 122.92, 122.83, 122.69, 122.38, 121.98, 121.53, 119.18, 117.98, 104.02, 100.10, 95.07, 95.00, 94.71, 93.03, 91.20, 90.27, 90.04, 89.94, 82.55, 82.48, 35.41, 35.10, 34.75, 32.36, 31.26, 31.08, 30.99, 30.88, 30.15, 30.14, 30.13, 30.11, 30.08, 30.06, 30.02, 30.00, 29.96, 29.89, 29.81, 29.78, 23.12, 14.53, 0.37.

Dodecyl-containing iodide octamer 42 from 40. **40** (0.81 g, 0.35 mmol) and CH$_3$I (5 mL) afforded 0.82 g (100%) of the title compound as a yellow solid. R$_f$ = 0.29 (9:1 hexane:CH$_2$Cl$_2$). IR (KBr) 3028, 2924, 2853, 2201, 2151, 1923, 1849, 1775, 1744, 1708, 1656, 1588, 1502, 1461, 1382, 1307, 1255, 1084, 886, 839, 723 cm^{-1}. ^{1}H NMR (500 MHz, CDCl$_3$) δ 7.59 (d, J = 1.0 Hz, 1 H), 7.51 (d, J = 1.3 Hz, 1 H), 7.49-7.46 (m, 5 H), 7.42-7.25 (m, 16 H), 7.19 (d, J = 8.1 Hz, 1 H), 2.90-2.70 (m, 16 H), 1.74-1.66 (m, 16 H), 1.45-1.25 (m, 144 H), 0.89-0.86 (m, 24 H), 0.28 (s, 9 H). ^{13}C NMR (125 MHz, CDCl$_3$) δ 147.52, 147.47, 146.11, 145.62, 145.55, 145.31, 143.57, 138.18, 138.14, 135.23, 135.18, 133.85, 133.60, 132.77, 132.56, 132.14, 132.03, 131.91, 131.86, 129.16, 129.12, 128.98, 128.80, 124.09, 123.81, 123.70, 123.64, 123.52, 123.50, 123.38, 123.20, 123.09, 123.06, 122.65, 122.54, 104.04, 100.10, 95.20, 95.10, 95.02, 94.97, 94.78, 94.67, 94.58, 91.43, 90.06, 89.96, 89.88, 88.74, 88.53, 87.90, 39.82, 39.50, 39.44, 39.28, 37.53, 37.06, 36.91, 35.50, 35.11, 34.94, 34.81, 34.77, 34.71, 33.21, 32.83, 32.70, 32.51, 32.38, 32.34, 31.77, 31.41, 31.34, 31.08, 31.02, 31.00, 30.82, 30.60, 30.42, 30.17, 30.13, 30.08, 30.01, 29.93, 29.91, 29.83, 29.80, 29.67, 29.41, 28. 91, 28.61, 28.40, 28.31, 27,85, 27.74, 27.55, 27.43, 27.14, 23.49, 23.26, 23.13, 23.09, 22.99, 20.57, 20.11, 19.65, 14.82, 14.68, 14.55, 14.40, 11.83, 4.96, 0.38. Anal. calcd for C$_{163}$H$_{233}$ISi: C, 83.39; H, 10.00. Found: C, 81.43; H, 9.32. M$_n$ = 3625, M$_w$ =4350, M$_w$/M$_n$ = 1.20. λ (THF) = 242, $\underline{370}$ (ε 1.72 × 10^5 M^{-1} cm^{-1}) nm. Emiss (THF) = $\underline{411}$, 433 nm.

Dodecyl-containing terminal alkynyl octamer 44. **40** (0.18 g, 0.11 mmol), MeOH (5 mL), CH$_2$Cl$_2$ (15 mL) and K$_2$CO$_3$ (0.15 g, 1.05 mmol) at 65 °C afforded 0.13 g (100%) of the title compound as a yellow oil which solidified upon standing. IR (KBr) 3310, 2924, 2854, 2198, 1593, 1502, 1460, 1403, 1343, 1234, 1096, 889, 837, 723, 612 cm^{-1}. ^{1}H NMR (400 MHz, CDCl$_3$) δ 7.50-7.43 (m, 9 H), 7.38-7.26 (m, 15 H), 3.76 (q, J = 7.0 Hz, 4 H), 3.32 (s, 1 H), 2.83 (m, 16 H), 1.71 (m, 16 H), 1.34 -1.23 (m, 152 H), 0.85-0.83 (m, 24 H).

Dodecyl-containing 16-mer 46. **42** (0.82 g, 0.35 mmol), **44** (0.68 g, 0.30 mmol), THF (10 mL), bis(dibenzylideneacetone)palladium(0) (0.020 g, 0.035 mmol), (C$_6$H$_5$)$_3$P (0.046 g, 0.18 mmol), CuI (0.0070 g, 0.035 mmol), N,N-diisopropylethylamine (0.49 mL, 2.80 mmol), for 4 d at 65 °C afforded

1.20 g (90%) of the title compound as.yellow solid after flash liquid chromatography (3:1 hexane:CH_2Cl_2). R_f = 0.64 (3:1 hexane:CH_2Cl_2). IR (KBr) 2924, 2854, 2199, 2150, 1593, 1502, 1460, 1402, 1345, 1235, 1096, 888, 838, 723, 658, 612 cm^{-1}. ^{1}H NMR (400 MHz, CDCl$_3$) δ 7.50-7.46 (m, 16 H), 7.38 (m, 16 H), 7.31-7.27 (m, 16 H), 3.77 (q, J = 7.1 Hz, 4 H), 2.84 (m, 32 H), 1.71 (m, 32 H), 1.23 (m, 288 H), 0.85-0.83 (m, 48 H), 0.25 (s, 9 H). ^{13}C NMR (125 MHz, CD$_2$Cl$_2$) δ 151.77, 146.44, 146.23, 145.75, 145.44, 136.24, 135.28, 134.84, 134.72, 133.80, 133.15, 132.54, 132.15, 131.23, 130.74, 130.45, 130.14, 129.87, 129.17, 128.97, 128.74, 128.63, 127.61, 127.48, 127.14, 126.58, 126.09, 125.85, 124.96, 124.40, 124.24, 123.72, 123.11, 122.68, 122.48, 121.50, 119.05, 118.02, 95.14, 93.07, 90.05, 41.17, 39.88, 39.71, 39.48, 39.31, 38.51, 38.12, 37.84, 37.55, 37.24, 37.09, 36.96, 36.40, 36.16, 35.41, 35.11, 34.90, 34.75, 34.14, 33.85, 33.51, 33.25, 32.86, 32.43, 31.82, 31.38, 31.15, 30.87, 30.19, 30.07, 29.88, 29.41, 29.02, 28. 67, 28.45, 28.35, 27.89, 26.78, 27.60, 27.23, 26.81, 26.35, 25.19, 25.70, 24.24, 24.13, 23.17, 22.91, 20.60, 20.31, 19.95, 19.48, 18.47, 14.66, 14.39, 11.67, 11.23, 0.10.

   N-Ethyl,N-hydroxyethyl-4-triazenyl-2-dodecylacetophenone. To a 3000 mL round-bottom flask equipped with a mechanical stirrer, N$_2$ inlet, and 0 °C condenser, containing boron trifluoride etherate (102.35 mL, 832.20 mmol) at 0 °C, was added **11** (129.31 g, 426.77 mmol) in CH$_2$Cl$_2$ (1.5 L) followed by t-butyl nitrite (83.75 mL, 704.17 mmol). The reaction was stirred for 30 min at 0 °C before K$_2$CO$_3$ (277.22 g, 2.06 mol) and 2-(ethylamino)ethanol (262.22 mL, 2.69 mol) were added sequentially. The reaction was then warmed to room temperature over 2 d before water (450 mL) was added in 3 portions. The K$_2$CO$_3$ was then filtered and the solvent removed in vacuo before the organic layer was extracted with CH$_2$Cl$_2$ and dried over MgSO$_4$. The sample was concentrated in vacuo yielding a crude oil that was purified by flash liquid chromatography (4:1 hexane:EtOAc) to afford 146.43 g (85%) of the title compound as brown oil. R_f = 0.22 (3:1 hexane:EtOAc). IR (neat) 3439, 2927, 2853, 1678, 1600, 1557, 1454, 1430, 1395, 1349, 1243, 1130, 1074, 951, 896, 825, 721, 624 cm^{-1}. ^{1}H NMR (400 MHz, CDCl$_3$) δ 7.65 (d, J = 8.9 Hz, 1 H), 7.26-7.25 (m, 2 H), 3.88-3.86 (m, 5 H), 2.87 (t, J = 8.0 Hz, 2 H), 2.55 (s, 3 H), 1.55 (m, 2 H), 1.40-1.20 (m, 23 H), 0.86 (t, J = 7.0 Hz, 3 H). ^{13}C NMR (100 MHz, CDCl$_3$) δ 200.66, 153.10, 145.14, 133.74, 131.10, 123.35, 117.00, 60.80, 51.13, 34.59, 31.87, 31.73, 29.85, 29.64, 29.51, 29.39, 29.31, 22.63, 14.07. HRMS calcd for C$_{24}$H$_{41}$N$_3$O$_2$: 403.3199. Found: 403.3206.

   N-Ethyl,N-(trimethylsiloxyethyl)-4-triazyl-2-dodecylacetophenone (47). To a solution of N-ethyl,N-hydroxyethyl-4-triazenyl-2-dodecynylacetophenone (1.99 g, 4.94 mmol) in THF (20 mL) was added dropwise Et$_3$N (4.13 mL, 29.64 mmol) at room temperature. The mixture was allowed to stir for 10 min before chlorotrimethylsilane (1.88 mL, 14.82 mmol) was added slowly. The resulting reaction mixture was allowed to stir for 1 d before being poured into water. The organic layer was extracted with ether and dried over MgSO$_4$. The sample was concentrated in vacuo to afford 2.23 g (95%) of the title compound as brown oil that was taken on to the next step without purification. R$_f$ = 0.61 (9:1 hexane:ether). IR (neat) 2927, 2857, 1679, 1598, 1559, 1454, 1532, 1388, 1348, 1246, 1200, 1101, 944, 843, 752, 720, 620 cm^{-1}. ^{1}H NMR (400 MHz, CDCl$_3$) δ 7.66 (d, J = 9.0 Hz, 1 H), 7.25-7.23 (m, 2 H), 3.84-3.82 (m, 5 H), 2.87 (t, J = 7.9 Hz, 2 H), 2.55 (s, 3 H), 1.56 (p, J = 7.0 Hz, 2 H), 1.26-1.23 (m, 22 H), 0.86 (t, J = 7.0 Hz, 3 H), 0.095 (s, 9 H). ^{13}C NMR (100 MHz, CDCl$_3$) δ 200.32, 153.14, 145.10, 133.72, 131.04, 123.45, 117.09, 60.21, 58.33, 45.80, 34.63, 31.91, 31.75, 29.87, 29.68, 29.66, 29.64, 29.55, 29.49, 29.35, 22.67, 14.09, -0.60. HRMS calcd for C$_{27}$H$_{49}$N$_3$O$_2$Si: 475.3594. Found: 475.3589.

N-Ethyl,N-(trimethylsiloxyethyl)-1-triazyl-3-dodecyl-4-
(trimethylsilylethynyl)benzene (48). To a stirred solution of diisopropylamine (0.88 mL, 6.70 mmol) in THF (5 mL) was added at -78 °C under N$_2$ n-BuLi (3.90 mL, 6.08 mmol, 1.56 M in hexanes). The reaction mixture was allowed to warm to 0 °C for 10 min then cooled to -78 °C. To the LDA solution was added triazene **47** (2.31 g, 4.86 mmol) in THF (5 mL). The reaction mixture was stirred at -78 °C for 1 h before adding, at the same temperature, diethylchlorophosphate (0.88 mL, 6.08 mmol). The reaction mixture was allowed to warm to room temperature over 30 min before being recooled to -78 °C. The reaction solution was then transferred via cannula into 2.25 equivalents of LDA [made as above with diisopropylamine (1.58 mL, 12.04 mmol) and n-BuLi (7.01 mL, 10.94 mmol, 1.56 M in hexanes)] in THF (15 mL) at -78 °C. The ensuing reaction mixture was allowed to warm to 0 °C over 1 h and then recooled to -78 °C before chlorotrimethylsilane (1.54 mL, 12.15 mmol) was added. The solution was allowed to warm to room temperature overnight before being poured into water and extracted with ether (3×). The combined ether extracts were washed with brine and dried over MgSO$_4$. Filtration and removal of the solvent in vacuo afforded 2.83 g of the title compound as a brown oil that was used directly for the next step. IR (neat) 2926, 2855, 2150, 1602, 1454, 1401, 1378, 1348, 1289, 1251, 1198, 1159, 1094, 978, 940, 842, 758, 721, 697, 654 cm^{-1}. ^{1}H NMR (300 MHz, CDCl$_3$) δ 7.36 (d, J = 8.2 Hz, 1 H), 7.19 (s, 1 H), 7.16 (d, J = 8.2 Hz, 1 H), 3.80-3.78 (m, 4 H), 2.74 (t, J = 8.0 Hz, 2 H), 1.63 (m, 2 H), 1.24-1.14 (m, 23 H), 0.86 (t, J = 6.9 Hz, 3 H), 0.23 (s, 9 H), 0.090 (s, 9 H). HRMS calcd for C$_{30}$H$_{55}$N$_3$OSi$_2$: 529.3884. Found: 529.3892.

N-Ethyl,N-(hydroxyethylethyl)-1-triazyl-3-dodecyl-4-
(ethynyl)benzene (49). **48** (2.83 g, 4.86 mmol), MeOH (10 mL), CH$_2$Cl$_2$ (10 mL), K$_2$CO$_3$ (4.15 g, 30 mmol), and gravity liquid chromatography (3:1 hexane:EtOAc) afforded 0.94 g (50% over 2 steps) of the title compound as a brown oil. R$_f$ = 0.34 (3:1 hexane:EtOAc). IR (neat) 3415, 3313, 2925, 2855, 2100, 1740, 1600, 1560, 1452, 1434, 1400, 1378, 1345, 1239, 1181, 1152, 1101, 1078, 1048, 892, 829, 791, 721, 643 cm^{-1}. ^{1}H NMR (400 MHz, CDCl$_3$) δ

7.40 (d, J = 8.2 Hz, 1 H), 7.21 (d, J = 1.9 Hz, 1 H), 7.18 (dd, J = 8.2, 1.2 Hz, 1 H), 3.87-3.81 (m, 5 H), 3.22 (s, 1 H), 2.76 (t, J = 8.0 Hz, 2 H), 1.63 (p, J = 7.0 Hz, 2 H), 1.32-1.23 (m, 23 H), 0.86 (t, J = 7.0 Hz, 3 H). ^{13}C NMR (100 MHz, CDCl$_3$) δ 150.22, 146.53, 133.55, 120.98, 118.56, 117.43, 82.82, 80.44, 61.28, 51.59, 34.64, 31.96, 30.68, 29.73 (2 C), 29.69, 29.65, 29.56, 29.52, 29.40, 22.72, 14.15. HRMS calcd for C$_{24}$H$_{39}$N$_3$O: 385.3093. Found: 385.3087.

1-Iodo-3-dodecyl-4-(trimethylsilylethynyl)benzene (16) prepared from 48. 48 (6.62 g, 12.5 mmol) and CH$_3$I (10 mL) afforded 2.83 g (60% over 2 steps) of the title compound as brown oil that was analytically identical to the titled material prepared from **13**.

Polymer supported monomer 50. A 3-neck 2 L round-bottomed flask equipped with a mechanical stirrer and a water condenser was charged with chloromethyl polystyrene (50.00 g, 50.00 mmol, 1.00 ± 0.05 meq g^{-1}, 1% cross-linked with divinylbenzene, 200-400 mesh). The vessel was flushed thoroughly with N$_2$ before a solution of anchor **49** (57.75 g, 150 mmol) in dry THF (250 mL) was added via a cannula followed by additional dry THF (350 mL) to facilitate free stirring. The reaction mixture was cooled to -78 °C, and NaH (5.62 g, 140.50 mmol, 60% dispersion in oil) was added carefully in portions under a N$_2$ blanket. After the addition was complete, the reaction mixture was warmed to room temperature over 30 min and then heated at 75 °C for 48 h with gentle stirring. The mixture was then cooled to 0 °C, quenched with MeOH (60 mL) and water (60 mL), and filtered. The polymer was then transferred to a pre-weighed fritted filter using CH$_2$Cl$_2$, and washed sequentially (ca. 3 × 30 mL/g polymer) with the following: THF, H$_2$O, MeOH, EtOAc, n-hexane, CH$_2$Cl$_2$, and MeOH, and dried to constant mass in a vacuum oven at 40 °C to afford 67.95 g (ΔW$_{theor}$ = 17.41 g, ΔW = 17.95 g) of the title polymer as yellow beads. IR (CCl$_4$) 3305, 3025, 2937, 2869, 2504, 2403, 2289, 2211, 2098, 1944, 1873, 1803, 1745, 1593, 1558, 1446, 1366, 1237, 1197, 1081, 1023, 893, 775 cm^{-1}. Gel-phase ^{13}C NMR showed some remaining benzylic chloride peak at δ 46.77 ppm. Anal. calcd for the functionalized resin: C, 87.61; H, 8.21; Cl, 0.64; N, 2.55; O, 0.97. Found: C, 87.12; H, 8.30; Cl, 0.61; N, 2.56. This signifies that 77% of the chloride was substituted. All subsequent calculated elemental composition data reflect this initial substitution.

General procedure for the coupling of a polymer supported terminal alkyne with an aryl halide. A thick-walled oven dried screw cap tube was charged with the polymer supported terminal alkyne (1.0 eq) and the monomeric or oligomeric aryl iodide (2.0 eq). To the reaction flask was added, via cannula, the supernatant of a pre-made solution (6 mL/g of polymer) of bis(dibenzylideneacetone)palladium(0) (4 mol %), (C$_6$H$_5$)$_3$P (20 mol %), and CuI (4 mol %) in dry Et$_3$N which had been degassed and stirred at 70 °C for 2 h. The tube was flushed with N$_2$, capped and kept at 65 °C for 12-48 h without stirring, and the vessel was shaken periodically. The polymer was then poured onto a pre-weighed fritted filter using CH_2Cl_2 to wash the beads that were sticking to the sides of the flask. The beads were then washed sequentially (ca. 30 mL/g polymer) with the following: CH_2Cl_2, DMF, 0.05 M solution of sodium diethyldithiocarbamate in 99:1 DMF:diisopropylethylamine, DMF, CH_2Cl_2, MeOH, and dried to constant mass in a vacuum oven at 60 °C. The degree of substitution of each reaction was estimated by infrared analysis of the polymer bound substrate. Absorptions at 3311 cm^{-1} (strong) and 2109 cm^{-1} (weak) were characteristic of the terminal alkynyl carbon-hydrogen and carbon-carbon stretches, respectively, and an absorption at 2156 cm^{-1} (strong) was characteristic of the carbon-carbon stretch of the trimethylsilyl-protected terminal alkyne. The excess aryl halide was recovered from the washings and repurified before further use. The amount of aryl iodide recovered after purification was generally ca. 30% less than expected.

Polymer Supported Silyl Dimer 51. **16** (33.72 g, 72.05 mmol), **50** (59.07 g, 36.03 mmol, 0.61 mmol g^{-1}), bis(dibenzylideneacetone)palladium(0) (1.65 g, 2.88 mmol), $(C_6H_5)_3P$ (3.78 g, 14.41 mmol), CuI (0.55 g, 2.88 mmol) and Et$_3$N (500 mL) for 2 d afforded 69.05 g (ΔW_{theor} = 12.25 g, ΔW = 9.98 g) of the title polymer as yellow beads. IR (CCl$_4$) 3062, 3027, 2952, 2914, 2853, 2291, 2201, 2150, 2009, 1943, 1873, 1803, 1746, 1663, 1595, 1549, 1491, 1447, 1345, 1244, 1196, 1156, 1093, 1008, 981, 790, 707, 627 cm^{-1}. Gel-phase ^{13}C NMR showed the gain of a trimethylsilyl peak at δ 0.47 ppm. Anal. calcd for the functionalized resin: C, 86.51; H, 8.63; Cl, 0.53; N, 2.11; O, 0.80; Si, 1.41. Found: C, 86.12; H, 8.66; N, 2.31.

General Procedure for the liberation of polymer supported oligomers. A thick-walled oven dried screw cap tube was charged with a suspension of the polymer supported oligomer and CH$_3$I (7 mL/g of polymer). The tube was flushed with N$_2$, capped and heated to 120 °C for 12-24 h without stirring. The reaction mixture was cooled and passed through a fritted filter before the resin was introduced to hot CH_2Cl_2 (3×) to extract any residual product trapped in the polymer matrix. The combined filtrate was then passed through a plug of silica gel with CH_2Cl_2.

Dodecyl-containing iodide dimer 25 from 51. **51** (45.72 g, 23.31 mmol, 0.51 mmol g^{-1}) and CH$_3$I (320 mL) afforded 14.66 g (86%) of the title compound as light brown oil that was analytically identical to the titled material prepared from **22**.

General procedure for the desilylation of polymer supported silylated alkynes. To a suspension of polymer supported aryl trimethylsilylalkyne (1.0 eq) and THF (9 mL/g of polymer) in an Erlenmeyer flask was added a solution of TBAF (2.0 eq, 1.0 M in THF). The suspension was swirled periodically for 15 min. The polymer was then transferred to a pre-weighed fritted filter using THF, and washed sequentially (ca. 30 mL/g polymer) with THF followed by MeOH, and dried to constant mass in a vacuum oven at 60 °C. The efficiency of each reaction was estimated by infrared analysis of the polymer bound substrate. Absorptions at 3311 cm^{-1} (strong) and 2109 cm^{-1} (weak) are characteristic of the terminal alkynyl carbon-hydrogen and carbon-carbon stretches, respectively. An absorption at 2156 cm^{-1} (strong) is characteristic of the carbon-carbon stretch of the trimethylsilyl-protected alkyne.[50]

Polymer supported terminal alkynyl dimer 52. **51** (22.87 g, 11.66 mmol, 0.51 mmol g^{-1}), THF (206 mL), and TBAF (6.96 mL, 24.00 mmol, 1.00 M in THF) afforded 22.09 g (ΔW_{theor} = -0.84 g, ΔW = -0.78 g) of the title polymer as goldish yellow beads. IR (CCl4) 3307, 3083, 3061, 3027, 2935, 2851, 2201, 2102, 1941, 1869, 1802, 1746, 1663, 1600, 1548, 1493, 1452, 1401, 1377, 1345, 1238, 1194, 1154, 1094, 1027, 1005, 979, 892, 804, 699, 648, 609 cm^{-1}. Gel-phase ^{13}C NMR showed the loss of a trimethylsilyl peak at δ 0.47 ppm. Anal. calcd for the functionalized resin: C, 87.89; H, 8.53; Cl, 0.55; N, 2.19; O, 0.83. Found: C, 87.94; H, 8.49; N, 2.31.

Polymer supported silyl tetramer 53. **25** (13.73 g, 18.65 mmol), **52** (17.28 g, 8.99 mmol, 0.52 mmol g^{-1}), bis(dibenzylideneacetone)palladium(0) (0.43 g, 0.75 mmol), (C$_6$H$_5$)$_3$P (0.98 g, 3.75 mmol), CuI (0.14 g, 0.75 mmol) and Et$_3$N (200 mL) for 1 d afforded 21.71 g (ΔW_{theor} = 5.47 g, ΔW = 4.43 g) of the title polymer as mustard-colored beads. IR (CCl$_4$) 3061, 3026, 2919, 2853, 2198, 2151, 1940, 1867, 1800, 1746, 1660, 1600, 1548, 1493, 1453, 1402, 1378, 1344, 1250, 1194, 1154, 1093, 1028, 1005, 979, 891, 792, 700 cm^{-1}. Gel-phase ^{13}C NMR showed the gain of a trimethylsilyl peak at δ 0.40 ppm. Anal. calcd for the functionalized resin: C, 87.14; H, 9.03; Cl, 0.42; N, 1.66; O, 0.63; Si, 1.11. Found: C, 86.89; H, 8.97; N, 1.82.

Dodecyl-containing iodide tetramer 34 from 53. **53** (14.47 g, 5.79 mmol, 0.40 mmol g^{-1}) and CH$_3$I (100 mL) afforded 6.02 g (77%) of the title compound as light brown oil that solidified upon standing and was analytically identical to the titled material prepared from **31**.

Polymer supported terminal alkynyl tetramer 54. **53** (6.88 g, 2.75 mmol, 0.40 mmol g^{-1}), THF (62 mL), and TBAF (1.60 mL, 5.50 mmol, 1.00 M in THF) afforded 6.80 g (ΔW_{theor} = -0.20 g, ΔW = -0.08 g) of the title polymer as orange beads. IR (CCl$_4$) 3308, 3061, 3027, 2920, 2853, 2198, 2102, 1941, 1870, 1802, 1745, 1662, 1596, 1550, 1495, 1450, 1401, 1375, 1345, 1239, 1195, 1155, 1092, 1007, 980, 892, 795, 702 cm^{-1}. Gel-phase ^{13}C NMR showed the loss of a trimethylsilyl peak at δ 0.40 ppm. Anal. calcd for the functionalized resin: C, 88.24; H, 8.97; Cl, 0.43; N, 1.71; O, 0.65. Found: C, 86.13; H, 8.44; N, 1.88.

Polymer supported silyl octamer 55. **34** (5.70 g, 4.48 mmol), **54** (5.46 g, 2.24 mmol, 0.41 mmol g^{-1}), bis(dibenzylideneacetone)palladium(0) (0.10 g, 0.18 mmol), (C$_6$H$_5$)$_3$P (0.23 g, 0.90 mmol), CuI (0.034 g, 0.18 mmol), and Et$_3$N (60 mL) afforded 7.17 g (ΔW_{theor} = 2.56 g, ΔW = 1.71 g) of the title polymer as rusty orange beads. IR (CCl$_4$) 3061, 3027, 2917, 2198, 2151, 1941, 1871, 1802, 1745, 1660, 1596, 1548, 1498, 1451, 1402, 1345, 1247, 1195,

1154, 1094, 1006, 980, 892, 794, 703 cm^{-1}. Gel-phase ^{13}C NMR showed the gain of a trimethylsilyl peak at δ 0.36 ppm. Anal. calcd for the functionalized resin: C, 87.84; H, 9.47; Cl, 0.29; N, 1.17; O, 0.44; Si, 0.78. Found: C, 87.56; H, 9.00; N, 1.44.

Dodecyl-containing iodide octamer 42 from 55. 55 (4.78 g, 1.34 mmol, 0.28 mmol g^{-1}) and CH$_3$I (35 mL) afforded 2.51 g (80%) of the title compound as a waxy yellow solid that was analytically identical to the titled material prepared from **40**.

Polymer supported terminal alkynyl octamer 56. 55 (1.90 g, , 0.53 mmol, 0.28 mmol g^{-1}), THF (20 mL), and TBAF (0.31 mL, 1.06 mmol, 1.00 M in THF) afforded 1.93 g (ΔW$_{theor}$ = -0.04 g, ΔW = 0.03 g) of the title polymer as rusty orange beads. IR (CCl$_4$) 3308, 3061, 3027, 2923, 2855, 2198, 2103, 1943, 1870, 1799, 1772, 1740, 1696, 1652, 1594, 1547, 1498, 1455, 1399, 1343, 1242, 1197, 1154, 1090, 1007, 891, 793, 706 cm^{-1}. Gel-phase ^{13}C NMR showed the loss of a trimethylsilyl peak at δ 0.36 ppm. Anal. calcd for the functionalized resin: C, 88.62; H, 9.44; Cl, 0.30; N, 1.22; O, 0.45. Found: C, 87.93; H, 8.92; N, 1.49.

Polymer supported dodecyl-containing 16-mer 57. 42 (2.18 g, 0.93 mmol), **56** (1.53 g, 0.44 mmol, 0.29 mmol g^{-1}), bis(dibenzylideneacetone)palladium(0) (0.021 g, 0.037 mmol), (C$_6$H$_5$)$_3$P (0.050 g, 0.19 mmol), CuI (0.0070 g, 0.037 mmol) and Et$_3$N (40 mL) afforded 1.76 g (ΔW$_{theor}$ = 0.98 g, ΔW = 0.23 g) of the title polymer as rusty orange beads. IR (CCl$_4$) 3061, 3027, 2923, 2855, 2198, 2151, 1942, 1869, 1798, 1772, 1740, 1696, 1675, 1652, 1594, 1545, 1499, 1455, 1401, 1343, 1244, 1195, 1152, 1088, 1007, 889, 793, 706 cm^{-1}. Gel-phase ^{13}C NMR showed the gain of a trimethylsilyl peak at δ 0.36 ppm. Anal. calcd for the functionalized resin: C, 88.46; H, 9.86; Cl, 0.18; N, 0.73; O, 0.28; Si, 0.49. Found: C, 86.95; H, 9.08; N, 1.30.

Dodecyl-containing iodide 16-mer 58. 57 (1.37 g, 0.23 mmol, 0.17 mmol g^{-1}) and CH$_3$I (25 mL) afforded 0.84 g (79%) of the title compound as a waxy yellow solid. R$_f$ = 0.27 (9:1 hexane:CH$_2$Cl$_2$). IR (KBr) 2924, 2855, 2146, 1590, 1502, 1461, 1384, 1262, 1088, 888, 837, 724, 618 cm^{-1}. ^{1}H NMR (500 MHz, CDCl$_3$) δ 7.58 (br s, 2 H), 7.50-7.47 (m, 12 H), 7.40-7.30 (m, 32 H), 7.19 (d, J = 8.0 Hz, 2 H), 2.90-2.70 (m, 32 H), 1.80-1.60 (m, 32 H), 1.40-1.20 (m, 288 H), 0.88-0.86 (m, 48 H), 0.27 (s, 9 H). ^{13}C NMR (125 MHz, CDCl$_3$) δ 152.57, 147.51, 145.61, 145.41, 138.19, 135.60, 135.23, 133.84, 136.76, 132.56, 132.14, 131.90, 129.16, 128.78, 125.95, 125.46, 123.65, 123.54, 123.06, 122.54, 107.71, 95.19, 95.09, 95.08, 95.05, 95.02, 94.97, 94.60, 94.59, 94.58, 94.57, 94.55, 90.15, 90.05, 90.02, 89.95, 89.93, 89.39, 89.37, 46.73, 46.41, 44.43, 42.73, 41.80, 41.09, 40.62, 40.52, 39.92, 39.86, 39.81, 39.50, 39.45, 39.29, 38.85, 38.17, 37.93, 37.88, 37.84, 37.82, 37.80, 37.73, 37.53, 37.42, 37.38, 37.09, 36.92, 36.71, 36.57, 36.07, 36.02, 35.77, 35.51, 35.12, 34.94, 34.83, 34.14, 33.90, 33.84, 33.65, 33.53, 33.33, 33.20, 33.18, 32.91,

32.83, 32.79, 32.70, 32.64, 32.38, 32.37, 32.08, 31.42, 31.09, 30.59, 30.47, 30.18, 30.14, 30.10, 30.02, 29.93, 29.90, 29.83, 29.80, 29.41, 29.26, 29.00, 28.48, 28.40, 27.96, 27.86, 27.74, 27.52, 27.22, 27.18, 27.14, 26.90, 26.83, 26.57, 267.34, 25.88, 25.62, 25.23, 24.90, 24.29, 24.11, 23.92, 23.85, 23.60, 23.48, 23.13, 23.08, 23.04, 22.04, 20.71, 20.57, 20.22, 20.16, 20.11, 19.68, 19.64, 19.13, 18.11, 14.97, 14.83, 14.54, 12.86, 11.82, 11.30, 8.11, 7.94, 7.89, 0.37. Anal. calcd for $C_{323}H_{457}ISi$: C, 86.31; H, 10.25. Found: C, 85.92; H, 10.76. M_n = 8860, M_w = 10,940, M_w/M_n =1.23. λ (THF) = 240, <u>377</u> (ε 3.59 × 10^5 M^{-1} cm^{-1}) nm. Emiss (THF) = <u>414</u>, 436 nm.

1-Phenylethynyl-4-(thioacetyl)benzene (61). **59** (0.36 g, 1.30 mmol), phenylacetylene (0.16 mL, 1.43 mmol), bis(dibenzylideneacetone palladium(0) (37.31 mg, 0.070 mmol), $(C_6H_5)_3P$ (85.20 mg, 0.33 mmol), CuI (24.80 mg, 0.13 mmol), N,N-diisopropylethylamine (0.90 mL, 5.20 mmol) and THF (3 mL) for 2 d afforded 0.31 g (95%) of the title compound as a yellow solid after gravity liquid chromatography (24:1 hexane:ether). R_f = 0.26 (24:1 hexane:ether). IR (KBr) 1699, 1587, 1493, 1442, 1396, 1352, 1314, 1211, 1175, 1112, 1012, 944, 877, 826, 756, 690, 622 cm^{-1}. ^{1}H NMR (300 MHz, CDCl$_3$) δ 7.54 (d, J = 8.5 Hz, 2 H), 7.53 (dd, J = 5.5, 2.4 Hz, 2 H), 7.38 (d, J = 8.5 Hz, 2 H), 7.36-7.33 (m, 3 H), 2.42 (s, 3 H). ^{13}C NMR (125 MHz, CDCl$_3$) δ 193.31, 134.16, 132.12, 131.65, 128.50, 128.35, 128.03, 124.52, 122.89, 91.04, 88.60, 30.21. HRMS calcd for $C_{16}H_{12}OS$: 252.0609. Found: 252.0607. λ (CH$_2$Cl$_2$) = 228, <u>293</u>, 309 nm.

2-Ethyl-4-(phenylethynyl)-1-(trimethylsilyethynyl)benzene (62). **14** (0.98 g, 3 mmol), phenylacetylene (0.36 mL, 3.30 mmol), bis(dibenzylideneacetone)palladium(0) (86.10 mg, 0.15 mmol), $(C_6H_5)_3P$ (196.70 mg, 3.30 mmol), CuI (57.10 mg, 0.30 mmol), N,N-diisopropylethylamine (2.10 mL, 12 mmol) and THF (6 mL) for 6 d afforded 0.89 g (98%) of the title compound as a yellow oil after gravity liquid chromatography (hexane). R_f = 0.39 (hexane). IR (neat) 3061, 2967, 2874, 2246, 2152, 1602, 1496, 1458, 1444, 1407, 1380, 1250, 1222, 1115, 1068, 1026, 909, 860, 756, 734, 690, 626 cm^{-1}. ^{1}H NMR (500 MHz, CDCl$_3$) δ 7.53-7.50 (m, 2 H), 7.38 (d, J = 7.9 Hz, 1 H), 7.36 (s, 1 H), 7.35-7.32 (m, 3 H), 7.27 (dd, J = 7.9, 1.7 Hz, 1 H), 2.78 (q, 7.6 Hz, 2 H), 1.25 (t, J = 7.6 Hz, 3 H), 0.25 (s, 9 H). ^{13}C NMR (125 MHz, CDCl$_3$) δ 147.19, 132.91, 132.17, 131.57, 129.37, 128.91, 128.90, 124.07, 123.75, 122.84, 104.19, 100.24, 91.54, 90.06, 28.18, 14.99, 0.54.

3-Ethyl-1-(phenylethynyl)-4-ethynylbenzene. **62** (0.89 g, 2.95 mmol), MeOH (15 mL) and K_2CO_3 (1.24 g, 9 mmol) afforded 0.63 g (93%) of the title compound as yellow oil. IR (neat) 3293, 3059, 2969, 2933, 2874, 2100, 1603, 1495, 1458, 1443, 1241, 1066, 1026, 912, 894, 831, 756, 689, 600 cm^{-1}. ^{1}H NMR (500 MHz, CDCl$_3$) δ 7.53-7.50 (m, 2 H), 7.42 (d, J = 7.9 Hz, 1 H), 7.38 (s, 1 H), 7.36-7.32 (m, 3 H), 7.29 (dd, J = 7.9, 1.6 Hz, 1 H), 3.32 (s, 1 H), 2.81 (q, J = 7.6 Hz, 2 H), 1.25 (t, J = 7.6 Hz, 3 H). ^{13}C NMR (125 MHz,

CDCl₃) δ 147.32, 133.34, 133.02, 132.16, 131.57, 129.33, 128.91, 124.31, 123.62, 121.82, 91.53, 89.87, 82.82, 82.55, 27.95, 15.12. HRMS calcd for $C_{18}H_{14}$: 230.1096. Found: 230.1089.

Trimer thioacetate 63. **59** (0.70 g, 2.50 mmol), 3-ethyl-1-(phenylethynyl)-4-ethynylbenzene (0.61 g, 2.67 mmol), THF (5 mL), bis(dibenzylideneacetone)palladium(0) (71.80 mg, 0.13 mmol), $(C_6H_5)_3P$ (164 mg, 0.63 mmol), CuI (47.60 mg, 0.25 mmol), N,N-diisopropylethylamine (1.73 mL, 10 mmol) for 5 d afforded 0.67 g (70%) of the title compound as yellow solid after gravity liquid chromatography (9:1 hexane:EtOAc). R_f = 0.49 (9:1 hexane:EtOAc). IR (KBr) 2962, 2930, 2874, 2201, 1691, 1589, 1501, 1459, 1442, 1396, 1356, 1304, 1266, 1118, 1014, 953, 887, 827, 753, 684, 668, 628 cm⁻¹. ¹H NMR (300 MHz, CDCl₃) δ 7.53 (d, J = 8.4 Hz, 2 H), 7.52 (dd, J = 7.5, 2.2 Hz, 1 H), 7.46 (d, J = 7.9 Hz, 1 H), 7.42 (d, J = 2 Hz, 1 H), 7.38 (d, J = 8.4 Hz, 2 H), 7.36-7.33 (m, 5 H), 2.85 (q, J = 7.6 Hz, 2 H), 2.43 (s, 3 H), 1.30 (t, J = 7.5 Hz, 3 H). ¹³C NMR (100 MHz, CDCl₃) δ 193.33, 146.40, 134.33, 132.23, 132.10, 131.70, 131.20, 129.00, 128.48, 128.46, 128.25, 124.61, 123.60, 123.17, 122.07, 93.87, 91.10, 89.70, 89.52, 30.30, 27.71, 14.70. HRMS calcd for $C_{26}H_{20}O_2$: 380.1235. Found: 380.1224. λ (CH₂Cl₂) = 230, 234, <u>328</u> nm.

General procedures for the hydrolysis of thioesters to thiols.[51] (a) A mixture of the thioester, NaOH (2 equiv), Zn dust (5 mg), and water (5 mL/mmol of thioester) were deoxygenated with argon gas for 1 h before THF (5 mL/mmol of thioester) was added. The reaction was allowed to stir overnight at room temperature while being purged with argon gas. 3 M HCl was to permit neutralization. The resulting precipitate was dissolved in degassed CHCl₃ (5 mL/mmol of thioester) and the organic layer was withdrawn via syringe and transferred to a Schlenk flask without exposure to air. Chloroform was added to the aqueous layer and withdrawn another 3× to permit complete extraction before the combined organic solvent was removed in vacuo to afford a solid which was immediately taken into the dry box. Eluents and other slight modifications are described below for each material. (b) A mixture of the thioester, 30 % NH₄OH (10 equivalents) and THF (2.50 mL/mmol of thioester) were allowed to stir for 15 min at room temperature while being purged with argon gas. The reaction mixture was then worked up in the same manner as above. Eluents and other slight modifications are described below for each material.

Trimer thiol 64. **63** (0.19 g, 0.50 mmol), NaOH (40 mg, 1.0 mmol), zinc dust (5.0 mg, 0.076 meq), water (3 mL), THF (5 mL) and 3 M HCl (0.33 mL) afforded 0.16 g (92%) of the title compound as yellow solid with slight impurities. R_f = 0.44 (9:1 hexane:EtOAc). IR (neat) 2963, 2929, 2869, 2562, 2204, 1722, 1654, 1598, 1499, 1460, 1400, 1271, 1118, 1015, 893, 824, 756, 691, 619 cm⁻¹. ¹H NMR (300 MHz, CDCl₃) δ 7.52 (d, J = 7.8 Hz, 2 H), 7.51 (dd, J = 7.5, 2.1 Hz, 1 H), 7.43 (d, J = 7.9 Hz, 1 H), 7.40 (s, 1 H), 7.36 (d, J = 8.4 Hz, 2 H), 7.34 (dd, J = 8.1, 3.4 Hz, 1 H), 7.33 (d, J = 7.9 Hz, 2 H), 7.23 (d, J

= 7.5 Hz, 2 H), 3.52 (s, 1 H), 2.85 (q, J = 7.6 Hz, 2 H), 1.29 (t, J = 7.5 Hz, 3 H). ^{13}C NMR (125 MHz, CDCl$_3$) δ 168.18, 146.57, 132.92, 132.47, 132.45, 132.41, 132.07, 131.54, 131.32, 129.43, 129.34, 129.24, 129.12, 128.82, 94.55, 91.32, 89.94, 88.83, 28.06, 15.01. LRMS Calcd for C$_{24}$H$_{18}$S: 338. Found: 338. λ (CH$_2$Cl$_2$) = 228, 234, 282, 339 nm.

2-Ethyl-1,4-bis(trimethylsilylethynyl)benzene 65. **14** (0.98 g, 3.0 mmol), TMSA (0.47 mL, 3.30 mmol), bis(dibenzylideneacetone)palladium(0) (86.10 mg, 0.15 mmol), (C$_6$H$_5$)$_3$P (0.20 g, 0.75 mmol), CuI (57.10 mg, 0.30 mmol), N,N-diisopropylethylamine (2.10 mL, 12 mmol) and THF (6 mL) for 1 d afforded 0.83 g (93%) of the title compound as yellow oil after gravity liquid chromatography (hexane). R$_f$ = 0.55 (hexane). IR (neat) 2963, 2935, 2899, 2875, 2157, 1602, 1484, 1458, 1406, 1250, 1213, 1120, 916, 859, 760, 700, 628 cm^{-1}. ^{1}H NMR (300 MHz, CDCl$_3$) δ 7.32 (d, J = 7.9 Hz, 1 H), 7.28 (d, J = 1.1 Hz, 1 H), 7.19 (dd, J = 7.9, 1.6 Hz, 1 H), 2.74 (q, J = 7.5 Hz, 2 H), 1.21 (t, J = 7.5 Hz, 3 H), 0.23 (s, 9 H), 0.22 (s, 9 H). ^{13}C NMR (100 MHz, CDCl$_3$) δ 146.57, 132.19, 131.37, 129.07, 123.17, 122.50, 104.97, 103.38, 99.75, 95.60, 27.52, 14.35, -0.056, -0.074. HRMS calcd for C$_{18}$H$_{26}$Si$_2$: 298.1573. Found: 298.1569.

2-Ethyl-1,4-diethynylbenzene 68. **65** (0.60 g, 2 mmol), MeOH (10 mL) and K$_2$CO$_3$ (1.66 g, 12 mmol) afforded 0.27 g (88%) of the title compound as yellow oil. R$_f$ = 0.27 (hexane). IR (neat) 3294, 2969, 2935, 2875, 2107, 1603, 1485, 1460, 1243, 896, 832, 735 cm^{-1}. ^{1}H NMR (300 MHz, CDCl$_3$) δ 7.39 (d, J = 7.9 Hz, 1 H), 7.33 (d, J = 1.1 Hz, 1 H), 7.25 (dd, J = 7.9, 1.6 Hz, 1 H), 3.31 (s, 1 H), 3.12 (s, 1 H), 2.78 (q, J = 7.6 Hz, 2 H), 1.23 (t, J = 7.6 Hz, 3 H). ^{13}C NMR (100 MHz, CDCl$_3$) δ 146.83, 132.74, 131.60, 129.26, 122.53, 121.89, 83.39, 82.29, 81.74, 78.53, 27.33, 14.51.

Diphenylthioester monomer 71. **59** (0.70 g, 2.50 mmol), **68** (0.15 g, 1.0 mmol), THF (3 mL), bis(dibenzylideneacetone)palladium(0) (72 mg, 0.13 mmol), (C$_6$H$_5$)$_3$P (164 mg, 0.63 mmol), CuI (48 mg, 0.25 mmol), N,N-diisopropylethylamine (0.70 mL, 4 mmol) for 1 d afforded 0.32 g (70%) of the title compound as yellow solid after gravity liquid chromatography (9:1 hexane:ether). R$_f$ = 0.11 (9:1 hexane:ether). IR (KBr) 2964, 2928, 2874, 2206, 1912, 1695, 1590, 1500, 1396, 1354, 1118, 1095, 1014, 955, 825, 720, 694, 617 cm^{-1}. ^{1}H NMR (400 MHz, CDCl$_3$) δ 7.54 (d, J = 8.1 Hz, 4 H), 7.46 (d, J = 8.1 Hz, 1 H), 7.41 (d, J = 1.7 Hz, 1 H), 7.39 (d, J = 8.2 Hz, 2 H), 7.38 (d, J = 8.3 Hz, 2 H), 7.33 (dd, J = 8.1, 1.7 Hz, 1 H), 2.86 (q, J = 7.6 Hz, 2 H), 2.43 (s, 3 H), 2.42 (s, 3 H), 1.30 (t, J = 7.5 Hz, 3 H). ^{13}C NMR (100 MHz, CDCl$_3$) δ 193.39, 146.41, 134.29, 134.22, 132.18, 132.07, 131.22, 128.98, 124.54, 124.36, 123.11, 122.34, 93.93, 91.05, 90.15, 89.52, 30.29, 27.64, 14.60. HRMS calcd for C$_{28}$H$_{22}$O$_2$S$_2$: 454.1061. Found: 454.1059. λ (CH$_2$Cl$_2$) = 205, 230, 234, 334 nm.

Diphenylthiol monomer 74. **71** (0.19 g, 0.40 mmol), NaOH (64 mg, 1 mmol), zinc dust (5.0 mg, 0.076 meq), water (1 mL), THF (2 mL) and 3 M

HCl (0.53 mL) afforded 0.14 g (94%) of the title compound as a yellow solid with slight impurities. R_f = 0.12 (4:1 hexane:ether). IR (KBr) 2962, 2927, 2869, 2561, 2204, 1902, 1712, 1656, 1587, 1498, 1461, 1401, 1266, 1099, 1016, 892, 823 cm^{-1}. ^{1}H NMR (300 MHz, CDCl$_3$) δ 7.42 (d, J = 8.0 Hz, 1 H), 7.37 (d, J = 8.2 Hz, 4 H), 7.36 (s, 1 H), 7.29 (dd, J = 8.0, 1.6 Hz, 1 H), 7.23 (d, J = 8.2 Hz, 2 H), 7.22 (d, J = 8.4 Hz, 2 H), 3.51 (s, 1 H), 3. 50 (s, 1 H), 2.84 (q, J = 7.5 Hz, 2 H), 1.29 (t, J = 7.5 Hz, 3 H). ^{13}C NMR (125 MHz, CDCl$_3$) δ 146.56, 132.56, 132.43, 131.44, 129.39, 129.37, 129.25, 123.51, 122.76, 120.88, 120.70, 94.57, 90.78, 90.33, 88.78, 28.02, 14.97. LRMS calcd for C$_{24}$H$_{18}$S$_2$: 370. Found: 370. λ (CH$_2$Cl$_2$) = 229, 234, 340 nm.

Bis(trimethylsilyl) dimer 66. **23** (0.91 g, 2.0 mmol), TMSA (0.31 mL, 2.20 mmol), bis(dibenzylideneacetone)palladium(0) (57.40 mg, 0.10 mmol), (C$_6$H$_5$)$_3$P (0.11 g, 0.40 mmol), CuI (38.10 mg, 0.20 mmol), N,N-diisopropylethylamine (1.38 mL, 8 mmol) and THF (4 mL) for 5 d afforded 0.65 g (76%) of the title compound as yellow oil after gravity liquid chromatography (hexane). R_f = 0.21 (hexane). IR (neat) 2964, 2899, 2875, 2154, 1598, 1495, 1458, 1407, 1250, 1222, 857, 760, 701, 636 cm^{-1}. ^{1}H NMR (500 MHz, CDCl$_3$) δ 7.39 (d, J = 7.9 Hz, 1 H), 7.38 (d, J = 7.9 Hz, 1 H), 7.33 (d, J = 1.2 Hz, 1 H), 7.32 (d, J = 1.1 Hz, 1 H), 7.25 (dd, 8.0, 1.3 Hz, 2 H), 2.82 (q, J = 7.6 Hz, 2 H), 2.78 (q, J = 7.6 Hz, 2 H), 1.27 (t, J = 7.5 Hz, 3 H), 1.25 (t, J = 7.5 Hz, 3 H), 0.25 (s, 9 H), 0.24 (s, 9 H). ^{13}C NMR (125 MHz, CDCl$_3$) δ 147.20, 146.48, 132.77, 132.40, 131.86, 131.27, 129.65, 129.04, 123.80, 123.53, 122.91, 122.82, 105.44, 103.85, 100.27, 96.13, 94.99, 94.99, 89.65, 28.01, 27.97, 14.97, 14.82, 0.37, 0.35. HRMS calcd for C$_{28}$H$_{34}$Si$_2$: 426.2199. Found: 426.2202.

Diethynyl dimer 69. **66** (0.64 g, 1.50 mmol), MeOH (10 mL), CH$_2$Cl$_2$ (5 mL) and K$_2$CO$_3$ (1.24 g, 9.0 mmol) afforded 0.42 g (100%) of the title compound as yellow oil. IR (neat) 3297, 3091, 3071, 3036, 2968, 2933, 2874, 1960, 1815, 1727, 1603, 1527, 1479, 1394, 1036, 910, 832, 734, 676 cm^{-1}. ^{1}H NMR (500 MHz, CDCl$_3$) δ 7.43 (d, J = 7.9 Hz, 1 H), 7.42 (d, J = 7.9 Hz, 1 H), 7.360 (d, J = 1.3 Hz, 1 H), 7.355 (d, J = 1.1 Hz, 1 H), 7.28 (dd, J = 7.9, 1.4 Hz, 2 H), 3.33 (s, 1 H), 3.14 (s, 1 H), 2.87-2.79 (two quartets, 4 H), 1.28 (t, J = 7.6 Hz, 3 H), 1.27 (t, J = 7.6 Hz, 3 H). ^{13}C NMR (125 MHz, CDCl$_3$) δ 147.34, 146.63, 133.28, 132.49, 132.05, 131.34, 131.29, 129.82, 129.22, 127.14, 124.12, 123.25, 94.90, 89.54, 84.01, 82.72, 82.37, 78.98, 27.98, 27.86, 15.03, 14.95. HRMS calcd for C$_{22}$H$_{18}$: 282.1408. Found: 282.1407.

Diphenylthioester dimer 72. **59** (1.04 g, 3.75 mmol), **69** (42 g, 1.50 mmol), THF (10 mL), bis(dibenzylideneacetone)palladium(0) (0.11 g, 0.19 mmol), (C$_6$H$_5$)$_3$P (0.25 g, 0.94 mmol), CuI (72.40 mg, 0.38 mmol), N,N-diisopropylethylamine (2.60 mL, 15 mmol) for 5 d to afforded 0.44 g (50%) of the title compound as a yellow waxy solid after gravity liquid chromatography (9:1 hexane:EtOAc). R_f = 0.26 (9:1 hexane:EtOAc). IR (KBr) 2966, 2931, 2871, 2203, 1910, 1701, 1588, 1542, 1500, 1478, 1460, 1435, 1396, 1350,

1264, 1115, 1088, 1014, 947, 888, 826, 616 cm^{-1}. ^{1}H NMR (400 MHz, CDCl$_3$) δ 7.54 (d, J = 8.2 Hz, 2 H), 7.47 (d, J = 7.9 Hz, 2 H), 7.41 (s, 1 H), 7.39 (d, J = 8.4 Hz, 2 H), 7.38 (d, J = 8.4 Hz, 2 H), 7.35-7.32 (m, 3 H), 7.21 (d, J = 7.8 Hz, 2 H), 2.89-2.86 (two quartets, 4 H), 2.43 (s, 3 H), 2.42 (s, 3 H), 1.32 (t, J = 7.6 Hz, 3 H), 1.31 (t, J = 7.6 Hz, 3 H). ^{13}C NMR (75 MHz, CDCl$_3$) δ 193.43, 146.49, 146.32, 134.32, 134.26, 134.09, 132.24, 132.20, 132.09, 131.20, 131.04, 129.69, 129.01, 128.87, 128.40, 128.22, 124.58, 124.41, 123.54, 122.60, 122.19, 94.77, 93.91, 91.13, 90.13, 89.60, 89.46, 30.33, 27.71, 27.65, 14.68, 14.63. HRMS calcd for C$_{38}$H$_{30}$O$_2$S$_2$: 582.1687. Found: 582.1684. λ (CH$_2$Cl$_2$) = <u>229</u>, 234, 272, 350 nm.

Diphenylthiol dimer 75. **72** (0.17 g, 0.30 mmol), NaOH (0.10 g, 2.50 mmol), zinc dust (5.0 mg, 0.076 meq), water (5 mL), THF (5 mL) and 3 M HCl (0.83 mL) afforded 0.13 g (87%) of the title compound as a yellow solid with slight impurities. R$_f$ = 0.17 (9:1 hexane:EtOAc). IR (KBr) 2963, 2929, 2868, 2561, 2201, 1899, 1719, 1654, 1589, 1542, 1499, 1459, 1400, 1265, 1181, 1116, 1096, 1013, 954, 890, 823, 744, 700, 620 cm^{-1}. Impurities, presumably due from disulfide formation, prevented complete assignment of the aromatic region. ^{1}H NMR (300 MHz, CDCl$_3$) δ 7.46 (d, J = 8.0 Hz, 1 H), 7.45 (d, J = 8.0 Hz, 1 H), 7.37 (d, J = 8.3 Hz, 4 H), 7.32 (d, J = 7.9 Hz, 1 H), 7.31 (d, J = 7.9 Hz, 2 H), 7.30 (d, J = 7.9 Hz, 1 H), 7.25 (d, J = 8.4 Hz, 2 H), 7.24 (d, J = 8.5 Hz, 2 H), 3.52 (s, 1 H), 3.51 (s, 1 H), 2.90-2.80 (two quartets, 4 H), 1.31 (t, J = 7.5 Hz, 3 H), 1.30 (t, J = 7.5 Hz, 3 H). ^{13}C NMR (125 MHz, CDCl$_3$) δ 146.64, 132.55, 132.54, 132.48, 132.43, 132.40, 131.44, 131.36, 131.27, 129.40, 129.37, 129.25, 129.22, 123.66, 123.62, 122.86, 122.73, 120.89, 120.72, 95.12, 94.61, 90.82, 90.32, 89.78, 88.78, 28.06, 28.02, 15.00, 14.99. LRMS calcd for C$_{34}$H$_{26}$S$_2$: 498. Found: 498. λ (CH$_2$Cl$_2$) = 207, 229, <u>352</u> nm.

Bis(trimethylsilyl) tetramer 67. **32** (0.56 g, 0.78 mmol), TMSA (TMSA) (0.12 mL, 0.86 mmol), bis(dibenzylideneacetone)palladium(0) (22.40 mg, 0.04 mmol), (C$_6$H$_5$)$_3$P (51.10 mg, 0.20 mmol), CuI (15 mg, 0.08 mmol), N,N-diisopropylethylamine (0.54 mL, 3.12 mmol) and THF (5 mL) for 4 d afforded 0.48 g (90%) of the title compound as a yellow oil after flash liquid chromatography (9:1 hexane:CH$_2$Cl$_2$). R$_f$ = 0.74 (9:1 hexane:CH$_2$Cl$_2$). IR (neat) 2963, 2930, 2868, 2152, 1919, 1592, 1501, 1459, 1409, 1248, 1117, 1062, 842, 759, 698, 666, 624 cm^{-1}. ^{1}H NMR (500 MHz, CDCl$_3$) δ 7.46 (dd, J = 7.9, 1.9 Hz, 2 H), 7.42 -7.39 (m, 4 H), 7.35-7.31 (m, 4 H), 7.29-7.26 (m, 2 H), 2.91-2.78 (m, 8 H), 1.33-1.24 (m, 12 H), 0.25 (s, 9 H), 0.24 (s, 9 H). ^{13}C NMR (125 MHz, CDCl$_3$) δ 147.24, 146.71, 146.51, 132.79, 132.58, 132.41, 131.89, 131.36, 131.29, 129.67, 129.23, 129.07, 123.81, 123.76, 123.54, 122.93, 122.83, 105.45, 103.86, 100.34, 96.18, 95.16, 95.11, 95.08, 89.82, 89.77, 89.74, 30.13, 28.08, 28.03, 28.00, 15.05, 14.99, 14.84, 0.38, 0.37. HRMS calcd for C$_{48}$H$_{50}$Si$_2$: 682.3451. Found: 682.3474.

Diethynyl tetramer 70. **67** (0.48 g, 0.70 mmol), MeOH (25 mL), CH$_2$Cl$_2$ (50 mL), and K$_2$CO$_3$ (0.58 g, 4.20 mmol) afforded 0.33 g (88%) of the title compound as a yellow solid. IR (KBr) 3300, 2964, 2929, 2867, 1593, 1500, 1459, 1409, 1373, 1116, 1062, 888, 838, 745, 642, 616 cm^{-1}. ^{1}H NMR (300 MHz, CDCl$_3$) δ 7.49-7.46 (m, 3 H), 7.43 (s, 1 H), 7.40-7.37 (m, 4 H), 7.35-7.28 (m, 4 H), 3.33 (s, 1 H), 3.14 (s, 1 H), 2.90-2.81 (m, 8 H), 1.35-1.24 (m, 12 H). ^{13}C NMR (125 MHz, CDCl$_3$) δ 147.35, 146.75, 146.73, 146.61, 133.25, 132.58, 132.56, 132.54, 132.45, 132.04, 131.37, 131.32, 129.78, 129.23, 129.10, 123.32, 122.49, 95.12, 94.87, 89.79, 89.76, 89.52, 84.00, 82.62, 82.36, 78.85, 30.10, 28.05, 27.95, 27.83, 15.02, 14.93.

Diphenylthioester tetramer 73. **59** (55.60 mg, 0.20 mmol), **70** (35.97 mg, 0.07 mmol), bis(dibenzylideneacetone)palladium(0) (5.74 mg, 0.01 mmol), (C$_6$H$_5$)$_3$P (13.11 mg, 0.05 mmol), CuI (3.80 mg, 0.02 mmol), N,N-diisopropylethylamine (0.14 mL, 0.80 mmol) and THF (3 mL) for 5 d to afforded 17.30 mg (32%) of the title compound as a yellow waxy solid with slight impurities after flash liquid chromatography (9:1 hexane:EtOAc). R_f = 0.20 (9:1 hexane:EtOAc). Impurities, presumably from disulfide formation, prevented complete assignment of the aromatic region. ^{1}H NMR (400 MHz, CDCl$_3$) δ 7.56-7.32 (m, 20 H), 3.00-2.80 (four quartets, 8 H), 2.43 (s, 3 H), 2.42 (s, 3 H), 1.40-1.30 (four triplets, 12 H). LRMS Calcd for C$_{58}$H$_{46}$O$_2$S$_2$: 838 (100%), 839 (68%), 840 (32%). Found: 838 (100%), 839 (60%), 840 (40%). λ (CH$_2$Cl$_2$) = 197, 209, 218, 228, <u>359</u> nm.

Thioester silane 76. **16** (0.35 g, 0.74 mmol), **60** (0.14 mL, 0.81 mmol), bis(triphenylphosphino)palladium(II) chloride (0.0052 g, 0.0074 mmol), (C$_6$H$_5$)$_3$P (0.0039 g, 0.015 mmol), CuI (0.0029 g, 0.015 mmol) and Et$_3$N (1 mL) for 2 d afforded 0.34 g (89%) of the title compound as a brown oil after gravity liquid chromatography (9:1 hexane:ether). R_f = 0.62 (9:1 hexane:ether). IR (neat) 3054, 2926, 2853, 2209, 2152, 1904, 1715, 1601, 1541, 1497, 1466, 1435, 1397, 1352, 1249, 1224, 1118, 1090, 1016, 949, 842, 760, 744, 721, 697, 638, 617 cm^{-1}. ^{1}H NMR (400 MHz, CDCl$_3$) δ 7.52 (d, J = 8.6 Hz, 2 H), 7.38 (d, J = 7.9 Hz, 1 H), 7.37 (d, J = 8.5 Hz, 2 H), 7.33 (d, J = 1.3 Hz, 1 H), 7.26 (dd, J = 8.0, 1.6 Hz, 1 H), 2.74 (t, J = 7.8 Hz, 2 H), 2.42 (s, 3 H), 1.62 (p, J =7.2 Hz, 2 H), 1.32-1.24 (m, 18 H), 0.86 (t, J = 7.0 Hz, 3 H), 0.24 (s, 9 H). HRMS calcd for C$_{33}$H$_{44}$OSSi: 516.2882. Found: 516.2878.

Thioester alkyne 80. **76** (0.32 g, 0.62 mmol), 49% concentrated HF (0.021 mL, 1.36 mmol), TBAF (0.20 mL, 0.68 mmol, 1.0 M in THF) and pyridine (1.24 mL) at room temperature for 20 min afforded 0.22 g (81%) of the title compound as a brown oil after gravity liquid chromatography (9:1 hexane:ether). R_f = 0.55 (9:1 hexane:ether) IR (neat) 3298, 2924, 2853, 2101, 1712, 1601, 1494, 1461, 1120, 1016, 949, 828 cm^{-1}. ^{1}H NMR (400 MHz, CDCl$_3$) δ 7.53 (d, J = 8.5 Hz, 2 H), 7.42 (d, J = 7.9 Hz, 1 H), 7.37 (d, J = 8.6 Hz, 2 H), 7.35 (d, J = 1.3 Hz, 1 H), 6.38 (dd, J = 7.9, 1.6 Hz, 1 H), 3.31 (s, 1 H), 2.75 (t, J = 7.8 Hz, 2 H), 2.42 (s, 3 H), 1.63 (p, J = 7.2 Hz, 2 H), 1.32-1.24 (m, 18 H), 0.86 (t, J = 7.1 Hz, 3 H).

Bisthioester 83. **59** (0.26 g, 0.92 mmol), **80** (0.21 g, 0.46 mmol), Et$_3$N (1 mL), bis(triphenylphosphino)palladium(II) chloride (0.0065 g, 0.0092 mmol), (C$_6$H$_5$)$_3$P (0.0047 g, 0.018 mmol) and CuI (0.0034 g, 0.018 mmol) for 1 d afforded 0.16 g (59%) of the title compound as a yellow solid after gravity liquid chromatography (9:1, hexane:ether). R$_f$ = 0.20 (9:1, hexane:ether). IR (KBr) 3032, 2924, 2855, 2677, 2254, 2211, 1907, 1705, 1594, 1541, 1498, 1461, 1400, 1354, 1305, 1267, 1116, 1012, 950, 908, 829, 731, 615 cm^{-1}. ^{1}H NMR (400 MHz, CDCl$_3$) δ 7.54 (d, J = 8.3 Hz, 2 H), 7.53 (d, J = 8.2 Hz, 2 H), 7.46 (d, J = 7.9 Hz, 1 H), 7.41-7.38 (m, 5 H), 7.33 (dd, J = 8.0, 2.5 Hz, 1 H), 2.83 (t, J = 7.6 Hz, 2 H), 2.42 (s, 3 H), 2.41 (s, 3 H), 1.69 (p, J = 7.5 Hz, 2 H), 1.38-1..25 (m, 18 H), 0.89 (t, J = 7.0 Hz, 3 H). ^{13}C NMR (100 MHz, CDCl$_3$) δ 193.27 (2C), 145.28, 134.29, 134.23, 132.24, 132.19, 132.05, 131.98, 129.00, 128.29, 128.24, 124.56, 124.36, 122.98, 122.56, 93.85, 91.13, 90.21, 89.79, 34.64, 31.98, 30.64, 30.29, 29.75, 29.75, 29.72, 29.69, 29.63, 29.61, 29.43, 22.76, 14.21. HRMS calcd for C$_{38}$H$_{42}$O$_2$S$_2$: 594.2626. Found: 594.2618.

Thioester silane 77. **34** (0.38 g, 0.29 mmol), **60** (0.13 mL, 0.75 mmol), bis(dibenzylideneacetone)palladium(0) (0.0086 g, 0.015 mmol), (C$_6$H$_5$)$_3$P (0.020 g, 0.073 mmol), CuI (0.0055 g, 0.029 mmol), N,N-diisopropylethylamine (0.40 mL, 2.32 mmol), and THF (4 mL) for 3 d afforded 0.28 g (73%) of the title compound as a waxy yellow solid after gravity liquid chromatography (9:1 hexane:ether). R$_f$ = 0.61 (9:1, hexane:ether). IR (KBr) 2924, 2853, 2206, 2150, 1710, 1594, 1500, 1460, 1403, 1250, 1118, 950, 835, 758, 617 cm^{-1}. ^{1}H NMR (400 MHz, CDCl$_3$) δ 7.54 (d, J = 8.4 Hz, 2 H), 7.46 (d, J = 7.9 Hz, 2 H), 7.45 (d, J = 7.9 Hz, 2 H), 7.39 (d, J = 8.2 Hz, 2 H), 7.38 (d, J = 8.3 Hz, 2 H), 7.37-7.26 (m, 6 H), 2.83 (t, J = 7.4 Hz, 6 H), 2.75 (t, J = 6.3 Hz, 2 H), 2.42 (s, 3 H), 1.70-1.20 (m, 8 H), 1.41-1.22 (m, 72 H), 0.86-0.83 (m, 12 H), 0.25 (s, 9 H). ^{13}C NMR (125 MHz, CDCl$_3$) δ 193.37, 146.11, 145.60, 134.73, 134.70, 134.44, 132.92, 132.88, 132.65, 132.57, 132.48, 132.43, 132.27, 132.22, 132.00, 131.90, 129.60, 129.38, 129.22, 129.07, 128.86, 128.70, 128.65, 124.85, 123.80, 123.73, 123.71, 123.30, 123.25, 123.08, 104.14, 100.08, 95.25, 95.21, 95.12, 91.65, 91.40, 90.57, 90.12, 90.08, 35.51, 35.18, 34.83, 34.63, 34.45, 31.27, 31.15, 31.06, 30.96, 30.87, 30.70, 30.63, 30.58, 30.25, 30.21, 30.08, 29.92, 29.34, 28.54, 28.38, 23.33, 23.21, 23.10, 14.74, 14.66, 14.60, 14.52. LRMS calcd for C$_{93}$H$_{128}$OSSi: 1321 (88%), 1322 (100%), 1323 (63%). Found: 1321 (95%), 1322 (100%), 1323 (70%).

Thioester alkyne 81. **77** (0.15 g, 0.11 mmol), 49% concentrated hydrofluoric acid (0.0074 mL, 0.48 mmol), TBAF (0.070 mL, 0.24 mmol, 1.0 M in THF), and pyridine (1.24 mL) at room temperature for 30 min afforded 0.12 g (87%) of the title compound as a waxy yellow solid after flash liquid chromatography (9:1 hexane:ether). R$_f$ = 0.57 (9:1 hexane:ether) IR (KBr) 3302, 2926, 2856, 2206, 2101, 1906, 1713, 1597, 1498, 1459, 1116, 827, 616 cm^{-1}. ^{1}H NMR (400 MHz, CDCl$_3$) δ 7.54 (d, J = 8.4 Hz, 2 H), 7.46 (d, J = 7.8 Hz, 2 H), 7.43 (d, J = 7.9 Hz, 2 H), 7.38 (d, J = 8.5 Hz, 2 H), 7.37-7.27 (m, 8

H), 3.32 (s, 1 H), 2.83 (t, J = 7.4 Hz, 6 H), 2.77 (t, J = 7.3 Hz, 2 H), 2.43 (s, 3 H), 1.70-1.60 (m, 8 H), 1.34-1.22 (m, 72 H), 0.86-0.83 (m, 12 H). LRMS calcd for $C_{90}H_{120}OS$: 1249 (98%), 1250 (100%), 1251 (55%). Found: 1249 (96%), 1250 (100%), 1251 (63%).

Thioester silane 78. **42** (0.13 g, 0.055 mmol), **60** (0.020 g, 0.75 mmol), bis(dibenzylideneacetone)palladium(0) (0.0016 g, 0.0028 mmol), $(C_6H_5)_3P$ (0.0037 g, 0.014 mmol), CuI (0.00053 g, 0.0028 mmol), N,N-diisopropylethylamine (2 mL), and THF (3 mL) for 3 days afforded 0.077 g (59%) of the title compound as a waxy yellow solid after gravity liquid chromatography (2:1 hexane:CHCl₃). R_f = 0.56 (2:1 hexane:CHCl₃). IR (KBr) 2928, 2858, 2198, 2154, 1716, 1510, 1468, 1398, 1262, 1126, 960, 840, 732 cm⁻¹. ¹H NMR (400 MHz, CDCl₃) δ 7.54 (d, J = 8.4 Hz, 2 H), 7.48-7.46 (m, 8 H), 7.40-7.32 (m, 18 H), 2.85 - 2.75 (m, 16 H), 2.42 (s, 3 H), 1.75-1.65 (m, 16 H), 1.40-1.20 (m, 144 H), 0.90-0.80 (m, 24 H), 0.25 (s, 9 H). ¹³C NMR (125 MHz, CDCl₃) δ 193.77, 146.11, 145.62, 145.32, 143.56, 134.62, 134.60, 133.37, 132.76, 132.57, 132.47, 132.36, 132.34, 132.32, 132.14, 132.02, 131.90, 131.84, 129.36, 129.16, 128.95, 128.85, 128.80, 128.70, 128.67, 128.60, 128.31, 124.83, 123.63, 123.37, 123.24, 123.16, 123.04, 104.01, 100.11, 95.07, 91.52, 90.46, 90.05, 39.80, 29.42, 39.26, 37.50, 37.04, 36.90, 36.18, 36.14, 35.48, 35.09, 34.75, 34.70, 33.20, 32.81, 32.50, 32.36, 32.22, 32.00, 31.72, 31.30, 31.07, 31.03, 30.98, 30.81, 30.67, 30.14, 30.13, 30.10, 30.05, 30.01, 30.00, 29.93, 29.90, 29.80, 29.78, 29.72, 29.68, 29.54, 29.47, 29.34, 28.88, 28.60, 28.39, 28.30, 27.82, 27.72, 27.51, 25.69, 23.24, 23.11, 23.07, 22.97, 21.11, 20.55, 20.10, 19.64, 14.81, 14.53, 14.39, 11.82, 4.96, 0.37.

Thioester silane 79. **46** (0.49 g, 0.11 mmol), **60** (0.039 mL, 0.22 mmol), bis(dibenzylideneacetone)palladium(0) (0.0032 g, 0.0055 mmol), $(C_6H_5)_3P$ (0.0073 g, 0.028 mmol), CuI (0.0010 g, 0.0055 mmol), N,N-diisopropylethylamine (2 mL), and THF (4 mL) for 3 days afforded 0.33 g (66%) of the title compound as a waxy yellow solid after gravity liquid chromatography (2:1 hexane:CHCl₃). R_f = 0.53 (2:1 hexane:CHCl₃). IR (KBr) 2924, 2855, 2203, 2151, 1710, 1594, 1502, 1460, 1386, 1261, 1094, 1027, 806 cm⁻¹. ¹H NMR (400 MHz, CDCl₃) δ 7.54 (d, J = 8.0 Hz, 2 H), 7.50-7.46 (m, 14 H), 7.38-7.31 (m, 36 H), 2.90-2.80 (m, 32 H), 2.42 (s, 3 H), 1.80-1.70 (m, 32 H), 1.40-1.20 (m, 288 H), 0.90-0.80 (m, 48 H), 0.25 (s, 9 H). ¹³C NMR (125 MHz, CDCl₃) δ 193.68, 145.62, 145.56, 145,32, 143.57, 143.55, 140.46, 134.61, 133.42, 132.88, 132.58, 132.48, 132.37, 132,34, 132.32, 132.15, 132.06, 131.92, 131.86, 129.80, 129.60, 129.39, 129.35, 129.28, 129.19, 129.00, 128.986, 128.86, 128.81, 128.76, 128.71, 128.68, 128.63, 128.33, 125.13, 124.84, 124.11, 123.82, 123.64, 123.51, 123.40, 123.26, 123.20, 123.08, 122.68, 95.41, 95.14, 91.88, 91.57, 91.45, 90.50, 90.41, 90.08, 89.42, 89.36, 88.66, 88.55, 78.78, 39.83, 39.52, 39.44, 39.29, 37.53, 37.08, 36.93, 36.21, 36.17, 35.52, 35.13, 35.08, 34.83, 34.78, 34.72, 33.23, 33.04, 33.00, 32.84, 32.53, 32.40, 32.26, 31.78, 31.75, 31.44, 31.36, 31.10, 31.06, 30.84,

30.66, 30.19, 30.14, 30.10, 30.06, 30.03, 29.93, 29.85, 29.76, 29.72, 28.91, 28.77, 28.62, 28.41, 28.32, 27.87, 27.76, 27.56, 27.45, 27.15, 23.50, 23.27, 23.24, 23.15, 23.11, 23.01, 20.59, 20.13, 19.70, 19.66, 14.84, 14.70, 14.57, 14.43, 11.85, 4.97, 0.40. Anal. calcd for $C_{333}H_{464}OSSi$: C, 88.03; H, 10.30. Found: C, 86.71; H, 10.80.

Thioester alkyne 82. 79 (0.27 g, 0.060 mmol), 49% HF (0.40 mL, 25.95 mmol), TBAF (3.84 mL, 13.26 mmol, 1.0 M in THF), pyridine (2 mL) and THF (15 mL) overnight afforded 0.14 g (52%) of the title compound as a waxy yellow solid after flash liquid chromatography (2:1 hexane:CHCl₃). R_f = 0.50 (2:1 hexane:CHCl₃). IR (KBr) 3302, 2926, 2856, 2204, 2096, 1714, 1655, 1590, 1504, 1466, 1386, 1262, 1116, 1030, 960, 896, 836, 798, 724 cm⁻¹. ¹H NMR (400 MHz, CDCl₃) δ 7.54 (d, J = 8.0 Hz, 2 H), 7.49-7.46 (m, 14 H), 7.38-7.31 (m, 36 H), 3.36 (s, 1 H), 2.90-2.80 (m, 32 H), 2.42 (s, 3 H), 1.80-1.70 (m, 32 H), 1.40-1.20 (m, 288 H), 0.90-0.80 (m, 48 H).

Bisthioester 84. 59 (0.011g, 0.039 mmol), **82** (0,12 g, 0.026 mmol), bis(dibenzylideneacetone)palladium(0) (0.0012g, 0.0020 mmol), $(C_6H_5)_3P$ (0.0026g, 0.0098 mmol), CuI (0.00038 g, 0.0020 mmol), N,N-diisopropylethylamine (2 mL, 11.48 mmol), and THF (3 mL) for 3 days afforded 0.070 g (58%) of the title compound as a waxy yellow solid after gravity liquid chromatography (3:1 hexane:CHCl₃). R_f = 0.43 (3:1 hexane:CHCl₃). IR (KBr) 2923, 2853, 2204, 1716, 1662, 1556, 1502, 1464, 1388, 1314, 1260, 1120, 1018, 890, 826, 692 cm⁻¹. ¹H NMR (400 MHz, CDCl₃) δ 7.54 (d, J = 8.4 Hz, 2 H), 7.48-7.46 (m, 14 H), 7.40-7.32 (m, 40 H), 2.90-2.80 (m, 32 H), 2.43 (s, 3 H), 2.41 (s, 3 H), 1.75-1.65 (m, 32 H), 1.4 -1.20 (m, 288 H), 0.90-0.80 (m, 48 H). ¹³C NMR (75 MHz, CDCl₃) δ 193.80, 145.62, 138.76, 136.34, 135.42, 134.61, 132.80, 132.56, 132.54, 132.46, 132.36, 132.32, 132.14, 131.89, 131.16, 129.34, 129.14, 128.94, 128.84, 128.80, 128.69, 128.59, 124.82, 124.06, 123.61, 123.46, 123.23, 123.14, 123.04, 95.06, 91.51, 90.44, 90.02, 36.17, 35.07, 34.72, 32.33, 31.74, 31.05, 31.01, 30.79, 30.67, 30.59, 30.12, 30.11, 30.08, 30.03, 29.99, 29.97, 29.88, 29.78, 28.36, 28.11, 27.80, 27.70, 23.24, 23.09, 20.09, 14.51. Anal. calcd for C338H462O2S2: C, 87.84; H, 10.08. Found: C, 83.52; H, 9.70.[52]

3.3 Shorter Phenylene-Based Molecular Wires and Devices and Experimental Details

3.3.1. *Synthesis of Molecular Scale Wires*

The syntheses of molecular wires that exhibit nearly linear *I(V)* curves are described in this section. Many of the molecular wires have previously been tested. In each case, the molecular wires bear one or more thioacetyl end groups that can be deprotected, during the self-assembled monolayer (SAM) formation step, to afford the free thiols for attachment to metallic probes.

3.3.1.1. Synthesis of One-Terminal Oligo(phenylene ethynylene) Molecular Wires

The synthesis of a simple wire, **4**, from readily available 1-bromo-4-iodobenzene (1) is shown (Scheme 3.20). The starting material was monocoupled to TMSA using typical Sonogashira coupling procedures.[29,40a,53] The reaction proceeded with good chemoselectivity due to the greater reactivity of the aryl iodide. The resulting aryl bromide was then coupled to phenylacetylene using similar conditions yet higher temperatures to enhance coupling to the aryl bromide. The terminal alkyne **2** was deprotected using K_2CO_3 and MeOH and then coupled to 1-iodo-4-thioacetylbenzene[54] (**3**) to form molecular wire **4**.

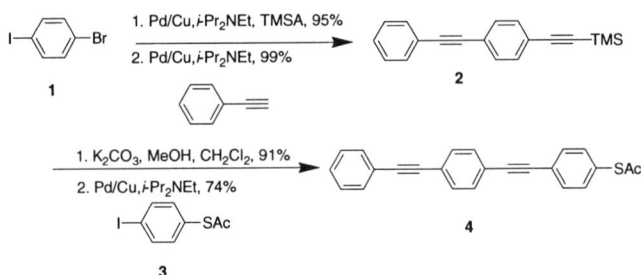

Scheme 3.20 Synthesis of unfunctionalized wire.

X-ray diffraction crystallography of thiol derived from **4** attached to an Os cluster has shown that this oligo(phenylene ethynylene) exists predominantly in a planarized form; the phenyl rings being nearly parallel.[55] It is hypothesized that the conductivity of these systems arise through the extended π-orbital overlap which is maximized while the molecule is planar. If the phenyl rings are skewed from planarity, the π-orbital overlap is diminished, and then conduction is decreased.[56]

The solubility of unsubstituted **4** is moderately low in most organic solvents; therefore, it was necessary to place n-alkyl side chain moieties on the phenylene ethynylene oligomers when there are more than three phenyl units. Although a long alkyl chain is important to retain solubility of a molecular wire in common organic solvents, it could sterically retard self-assembly or inhibit formation of a well-ordered and densely packed monolayer

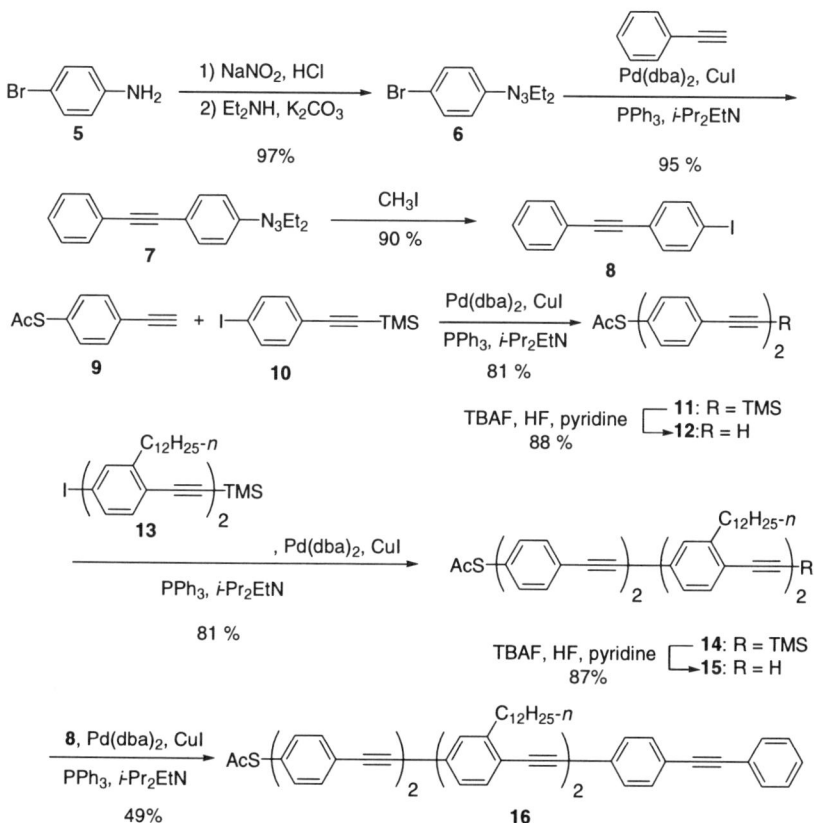

Scheme 3.21   Synthesis of molecular wire **16**.   dba = trans,trans-dibenzylideneacetone; TBAF = tetrabutylammonium fluoride.

In order to make more soluble systems, we prepared molecular wire **16** by Pd/Cu-catalyzed coupling reactions of **8**, **12**, and **13** (Scheme 3.21). **8** was synthesized by the coupling reaction of **6**[57] with phenylacetylene followed by an iodination with CH$_3$I. **9**[54] was coupled with **10** to afford dimer **11**. Deprotection of the terminal alkyne with TBAF provided intermediate **12** that was coupled with **13**[58] to afford tetramer **14** which, upon deprotection and coupling with aryl iodide **8**, afforded **16**. **16** has dodecyl chains on the two central units that allow this system to be soluble in many organic solvents but the chains point in the opposite direction of thiol group that serves as a molecular alligator clip; therefore, it does not impede the formation of the SAM.[59]

The synthesis of wires with central conduction units and terminal conducting barrier units are shown in Scheme 3.22. These were prepared to study the effects of imbedding the molecular system in a mildly insulation terminal framework. Monolithiation on 4,4'-dibromobiphenyl (17) followed by treatment with iodoethane afforded 18 that was then converted to the alligator clip-bearing molecular scale wire 19 with one ethyl end group barrier. The two-barrier system 22 was synthesized by conversion of 4-bromo-4'-propylbiphenyl (20) to 4-allyl-4'-propylbiphenyl (21). Radical thioacetyl formation[60] afforded the thiol-protected molecular scale wire with an imbedded conductive portion that could be further converted to the alkylthiol 22.

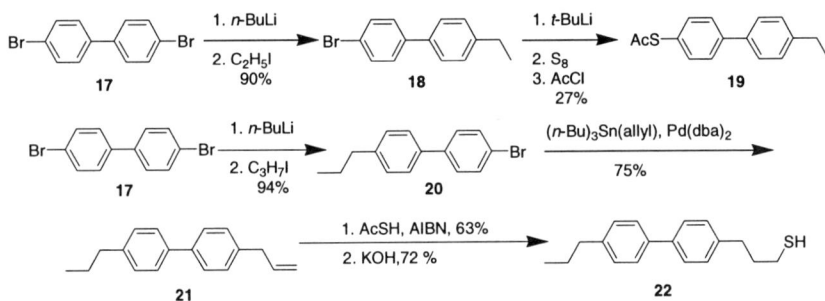

Scheme 3.22 Synthesis of two-barrier system 22. AIBN = 2,2'-azobisisobutronitrile.

3.3.1.2. Synthesis of Two-Terminal Oligo(phenylene ethynylene) Molecular Wires

Several syntheses of oligo(phenylene ethynylene)s with α,ω-dithioacetyl moieties, used as protected alligator clips, have been executed. These compounds will permit molecular scale wires to perform as interconnects between metallic probes (Scheme 3.23). Specifically, Pd/Cu-catalyzed cross couplings of 1,4-diiodobenzene (23) with two equivalents of alligator clip 9[54] afforded the rigid rod molecular scale wire 24. Due to the poor solubility of the deprotected dithiol made from 24, the more soluble diethyl-containing wire, 28, was synthesized. Iodination of 1,4-diethylbenzene followed by a series of Pd/Cu-catalyzed couplings led to the formation of 27. Removal of the acetyl protecting groups with NaOH in THF/H$_2$O and rapid workup produced soluble 28 with free thiol end groups. However, it is recommended that the end groups remain protected until the SAM formation step. In this way, oligomerization via oxidative disulfide formation is inhibited.

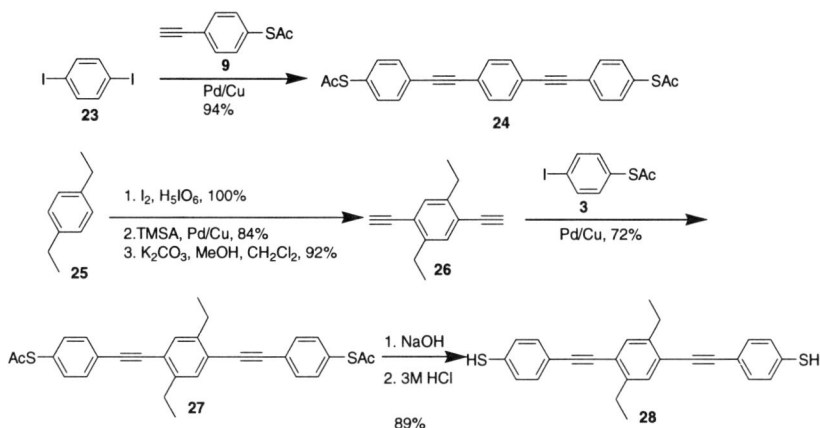

Scheme 3.23 Synthesis of two-dimensional molecular wires **24** and **28**.

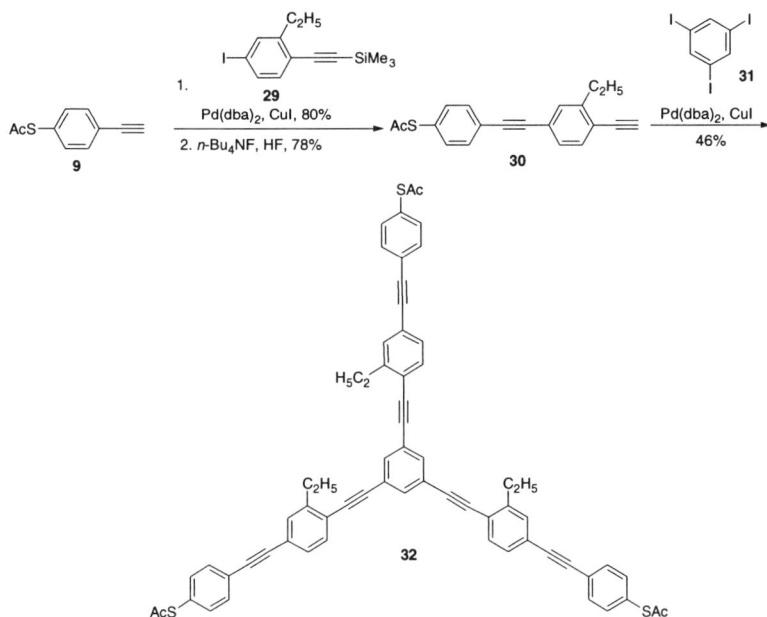

Scheme 3.24 Synthesis of three-terminal molecular wire **32**.

3.3.1.3. Syntheses of Three-Terminal Molecular Scale Wires

Three-terminal interconnects were prepared for branched interconnect locations (Scheme 3.24).[61] Alligator clip **9**[54] was cross-coupled with **29**[58] followed by subsequent deprotection of the terminal alkyne to afford **30**. Three equivalents of intermediate **30** were coupled with 1,3,5-triiodobenzene (**31**) to afford the desired **32**.

3.3.1.4. Molecular Wires with Internal Methylene and Ethylene Transport Barriers

Molecular scale wires with internal methylene and ethylene conduction barriers have been synthesized. These alkyl conduction barriers are positioned in the rigid rod phenylene ethynylene backbone to disrupt the electronic characteristics of the wires. It was hoped that the use of these methylene and ethylene conduction barriers in molecular wires might allow for the development of nanoscale molecular devices, i.e. resonant tunneling diodes (RTDs). Monolithiation of 1,4-dibromobenzene and subsequent quenching with p-bromobenzaldehyde gave 2,2-diarylethanol **34** that was subsequently converted to the diarylmethane **35** by reduction with NaBH$_4$.[62] **36**, with one central methylene conduction barrier, was easily synthesized from **35** by lithium-halogen exchange followed by quenching with sulfur and subsequent addition of acetyl chloride (Scheme 3.25).

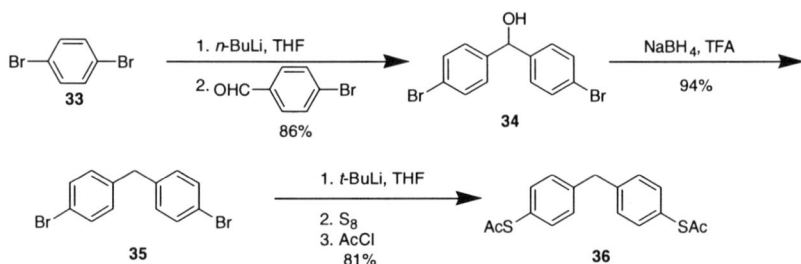

Scheme 3.25 Synthesis of two-terminal wire **36** with a methylene barrier.

Compounds **41** and **45** are molecular wires with a tunnel barrier to study the effects of asymmetric and symmetric barrier placement on the electronic properties. The synthesis of **41**, a 3-phenyl ring molecular scale wire with a methylene conduction barrier, is described in Scheme 3.26. 1,4-Diiodobenzene was monolithiated and quenched with 4-bromobenzaldehyde to form intermediate **38** followed by reduction of the secondary alcohol to form **39** in high yield. Coupling to the more labile aryl iodide gave compound **40**. Lithium-halogen exchange followed by quenching with sulfur and subsequent addition of acetyl chloride afforded the molecular scale wire **41** containing a methylene conduction barrier.

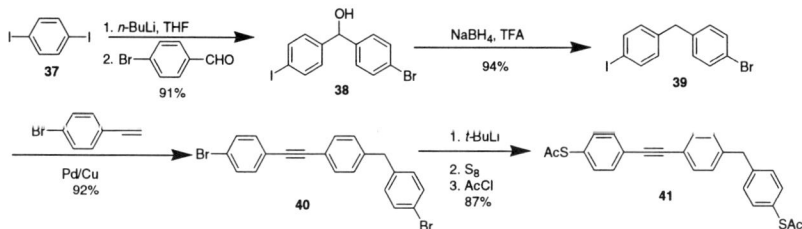

Scheme 3.26 Synthesis of molecular-scale wire **41** with one methylene barrier.

The synthesis of a symmetric molecular wire with a methylene conduction barrier is described in Scheme 3.27. Conversion of 4,4′-diaminodiphenylmethane (**42**) to the diiodide **44** through the formation of the bistriazene **43** proceeded in moderate yields. Intermediate **44** was coupled with the molecular alligator clip **9**[54] to afforded molecular wire **45** with the desired central methylene transport barrier.[61]

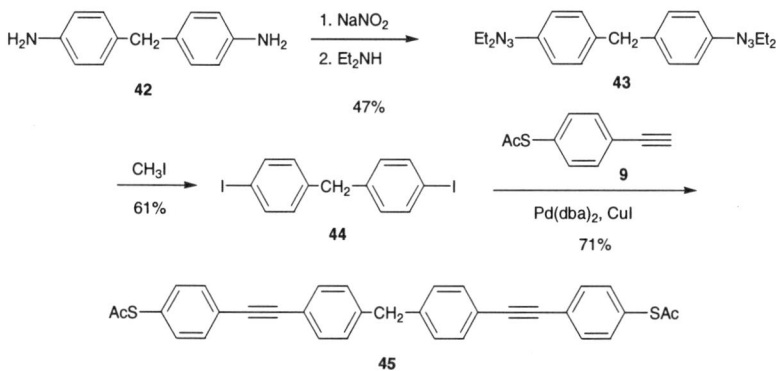

Scheme 3.27 Synthesis of two-terminal wire **45** with one methylene barrier.

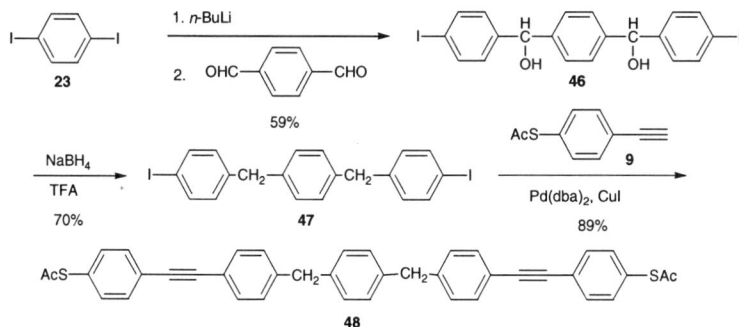

Scheme 3.28 Synthesis of two-terminal wire **48** with two methylene barriers.

Compound **48** is a more sophisticated device with two barriers that resembles a linear quantum dot or a RTD.[61] **46** was synthesized from terephthaldehyde and 1-iodo-4-lithiobenzene (Scheme 3.28). Reduction of the two hydroxyl moieties on **46** afforded **47** that was further coupled with two equivalents of alligator clip **9** to afford the desired **48**. This compound did indeed respond as a room temperature RTD when placed in the nanopore configuration.[63]

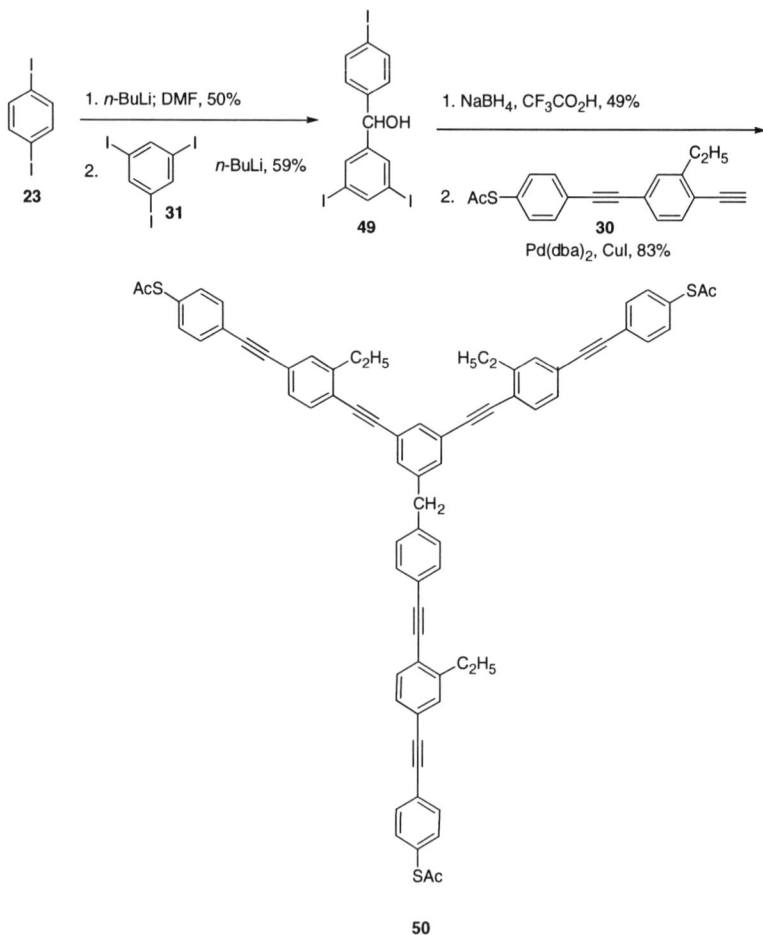

50

Scheme 3.29 Synthesis of three-terminal molecular wire **50** with one methylene barrier.

A three-terminal system with one barrier could be reminiscent of a molecular-sized field effect transistor (FET) or switch in which there is a source, drain and gate (Scheme 3.29).[61] 4-Iodobenzaldehyde was treated with 1,3-diiodo-5-lithiobenzene to afford the alcohol **49**. Reduction and Pd/Cu-catalyzed coupling with **30** yielded **50**, the desired three-terminal system with one methylene transport barrier.

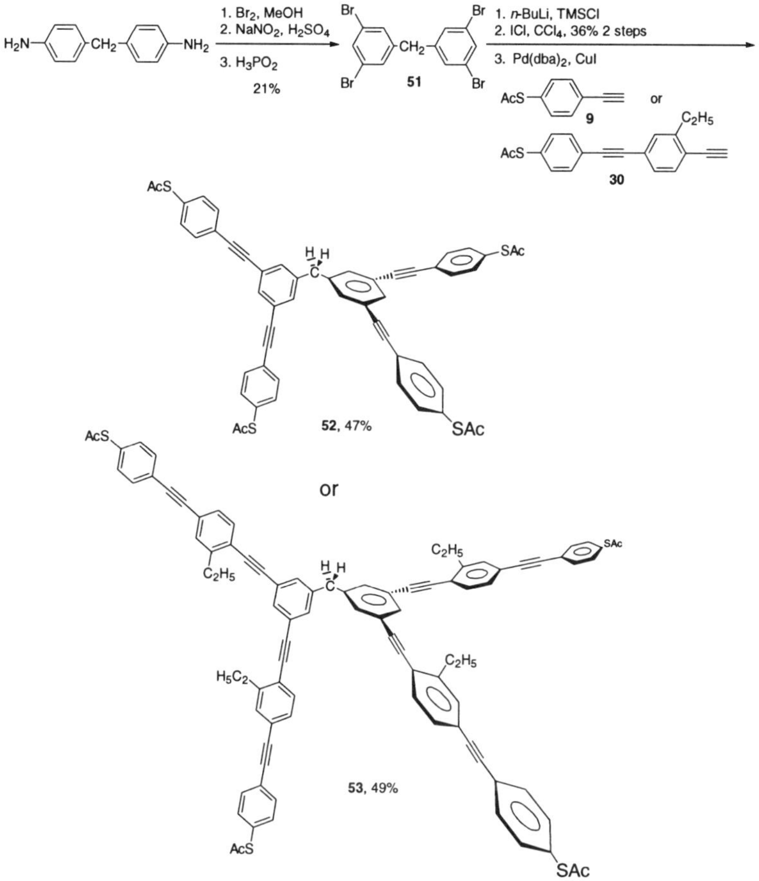

Scheme 3.30 Synthesis of four-terminal wires **52** and **53**, both with one methylene barrier.

Four terminal systems were synthesized according to Scheme 3.30. 4,4'-Diaminodiphenylmethane was treated with bromine followed by removal of the amino groups to afford 3,3',5,5'-tetrabromodiphenylmethane (**51**). Compound **51** proved to be too unreactive toward Pd/Cu-coupling; therefore, conversion of the bromides to the iodides was necessary. Lithium-halogen exchanges on **51** followed by quenching with molecular iodine resulted in mono-iodination on each ring. Complete halogen exchange reaction on **51** was achieved via conversion to the tetra(trimethylsilyl) system by addition of n-BuLi and chlorotrimethylsilane followed by treatment with iodine monochloride. The Pd/Cu-catalyzed coupling reaction of **9** or **30** with the tetraiodide intermediate afforded the four terminal systems **52** and **53**, respectively. Compounds **52** and **53** could be viewed as molecular logic devices as described previously.[61]

The synthesis of two four-terminus systems with two methylene conduction barriers is shown in Scheme 3.31. The dibromoxylene was oxidized and converted to the di(acid chloride) **54**. Friedel-Crafts acylation of **54** with bromobenzene was sluggish and low yielding. However, the tetrabromobis(arylketone) **56** was conveniently prepared by treatment of **54** with 1-bromo-4-trimethylsilylbenzene.[64] The reduction of the tetrabromobis(arylketone) was successfully carried out using triethylsilane and trifluoromethanesulfonic acid.[65] Conversion of the bromides to the iodides was achieved by lithium-halogen exchange with tert-BuLi followed by quenching with iodine. Pd/Cu-couplings of **57** with the alligator clip **9** or **30** afforded the four terminal systems **58** and **59**, respectively.

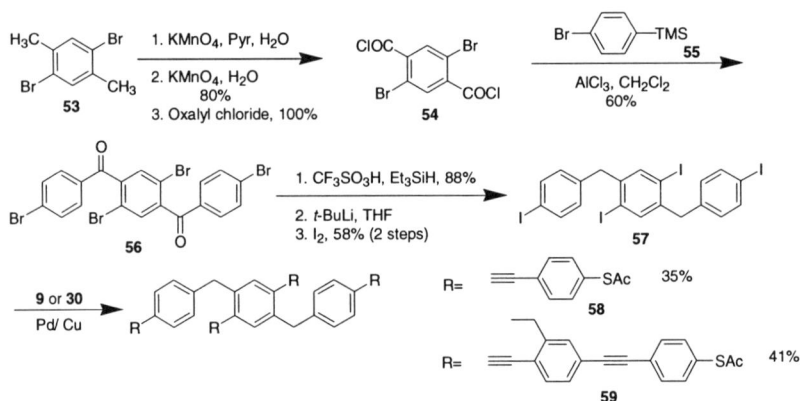

Scheme 3.31 Syntheses of molecular wires **58** and **59**, each containing two methylene barriers.

A two-terminal system with a lengthened resistive section was sought. Conversion of 1,2-(4,4'-dinitrodiphenyl)ethane (**60**) to the diiodide **62** followed by Pd/Cu-catalyzed coupling with alligator clip **9** afforded **63** with the desired central ethylene transport barrier (Scheme 3.32).

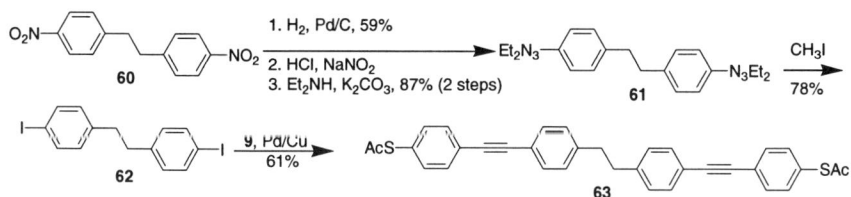

Scheme 3.32 Synthesis of molecular wire **63** with an ethylene barrier.

The syntheses of two ethylene-barrier containing systems, **66** and **67**, are described in Scheme 3.33. **64** was synthesized in three steps from 1,4-diiodobenzene. Hydrogenation of **64** was achieved over Pd/C in the presence of a small amount of HCl. Without an acid additive, no reduced products were isolated in a range of solvents and temperatures. The intermediate was then converted to **65** by treating with ICl in carbon tetrachloride. Pd/Cu couplings of **65** with two equivalents of the alligator clip **9** produced wire **66**. Alternatively, coupling with one equivalent of phenylacetylene followed by one equivalent of the alligator clip **9** afforded **67** with one thioacetyl terminal group.

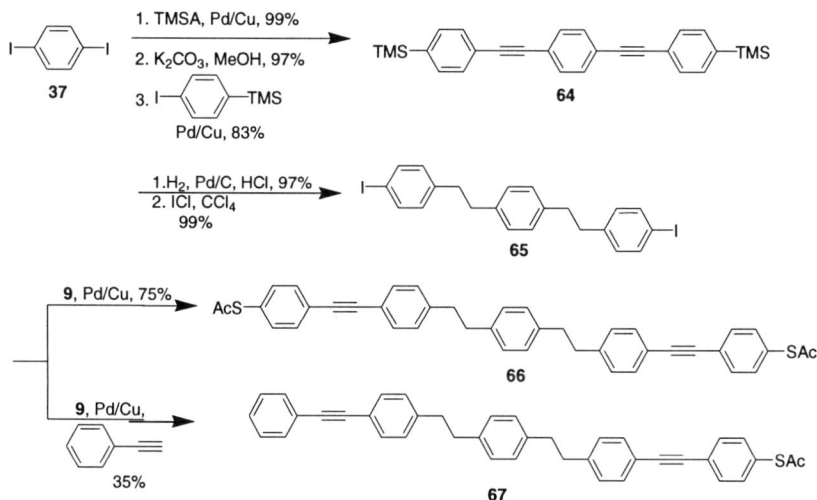

Scheme 3.33 Synthesis of one-terminal and two-terminal molecular wires **66** and **67**, both containing two ethylene barriers.

3.3.2. *Synthesis of Molecular Scale Devices with Heteroatomic Functionalities*

Described here are the syntheses of functionalized molecular scale devices which are designed to have nonlinear I(V) responses by adding heterofunctionalities to modulate the π-electron system. Some of the systems have been shown to possess NDR and memory properties. The majority of these molecules are based on functionalized oligo(phenylene ethynylene)s which are substituted with electron withdrawing and donating groups and are terminated with thioacetyl alligator clips.[54,58]

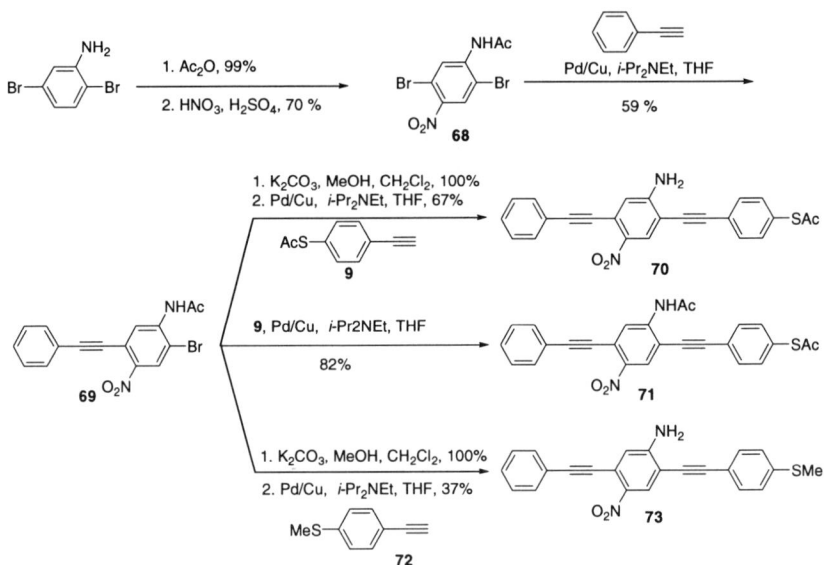

Scheme 3.34　Synthesis of compounds **70**, **71**, and **73** for cyclic voltammetry experiments.

The synthesis of a molecular scale device with amino and nitro moieties is described in Scheme 3.34. The formation of 2,5-dibromo-4-nitroacetanilide (**68**) proceeded according to a literature procedure.[64b,66] Caution must be used during the synthesis of **68** due to the possibility of multiple nitrations on the phenyl ring which could generate polynitrated compounds; on one occasion the compound exploded violently upon drying.[64b] The Pd/Cu-catalyzed coupling of phenylacetylene to the substituted dibromobenzene gave a moderate yield of the product due to the expected mixture of the mono and dicoupled products. The coupling of **68** was expected to proceed faster at the bromide ortho to the nitro (electron withdrawing) moiety since it is more active toward the electron rich late-transition metal catalyst system. X-ray analysis confirmed the assigned regiochemistry. The acetyl-protecting group was removed during the deprotection of the terminal alkynes in the presence of K_2CO_3 and MeOH. The electron withdrawing ability of the nitro moiety allowed for the removal of the acetyl-protecting group under such mild conditions. Finally, intermediate **69** was coupled by Pd/Cu-catalysis to alligator clip **9**[54] to afford molecular scale device **70**. An additional method for the synthesis of **70** has been developed. Intermediate **69** was coupled with TMSA, then deprotection of the terminal acetylene and the amine with K_2CO_3, and finally coupling with **3** afforded **70** in slightly lower yields than described in Scheme 3.34. The dipole moment of the interior phenyl ring in **70**, which is directed away from the thioacetyl group, was calculated to be 5.8 Debye.[56]

Compound **71** differs from **70** in that it possesses an acetamide rather than an amine moiety. The Pd/Cu-catalyzed coupling reaction to form **71** proceeded at a faster rate than the coupling to the amine/nitro compound due to the diminished electron donating potential of the acetamide allowing for faster Pd oxidative addition across the aryl bromide bond. The overall net dipole moment of this compound has been calculated to be 2.7 Debye, substantially lower than that for **70**.[56]

The cyclic voltammetry characteristics of the nitroaniline-containing molecular scale devices were determined to help elucidate the transport mechanism. It was therefore necessary to synthesize thioether **73** that is more stable to hydrolysis and subsequent oxidation than the thioacetate-terminated system (Scheme 3.34). For the synthesis of thioether **73**, intermediate **69** was deprotected and Pd/Cu-catalyzed coupled to **72** to form thioether terminated **73**. This compound was subjected to cyclic voltammetry that confirmed that the compound was being reduced at -1.7 V and again reduced at -2.3 V (Ag/AgNO$_3$ reference electrode, 1.0 M n-tetrabutylammonium tetrafluoroborate in DMF at a scan rate of 100 mV/sec). Of course, there can be no correlation of absolute reduction potentials between the solution-phase and SAM experiments since the environments are grossly different. However, that **73** could undergo a reversible 2-electron reduction was useful in the development of a hypothesis of a mechanism of the transport effect.[56]

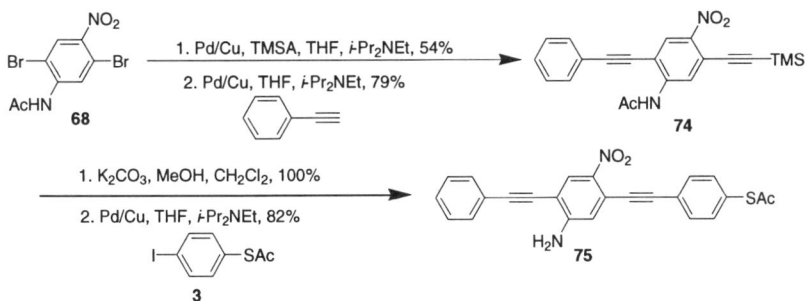

Scheme 3.35 Synthesis of molecular device **75**.

To determine the effect of the direction, if any, of the dipole moment on I(V) properties, compound **75** was synthesized, according to Scheme 3.35. It possesses a dipole that is directed toward the thioacetyl terminus, a direction opposite that of the dipole in **70**. With a deficient amount of TMSA, the coupling with intermediate **68** proceeded at the more labile bromide alpha to the nitro group (vide supra). Subsequent coupling to phenylacetylene provided **74**. Deprotection of the amine and the terminal alkyne, followed by coupling to **3** afforded **75**.

To determine the effects of an electron withdrawing or donating moiety on the electrical properties of these compounds, materials with solely an amine, nitro, or acetamide moiety have been synthesized. 2,5-Dibromonitrobenzene was coupled with TMSA at the more reactive bromide, α to the nitro moiety, followed by coupling with phenylacetylene. Deprotection of the terminal alkyne afforded intermediate **77**. Coupling of **77** with **3** afforded product **78** (Scheme 3.36).

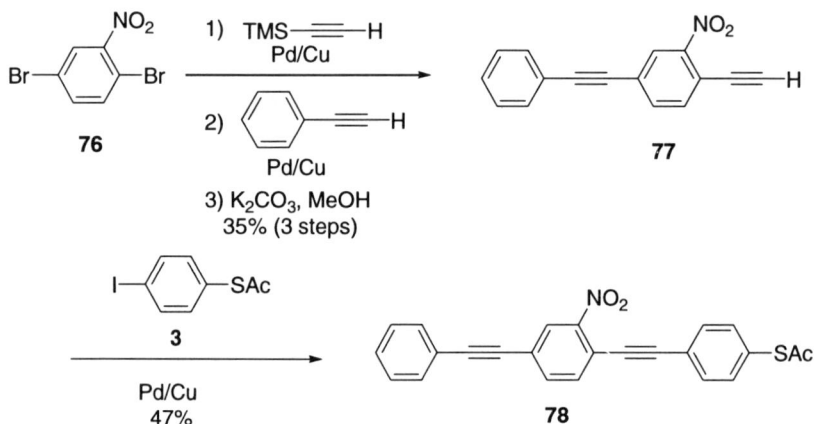

Scheme 3.36 Synthesis of mononitro molecular device **78**.

A system which possessed a amino moiety was synthesized according to Scheme 3.37. First, phenylacetylene and TMSA were sequentially coupled to **79** to form **80**. The deprotection of compound **80** with 3 M HCl afforded two compounds: the desired amine product and the amine cyclized bicyclic indole product.[67] The separation of these compounds was not attempted due to similar retention times on silica gel in most eluents. The terminal alkyne was revealed using K_2CO_3 and MeOH followed by Pd/Cu-catalyzed coupling to **3** to form **81** that, at this stage, could be separated from the by-products. The sequence of couplings to the bromo moieties on **79** was inferred based upon the electron donation of the acetamide; however, no crystallographic confirmation of the regioselectivity was obtained.

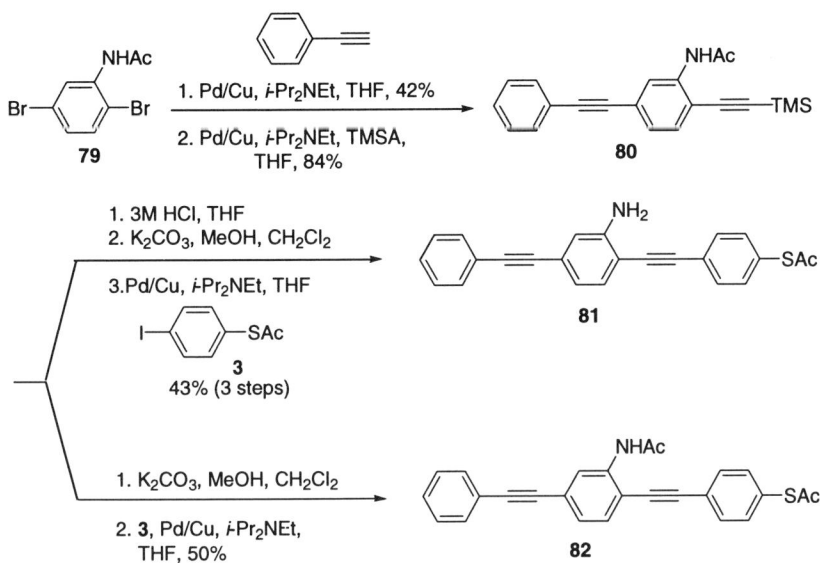

Scheme 3.37 Synthesis of monoamino compound **81** and molecular wire **82**.

Similar to **81**, the acetamide adduct **82** was synthesized according to Scheme 3.37. In this case, the deprotection of the terminal alkyne with K_2CO_3 and MeOH did not remove the acetyl-protecting group.

A two terminal molecular scale device **83** that is similar to compound **70** has been synthesized according to Scheme 3.38 although this bears α,ω-alligator clips.

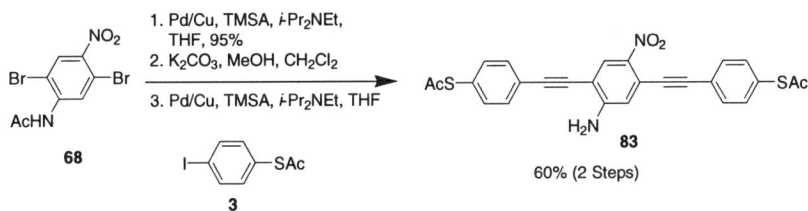

Scheme 3.38 Synthesis of two-terminal molecular device **83**.

To study the effects of other alligator clips on the impedance of molecular/metal junctions,[5a] compounds with isonitrile end groups were synthesized. The nitroaniline with an isonitrile terminus, **86**, was synthesized according to Scheme 3.39. The amine moiety in intermediate **69** was unmasked with K_2CO_3 and MeOH followed by Pd/Cu-catalyzed cross coupling with the formanilide-bearing end-group, **84**, to afford compound **85**. Although

85 had limited solubility, it was dehydrated in the presence of $(C_6H_5)_3P$ and Et_3N to afford the isonitrile **86**.[68]

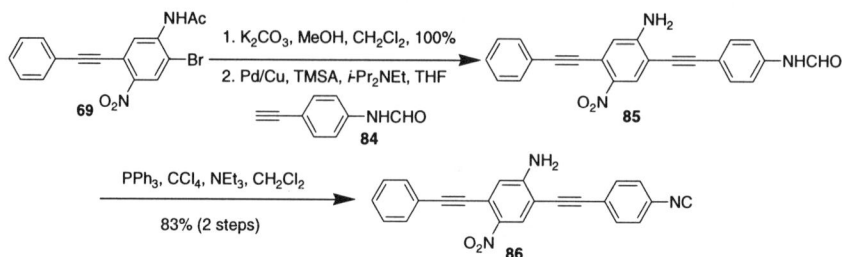

Scheme 3.39 Synthesis of molecular device **86** with an isonitrile alligator clip.

Currently, these molecular systems are studies as SAMs on a metal surface. An additional method of preparing ordered monolayers of molecular devices is the use of Langmuir-Blodgett (LB) films.[69] Therefore, a compound with hydrophilic and hydrophobic subunits with the central nitroaniline core similar to **70** was synthesized as in Scheme 3.40.[69] n-Hexylbenzene was easily brominated on neutral alumina[70] and coupled to TMSA followed by silyl removal and coupling to the nitroacetanilide core intermediate, **68**, to afford **88**. The methyl ester, intermediate **90**, was synthesized by the coupling of methyl 4-ethynylbenzoate (**89**) to **88**. The amine was unmasked and the methyl ester was saponified with lithium hydroxide to afford molecular scale device **91**.[71] Compound **91** is suitable for the formation of a LB film due to its hydrophilic carboxylic acid end-group and the hydrophobic n-hexyl end-group.

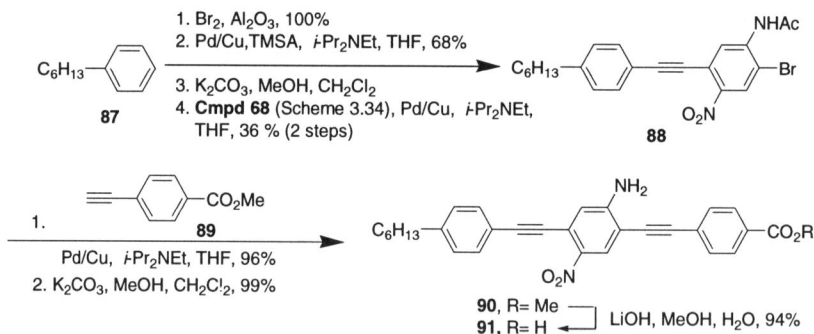

Scheme 3.40 Synthesis of compound **91**.

Other compounds with substituted biphenyl and bipyridyl core units have been sought (Schemes 3.41 and 3.42). 2,2´-Dinitrobiphenyl (**92**) was brominated at the 4 and 4´-positions on the biphenyl core using bromine, silver acetate, and acid.[72] The brominated biphenyl was coupled to TMSA to afford **93** that was then mono-reduced to the nitroamine **94** in the presence of iron and acetic acid.[73] Finally, the terminal alkynes were revealed and coupled to two equivalents of alligator clip **3** to afford compound **95**.

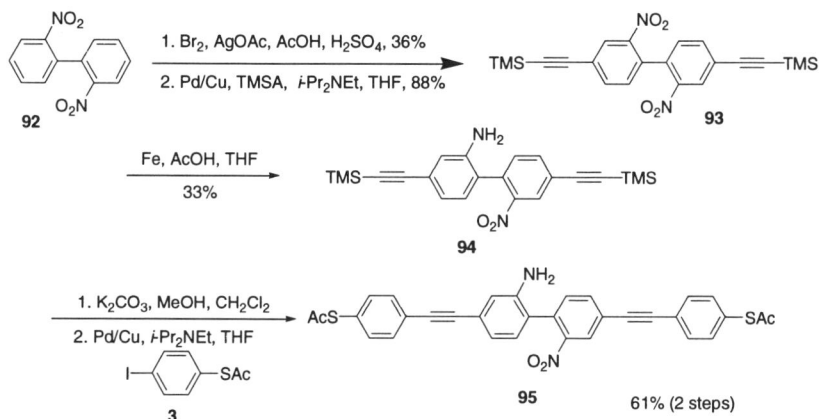

Scheme 3.41 Synthesis of nitro-amine biphenyl compound **95**.

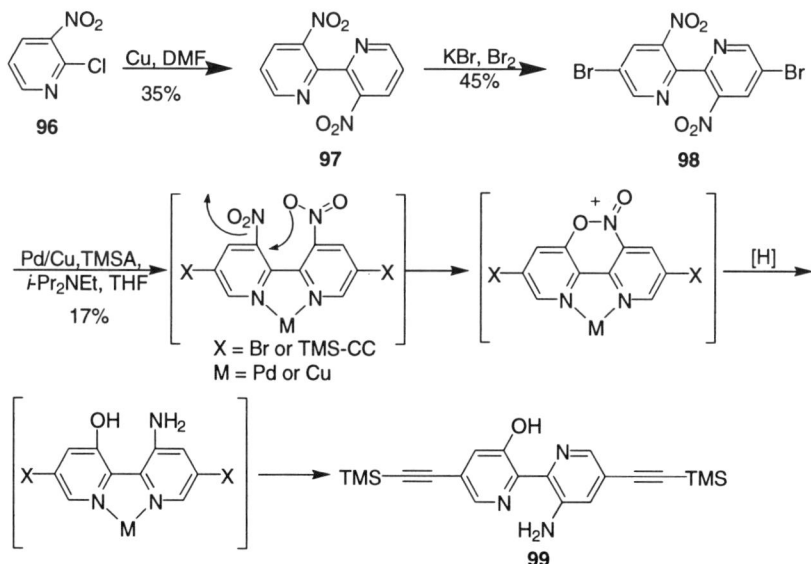

Scheme 3.42 Synthesis of bipyridyl compound **99**.

A similar compound with a bipyridyl central core was sought according to Scheme 3.42. In this manner, a greater degree of planarity could be achieved due to reduced interactions in the absence of 2- and 2'-steric interactions. To that end, 2-chloro-3-nitropyridine was homocoupled in the presence of copper/bronze and dimethylformamide.[74] The bipyridine ring system was brominated at the 5-and 5′-position under harsh conditions[75] (due to its electrophilicity) to afford intermediate **98** that was then coupled with two equivalents of TMSA. These coupling conditions unfortunately afforded the hydroxyamine and a very small amount of the dinitro-coupled product. The electron deficient **98** presumably underwent nitro loss and Pd-catalyzed reduction by the hydridopalladium species that are present in the coupling catalytic cycle to afford the undesired **99** (Scheme 3.42).[75b]

3.3.3. Porphyrin Containing Molecular Scale Wires

Initial efforts directed toward the porphyrin targets involved the preparation of dipyrromethane or aryl-substituted-dipyrromethanes with the intent of subsequent Pd/Cu-catalyzed coupling[76] to the aryl halides for preparation of the final compounds.[77] The porphyrin syntheses are shown in Scheme 3.43.

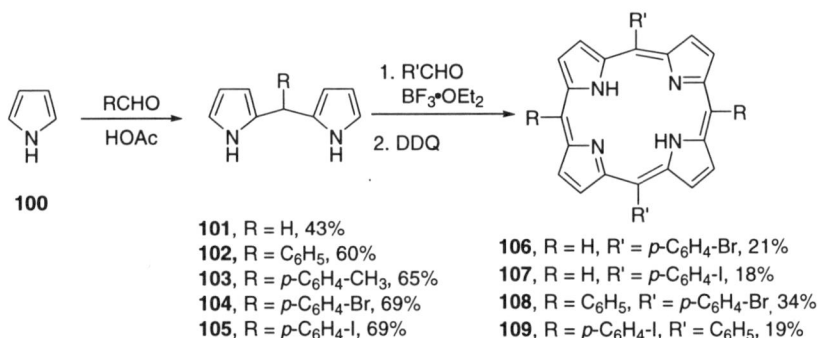

100

101, R = H, 43%
102, R = C$_6$H$_5$, 60%
103, R = p-C$_6$H$_4$-CH$_3$, 65%
104, R = p-C$_6$H$_4$-Br, 69%
105, R = p-C$_6$H$_4$-I, 69%

106, R = H, R' = p-C$_6$H$_4$-Br, 21%
107, R = H, R' = p-C$_6$H$_4$-I, 18%
108, R = C$_6$H$_5$, R' = p-C$_6$H$_4$-Br, 34%
109, R = p-C$_6$H$_4$-I, R' = C$_6$H$_5$, 19%

Scheme 3.43 Synthesis of various porphyrin compounds.

The dipyrromethanes could be prepared in reasonable yield, and further condensed with the complementary benzaldehyde component to generate the trans-(halophenyl)porphyrins.[78] Unfortunately, further attempts to elaborate the halogenated positions via Pd/Cu-catalyzed cross coupling or lithium-halogen exchange and subsequent conversion directly to thioacetyl moieties (excess BuLi, sequential quenching with S$_8$ and AcCl)[54] were unsuccessful; all reactions afforded only small amounts of mono-substituted products, if any. Additionally, complexing the porphyrin with zinc did not change the unsuccessful course of the subsequent derivatizations of **106-109**.

The strategy was therefore modified by preparing the aldehyde-bearing protected thiol using Pd/Cu-catalyzed coupling of 4-iodobenzaldehyde with TMSA, subsequent deprotection, and another Pd/Cu-catalyzed coupling with 4-iodo-1-thioacetylbenzene (**8**) to afford aldehyde **110** (Scheme 3.44).[79] Protected thiol **110** was then condensed with the substituted dipyrromethanes (**102-105**) and oxidized to form porphyrins **111-114**, respectively (Scheme 3.44). Likewise, **110** was condensed with pyrrole to form **115** and then further condensed with benzaldehyde and oxidized to form **111**. Accordingly, no further functionalization of the porphyrin was needed. Furthermore, the thioacetyl moieties did not inhibit the reaction neither were they affected to a significant extent; the yields were similar to those obtained in reactions that did not have these thioacetyl functionalities. In a less controlled manner, the three component system involving pyrrole, benzaldehyde, and **110** could be used to prepare **111** in 8 % yield after oxidation with p-chloranil. Similarly, the tetra(alligator clip) substituted system **116** could be prepared from **110** and pyrrole (Scheme 3.45).[76-78]

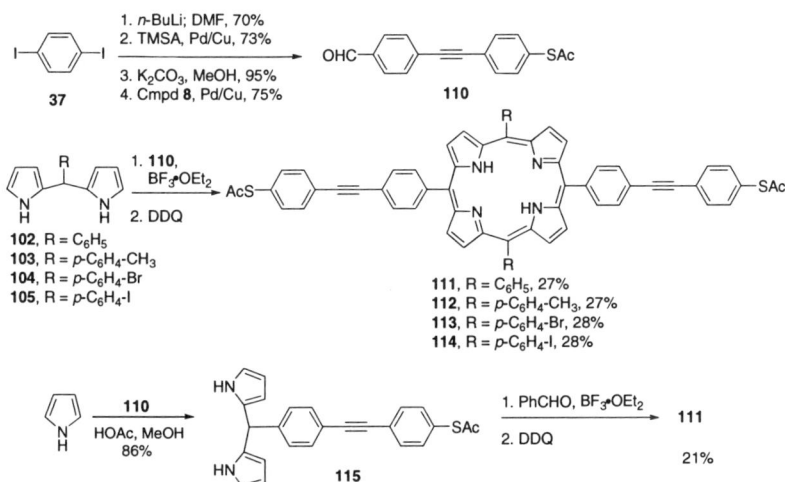

Scheme 3.44 Synthesis of porphyrin intermediate **110**, and porphyrins **111**, **112**, **113**, and **114**.

Finally, we have demonstrated efficient removal of the acetyl groups in **111** using NH$_4$OH.[23a] Metal incorporation into **111**, specifically Zn (91%), Cu (95%), and Co (90%), using the corresponding hydrated metal acetates, followed by NH$_4$OH-promoted thiol generation,[23a] proceeded without metal loss as indicated by ^{1}H NMR analysis.

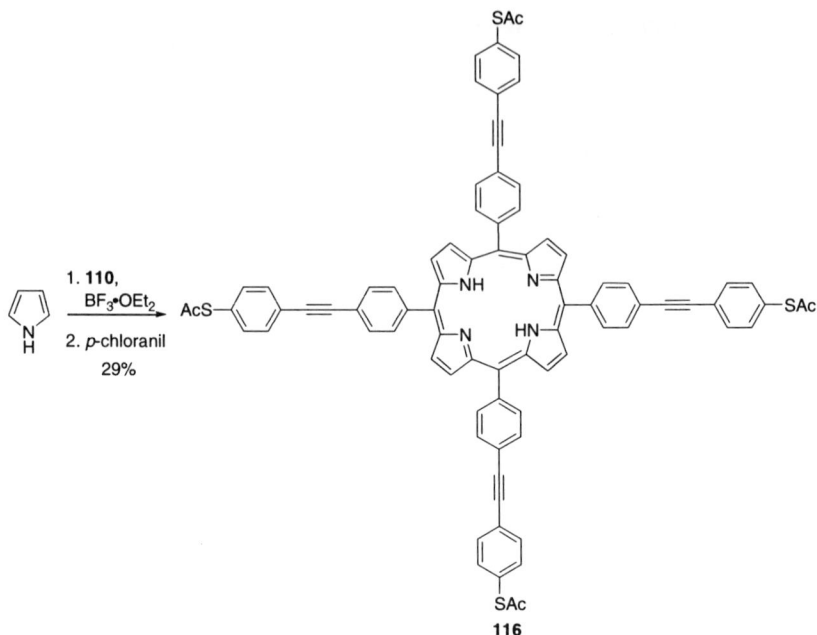

Scheme 3.45 Synthesis of four-terminal porphyrin **116**.

3.3.4. *Synthesis of Dipole-Possessing Molecular Wire SAMs to Control Schottky Barriers in Organic Electronic Devices*

Concurrent with our efforts to build molecules for SAMs to be used for molecular electronic devices, we are considering compounds that would form SAMs at metal interfaces in organic polymer-based LEDs. Similar issues that affect the efficiency of the metal's Fermi level overlap with the molecule's LUMO in molecular electronics[5a] will affect the electron injection at LED interfaces. Therefore, we are currently synthesizing molecules to act as SAM interfaces between the metal contacts and the organic substrates in LEDs. By tailoring the Schottky barrier of the metal/organic interface, we are hoping to improve the efficiency of the LEDs. The Cu/SAM injection of holes at low voltage could also improve ohmic contact.

We envisioned conjugated phenylene-ethynylene compounds that possess electron deficient units or electron rich units to be good candidates for lowering or raising the LUMO energies, respectively, as needed for the electron or hole injecting interfaces. Again, these compounds need alligator clips to provide the SAM formation.

Compounds **118** and **120** were synthesized by Pd/Cu-catalyzed cross couplings reactions (Scheme 3.46). Surprisingly, each of these compounds was extremely difficult to separate by column chromatography and recrystallization. They were finally purified by multiple cold hexanes washes.

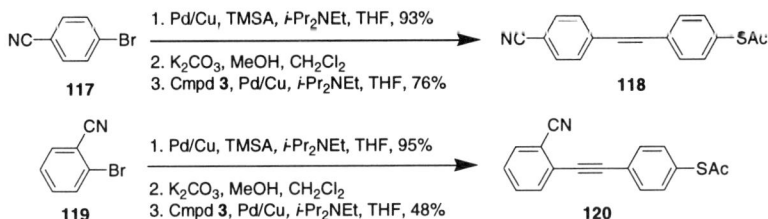

Scheme 3.46 Synthesis of cyano-containing systems **118** and **120**.

Likewise, compounds **122** and **124** were synthesized (Scheme 3.47). Again, these compounds were produced in a straightforward fashion by Pd/Cu-catalyzed cross couplings.

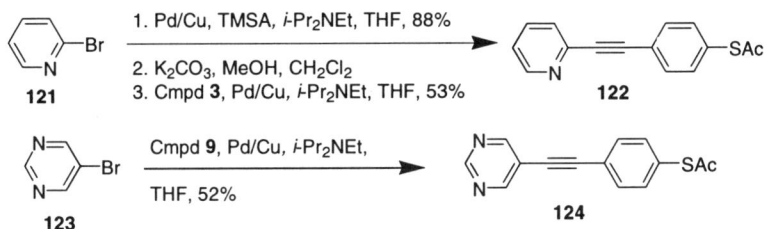

Scheme 3.47 Syntheses of pyridine system **122** and pyrimidine system **124**.

A three-aryl system with a pyridine interior was synthesized for the LED interfaces (Scheme 3.48). 2,5-Dibromopyridine was coupled to TMSA followed by phenylacetylene under Pd/Cu-catalyzed conditions. The first coupling reaction occurred at the more labile bromide at the 2-position in the pyridine ring. The alkyne in intermediate **126** was unmasked and then was coupled to alligator clip **3** to form the desired **127**.

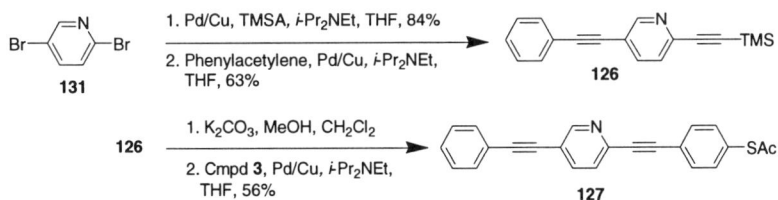

Scheme 3.48 Synthesis of pyrdine system **127**.

To decrease to the Schottky barrier for electron injection in LEDs, compounds with electron donating moieties and carboxylic acid alligator clips were synthesized for the formation of SAMs on aluminum oxide contacts.[80] In separate reactions, compounds **128** and **131** were coupled to **89** to afford methyl ester intermediates, **129** and **132**, respectively. The methyl ester moieties were saponified in the presence of lithium hydroxide to afford compounds **130** and **133** (Scheme 3.49).[71] These compounds are currently being tested for their ability to lower the electron injection barrier between the aluminum oxide contact and the organic polymer in organic LEDs and provide corroborating evidence for impedance lowering in molecular electronic devices.

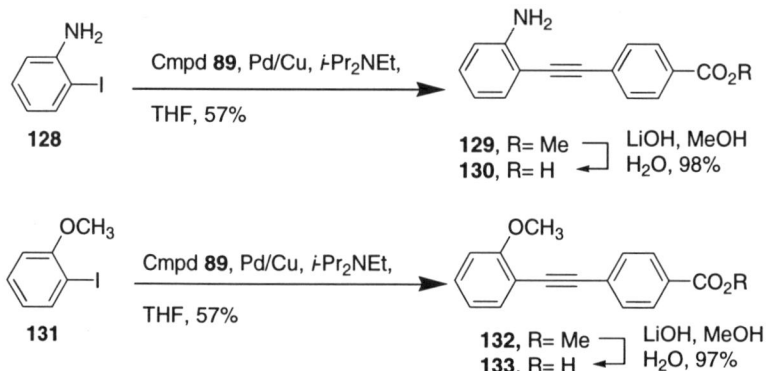

Scheme 3.49 Syntheses of compounds **130** and **133**.

3.3.5. *Experimental*

General Procedure for the Coupling of a Terminal Alkyne with an Aryl Halide Using the Palladium-Copper Cross-Coupling (Castro-Stephens/Sonogashira Protocol).[53] To an oven-dried round bottom flask equipped with a water cooled West condenser and magnetic stir bar or to a screw cap pressure tube with a magnetic stir bar were added the aryl halide, a palladium catalyst such as $((C_6H_5)_3P)_2PdCl_2$ (3-5 mol % per halide), and CuI (6-10 mol % per halide). $(C_6H_5)_3P$ was used in some reactions to keep the Pd in solution. The vessel was then sealed with a rubber septum (flask) or capped (tube) under a N_2 atmosphere. A solvent system of THF and/or benzene and/or CH_2Cl_2 was added depending on the solubility of the aryl halide. Then base, Et_3N or diisopropylethylamine, was added. Finally, the terminal alkyne (1-1.5 mol % per halide) was added and the reaction was heated until complete. Upon completion of the reaction, the reaction mixture was quenched with water, a saturated solution of NH_4Cl, or brine. The organic layer was diluted with CH_2Cl_2 or Et_2O and washed with water, a saturated solution of NH_4Cl, or brine (3×). The combined aqueous layers were extracted with CH_2Cl_2 or Et_2O (2×). The combined organic layers were dried over $MgSO_4$ and the solvent removed in vacuo to afford the crude product that was purified by column chromatography (silica gel). Eluents and other slight modifications are described below for each material.

General Procedure for the Iodination of Triazenes.[81] To an oven-dried screw cap tube was added the corresponding triazene and CH$_3$I. The mixture was degassed by slowly bubbling N$_2$ for more than 15 min. After flushing with N$_2$, the tube was capped and heated at 120 °C overnight. The reaction mixture was cooled and diluted with hexane. The mixture was passed through a plug of silica gel. After evaporation of the solvent in vacuo, purified product was obtained by chromatography. Eluents and other slight modifications are described below for each material.

General Procedure for the Deprotection of Trimethylsilyl-Protected Alkynes. (Method A) The silylated alkyne was dissolved in MeOH and often a co-solvent, and K$_2$CO$_3$ was added. The mixture was stirred at room temperature before being poured into water. The solution was extracted with ether or EtOAc and washed with brine. After drying over MgSO$_4$, the solvent was evaporated in vacuo to afford the products that generally required no purification. (Method B) The silylated alkyne was dissolved in pyridine in a plastic vessel. A mixed solution of 49% HF and 1.0 M tetrabutylammonium fluoride (TBAF) in THF was added at room temperature. The solution was stirred for 15 min and quenched with silica gel. The mixture was poured into water and extracted with ether. The extract was washed with brine and dried over MgSO$_4$. After filtration the solvent was evaporated in vacuo. The crude product was purified by a flash chromatography on silica gel. Eluents and other slight modifications are described below for each material.

General Procedure for the Conversion of Aryl Halides to Arylthioacetates. To tert-BuLi (2 equiv per halide) in ether or THF at -78 °C was added a solution of the aryl halide in THF. After stirring for 40 min, sulfur powder was added as a solid or via cannula as a slurry in THF. The resulting green slurry was stirred for 1 h and then warmed to 0 °C. The mixture was re-cooled to -78 °C and acetyl chloride (1.2 equiv per halide) was added. The resultant yellow solution was allowed to warm to room temperature and stirred for 1 h before quenching with water. The mixture was extracted with ether (3×). The combined organic fractions were washed with water (2 ×) and dried over MgSO$_4$. Removal of solvents in vacuo followed by flash chromatography afforded the desired material. Eluents and other slight modifications are described below for each material.

1-Bromo-4-(trimethysilylethynyl)benzene.[54] See the general procedure for the Pd/Cu coupling reaction except that amine was added at 0 °C. The compounds used were 1-bromo-4-iodobenzene (2.83 g, 10.0 mmol), TMSA (1.47 mL, 10.4 mmol), ((C$_6$H$_5$)$_3$P)$_2$PdCl$_2$ (0.21 g, 0.30 mmol), CuI (0.11 g, 0.60 mmol), benzene (13 mL), and Et$_3$N (5.6 mL, 40 mmol). The mixture was stirred at room temperature for 10 h. Flash chromatography (silica gel, hexane) afforded 2.37 g (95 %) of the title compound as yellow oil with slight impurities. The compound was used for the next step without further purification. Spectral data were identical to that reported earlier.[54]

1-Ethynylphenyl-4-(trimethylsilylethynyl)benzene (2). See the general procedure for the Pd/Cu coupling reaction. The compounds used were CuI (78 mg, 0.41 mmol), $((C_6H_5)_3P)_2PdCl_2$ (0.14 g, 0.20 mmol), 1-bromo-4-(trimethysilylethynyl)benzene (1.0 g, 4.0 mmol), phenylacetylene (0.60 mL, 5.5 mmol), Et_3N (2.0 mL, 14 mmol), and benzene (2 mL) at 80 °C overnight. The resulting brown solid was eluted 2× through a 4 × 20-cm column of silica gel using hexanes as the eluent. The product was obtained as crystalline white solid (1.08 g, 99 %). TLC R_f=0.28 (hexanes). IR (KBr) 3053, 2957, 2897, 2153, 1602, 1509, 1441, 1406, 1249, 866, 844, 757, 692, 628, 550, 529 cm^{-1}. ^1H NMR (CDCl$_3$) δ 7.512 (m, 2 H), 7.441 (m, 4 H), 7.336 (m, 3 H), 0.253 (s, 9 H), ^{13}C NMR (CDCl$_3$) δ 131.87, 131.60, 131.37, 128.45, 128.36, 123.34, 122.99, 122.89, 104.64, 96.21, 91.28, 89.01, -0.07. LRMS calcd for $C_{19}H_{18}Si$: 274 m/e. Found 274 (M$^+$), 259 [(M-CH$_3$)$^+$], 202 [(M-C$_3$H$_{10}$Si)$^+$].

1-Ethynyl-4-(ethynylphenyl)benzene. See the general procedure for the deprotection of a trimethylsilyl-protected alkyne. The compounds used were **2** (0.94 g, 3.4 mmol), K_2CO_3 (1.9 g, 14 mmol), MeOH (2.5 mL), and CH_2Cl_2 (4 mL). The product was obtained as a pale yellow solid (0.63 g, 91 %). IR (KBr) 3278, 3079, 3062, 3053, 3033, 3017, 1602, 1500, 1440, 1406, 1265, 1249, 1181, 1111, 1101, 1070, 1025, 922, 842, 834, 759, 690, 666, 629, 548, 527, 460 cm^{-1}. ^1H NMR (CDCl$_3$) δ 7.515 (m, 2 H), 7.462 (m, 4 H), 7.341 (m, 3 H), 3.159 (s, 1 H). ^{13}C NMR (CDCl$_3$) δ 132.06, 131.64, 131.46, 128.52, 128.38, 123.79, 122.94, 121.86, 91.36, 88.82, 83.28, 78.85. LRMS calcd for $C_{16}H_{10}$: 202 m/e. Found 202 (M$^+$), 176 [(M-C$_2$H$_2$)$^+$], 150 [(M-2C$_2$H$_2$)$^+$], 101 [(M-C$_8$H$_5$)$^+$].

4,4′-Di(ethynylphenyl)-1-(thioacetyl)benzene (4). See the general procedure for the Pd/Cu coupling reaction. The compounds used were CuI (0.042 g, 0.22 mmol), bis(dibenzylideneacetone)palladium(0) (0..063 g, 0.11 mmol), $(C_6H_5)_3P$ (0.115 g, 0.44 mmol), **3** (0.64 g, 2.3 mmol) 1-ethynyl-4-(ethynylphenyl)benzene (0.44 g, 2.2 mmol), diisopropylethylamine (1.7 mL, 10.0 mmol), and THF (10 mL) at 50° C for 3 h. The residue purified by flash liquid chromatography using silica gel (1:1 hexanes:CH$_2$Cl$_2$) yielding 0.57 g (74%) of the titled compound. IR (KBr) 3435.9, 3138.5, 2215.4, 1697.4, 1656.4, 1507.7, 1379.5, 1353.8, 1128.2, 1107.7, 1015.4, 943.6, 838.6, 828.1, 759.0, 756.7, 692.0, 620.5 cm^{-1}. ^1H NMR (300 MHz, C$_6$D$_6$) δ 7.54-7.50 (m, 2 H), 7.39 (d, J=8.5 Hz, 2 H), 7.34 (d, J=2 Hz, 3 H), 7.24 (d, J=8.5 Hz, 2 H), 7.16 (br s, 1 H), 7.03-6.98 (m, 3 H), 1.81 (s, 3 H). ^{13}C NMR (400 MHz, C$_6$D$_6$) δ 190.94, 134.24, 132.01, 131.62, 131.58, 128.91, 128.35, 127.21, 126.96, 124.12, 123.60, 123.28, 122.93, 91.87, 91.01, 90.90, 89.52, 29.55. HRMS calcd for $C_{23}H_{16}SO$: 352.0922. Found 352.0921.

1-Diethyltriazenyl- 4-ethynylphenylbenzene (7). See the general procedure for the Pd/Cu coupling reaction. **6** (2.56 g, 10.0 mmol), phenylacetylene (1.21 mL, 11.0 mmol), bis(dibenzylideneacetone)palladium(0) (0.26 g, 0.280 mmol), CuI (0.21 g, 11.0 mmol), $(C_6H_5)_3P$ (0.83 g, 2.75 mmol), and diisopropylethylamine (7.65 mL, 44.0 mmol) were reacted in THF (10 mL) at room temperature for 2 d and 70 °C for 3 d. An additional portion of phenylacetylene (0.60 mL, 5.5 mmol) was added and the mixture was stirred at 70 °C for 1 d. The crude product was purified by flash chromatography on silica gel (hexane:ether 19:1) to afford desired product (2.64 g, 95%) as yellow oil. FTIR (neat) 2976, 2359, 2213, 1594, 1393, 1237, 1201, 1162, 1097, 841, 756, 690 cm^{-1}. ^{1}H NMR (CDCl$_3$) δ 7.51 (dd, J = 7.7, 1.7 Hz, 2 H), 7.48 (dt, J = 8.5, 1.6 Hz, 2 H), 7.38 (dt, J = 8.5, 1.6 Hz, 2 H), 7.36-7.26 (m, 3 H), 3.76 (q, J = 7.2 Hz, 2 H), 1.26 (br t, 3 H). ^{13}C NMR (CDCl$_3$) δ 151.1, 132.3, 131.5, 128.3, 128.0, 123.6, 120.4, 119.4, 90.1, 89.1. (Two carbons are missing due to the quadropolar effect of nitrogen.) HRMS calcd for $C_{18}H_{19}N_3$: 277.1579. Found: 277.1582.

1-(Ethynylphenyl)-4-iodobenzene (8). See the general procedure for the iodination of triazenes. **7** (2.51 g, 9.06 mmol) was stirred in CH$_3$I (10 mL) to afford **8** (2.46 g, 90%) as a white solid. The solid was recrystallized from ethanol to afford purified product (2.06 g, 75%) as white crystals. Mp 104-105 °C. FTIR (KBr) 1493, 1385, 1004, 821, 758, 750, 690 cm^{-1}. ^{1}H NMR (CDCl$_3$) δ 7.67 (dt, J = 8.5, 1.9 Hz, 2 H), 7.52-7.47 (m, 2 H), 7.36-7.30 (m, 3 H), 7.23 (dt, J = 8.5, 1.9 Hz, 2 H). ^{13}C NMR (CDCl$_3$) δ 137.5, 133.1, 131.6, 128.5, 128.4, 122.9, 122.8, 94.1, 90.8, 88.5. HRMS calcd for $C_{14}H_9I$: 303.9749. Found: 303.9738.

11. See the general procedure for the Pd/Cu coupling reaction. **10**[58] (528 mg, 3.0 mmol), **9**[54] (990 mg, 3.3 mmol), bis(dibenzylideneacetone)palladium(0) (73 mg, 0.080 mmol), CuI (63 mg, 0.33 mmol), $(C_6H_5)_3P$ (256 mg, 0.85 mmol), and diisopropylethylamine (2.29 mL, 13.2 mmol) were reacted in THF (10 mL) at room temperature for 2.5 d. The crude product was purified by flash chromatography on silica gel (hexane:CH$_2$Cl$_2$ 7:3) to afford desired product (841 mg, 81%) as white solid. Mp 114 °C. FTIR (KBr) 2150, 1694, 1504, 1384, 1248, 841 cm^{-1}. ^{1}H NMR (CDCl$_3$) δ 7.52 (d, J = 8.3 Hz, 2 H), 7.43 (s, 4 H), 7.38 (d, J = 8.3 Hz, 2 H), 2.42 (s, 3 H), 0.24(s,9H).^{13}C NMR (CDCl$_3$) δ 193.2, 134.2, 131.9, 131.5, 124.2, 123.0, 104.6, 96.5, 90.7, 90.5, 30.3, -0.1. HRMS calcd for $C_{21}H_{20}OSSi$: 348.1004. Found: 348.1004.

12. See the general procedure for the deprotection of a trimethylsilyl-protected alkyne. **11** (230 mg, 0.65 mmol) was desilylated with TBAF (1.0 M solution in THF, 0.72 mL, 0.72 mmol) and 48% hydrofluoric acid (0.045 mL, 1.40 mmol) in pyridine (4.0 mL) for 10 min according to Method B. The crude product was purified by flash chromatography on silica gel (hexane:EtOAc 9:1) to afford desired product (157 mg, 88%) as a pale yellow solid. Mp 113-115 °C. FTIR (KBr) 3272, 1670, 1508, 1384, 1125, 833 cm^{-1}. ^{1}H NMR (CDCl$_3$) δ 7.53 (dt, J = 8.5, 1.9 Hz, 2 H), 7.46 (s, 4 H), 7.38 (dt, J = 8.5, 1.9 Hz, 1 H), 3.17

(s, 1 H), 2.42 (s, 3 H). ^{13}C NMR (CDCl$_3$) δ 193.4, 134.2, 132.2, 132.1, 131.5, 128.4, 124.2, 123.4, 122,2, 90.5, 90.4, 83.2, 79.1, 30.3. HRMS calcd for C$_{18}$H$_{12}$OS: 276.0609. Found: 276.0615.

14. See the general procedure for the Pd/Cu coupling reaction. **12** (319 mg, 0.43 mmol), **13**[58] (144 mg, 0.52 mmol), bis(dibenzylideneacetone)palladium(0) (13 mg, 0.023 mmol), CuI (8 mg, 0.042 mmol), (C$_6$H$_5$)$_3$P (33 mg, 0.11 mmol), and diisopropylethylamine (0.30 mL, 1.73 mmol) were stirred in THF (2.0 mL) at room temperature for 2 d. An additional portion of bis(dibenzylideneacetone)palladium(0) (13 mg, 0.023 mmol), CuI (8 mg, 0.042 mmol) and (C$_6$H$_5$)$_3$P (33 mg, 0.11 mmol) were then added. The mixture was stirred at room temperature another 2.5 d. The crude product was purified by flash chromatography on silica gel (hexane:EtOAc 8:2) to afford titled compound (307 mg, 81%) as a yellow solid. Mp 98-101 °C. FTIR (KBr) 2922, 2852, 2150, 1700, 1511, 1249, 1119, 862, 839 cm^{-1}. ^{1}H NMR (CDCl$_3$) δ 7.54 (dt, J = 8.4, 1.6 Hz, 2 H), 7.50 (s, 4 H), 7.45 (d, J = 7.9 Hz, 1 H), 7.42-7.37 (m, 4 H), 7.32 (dd, J = 7.9, 1.6, Hz, 1 H), 7.32 (d, J = 1.5 Hz, 1 H), 7.26 (dd, J = 7.9, 1.5 Hz, 1 H), 2.82 (t, J = 7.8, 2 H), 2.75 (t, J = 7.8, 2 H), 2.42 (s, 3 H), 1.70 (pent, J = 7.9 Hz, 2 H), 1.64 (pent, J = 8.0 Hz, 2 H), 1.45-1.13 (m, 36 H), 0.86 (t, J = 6.5 Hz, 6 H), 0.25 (s, 9 H). ^{13}C NMR (CDCl$_3$) δ 193.3, 145.7, 145.2, 1342., 132.4, 132.2, 132.1, 131.9, 131.7, 131.6, 128.9, 128.6, 128.0, 124.3, 123.3, 122.8, 122.6, 103.6, 99.7, 94.5, 91.5, 90.8, 90.6, 90.5, 89.5, 34.7, 34.6, 32.0, 30.6, 30.3, 29.7, 29.6, 29.4, 22.7, 14.2, -0.01. HRMS calcd for C$_{61}$H$_{76}$OSSi: 884.5386. Found: 884.5386.

15. See the general procedure for the deprotection of a trimethylsilyl-protected alkyne. **14** (119 mg, 0.13 mmol) was desilylated with TBAF (1.0 M solution in THF, 0.14 mL, 0.14 mmol) and 48% HF (0.009 mL, 0.29 mmol) in pyridine (1.5 mL) for 10 min according to Method B described above. The crude product was purified by flash chromatography on silica gel (hexane:EtOAc 19:1) to afford desired product (83 mg, 79%) as pale yellow solid. ^{1}H NMR (CDCl$_3$) δ 7.54 (dt, J = 8.4, 1.9 Hz, 2 H), 7.50 (s, 4 H), 7.46 (d, J = 8.0 Hz, 1 H), 7.43 (d, J = 8.0 Hz, 1 H), 7.39 (d, J = 1.6 Hz, 1 H), 7.39 (dt, J = 8.4, 1.9 Hz, 1 H), 7.34 (br s, 1 H), 7.33 (dd, J = 8.0, 1.6 Hz, 1 H), 7.27 (dd, J = 8.0, 1.6 Hz, 1 H), 3.32 (s, 1 H), 2.82 (t, J = 7.4 Hz, 2 H), 2.77 (t, J = 7.6 Hz, 2 H), 2.43 (s, 3 H), 1.69 (pent, J = 7.4 Hz, 2 H), 1.64 (pent, J = 7.6 Hz, 2 H), 1.46-1.14 (m, 36 H), 0.86 (t, J = 6.1 Hz, 3 H). ^{13}C NMR (CDCl$_3$) δ 193.3, 145.7, 145.2, 134.2, 132.9, 132.2, 132.1, 131.9, 131.6, 131.6, 128.9, 128.6, 128.3, 124.3, 123.6, 123.3, 122.9, 122.8, 122.7, 121.6, 94.4, 91.5, 90.7, 90.6, 90.5, 89.6, 82.1, 34.6, 34.3, 31.9, 31.6, 30.6, 30.5, 30.3, 29.7, 29.7, 29.6, 29.6, 29.6, 29.5, 29.4, 22.7, 14.1. LRMS calcd for C$_{58}$H$_{68}$OS: 812. Found: 812. (This compound is too unstable to afford a HRMS.)

16. See the general procedure for the Pd/Cu coupling reaction. **15** (464 mg, 0.57 mmol), **8** (207 mg, 0.68 mmol), bis(dibenzylideneacetone)palladium(0) (17 mg, 0.029 mmol), CuI (11 mg,

0.057 mmol), $(C_6H_5)_3P$ (38 mg, 0.145 mmol), and diisopropylethylamine (0.47 mL, 2.72 mmol) were stirred in THF (2.0 mL) at room temperature for 2 d. More bis(dibenzylideneacetone)palladium(0) (17 mg, 0.029 mmol), CuI (11 mg, 0.057 mmol) and $(C_6H_5)_3P$ (38 mg, 0.145 mmol) were then added. The mixture was stirred at room temperature another 4 d. The crude product was purified by recrystallization from hexane to afford desired product (369 mg, 66%) as yellow solid. Mp 124-125 °C. FTIR (KBr) 2922, 1709, 1511, 1384, 836 cm^{-1}. ^{1}H NMR (CDCl$_3$) δ 7.57-7.43 (m, 13 H), 7.42-7.30 (m, 10 H), 2.84 (t, J = 7.4 Hz, 4 H), 2.42 (s, 3 H), 1.70 (p, J = 7.4 Hz, 4 H), 1.47-1.14 (m, 36 H), 0.90-0.78 (m, 6 H). ^{13}C NMR (CDCl$_3$) δ 193.3, 145.2, 137.5, 134.2, 133.1, 132.2, 131.9, 131.8, 131.7, 131.6, 131.6, 131.4, 129.0, 128.8, 128.4, 128.4, 124.3, 123.3, 123.3, 123.2, 123.1, 122.9, 122.8, 122.8, 122.5, 94.8, 94.4, 91.6, 91.4, 90.8, 90.6, 90.1, 89.7, 89.2, 34.7, 32.0, 30.6, 30.3, 29.8, 29.7, 29.7, 29.6, 29.4, 22.8, 14.2. HRMS calcd for $C_{72}H_{76}OS$: 988.5613. Found: 988.5630.

4-Bromo-(4'-ethyl)biphenyl (18). To a solution of 4,4'-dibromobiphenyl **(17)** (6.24 g, 20.0 mmol) in THF (100 mL) at -78 °C was added n-BuLi (12.4 mL, 20.0 mmol, 1.61 M in hexane) dropwise. The yellow slurry was stirred for 1 h and iodoethane was added. The mixture was allowed to warm to room temperature and stirred for 20 h. The mixture was poured into water. The organic layer was separated and washed with water (2×) and brine (1×). The combined aqueous solution was extracted with ether (2×). The combined organic fractions were dried over MgSO$_4$. Removal of solvent followed by flash chromatography (silica gel, hexane) gave desired product as white solid (4.70 g, 90%). FTIR (neat) 2964, 2923, 2872, 1482, 1390, 1267, 1133, 1072, 1000, 815, 739 cm^{-1}. ^{1}H NMR (300 MHz, CDCl$_3$) δ 7.53 (d, J = 8.4 Hz, 2 H), 7.46 (d, J = 8.2 Hz, 2 H), 7.43 (d, J = 8.5 Hz, 2 H), 7.26 (d, J = 8.8 Hz, 2 H), 2.68 (q, J = 7.6 Hz, 2 H), 1.26 (t, J = 7.6 Hz, 3 H). ^{13}C NMR (75 MHz, CDCl$_3$) δ 143.92, 140.14, 137.40, 131.89, 128.63, 128.51, 126.93, 121.28, 28.62, 15.65. HRMS calcd for $C_{14}H_{13}Br$: 260.0201. Found: 260.0204.

4-Ethyl-4'-thioacetylbiphenyl (19). See the general procedure for the conversion of aryl halide to arylthioacetate. The compounds used were **18** (0.784 g, 3.00 mmol) in ether (10 mL), tert-BuLi (2.6 mL, 6.0 mmol, 2.30 M in pentane) in ether (10 mL), sulfur powder (0.16 g, 5.0 mmol) in ether (5 mL), and acetyl chloride (0.43 mL, 6.0 mmol). Gravity chromatography (silica gel, hexane:ether 9:1) afforded desired material as white solid (0.21 g, 27%). Mp 84-86 °C. FTIR (neat) 2964, 2923, 2872, 1703, 1482, 1395, 1354, 1123, 1097, 1005, 954, 815 cm^{-1}. ^{1}H NMR (300 MHz, CDCl$_3$) δ 7.60 (d, J = 8.1 Hz, 2 H), 7.50 (d, J = 8.1 Hz, 2 H), 7.44 (d, J = 8.1 Hz, 2 H), 7.27 (d, J = 8.0 Hz, 2 H), 2.68 (q, J = 7.6 Hz, 2 H), 2.43 (s, 3 H), 1.26 (t, J = 7.6 Hz, 3 H). ^{13}C NMR (75 MHz, CDCl$_3$) δ 194.30, 144.04, 142.40, 137.54, 134.77, 128.43, 127.80, 127.15, 126.39, 30.24, 28.57, 15.57. HRMS calcd for $C_{16}H_{16}OS$: 256.0922. Found: 256.0918.

4-Bromo-(4'-propyl)biphenyl (20). To a solution of **17** (9.36 g, 30.0 mmol) in THF (150 mL) at -78 °C was added n-BuLi (18.8 mL, 30.0 mmol, 1.60 M in hexane) dropwise. The yellow slurry was stirred for 1.5 h and iodopropane was added. The mixture was allowed to warm to room temperature and stirred for 5 h. The mixture was poured into water and extracted with ether (2×). The organic fractions were dried over $MgSO_4$. Removal of solvent followed by flash chromatography (silica gel, hexane) gave product as white solid (7.80 g, 94%). Mp 103-104 °C. FTIR (KBr) 2954, 2933, 2872, 1482, 1462, 1390, 1262, 1133, 1077, 1005, 826, 800, 739 cm^{-1}. ^{1}H NMR (300 MHz, CDCl$_3$) δ 7.53 (d, J = 8.4 Hz, 2 H), 7.46 (d, J = 7.9 Hz, 2 H), 7.43 (d, J = 8.4 Hz, 2 H), 7.24 (d, J = 8.1 Hz, 2 H), 2.62 (t, J = 7.6 Hz, 2 H), 1.67 (sext, J = 7.5 Hz, 2 H), 0.96 (t, J = 7.3 Hz, 3 H). ^{13}C NMR (75 MHz, CDCl$_3$) δ 142.37, 140.13, 137.38, 131.86, 129.09, 128.62, 126.81, 121.24, 37.76, 24.61, 13.96. HRMS calcd for $C_{15}H_{15}Br$: 274.0357. Found: 274.0350.

4-Allyl-4'-(propyl)biphenyl (21). A mixture of **20** (3.96 g, 14.4 mmol), tributylallyltin (4.97 g, 15.0 mmol), Pd(dba)$_2$ (0.248 g, 0.432 mmol), PPh$_3$ (0.453 g, 1.73 mmol), and BHT (four crystals) in toluene (20 mL) was heated to reflux for 21 h. The mixture was cooled to room temperature, filtered and concentrated in vacuo. The residue was diluted with ether and aqueous KF (8 mL, 3 M) was added. The mixture was stirred for 15 min and filtered through a pad of celite. The filtrate was washed with water (2×) and dried over $MgSO_4$. Removal of solvent in vacuo followed by flash chromatography (silica gel, hexane) gave product as white solid (2.55 g, 75 %). Mp 44-46 °C. FTIR (KBr) 3077, 3026, 2964, 2923, 2872, 1641, 1497, 1456, 1431, 1400, 1267, 1118, 1005, 995, 913, 831, 795, 739 cm^{-1}. ^{1}H NMR (300 MHz, CDCl$_3$) δ 7.52 (d, J = 8.0 Hz, 2 H), 7.49 (d, J = 8.0 Hz, 2 H), 7.25 (d, J = 8.0 Hz, 2 H), 7.24 (d, J = 8.0 Hz, 2 H), 6.07-5.94 (ddt, J = 17.0, 8.2, 6.7 Hz, , 1 H), 5.12 (dd, J = 17.0, 1.5 Hz, 1 H), 5.09 (dd, J = 8.2, 1.3 Hz, 1 H), 3.42 (d, J = 6.7 Hz, 2 H), 2.62 (t, J = 7.6 Hz, 2 H), 1.67 (sext, J = 7.5 Hz, 2 H), 0.97 (t, J = 7.3 Hz, 3 H). ^{13}C NMR (75 MHz, CDCl$_3$) δ 141.85, 139.37, 139.09, 138.77, 137.74, 129.33, 129.23, 127.35, 127.19, 116.22, 40.26, 38.09, 24.96, 14.30. HRMS calcd for $C_{18}H_{20}$: 236.1565. Found: 236.1564.

1-(4'-Propylbiphenyl)-3-thioacetylpropane. A mixture of **21** (0.90 g, 3.8 mmol), thioacetic acid (0.44 mL, 6.0 mmol) and AIBN (0.005 g) in benzene (5 mL) was heated to reflux overnight.[60] After cooling to room temperature, the mixture was poured into water, extracted with ether (2×) and the extracts were dried over $MgSO_4$. Removal of solvent followed by flash chromatography (silica gel, hexane:ether = 20/1) gave the title compound as white solid (0.75 g, 63%). FTIR (KBr) 2954, 2933, 2862, 1677, 1497, 1451, 1421, 1400, 1354, 1144, 1118, 954, 831, 790, 636 cm^{-1}. ^{1}H NMR (300 MHz, CDCl$_3$) δ 7.49 (d, J = 8.0 Hz, 2 H), 7.48 (d, J = 8.0 Hz, 2 H), 7.23 (d, J = 8.3 Hz, 2 H), 7.22 (d, J = 8.1 Hz, 2 H), 2.91 (t, J = 7.3 Hz, 2 H), 2.71 (t, J = 7.6 Hz,

2 H), 2.61 (t, J = 7.6 Hz, 2 H), 2.33 (s, 3 H), 1.92 (p, J = 7.5 Hz, 2 H), 1.67 (sext, J = 7.4 Hz, 2 H), 0.96 (t, J = 7.3 Hz, 3 H). ^{13}C NMR (100 MHz, CDCl$_3$) δ 195.55, 141.46, 139.75, 138.79, 138.20, 128.71, 128.67, 126.85, 126.66, 37.72, 34.50, 31.15, 30.71, 28.66, 24.62, 13.98. HRMS calcd for C$_{20}$H$_{24}$OS: 312.1548. Found: 312.1539.

1-(4′-Propylbiphenyl)-3-propanethiol (22). To a solution of 1-(4′-propylbiphenyl)-3-thioacetylpropane (0.50 g, 1.6 mmol) in ethanol (4 mL) was added water (4 mL) and KOH (0.45 g, 8.0 mmol). The mixture was heated to reflux for 15 min, then cooled to room temperature. The solution was acidified with 3 N HCl and extracted with ether (3×). The extracts were dried over MgSO$_4$. Removal of solvent in vacuo followed by flash chromatography (silica gel, hexane) gave desired product as white solid (0.31 g, 72%). Mp 32-33 °C. FTIR (KBr) 3026, 2954, 2923, 2862, 1497, 1456, 1400, 1374, 1344, 1292, 1256, 1185, 1118, 1005, 800, 739 cm^{-1}. ^{1}H NMR (300 MHz, CDCl$_3$) δ 7.50 (d, J = 8.1 Hz, 2 H), 7.48 (d, J = 8.1 Hz, 2 H), 7.23 (d, J = 8.5 Hz, 4 H), 2.75 (t, J = 7.5 Hz, 2 H), 2.61 (t, J = 7.6 Hz, 2 H), 2.56 (q, J = 7.4 Hz, 2 H), 1.96 (p, J = 7.3 Hz, 2 H), 1.67 (sext, J = 7.5 Hz, 2 H), 1.37 (t, J = 7.8 Hz, 2 H), 0.96 (t, J = 7.3 Hz, 3 H). ^{13}C NMR (100 MHz, CDCl$_3$) δ 141.69, 140.10, 138.98, 138.41, 128.92, 128.91, 127.04, 126.86, 37.76, 35.51, 34.04, 24.63, 24.09, 13.98. HRMS calcd for C$_{18}$H$_{22}$S: 270.1442. Found: 270.1437.

24. See the general procedure for the Pd/Cu coupling reaction. The compounds used were 1,4-diiodobenzene (0.165 g, 0.500 mmol), **9**[54] (0.247 g, 1.40 mmol), Pd(dba)$_2$ (0.029 g, 0.05 mmol), CuI (0.019 g, 0.10 mmol), PPh$_3$ (0.052 g, 0.20 mmol), THF (13 mL), and diisopropylethylamine (0.70 mL, 4.0 mmol). Flash chromatography (silica gel, hexane, then hexane:CH$_2$Cl$_2$ 1:1) afforded desired product as a light brown solid (0.20 g, 94%). Mp 199-120 °C. FTIR (KBr) 1692, 1590, 1513, 1385, 1354, 1118, 1092, 1010, 964, 826, 621 cm^{-1}. ^{1}H NMR (300 MHz, CDCl$_3$) δ 7.54 (d, J = 8.5 Hz, 4 H), 7.50 (s, 4 H), 7.39 (d, J = 8.5, 4 H), 2.43 (s, 6 H). ^{13}C NMR (75 MHz, CDCl$_3$) δ 193.40, 134.22, 132.16, 131.62, 128.30, 124.21, 122.99, 90.65, 90.57, 30.29. HRMS calcd for C$_{26}$H$_{18}$O$_2$S$_2$: 426.0748. Found: 426.0740.

1,4-Diethyl-2,5-diiodobenzene. A mixture of 1,4-diethylbenzene (2.43 g, 18.1 mmol), iodine (6.13 g, 24.1 mmol), periodic acid (2.74 g, 12.0 mmol), acetic acid (12 mL) water (2.4 mL) and concentrated sulfuric acid (0.4 mL) was heated to 95 °C for 1 d. The mixture was cooled to room temperature and poured into water. The mixture was neutralized carefully with saturated aqueous NaHCO$_3$. The precipitate was collected by filtration and re-dissolved in ether. The ether solution was washed with aqueous sodium thiosulfate (1×), water (1×), brine (1×) and dried over MgSO$_4$. After filtration, removal of solvent in vacuo gave the title compound as white solid (6.92 g, 99%). Mp 68-69 °C. FTIR (neat) 2964, 2933, 2862, 1462, 1380, 1349, 1318, 1046, 1036, 980, 882, 713, 667 cm^{-1}. ^{1}H NMR (300 MHz, CDCl$_3$) δ 7.60 (s, 2 H), 2.62 (q, J

= 7.5 Hz, 4 H), 1.16 (t, J= 7.5 Hz, 6 H). ^{13}C NMR (75 MHz, CDCl$_3$) δ 145.93, 138.70, 100.52, 33.26, 14.61.

1,4-Diethyl-2,5-bis(trimethysilylethynyl)benzene. See the general procedure for the Pd/Cu coupling reaction. The compounds used were 1,4-diethyl-2,5-diiodobenzene (3.86 g, 10.0 mmol), TMSA (3.53 mL, 25.0 mmol), ((C$_6$H$_5$)$_3$P)$_2$PdCl$_2$(0.35 g, 0.50 mmol), CuI (0.19 g, 1.0 mmol), diisopropylethylamine (7 mL, 40 mmol), and THF (10 mL). Flash chromatography (silica gel, hexane) gave the title compound as yellow crystals (2.73 g, 84%). FTIR (neat) 2964, 2872, 2154, 1487, 1456, 1400, 1251, 1195, 1062, 897, 867, 841, 764, 708, 626 cm^{-1}. ^{1}H NMR (300 MHz, CDCl$_3$) δ 7.26 (s, 2 H), 2.73 (q, J = 7.5 Hz, 4 H), 1.21 (t, J = 7.5 Hz, 6 H), 0.25 (s, 18 H). ^{13}C NMR (75 MHz, CDCl$_3$) δ 143.82, 131.76, 122.50, 103.78, 99.12, 27.10, 14.50, -0.02.

1,4-Diethyl-2,5-(diethynyl)benzene (26). See the general procedure for the deprotection of a trimethylsilyl-protected alkyne. The compounds used were 1,4-diethyl-2,5-bis(trimethysilylethynyl)benzene (2.52 g, 7.70 mmol) and K$_2$CO$_3$ (6.40 g, 46.2 mmol) in MeOH (50 mL) for 1 d to afford titled compound as yellow oil (1.29 g, 92%). FTIR (neat) 3290, 2971, 2933, 2875, 1491, 1457, 1239, 1061, 896 cm^{-1}. ^{1}H NMR (300 MHz, CDCl$_3$) δ 7.31 (s, 2 H), 3.29 (s, 2 H), 2.75 (q, J = 7.6 Hz, 4 H), 1.22 (t, J = 7.6 Hz, 6 H). ^{13}C NMR (75 MHz, CDCl$_3$) δ 143.99, 132.27, 121.88, 82.10, 81.68, 26.89, 14.63. HRMS calcd for C$_{14}$H$_{14}$: 182.1096. Found: 182.1088.

27. See the general procedure for the Pd/Cu coupling reaction. The compounds used were **26** (0.624 g, 3.42 mmol), **3** (1.95 g, 7.00 mmol), di(benzylidineacetone)palladium(0) (0.20 g, 0.35 mmol), (C$_6$H$_5$)$_3$P (0.37 g, 1.4 mmol), CuI (0.13 g, 0.70 mmol), diisopropylethylamine (4.9 mL, 28 mmol) and THF (10 mL). Flash chromatography (silica gel, hexane:CH$_2$Cl$_2$:ether = 12:6:1) gave desired product as light yellow crystalline solid (1.19 g, 72%). Mp 154-158 °C. FTIR (neat) 2954, 2933, 2872, 1692, 1590, 1497, 1395, 1123, 949, 887, 826 cm^{-1}. ^{1}H NMR (300 MHz, CDCl$_3$) δ 7.55 (d, J = 8.5 Hz, 4 H), 7.39 (d, J = 8.4 Hz, 4 H), 7.38 (s, 2 H), 2.83 (q, J = 7.5 Hz, 4 H), 2.42 (s, 6 H), 1.29 (t, J = 7.5 Hz, 6 H). ^{13}C NMR (75 MHz, CDCl$_3$) δ 193.42, 143.66, 134.29, 132.08, 131.69, 128.11, 124.68, 122.42, 93.44, 89.88, 30.30, 27.20, 14.73. HRMS calcd for C$_{30}$H$_{26}$O$_2$S$_2$: 482.1374. Found: 482.1373.

28. To **27** (0.15 g, 0.31 mmol) and NaOH (0.740 g, 18.5 mmol) was added THF (20 mL) and water (4 mL). The mixture was stirred for 12 h. The solvent was decanted and the precipitate was washed with ether (5×). The solid was suspended in ether and acidified with 3 N HCl (10 mL). The organic layer was separated and washed with water (1×), brine (1×) and dried over MgSO$_4$. Removal of solvent in vacuo gave desired product as yellow solid (0.11 g, 89%). FTIR (KBr) 2964, 2933, 2872, 2359, 2339, 1590, 1497, 1456, 1400, 1097, 1015, 897, 821 cm^{-1}. ^{1}H NMR (300 MHz, CDCl$_3$) δ 7.38 (d, J = 8.2 Hz, 4 H), 7.35 (s, 2 H), 7.23 (d, J = 8.4 Hz, 4 H), 3.52 (s, 2 H), 2.82 (q, J = 7.5 Hz,

4 H), 1.28 (t, J = 7.5 Hz, 6 H). ^{13}C NMR (75 MHz, CDCl$_3$) δ 144.28, 132.94, 132.73, 132.42, 129.91, 123.33, 121.55, 94.59, 89.66, 28.27, 15.82. HRMS calcd for C$_{26}$H$_{22}$S$_2$: 398.1163. Found: 398.1167.

4-Ethynylphenyl-3´-ethyl-4´-trimethylsilylethynyl-1-thioacetylbenzene. See the general procedure for the Pd/Cu coupling reaction. **29**[58] (3.28 g, 10.0 mmol), **9**[54] (2.23 g, 12.7 mmol), bis(dibenzylideneacetone)palladium(0) (288 mg, 0.50 mmol), CuI (0.190 g, 0.042 mmol), (C$_6$H$_5$)$_3$P (655 mg, 2.50 mmol), and diisopropylethylamine (7.0 mL, 40.0 mmol) were stirred in THF (20.0 mL) at room temperature for 1 d. The crude product was purified by flash chromatography on silica gel (hexane:EtOAc 19:1) to afford desired product (3.00 g, 80%) as an orange oil. FTIR 2965, 2152, 1713, 1601, 1495, 1250, 1114, 864 cm^{-1}. ^{1}H NMR (CDCl$_3$) δ 7.52 (dt, J = 8.2, 1.6 Hz, 2 H), 7.39 (d, J = 7.8 Hz, 1 H), 7.38 (dt, J = 8.2, 1.6 Hz, 2 H), 7.35 (d, J = 1.4 Hz, 1 H), 7.27 (dd, J = 7.8, 1.4 Hz, 1 H), 2.78 (q, J = 7.6 Hz, 2 H), 2.42 (s, 3 H), 1.24 (t, J = 7.6 Hz, 3 H), 0.24 (s, 9 H). ^{13}C NMR (CDCl$_3$) δ 193.3, 146.8, 134.2, 312.2, 131.1, 128.8, 128.2, 124.4, 123.0, 122.6, 103.4, 100.0, 91.1, 90.0,b 30.3, 27.6, 14.4, 0.0. HRMS calcd for C$_{23}$H$_{24}$OSSi: 376.1317. Found: 376.1308.

30. See the general procedure for the deprotection of a trimethylsilyl-protected alkyne. The compounds used were (**31**) (940 mg, 2.5 mmol), pyridine (5.0 mL), concentrated hydrofluoric acid (48% in water, 0.18 mL, 5.60 mmol) and TBAF (1.0 M in THF, 2.75 mL, 2.75 mmol) at room temperature for 10 min. The crude product was then purified by flash column chromatography on silica gel (hexane:ether 19:1) to afford desired product (629 mg, 83%). Mp 97-98°C. FTIR (KBr) 3255, 2966, 1702, 1491, 1123, 956, 825 cm^{-1}. ^{1}H NMR (CDCl$_3$) δ7.53 (dt, J = 8.4, 1.9 Hz, 2 H), 7.43 (d, J = 7.9 Hz, 1 H), 7.38 (dt, J = 8.4, 1.9 Hz, 2 H), 7.38 (d, J = 1.6 Hz, 1 H), 7.30 (dd, J = 7.9, 1.6 Hz, 1 H), 3.33 (s, 1 H), 2.81 (q, J = 7.6 Hz, 2 H), 2.42 (s, 3 H), 1.25 (t, J = 7.6 Hz, 3 H) ^{13}C NMR (CDCl$_3$) δ 193.4, 146.9, 134.2, 132.8, 131.1, 128.9, 128.3, 123.4, 121.9, 90.0, 90.1, 82.4, 81.9, 30.3, 27.4, 14.5. HRMS calcd for C$_{20}$H$_{16}$OS: 304.0922. Found: 304.0920.

32. See the general procedure for the Pd/Cu coupling reaction. 1,3,5-Triiodobenzene (228 mg, 0.50 mmol), **30** (547 mg, 1.80 mmol), bis(dibenzylideneacetone)palladium(0) (43 mg, 0.075 mmol), CuI (29 mg, 0.15 mmol), (C$_6$H$_5$)$_3$P (98 mg, 0.37 mmol), and diisopropylethylamine (1.0 mL, 6.0 mmol) were stirred in THF (5.0 mL) at room temperature for 65 h. The crude product was washed with a small amount of EtOAc to afford a pale yellow solid (107 mg). The washings were combined, evaporated to dryness, and purified by flash chromatography on silica gel (hexane:EtOAc 85:15) to afford another 219 mg yielding a total of 326 mg (66%) of titled compound. Mp 87-88 °C. FTIR (KBr) 1700, 1578, 1498, 1384, 1115, 827, 619 cm^{-1}. ^{1}H NMR (CDCl$_3$) δ 7.63 (s, 3 H), 7.55 (dt, J = 8.5, 1.9 Hz, 6 H), 7.49 (d, J = 7.9 Hz, 3 H), 7.43 (d, J = 1.6 Hz, 3 H), 7.39 (dt, J = 8.5, 1.9 Hz, 6 H), 7.36 (dd, J = 7.9,

1.6 Hz, 3 H), 2.90 (q, J = 7.5 Hz, 6 H), 2.43 (s, 9 H), 1.33 (t, J = 7.5 Hz, 9 H). ^{13}C NMR (CDCl$_3$) δ 193.4, 146.5, 134.3, 133.9, 132.2, 131.2, 129.1, 128.3, 124.3, 124.2, 123.3, 122.1, 93.0, 91.0, 90.3, 89.2, 30.3, 27.7, 14.7. HRMS calcd for C$_{66}$H$_{48}$O$_3$S$_3$: 984.2766. Found: 984.2717.

Bis(4-bromophenyl)methanol (35). To a solution of 1,4-dibromobenzene (5.66 g, 24.0 mmol) in THF (50 mL) at -78 °C was added n-BuLi (14.6 mL, 22.0 mmol, 1.51 M in hexane) dropwise. The slurry was stirred for 40 min and added to a solution of 4-bromobenzylaldehyde (3.7 g, 20 mmol) in THF (40 mL) which was cooled at -78 °C. The yellow solution was allowed to warmed to room temperature and stirred for 2 h before pouring into water. The mixture was extracted with EtOAc (3×). The combined organic fractions were washed with water (2×) and dried over MgSO$_4$. Removal of solvents in vacuo followed by washing with hexane afforded titled compound as a white solid (5.89 g, 86%). Mp 112-113 °C. FTIR (KBr) 3323, 2903, 1590, 1482, 1400, 1328, 1190, 1113, 1072, 1041, 1010, 862, 810, 795 cm^{-1}. ^{1}H NMR (300 MHz, CDCl$_3$) δ 7.45 (d, J = 8.5 Hz, 4 H), 7.21 (d, J = 8.3 Hz, 4 H), 5.74 (d, J = 3.3 Hz, 1 H), 2.21 (d, J = 3.4 Hz, 1 H). ^{13}C NMR (75 MHz, CDCl$_3$) δ 142.23, 131.74, 128.21, 121.78, 75.03. HRMS calcd for C$_{13}$H$_{10}$Br$_2$O: 339.9098. Found: 339.9084.

Bis(4-bromophenyl)methane (36). To a solution of **35** (1.71 g, 5.00 mmol) in TFA (40 mL) was added sodium borohydride (1.89 g, 50.0 mmol) in small portions at room temperature over 10 min.[62] The resulting white slurry was stirred for 40 min before pouring into water. The suspension was carefully made alkaline with aqueous NaOH solution. The mixture was extracted with ether (3×). The combined organic fraction was washed with water (2×), brine (1×), and dried over MgSO$_4$. Removal of solvents followed by filtering through a short silica gel column (hexane) afforded desired product as a white solid (1.53 g, 94%). Mp 62-62.5 °C. FTIR (KBr) 2923, 2851, 1482, 1436, 1400, 1200, 1113, 1067, 1010, 856, 805, 780, 621 cm^{-1}. ^{1}H NMR (300 MHz, CDCl$_3$) δ 7.39 (d, J = 8.4 Hz, 4 H), 7.01 (d, J = 8.3 Hz, 4 H), 3.86 (s, 2 H). ^{13}C NMR (75 MHz, CDCl$_3$) δ 139.46, 131.69, 130.64, 120.26, 40.71. HRMS calcd for C$_{13}$H$_{10}$Br$_2$: 323.9149. Found: 323.9147.

Bis(4-thioacetylphenyl)methane (37). See the general procedure for the conversion of aryl halides to arylthioacetates. The compounds used were **36** (0.978 g, 3.00 mmol) in THF (15 mL), tert-BuLi (8.7 mL, 15 mmol, 1.72 M in pentane) in ether (5 mL), sulfur powder (0.39 g, 12 mmol) in THF (15 mL), and acetyl chloride (1.07 mL, 15.0 mmol). Gravity chromatography (silica gel, hexane:ether 4/1) afforded desired product as a colorless oil (0.764, 81%). Mp 54-55 °C. FTIR (neat) 3395, 3026, 2923, 1703, 1595, 1492, 1431, 1405, 1354, 1118, 1092, 1015, 949, 805, 790, 610 cm^{-1}. ^{1}H NMR (300 MHz, CDCl$_3$) δ 7.32 (d, J = 8.2 Hz, 4 H), 7.21 (d, J = 8.1 Hz, 4 H), 4.00 (s, 2 H), 2.39 (s, 6 H). ^{13}C

NMR (75 MHz, CDCl$_3$) δ 194.10, 141.98, 134.70, 129.95, 125.81, 41.37, 30.24. HRMS calcd for C$_{17}$H$_{16}$O$_2$S$_2$: 316.0592. Found: 316.0583.

4'-Bromo-(4''-iodo)diphenylmethanol (38). See the preparation of **34**. The compounds used were 1,4-diiodobenzene (4.29 g, 13.0 mmol) in THF (50 mL), n-BuLi (8.0 mL, 12 mmol, 1.51 M in hexane), and 4-bromobenzylaldehyde (1.85 g, 10.0 mmol) in THF (40 mL). After workup, the solvent was removed in vacuo followed by washing with hexane to give desired compound as a white solid (3.53 g, 91%). Mp 119-120 °C. FTIR (KBr) 3333 (broad), 2903, 1590, 1482, 1400, 1328, 1292, 1190, 1113, 1072, 1036, 1005, 862, 831, 810, 790 cm^{-1}. ^{1}H NMR (300 MHz, CDCl$_3$) δ 7.65 (d, J = 8.4 Hz, 2 H), 7.44 (d, J = 8.5 Hz, 2 H), 7.20 (d, J = 8.7 Hz, 2 H), 7.08 (d, J = 8.5 Hz, 2 H), 5.73 (d, J = 3.2 Hz, 1 H), 2.20 (d, J = 3.5 Hz, 1 H). ^{13}C NMR (75 MHz, CDCl$_3$) δ 142.92, 142.21, 137.70, 131.75, 128.44, 128.22, 121.79, 93.48, 75.12. HRMS calcd for C$_{13}$H$_{10}$BrIO: 389.8939. Found: 389.8930.

4'-Bromo-(4''-iodo)diphenylmethane (39). See the preparation of **35**. The compounds used were **38** (1.36 g, 3.50 mmol), sodium borohydride (1.32 g, 35.0 mmol), and TFA (30 mL). Flash chromatography (silica gel, hexane) afforded desired product as white needle-like crystals (1.23 g, 94%). Mp 68-70 °C. FTIR (KBr) 3036, 2923, 2851, 1482, 1436, 1395, 1200, 1108, 1067, 1010, 856, 800, 774 cm^{-1}. ^{1}H NMR (300 MHz, CDCl$_3$) δ 7.59 (d, J = 8.4 Hz, 2 H), 7.39 (d, J = 8.4 Hz, 2 H), 7.01 (d, J = 8.5 Hz, 2 H), 6.89 (d, J = 8.4 Hz, 2 H), 3.85 (s, 2 H). ^{13}C NMR (75 MHz, CDCl$_3$) δ 140.15, 139.42, 137.68, 131.71, 131.00, 130.68, 120.28, 91.71, 40.84. HRMS calcd for C$_{13}$H$_{10}$BrI: 371.9011. Found: 371.8996.

40. See the general procedure for the Pd/Cu coupling reaction. The compounds used were **39** (1.12 g, 3.00 mmol), 1-bromo-4-ethynylbenzene (0.58 g, 3.2 mmol), bis(dibenzylidineacetone)palladium(0) (0.086 g, 0.15 mmol), CuI (0.057 g, 0.30 mmol), (C$_6$H$_5$)$_3$P (0.157 g, 0.600 mmol), THF (20 mL), and diisopropylethylamine (2.1 mL, 12 mmol) at room temperature for 2 d. Flash chromatography (silica gel, hexane) afforded desired product as white crystals (1.17 g, 92%). Mp 151-153 °C. FTIR (KBr) 2215, 1508, 1482, 1385, 1067, 1005, 867, 826, 810, 780, 621 cm^{-1}. ^{1}H NMR (300 MHz, CDCl$_3$) δ 7.46 (d, J = 8.8 Hz, 2 H), 7.43 (d, J = 9.0 Hz, 2 H), 7.40 (d, J = 8.7 Hz, 2 H), 7.35 (d, J = 8.3 Hz, 2 H), 7.12 (d, J = 8.0 Hz, 2 H), 7.03 (d, J = 8.2 Hz, 2 H), 3.92 (s, 2 H). ^{13}C NMR (75 MHz, CDCl$_3$) δ 141.10, 139.48, 133.03, 131.85, 131.66, 131.64, 130.70, 129.01, 122.45, 122.31, 120.94, 120.22, 90.43, 88.23, 41.22. HRMS calcd for C$_{21}$H$_{14}$Br$_2$: 423.9462. Found: 423.9465.

41. See the general procedure for the conversion of aryl halide to arylthioacetate. The compounds used were **40** (0.852 g, 2.00 mmol) in THF (20 mL), tert-BuLi (6.0 mL, 10 mmol, 1.67 M in pentane) in ether (5 mL), sulfur powder (0.257 g, 8.00 mmol) in THF (10 mL), and acetyl chloride (0.71 mL, 10 mmol). Flash chromatography (silica gel, hexane:ether 4/1, then hexane:CH$_2$Cl$_2$ 1/1) afforded desired product as a white solid (0.724 g, 87%).

Mp 99-100 °C. FTIR (KBr) 1697, 1513, 1385, 1354, 1123, 1015, 959, 826, 785, 621 cm^{-1}. ^{1}H NMR (300 MHz, CDCl$_3$) δ 7.53 (d, J = 8.5 Hz, 2 H), 7.45 (d, J = 8.3 Hz, 2 H), 7.37 (d, J = 8.6 Hz, 2 H), 7.32 (d, J = 8.4 Hz, 2 H), 7.20 (d, J = 8.4 Hz, 2 H), 7.16 (d, J = 8.4 Hz, 2 H), 4.00 (s, 2 H), 2.42 (s, 3 H), 2.40 (s, 3 H). ^{13}C NMR (75 MHz, CDCl$_3$) δ 194.29, 193.51, 142.13, 140.98, 134.65, 134.23, 132.17, 131.93, 129.83, 129.15, 127.95, 125.66, 124.64, 120.93, 91.03, 88.52, 41.55, 30.30, 30.19. HRMS calcd for C$_{25}$H$_{20}$O$_2$S$_2$: 416.0905. Found: 416.0919.

Bis(4-diethyltriazenylphenyl)methane **(43)**. To 4,4'-methylenedianiline **(42)** (19.83 g, 100 mmol) in water (80 mL) and concentrated HCl (30 mL) was added sodium nitrite (15.18 g, 220 mmol) in water (120 mL) at 0 °C. The reaction was stirred at 0 °C for 30 min and then poured into a solution of K$_2$CO$_3$ (165.85 g, 1200 mmol) and diethylamine (22.76 mL, 220 mmol) in water (500 mL) at 0 °C. The reaction was stirred for 30 min at 0 °C and then poured into water. The aqueous layer was extracted with diethyl ether (3 × 25 mL) and the organic layer was dried over MgSO$_4$ and the product concentrated in vacuo to afford 17.30 g (47%) of the title compound as a viscous brown liquid. IR (neat) 3083, 3024, 2931, 1905, 1601, 1502, 1090, 1014, 854, 821, 787, 736, 700, 624 cm^{-1}. ^{1}H NMR (400 MHz, CDCl$_3$) δ 7.31 (d, J = 8.3 Hz, 4 H), 7.12 (d, J = 8.4 Hz, 4 H), 3.94 (s, 2 H), 3.72 (q, J = 7.2 Hz, 8 H), 1.23 (t, J = 7.1 Hz, 12 H). ^{13}C NMR (125 MHz, CDCl$_3$) δ 150.03, 138.76, 129.85, 120.98, 54.03, 41.60, 13.48.

Bis(4-iodophenyl)methane (44). See the general procedure for the iodination of triazenes. The title compound was prepared as above from **43** (9.15 g, 25 mmol) and CH$_3$I (25 mL) to yield 6.36 g (61%) of the title compound as a fluffy white solid. IR (KBr) 3025, 2919, 2848, 1898, 1477, 1394, 1196, 1105, 1056, 1003, 854, 798, 772, 617 cm^{-1}. ^{1}H NMR (400 MHz, CDCl$_3$) δ 7.58 (d, J = 8.3 Hz, 4 H), 6.88 (d, J = 8.4 Hz, 4 H), 3.83 (s, 2 H). ^{13}C NMR (100 MHz, CDCl$_3$) δ 140.15, 137.76, 131.15, 92.04, 41.07. LRMS calcd for C$_{13}$H$_{10}$I$_2$: 420. Found: 420.

45. See the general procedure for the Pd/Cu coupling reaction. **44** (84 mg, 0.20 mmol), **9**[54] (84 mg, 0.48 mmol), bis(dibenzylideneacetone)palladium(0) (12 mg, 0.021 mmol), CuI (8 mg, 0.042 mmol), (C$_6$H$_5$)$_3$P (30 mg, 0.10 mmol), and diisopropylethylamine (0.33 mL, 1.92 mmol) were stirred in THF (2.0 mL) at room temperature for 80 h. The crude product was purified by flash chromatography on silica gel (hexane:EtOAc 9:1) to afford titled compound (73 mg, 71%) as a yellow solid. Mp 173-174 °C. FTIR (KBr) 1701, 1508, 1385, 1118, 828 cm^{-1}. ^{1}H NMR (CDCl$_3$) δ 7.52 (dt, J = 8.3, 1.8 Hz, 4 H), 7.44 (d, J = 8.2 Hz, 4 H), 7.37 (dt, J = 8.3, 1.8 Hz, 4 H), 7.15 (d, J = 8.2 Hz, 4 H), 3.98 (s, 2 H), 2.41 (s, 6 H). ^{13}C NMR (CDCl$_3$) δ 193.5, 141.2, 134.2, 132.1, 131.9, 129.0, 127.9, 120.9, 91.0, 88.5, 41.7, 30.3. HRMS calcd for C$_{33}$H$_{24}$O$_2$S$_2$: 516.1218. Found: 516.1207.

46. To a solution of 1,4-diiodobenzene (726 mg, 2.2 mmol) in dry THF (10 mL) at -78 °C was added n-BuLi (1.53 M in hexane, 1.37 mL, 2.1 mmol). The yellow suspension was stirred at -78 °C for 30 min and terephthaldehyde (134 mg, 1.0 mmol) in dry THF (5.0 mL) was added. After stirring at room temperature for 30 min, the suspension was poured into water. The solution was extracted with ether and dried over MgSO$_4$. After filtration, the solvent was evaporated in vacuo to afford colorless oil. The oil was separated by flash chromatography on silica gel (hexane:EtOAc 7:3) to afford the desired product as white solid (317 mg, 59%). The product was a 1:1 mixture of diastereomers. Mp 160-164 °C. FTIR (KBr) 3355, 1482, 1397, 1192, 1038, 1006, 799, 774 cm^{-1}. ^{1}H NMR (CDCl$_3$) δ 7.63 (d, J = 8.4 Hz, 4 H), 7.29 (s, 4 H), 7.09 (d, J = 8.4 Hz, 4 H), 5.75 (s, 1 H), 5.74 (s, 1 H), 2.16 (s, 1 H), 2.15 (s, 1 H). ^{13}C NMR (CDCl$_3$) δ 143.2, 143.0, 137.6, 128.4, 126.8, 93.1, 75.4. HRMS calcd for C$_{20}$H$_{16}$O$_2$I$_2$: 541.9240. Found: 541.9216.

47. To trifluoroacetic acid (20 mL) was added under N$_2$ at 0 °C a mixture of **46** (542 mg, 1.0 mmol) and NaBH$_4$ (760 mg, 20.0 mmol).[62] The mixture was stirred at 0 °C for 1.5 h and poured into water. The solution was extracted with CH$_2$Cl$_2$ and washed with a saturated solution of NaHCO$_3$ and brine. The solution was dried over MgSO$_4$. After filtration, the solvent was evaporated in vacuo to afford a white solid. The solid was crystallized from cyclohexane and purified by flash chromatography on silica gel to afford the desired product as a white solid (359 mg, 70%). Mp 141-142 °C. FTIR (KBr) 1511, 1480, 1426, 1398, 1181, 1003, 797, 752, 627, 471 cm^{-1}. ^{1}H NMR (CDCl$_3$) δ 7.57 (d, J = 8.2 Hz, 4 H), 7.05 (s, 4 H), 6.91 (d, J = 8.2 Hz, 4 H), 3.86 (s, 4 H). ^{13}C NMR (CDCl$_3$) δ 140.8, 138.4, 137.5, 131.0, 129.0, 91.3, 41.0. HRMS calcd for C$_{20}$H$_{16}$I$_2$: 509.9342. Found: 509.9331.

48. See the general procedure for the Pd/Cu coupling reaction. **47** (255 mg, 0.50 mmol), **9** (201 mg, 1.1 mmol), bis(dibenzylideneacetone)palladium(0) (23 mg, 0.040 mmol), CuI (20 mg, 0.10 mmol), (C$_6$H$_5$)$_3$P (75 mg, 0.25 mmol), and diisopropylethylamine (0.70 mL, 4.0 mmol) were stirred in THF (4.0 mL) at room temperature for 65 h. The crude product was washed with a small amount of EtOAc to afford a yellow solid. The solid was dissolved in hot EtOAc and the solution was filtered. After the solvent was evaporated in vacuo, the title compound was afforded as yellow solid (269 mg, 89%). Mp 188-190 °C (EtOAc). FTIR (KBr) 1692, 1560, 1508, 1384, 1112, 826, 695 cm^{-1}. ^{1}H NMR (CDCl$_3$) δ 7.52 (dt, J = 8.5, 1.9 Hz, 4 H), 7.43 (d, J = 8.3 Hz, 4 H), 7.37 (dt, J = 8.5, 1.9 Hz, 4 H), 7.15 (d, J = 8.3 Hz, 4 H), 7.09 (s, 4 H), 3.95 (s, 4 H), 2.41 (s, 6 H). ^{13}C NMR (CDCl$_3$) δ 193.5, 141.9, 138.5, 134.2, 132.1, 131.8, 129.1, 129.0, 127.9, 124.7, 120.6, 91.1, 88.3, 41.5, 30.3. HRMS calcd for C$_{40}$H$_{30}$O$_2$S$_2$: 606.1687. Found: 606.1698.

4-Iodobenzaldehyde.[78] To a solution of 1,4-diiodobenzene (5.11 g, 0.015 mol) in diethyl ether (2.1 mL) at −78° C was added dropwise n-BuLi (6.14 mL, 1.50 M in hexanes) over a period of 30 min. The reaction was stirred for 1 h. To this solution was added dropwise dry DMF (1.19 mL) over a period of 30 min. The reaction mixture was gradually allowed to warm to room temperature. The reaction mixture was quenched with distilled water and the mixture extracted with CH_2Cl_2 (3 × 70 mL). It was dried over Na_2SO_4 and the solvents were removed in vacuo to yield yellow oil that solidified on cooling. The sample was purified by silica gel column chromatography using hexane:CH_2Cl_2 (1:1) to provide 2.50 g (70%) of the title compound as a white solid. ^{1}H NMR (300 MHz, $CDCl_3$) δ 9.95 (s, 1 H), 7.55 (d, J = 8.4 Hz, 2 H), 7.87 (d, J = 8.37 Hz, 2 H). FABMS Calcd for C_7H_5IO: 232. Found: 232. ^{13}C NMR (100 MHz, $CDCl_3$) δ 130.77, 135.56, 138.39, 191.23. Anal. Calcd for C_7H_5IO: C, 36.23; H, 2.17. Found. C, 36.46; H, 2.12.

3,4',5-TriiododiphenylMeOH (49). To a solution of 1,3,5-triiodobenzene (3.37 g, 7.41 mmol) in dry THF (90 mL) at -78 °C was added n-BuLi (1.57 M in hexane, 5.18 mL, 8.14 mmol). The solution was stirred for 30 min and transferred via cannula into 4-iodobenzaldehyde (2.06 g, 8.88 mmol) in dry THF (50 mL) at -78 °C. The solution was stirred for 10 min and temperature was gradually raised to room temperature. The solution was poured into water and extracted with ether and washed with brine. The extract was dried over $MgSO_4$. After filtration, the solvent was evaporated in vacuo to afford brown oil. The oil was separated by flash chromatography on silica gel (hexane:EtOAc 19:1 to 8:2) to give the desired product (2.32 g, 59%) as white crystals. Mp 146-147 °C. FTIR (KBr) 3416, 1567, 1540, 1410, 1384, 1167, 1038, 1005, 830 cm[-1]. ^{1}H NMR ($CDCl_3$) δ 7.93 (t, J = 1.5 Hz, 1 H), 7.67 (dt, J = 8.7, 2.0 Hz, 2 H), 7.63 (dd, J = 1.5, 0.6 Hz, 2 H), 7.06 (dtd, J = 8.7, 2.0, 0.3 Hz, 2 H), 5.63 (d, J = 2.4 Hz, 1 H), 2.25 (d, J = 3.3 Hz, 1 H). ^{13}C NMR ($CDCl_3$) δ 147.1, 144.4, 142.2, 137.9, 134.7, 128.4, 95.0, 93.9, 74.3. HRMS calcd for $C_{13}H_9I_3O$: 561.7789. Found: 561.7798.

3,4',5-Triiododiphenylmethane. The procedure by Gribble was modified as follows.[62] To trifluoroacetic acid (50 mL) was added under N_2 at room temperature $NaBH_4$ (1.05 g, 27.6 mmol). Before all of the $NaBH_4$ reacted with the trifluoroacetic acid, **49** (1.56 g, 2.78 mmol) in CH_2Cl_2 (50 mL) was added dropwise. The mixture was stirred for 1 h. Additional pieces of $NaBH_4$ (606 mg, 15.9 mmol) were added in portions over 6 h. The mixture was stirred for 1 h and then poured into ice water. The solution was neutralized by a careful addition of NaOH pellets. The solution was extracted with CH_2Cl_2 and washed with brine. The extract was dried over $MgSO_4$. After filtration, the solvent was evaporated in vacuo to afford a mixture of yellow oil and white solid. The mixture was washed with hexane:EtOAc (8:2) and filtered to afford white solid (599 mg). The washings were combined, evaporated to dryness, and again purified by flash chromatography on silica gel (hexane:EtOAc 8:2) to afford another 145 mg yielding a total of 744 mg (49%) of desired compound in addition to the recovery of **49** (338 mg, 22%). Mp 147-148 °C. FTIR (KBr) 1571, 1539, 1482, 1418, 1384, 1006, 858, 787, 706 cm[-1]. ^{1}H NMR ($CDCl_3$) δ

7.88 (t, J = 1.5 Hz, 1 H), 7.61 (dt, J = 8.6, 2.1 Hz, 2 H), 7.43 (d, J = 1.5 Hz, 2 H), 6.88 (dt, J = 8.6, 2.1 Hz, 2 H), 3.77 (s, 2 H). ^{13}C NMR (CDCl$_3$) δ 144.5, 143.1, 139.0, 137.8, 137.1, 130.9, 95.1, 92.1, 40.4. HRMS calcd for C$_{66}$H$_{48}$O$_3$S$_3$: 545.7838. Found. 545.7840.

50. See the general procedure for the Pd/Cu coupling reaction. **49** (295 mg, 0.54 mmol), **30** (587 mg, 1.93 mmol), bis(dibenzylideneacetone)palladium(0) (47 mg, 0.081 mmol), CuI (30 mg, 0.16 mmol), (C$_6$H$_5$)$_3$P (107 mg, 0.41 mmol), and diisopropylethylamine (1.13 mL, 6.5 mmol) were stirred in THF (5.0 mL) at room temperature for 3 d. The crude product was washed with a small amount of EtOAc to afford a pale brown solid (355 mg). The washings were combined, evaporated to dryness, and further purified by flash chromatography on silica gel (hexane:EtOAc 8:2) to afford yellow oil. The oil was crystallized from EtOAc to afford another 125 mg yielding a total of 480 mg (83%) of desired product. Mp 132-134 °C. FTIR (KBr) 2954, 2203, 1698, 1588, 1498, 1384, 1116, 827, 620 cm^{-1}. ^{1}H NMR (CDCl$_3$) δ 7.55-7.53 (m, 7 H), 7.48 (d, J = 8.2 Hz, 2 H), 7.46 (d, J = 7.9 Hz, 2 H), 7.46 (d, J = 8.0 Hz, 1 H), 7.41 (d, J = 1.4 Hz, 2 H), 7.41-7.37 (m, 7 H), 7.33 (dd, J = 7.9, 1.4 Hz, 2 H), 7.34-7.31 (m, 3 H), 7.21 (d, J = 8.2 Hz, 2 H), 4.00 (s, 2 H), 2.87 (q, J = 7.6 Hz, 4 H), 2.86 (q, J = 7.5 Hz, 2 H), 2.42 (s, 9 H), 1.31 (t, J = 7.6 Hz, 6 H), 1.30 (t, J = 7.5 Hz, 3 H). ^{13}C NMR (CDCl$_3$) δ 193.4, 146.4, 146.2, 141.3, 140.5, 134.4, 134.2, 132.4, 132.3, 132.2, 132.1, 131.9, 131.6, 131.2, 129.1, 129.0, 129.0, 128.4, 128.3, 128.3, 128.2, 124.4, 124.4, 124.0, 123.1, 122.8, 122.7, 122.4, 121.5, 94.7, 93.8, 91.2, 91.1, 90.2, 90.0, 88.6, 87.9, 41.4, 30.3, 27.7, 27.6, 14.6, 14.6. Anal. calcd for C$_{73}$H$_{54}$O$_3$S$_3$: C, 81.53; H, 5.06. Found: C, 81.48; H, 5.07.

Bis(3,5-dibromo-4-aminophenyl)methane. To 4,4'-diaminodiphenylmethylene (594 mg, 3.0 mmol) in a MeOH: CH$_2$Cl$_2$ (1:1) solution (20 mL) was added dropwise bromine (0.77 mL, 15.0 mmol) in a MeOH: CH$_2$Cl$_2$ (1:1) solution (20 mL). The mixture was stirred at room temperature for 3 h before poured into 1 N NaOH solution. The mixture was filtered to afford a white solid. The solid was washed with water and dried to give titled compound (1.45 g, 94%). Mp>250 °C. FTIR (KBr) 3424, 3315, 1618, 1472, 1060 cm^{-1}. ^{1}H NMR (CDCl$_3$) δ 7.14 (s, 4 H), 3.66 (s, 2 H). HRMS calcd for C$_{13}$H$_{10}$N$_2$Br$_4$: 509.7577, Found: 509.7600. Insolubility of the material inhibited obtaining other spectral characterization.

Bis(3,5-dibromophenyl)methane (51). To NaNO$_2$ (208 mg, 3.0 mmol) in sulfuric acid (5.0 mL) at 5 °C was added dropwise a suspension of bis(3,5-dibromo-4-aminophenyl)methane (514 mg, 1.0 mmol) in glacial AcOH (5.0 mL). During the addition, the temperature was maintained below 10 °C. The mixture was stirred at 5 °C for 30 min and a 50% aqueous solution of hypophosphorous acid (3.12 mL, 30 mmol) was added dropwise. After stirring for 30 min at 5 °C, the mixture was placed in a refrigerator for 1 d and then allowed to stand at room temperature overnight. The mixture was poured into water and extracted with EtOAc. The extract was washed with $NaHCO_3$ solution and brine and dried over $MgSO_4$. After filtration, the solvent was evaporated in vacuo to afford brown solid. The solid was crystallized from $CHCl_3$ to afford desired product (109 mg, 23%) as white solid. Mp 196 °C. FTIR (KBr) 3036, 1575, 1556, 1417, 1104, 849 cm^{-1}. ^{1}H NMR (CDCl$_3$) δ 7.54 (t, J = 1.7 Hz, 2 H), 7.21 (d, J = 1.7 Hz, 4 H), 3.82 (s, 2 H). ^{13}C NMR (CDCl$_3$) δ 143.1, 132.5, 130.7, 123.2, 40.4. HRMS calcd for $C_{13}H_8Br_4$: 479.7359. Found: 479.7357.

Bis(3,5-diiodophenyl)methane. To a solution of bis(3,5-dibromophenyl)methane (484 mg, 1.0 mmol) in dry THF (1.0 mL) was added under N_2 at -78 °C n-BuLi (1.58 M in hexane, 3.2 mL, 5.0 mmol). The solution was stirred at -78 °C for 1 h. After chlorotrimethylsilane (1.27 mL, 10.0 mmol) was added, the solution was stirred at -78 °C for 30 min and at room temperature overnight. The solution was poured into water and extracted with ether. The extract was dried over $MgSO_4$. After filtration, the solvent was evaporated in vacuo to afford a brown oil. The oil was separated by flash chromatography on silica gel (hexane:EtOAc 19:1) to afford bis(3,5-bistrimethylsilylphenyl)methane (377 mg) as a yellow oil. The oil contained a small amount of impurity but it was used for next reaction without further purification. To a solution of bis(3,5-bistrimethylsilylphenyl)methane (332 mg, 0.73 mmol) in CCl_4 (10 mL) was added at room temperature iodine monochloride (0.16 mL, 3.2 mmol) in carbon tetrachloride (5.0 mL). The solution was stirred at room temperature for 1 h and poured into an aqueous solution of sodium thiosulfate. The aqueous solution was extracted with CH_2Cl_2. The solution was dried over $MgSO_4$. After filtration, the solvent was evaporated in vacuo to afford brown oil. The oil was washed with a small amount of CH_2Cl_2 to afford the desired product (209 mg, 36%) as a white solid. Mp 219-221 °C. FTIR (KBr) 1560, 1542, 1412, 1384, 712 cm^{-1}. ^{1}H NMR (CDCl$_3$) δ 7.91 (s, 2 H), 7.42 (t, J = 1.5 Hz, 4 H), 7.42 (d, J = 1.5, 4 H), 3.71 (s, 2 H). ^{13}C NMR (CDCl$_3$) δ 143.6, 137.1, 94.8, 39.8. HRMS calcd for $C_{13}H_8I_4$: 671.6805. Found: 671.6802.

2. See the general procedure for the Pd/Cu coupling reaction. Bis(3,5-diiodophenyl)methane (170 mg, 0.25 mmol), **9**[54] (211 mg, 1.20 mmol), bis(dibenzylideneacetone)palladium(0) (29 mg, 0.050 mmol), CuI (19 mg, 0.10 mmol), $(C_6H_5)_3P$ (66 mg, 0.25 mmol), and diisopropylethylamine (0.70 mL, 4.0 mmol) were stirred in THF (4.0 mL) at room temperature for 2 d. The crude product was dissolved in EtOAc and passed through a plug of silica gel. Then the crude solid was washed with a small amount of EtOAc, dissolved in hot EtOAc and filtered to afford titled compound (102 mg, 47%) as a pale yellow solid. Mp 177-178 °C. FTIR (KBr) 1701, 1593, 1486, 1385, 1118, 828 cm^{-1}. ^{1}H NMR (CDCl$_3$) δ 7.59 (t, J = 2.5 Hz, 2 H), 7.53 (dt, J = 8.3, 1.7 Hz, 8 H), 7.38 (dt, J = 8.3, 1.7 Hz, 8 H), 7.34 (d, J = 1.5 Hz, 4 H), 3.96 (s, 2 H), 2.41 (s, 12 H). ^{13}C NMR (CDCl$_3$) δ 193.3, 140.7, 134.2, 133.0, 132.2, 132.2, 128.4,

124.1, 123.7, 90.1, 89.5, 30.0. HRMS calcd for $C_{53}H_{36}O_4S_4$: 864.1496. Found: 864.1453.

53. See the general procedure for the Pd/Cu coupling reaction. Bis(3,5-diiodophenyl)methane (108 mg, 0.16 mmol), **30** (220 mg, 0.72 mmol), bis(dibenzylideneacetone)palladium(0) (18 mg, 0.032 mmol), CuI (12 mg, 0.064 mmol), $(C_6H_5)_3P$ (42 mg, 0.16 mmol), and diisopropylethylamine (0.45 mL, 2.59 mmol) were stirred in THF (3.0 mL) at room temperature for 60 h. The crude product was dissolved in hexane:EtOAc (1:1) to afford a pale yellow solid. The solid was washed with a small amount of EtOAc to afford the desired product (107 mg, 49%) as a pale yellow solid. Mp 104-107 °C. FTIR (KBr) 1707, 1585, 1498, 1384, 1108, 826 cm^{-1}. ^{1}H NMR (CDCl$_3$) δ 7.57 (t, J = 1.3 Hz, 2 H), 7.54 (dt, J = 8.3, 1.7 Hz, 8 H), 7.48 (d, J = 8.0 Hz, 4 H), 7.41 (d, J = 1.3 Hz, 4 H), 7.38 (dt, J = 8.3, 1.7 Hz, 8 H), 7.35 (d, J = 1.3 Hz, 4 H), 7.33 (dd, J = 8.0, 1.3 Hz, 4 H), 4.00 (s, 2 H), 2.88 (q, J = 7.6 Hz, 8 H), 2.42 (s, 12 H), 1.31 (t, J = 7.6 Hz, 12 H). ^{13}C NMR (CDCl$_3$) δ 193.8, 146.8, 141.1, 134.6, 133.0, 132.6, 132.6, 132.3, 131.6, 129.4, 128.6, 124.8, 124.5, 123.5, 122.7, 94.1, 91.4, 90.6, 89.1, 41.3, 30.7, 28.0, 15.0 Anal. calcd for $C_{93}H_{68}O_4S_4$: C, 81.07; H, 4.97. Found: C, 81.16; H, 4.99.

2,5-Bis(p-bromobenzoyl)-1,4-dibromobenzene (56). To a suspension of AlCl$_3$ (2.67 g, 20.0 mmol) in CH$_2$Cl$_2$ (50 mL) at 0 °C was slowly added a solution of **54**[26] (3.25 g, 9.00 mmol) in CH$_2$Cl$_2$. The resultant yellow slurry was stirred for 10 min and a solution of 1-bromo-4-trimethylsilylbenzene (4.89 g, 21.3 mmol) in CH$_2$Cl$_2$ (15 mmol) was added. The mixture was stirred for 2 h at 0 °C and overnight at room temperature. The brown mixture was carefully poured into cold 1.5 N HCl solution. CH$_2$Cl$_2$ (100 mL) was added to dissolve the precipitate and the organic phase was separated. The aqueous phase was extracted with CH$_2$Cl$_2$ (2×). Combined organic fractions were washed with H$_2$O (1×) and dried over MgSO$_4$. After filtration, the solvent was concentrated to ca. 100 mL and filtered through a short silica gel column [CH$_2$Cl$_2$:hexane 1:1]. Removal of solvents followed by washing with hexane and ether afforded desired product as white solid (3.25 g, 60%). Mp 254-256 °C. FTIR (KBr) 3097, 1677, 1585, 1400, 1385, 1339, 1246, 1067, 1010, 928, 882, 841, 749 cm^{-1}. ^{1}H NMR (300 MHz, CDCl$_3$) δ 7.69 (d, J = 8.9 Hz, 4 H), 7.65 (d, J = 9.0 Hz, 4 H), 7.58 (s, 2 H). ^{13}C NMR (CDCl$_3$, 50 °C, 100 MHz) 192.16, 142.94, 134.02, 133.08, 132.29, 131.42, 129.86, 118.54. HRMS calcd for $C_{20}H_{10}Br_4O_2$: 597.7414. Found: 597.7400.

2,5-Bis(p-bromobenzyl)-1,4-dibromobenzene. To a suspension of **56** (2.11 g, 3.50 mmol) in CH$_2$Cl$_2$ (70 mL) was added dropwise trifluoromethanesulfonic acid (3.15 g, 21.0 mmol). The clear golden solution was cooled to 0 °C and a solution of triethylsilane (3.15 g, 17.5 mmol) in CH$_2$Cl$_2$ (10 mL) was added dropwise. The resulting light yellow solution was stirred at 0 °C for 10 min. Another portion of trifluoromethanesulfonic acid (3.15 g, 21.0 mmol) and triethylsilane (3.15 g, 17.5 mmol) was added by the above addition sequence at 0 °C. The obtained light yellow solution was allowed to warm to room temperature and stir for 3 h before pouring into saturated aqueous Na_2CO_3 (100 mL). The aqueous phase was separated and extracted with CH_2Cl_2 (2×). The combined organic fractions were washed with H_2O (2×) and dried over $MgSO_4$. Removal of solvents followed by washing with hexane afforded desired product as white solid (1.76 g, 88%). Mp 161-167 °C. FTIR (KBr) 1487, 1472, 1436, 1405, 1385, 1072, 1056, 1010, 897, 831, 774 cm^{-1}. ^1H NMR (300 MHz, $CDCl_3$) δ 7.41 (d, J = 8.4 Hz, 4 H), 7.28 (s, 2 H), 7.03 (d, J = 8.4 Hz, 4 H), 3.97 (s, 4 H). ^{13}C NMR (75 MHz, $CDCl_3$) δ 139.97, 137.63, 134.75, 131.76, 130.69, 123.74, 120.52, 40.59. HRMS calcd for $C_{20}H_{14}Br_4$: 569.7829. Found: 569.7834.

2,5-Bis(p-iodobenzyl)-1,4-diiodobenzene (57). To tert-BuLi (5.62 mL, 10.0 mmol, 1.78 M in pentane) in ether (5 mL) at -78 °C was added via cannula a solution of 2,5-bis(p-bromobenzyl)-1,4-dibromobenzene (0.574 g, 1.00 mmol) in THF (15 mL) dropwise. The brown slurry was stirred for 30 min and then warmed to 0 °C. The slurry was re-cooled to -78 °C and a solution of iodine (2.54 g, 10.0 mmol) in THF (10 mL) was added via cannula. The mixture was allowed to warmed to room temperature and stir for 1 h before pouring into an aqueous solution of sodium thiosulfate. The organic phase was separated. The aqueous layer was extracted with CH_2Cl_2 (2×). Combined organic fractions were washed with H_2O (2×) and dried over $MgSO_4$. Removal of solvents followed by washing with EtOAc afforded desired product as white solid (0.442 g, 58%). Mp 210-213 °C. FTIR (KBr) 3149, 1482, 1431, 1400, 1385, 1354, 1185, 1041, 1005, 897, 815, 774 cm^{-1}. ^1H NMR (300 MHz, $CDCl_3$) δ 7.61 (d, J = 8.4 Hz, 4 H), 7.53 (s, 2 H), 6.89 (d, J = 8.4 Hz, 4 H), 3.94 (s, 4 H). No ^{13}C could be obtained due to the limited solubility of **57**. HRMS calcd for $C_{20}H_{14}I_4$: 761.7274. Found: 761.7270.

58. See the general procedure for the Pd/Cu coupling reaction. The compounds used were **57** (0.38 g, 0.50 mmol), **9** (0.44 g, 2.5 mmol), di(benzylidineacetone)palladium(0) (0.058 g, 0.10 mmol), CuI (0.038 g, 0.20 mmol), $(C_6H_5)_3P$ (0.53 g, 0.20 mmol), THF (15 mL), and diisopropylethylamine (1.4 mL, 8.0 mmol) at room temperature. The mixture was stirred for 8 h. Another portion of di(benzylidineacetone)palladium(0) (0.029 g, 0.050 mmol) and PPh_3 (0.026 g, 0.10 mmol) in THF (5 mL) was added. The mixture was further stirred for 21 h. Flash chromatography (silica gel, hexane:CH_2Cl_2 1:1) gave desired product as a white solid (0.165 g, 35%). Mp 239-240 °C. FTIR (KBr) 1708, 1497, 1385, 1354, 1123, 1015, 854, 826, 621 cm^{-1}. ^1H NMR (300 MHz, $CDCl_3$) δ 7.52 (d, J = 8.5 Hz, 4 H), 7.46 (d, J = 7.9 Hz, 8 H), 7.37 (d, J = 8.5 Hz, 4 H), 7.36 (d, J = 8.6 Hz, 4 H), 7.35 (s, 2 H), 7.24 (d, J = 8.3 Hz, 4 H), 4.19 (s, 4 H), 2.42 (s, 6 H), 2.41 (s, 6 H). ^{13}C NMR (75 MHz, $CDCl_3$) δ 193.53, 193.37, 140.72, 140.69, 134.32, 134.24, 133.34, 132.19, 132.11, 131.94, 129.10, 128.53, 127.94, 124.67, 124.12, 123.23,

120.89, 94.61, 91.17, 89.64, 88.56, 39.80, 30.36, 30.32. HRMS calcd for $C_{60}H_{42}O_4S_4$: 954.1966. Found: 954.1999. Anal. calcd for $C_{60}H_{42}O_4S_4$: C, 75.44; H, 4.43. Found: C, 75.52; H, 4.51.

59. See the general procedure for the Pd/Cu coupling reaction. The compounds used were **57** (0.076 g, 0.10 mmol), **30** (0.157 g, 0.500 mmol), di(benzylidineacetone)palladium(0) (0.012 g, 0.020 mmol), CuI (0.0076 g, 0.040 mmol), $(C_6H_5)_3P$ (0.026 g, 0.10 mmol), THF (3 mL), and diisopropylethylamine (0.28 mL, 1.6 mmol) for 60 h at room temperature. Flash chromatography (silica gel, $CHCl_3$:hexane 1:1) afforded desired product as green/yellow solid (0.060 g, 41%). Mp 159-162 °C. FTIR (KBr) 2964, 2933, 2872, 2205, 1708, 1595, 1508, 1400, 1385, 1349, 1118, 1087, 1015, 949, 892, 826, 613 cm^{-1}. ^{1}H NMR (300 MHz, CDCl$_3$) δ 7.54 (d, J = 8.1 Hz, 8 H), 7.48-7.43 (m, 6 H), 7.40-7.36 (m, 16 H), 7.32 (dd, J = 8.2, 1.7 Hz, 4 H), 7.25 (d, J = 8.1 Hz, 4 H), 4.25 (s, 4 H), 2.86 (q, J = 7.6 Hz, 4 H), 2.76 (q, J = 7.6 Hz, 4 H), 2.42 (s, 12 H), 1.29 (t, J = 7.6 Hz, 6 H), 1.21 (t, J = 7.6 Hz, 6 H). ^{13}C NMR (100 MHz, CDCl$_3$) δ 193.31, 193.28, 146.19, 146.11, 140.53, 140.22, 134.21, 133.37, 132.27, 132.16, 132.04, 131.73, 131.16, 128.97, 128.94, 128.25, 128.14, 124.43, 124.32, 123.59, 123.15, 122.83, 122.68, 122.37, 121.26, 94.82, 93.72, 93.30, 91.21, 91.07, 90.32, 90.00, 87.77, 39.78, 30.45, 27.81, 27.75, 14.86, 14.77. LRMS calcd for $C_{100}H_{74}O_4S_4$: 1468. Found: 1468. Anal. calcd for $C_{100}H_{74}O_4S_4$: C, 81.82; H, 5.08. Found: C, 81.68; H, 5.13.

1,2-Bis(4'-aminophenyl)ethane. To a Parr flask was added **60** (5.45 g, 20.0 mmol), 10% palladium on activated carbon (274 mg), and ethanol (50 mL). The flask was purged with hydrogen and pressurized to 60 psi. The flask was shaken for 5 h at room temperature. After filtration, the solvent was evaporated in vacuo to afford desired compound (2.50 g, 59%) as a white solid. Mp >250 °C. ^{1}H NMR (CDCl$_3$) δ 6.95 (d, J = 8.2 Hz, 4 H), 6.61 (d, J = 8.2 Hz, 4 H), 3.47 (br, 4 H), 2.74 (s, 4 H).

1,2-Bis(4'-diethyltriazenylphenyl)ethane (61). To 1,2-bis(4'-aminophenyl)ethane (1.00 g, 4.72 mmol), HCl (15 mL), and water (50 mL) was added at 0 °C NaNO$_2$ (716 mg, 10.4 mmol) in water (2.0 mL). The solution was stirred for 30 min at 0 °C and poured into K_2CO_3 (10.4 g, 75.2 mmol), diethylamine (10 mL), and water (100 mL). An orange solid was removed by filtration and washed with water. After drying, the desired solid (1.59 g, 87%) was obtained. Mp 64-66°C. FTIR (KBr) 2980, 2935, 1433, 1402, 1384, 1351, 1235, 1089, 841 cm^{-1}. ^{1}H NMR (CDCl$_3$) δ 7.30 (dt, J = 8.3, 2.0 Hz, 4 H), 7.11 (dt, J = 8.3, 2.0 Hz, 4 H), 3.73 (q, J = 7.2 Hz, 8 H), 2.87 (s, 4 H), 1.24 (t, J = 7.2 Hz, 12 H). ^{13}C NMR (CDCl$_3$) δ 149.4, 138.8, 128.9, 120.3, 37.7, 13.1 (br) (one carbon is missing due to the quadropolar effect of nitrogen.). HRMS calcd for $C_{22}H_{32}N_6$: 380.2688. Found: 380.2696.

1,2-Bis(4'-iodophenyl)ethane (62). See the standard procedure. The compounds used were **61** (800 mg, 2.48 mmol) and CH$_3$I (15 mL) at 120 °C overnight. After cooling, the reaction was diluted with hexane:EtOAc 1:1 and passed through a plug of silica gel. The solvent was evaporated in vacuo to afford desired compound (834 mg, 78%) as yellow solid. Mp 150-151°C. FTIR (KBr) 1482, 1384, 1002, 816 cm^{-1}. ^{1}H NMR (CDCl$_3$) δ 7.56 (d, J = 8.3 Hz, 4 H), 6.86 (d, J = 8.3 Hz, 4 H), 2.81 (s, 4 H). ^{13}C NMR (CDCl$_3$) δ 140.8, 137.4, 130.6, 91.2, 37.1. HRMS calcd for C$_{14}$H$_{12}$I$_2$: 433.9028. Found: 433.9027.

63. See the general procedure for the Pd/Cu coupling reaction. **62** (304 mg, 0.7 mmol), **9** (296 mg, 1.68 mmol), bis(dibenzylideneacetone)palladium(0) (40 mg, 0.070 mmol), CuI (27 mg, 0.14 mmol), (C$_6$H$_5$)$_3$P (92 mg, 0.35 mmol), and diisopropylethylamine (0.97 mL, 5.6 mmol) were stirred in THF (10 mL) at room temperature for 2 d. The crude product was passed thorough a plug of silica gel (hexane:EtOAc 1:1) to afford a yellow solid. The solid was recrystallized from EtOAc to afford the desired compound (227 mg, 61%). FTIR (KBr) 1700, 1512, 1384, 1123, 828, 623 cm^{-1}. ^{1}H NMR (CDCl$_3$) δ 7.54 (dt, J = 8.5, 1.8 Hz, 4 H), 7.43 (dt, J = 8.3, 1.7 Hz, 4 H), 7.38 (dt, J = 8.5, 1.8 Hz, 4 H), 7.11 (dt, J = 8.3, 1.7 Hz, 4 H), 2.92 (s, 4 H), 2.42 (s, 6 H). ^{13}C NMR (CDCl$_3$) δ 193.5, 142.1, 134.2, 132.2, 131.7, 128.7, 127.9, 124.7, 120.5, 91.2, 88.4, 37.6, 30.3. HRMS calcd for C$_{34}$H$_{26}$O$_2$S$_2$: 530.1374. Found: 530.1366.

64. See the general procedure for the Pd/Cu coupling reaction. The compounds used were 1,4-diethynylbenzene (1.26 g, 10.0 mmol), 1-iodo-4-trimethylsilylbenzene (6.09 g, 22.0 mmol), di(benzylidineacetone)palladium(0) (0.57 g, 1.0 mmol), (C$_6$H$_5$)$_3$P (0.53 g, 2.0 mmol), CuI (0.38 g, 2.0 mmol), THF (40 mL), and diisopropylethylamine (13.9 mL, 80.0 mmol). The mixture was stirred at room temperature for 30 h. After workup, the residue was dissolved in CH$_2$Cl$_2$ and filtered through a silica gel column [hexane:CH$_2$Cl$_2$ 2:1]. Removal of the solvent in vacuo followed by crystallization from hexane gave pale yellow solid (2.86 g). The mother liquor was purified by flash chromatography to give another pale yellow solid (0.66 g). A total of 3.52 g (83%) of desired product was obtained. Mp 214-220 °C. FTIR (KBr) 3067, 3015, 2954, 2892, 1595, 1513, 1385, 1308, 1246, 1103, 846, 821, 754, 718, 682 cm^{-1}. ^{1}H NMR (300 MHz, CDCl$_3$) δ 7.49 (s, 12 H), 0.26 (s, 18 H). ^{13}C NMR (75 MHz, CDCl$_3$) δ 141.36, 133.31, 131.61, 130.74, 123.36, 123.18, 91.51, 89.59, -1.17. HRMS calcd for C$_{28}$H$_{30}$Si$_2$: 422.1886. Found: 422.1878.

1,4-Bis(2-(4′-trimethylsilylphenyl)ethyl)benzene. A mixture of **64** (1.27 g, 3.00 mmol) in ethanol (100 mL) and 37% HCl (10 drops) was hydrogenated over Pd on carbon (0.2 g, 10% of Pd on carbon) at 60 psi for 21 h. The mixture was filtered and the residue was washed with EtOAc. Removal of solvent in vacuo gave desired compound as white solid (1.26 g, 97%). Mp 168-173 °C. FTIR (KBr) 3067, 3015, 2954, 2923, 2851, 1600, 1513, 1451, 1395, 1246, 1108, 831, 754, 718, 692, 651 cm^{-1}. ^{1}H NMR (300 MHz, CDCl$_3$) δ 7.44 (d, J = 7.9 Hz, 4 H), 7.20 (d, J = 7.9 Hz, 4 H), 7.14 (s, 4 H), 2.89 (s, 8 H), 0.29 (s, 18 H). ^{13}C NMR (75 MHz, CDCl$_3$) δ 142.66, 139.52, 137.61, 133.52,

128.45, 127.96, 38.07, 37.50, -0.96. HRMS calcd for $C_{28}H_{38}Si_2$: 430.2512. Found: 430.2497.

1,4-Bis(2-(4′-iodophenyl)ethyl)benzene (65). To a suspension of 1,4 bis(2 (4′-trimethylsilylphenyl)ethyl)benzene (1.13 g, 2.62 mmol) in CCl_4 (60 mL) was added dropwise iodine monochloride (0.37 mL, 7.3 mmol). The mixture was stirred for 80 min and then decolorized with aqueous sodium thiosulfate. The mixture was extracted with CH_2Cl_2 (2×). The extracts were dried over $MgSO_4$. Removal of solvent in vacuo gave a white solid. The solid was re-dissolved in CH_2Cl_2 and passed through a short silica gel column to afford desired compound as white solid (1.39 g, 98%). Mp 147-160 °C. FTIR (KBr) 3026, 2944, 2923, 2851, 1513, 1482, 1451, 1400, 1385, 1200, 1139, 1087, 1062, 1005, 815, 790, 759, 703, 610 cm^{-1}. ^{1}H NMR (300 MHz, CDCl$_3$) δ 7.56 (d, J = 8.2 Hz, 4 H), 7.03 (s, 4 H), 6.88 (d, J = 8.2 Hz, 4 H), 2.83 (s, 8 H). ^{13}C NMR (75 MHz, CDCl$_3$) δ 141.39, 138.94, 137.35, 130.68, 128.50, 91.04, 37.41, 37.26. HRMS calcd for $C_{22}H_{20}I_2$: 537.9655. Found: 537.9634.

66. See the general procedure for the Pd/Cu coupling reaction. The compounds used were **65** (0.463 g, 0.860 mmol), **9**[54] (0.379 g, 2.15 mmol), di(benzylidineacetone)palladium(0) (0.049 g, 0.086 mmol), CuI (0.033 g, 0.17 mmol), $(C_6H_5)_3P$ (0.090 g, 0.34 mmol), THF (15 mL), and diisopropylethylamine (0.91 mL, 5.2 mmol) for 24 h at room temperature. Flash chromatography (silica gel, CH_2Cl_2:hexane 2:1) afforded desired compound as a white solid (0.41 g, 75%). Mp 182 °C (decompose). FTIR (KBr) 2913, 2851, 1703, 1513, 1385, 1129, 1092, 1015, 949, 831, 821 cm^{-1}. ^{1}H NMR (300 MHz, CDCl$_3$) δ 7.52 (d, J = 8.2 Hz, 4 H), 7.43 (d, J = 8.1 Hz, 4 H), 7.37 (d, J = 8.3 Hz, 4 H), 7.13 (d, J = 8.2 Hz, 4 H), 7.05 (s, 4 H), 2.89 (br s, 8 H), 2.42 (s, 6 H). ^{13}C NMR (100 MHz, CDCl$_3$) δ 193.28, 142.40, 138.83, 134.07, 131.99, 131.52, 128.51, 128.35, 127.69, 124.61, 120.24, 91.20, 88.16, 37.93, 37.25, 30.34. HRMS calcd for $C_{42}H_{34}O_2S_2$: 634.2000. Found: 634.1990.

67. See the general procedure for the Pd/Cu coupling reaction. A solution of bis(dibenzylidineacetone)palladium(0) (0.0770 g, 0.135 mmol) and $(C_6H_5)_3P$ (0.14 g, 0.54 mmol) in THF (5 mL) was added to a solution of **65** (0.724 g, 1.35 mmol), phenylacetylene (0.138 g, 1.35 mmol) and CuI (0.050 g, 0.27 mmol) in THF (10 mL). The mixture was stirred for 19 h at room temperature. A solution of **9**[54] (0.44 g, 2.5 mmol), bis(dibenzylidineacetone)palladium(0) (0.015 g, 0.027 mmol) and $(C_6H_5)_3P$ (0.028 g, 0.11 mmol) in THF (5 mL) was added. The mixture was stirred for 28 h at room temperature and then poured into water. The mixture was extracted with CH_2Cl_2 (2×). The filtrate was dried over $MgSO_4$. Removal of solvent followed by flash chromatography (silica gel, CH_2Cl_2:hexane 1:1) and recrystallization from cyclohexane:CH_2Cl_2 afforded desired compound as white solid (0.266 g, 35%). Mp 180-183 °C. FTIR (KBr) 2915, 2850, 2371, 2213, 1707, 1591, 1508, 1387, 1113, 1011, 946, 830 cm^{-1}. ^{1}H NMR (300 MHz,

CDCl$_3$) δ 7.52 (d, J = 8.5 Hz, 2 H), 7.51 (dd, J = 7.8, 2.0 Hz, 2 H), 7.43 (d, J = 8.1 Hz, 4 H), 7.38-7.30 (m, 3 H), 7.13 (d, J = 8.3 Hz, 2 H), 7.12 (d, J = 8.2 Hz, 2 H), 7.05 (s, 4 H), 2.89 (br s, 8 H), 2.42 (s, 3 H). ^{13}C NMR (100 MHz, CDCl$_3$) δ 193.12, 142.28, 141.92, 138.74, 138.68, 133.94, 131.87, 131.39, 131.31, 128.38, 128.33, 128.22, 128.07, 127.87, 127.57, 124.49, 123.19, 120.49, 120.12, 91.09, 89.34, 88.77, 88.04, 37.79, 37.12, 30.20. HRMS calcd for C$_{40}$H$_{32}$OS: 560.2174. Found: 560.2157.

2-Bromo-4-nitro-5-(phenylethynyl)acetanilide (69). See the general procedure for the Pd/Cu-catalyzed coupling reaction. The compounds used were 2,5-dibromo-4-nitroacetanilide **(68)**[64] (3.0 g, 8.88 mmol), phenylacetylene (0.98 mL, 8.88 mmol), CuI (0.17 g, 0.89 mmol), ((C$_6$H$_5$)$_3$P)$_2$PdCl$_2$ (0.25 g, 0.44 mmol), (C$_6$H$_5$)$_3$P (0.47 g, 1.78 mmol), diisopropylethylamine (6.18 mL, 35.52 mmol), and THF (25 mL) at room temperature for 1 d then 50 °C for 12 h. The resultant mixture was subjected to an aqueous workup as described above. The desired material was purified by gravity liquid chromatography using silica gel as the stationary phase and CH$_2$Cl$_2$ as the eluent. R$_f$ (product) = 0.60. The reaction afforded 1.79 g (56% yield) of the desired product. IR (KBr) 3261.5, 3097.4, 2215.4, 1671.8, 1553.8, 1533.3, 1502.6, 1379.5, 1333.3, 1261.5, 1092.3, 1020.5, 892.3, 851.3, 753.8, 687.2, 651.3 cm^{-1}. ^{1}H NMR (400 MHz, CDCl$_3$) δ 8.84 (s, 1 H), 8.39 (s, 1 H), 7.80 (br s, 1 H), 7.66-7.60 (m, 2 H), 7.43-7.36 (m, 3 H), 2.32 (s, 3 H). ^{13}C NMR (400 MHz, CDCl$_3$) δ 168.30, 139.81, 132.20, 129.49, 129.03, 128.49, 124.87, 122.21, 119.88, 117.49, 111.00, 98.64, 84.81, 25.33. HRMS calcd C$_{16}$H$_{11}$N$_2$O$_3$Br: 357.9953. Found: 357.9948.

2-Bromo-4-nitro-5-(phenylethynyl)aniline. To a 100 mL round bottom flask equipped with a magnetic stirbar, **69** (0.33 g, 0.92 mmol), K$_2$CO$_3$ (0.64 g, 4.6 mmol), MeOH (15 mL), and CH$_2$Cl$_2$ (15 mL) were added. The reaction was allowed to stir at room temperature for 1 h. The reaction mixture was quenched with water and extracted with CH$_2$Cl$_2$ (3×). The organic layers were combined and dried over MgSO$_4$. Solvents were removed in vacuo. No further purification needed. The reaction afforded 0.29 g (100% yield) of the titled compound as yellow solid. IR (KBr) 3476.9, 3374.4, 3159.0, 1656.4, 1615.4, 1559.0, 1379.5, 1307.7, 1138.5, 1102.6, 892.3, 748.7, 687.2 cm^{-1}. ^{1}H NMR (400 MHz, CDCl$_3$) δ 8.46 (s, 1 H), 7.74-7.68 (m, 2 H), 7.52-7.46 (m, 3 H), 7.06 (s, 1 H), 4.93 (br s, 2 H). ^{13}C NMR (400 MHz, CDCl$_3$) δ 148.55, 139.41, 132.02, 130.45, 129.25, 128.46, 122.46, 120.17, 118.38, 106.86, 96.94, 85.46. HRMS calcd: 317.9828. Found: 317.9841.

2′-Amino-4,4′-diphenylethynyl-5′-nitro-1-thioacetylbenzene (70). See the general procedure for the Pd/Cu-catalyzed coupling reaction. 2-Bromo-4-nitro-5-(phenylethynyl)aniline (0.10 g, 0.30 mmol) was coupled to **9** (0.10 g, 0.56 mmol) as described above using CuI (0.01 g, 0.03 mmol), ((C$_6$H$_5$)$_3$P)$_2$PdCl$_2$ (0.01 g, 0.02 mmol), (C$_6$H$_5$)$_3$P (0.02 g, 0.06 mmol), diisopropylethylamine (0.24 mL, 1.40 mmol), and THF (10 mL) in an oven dried round screw capped pressure tube equipped with a stirbar. The reaction mixture was allowed to react at 80 °C for 3 d. The resultant mixture was subjected to an aqueous workup as described above. The desired material was purified by gravity liquid chromatography using silica gel as the stationary phase and 3:1 CH_2Cl_2:hexanes as the eluent. R_f (product): 0.26. An additional hexanes wash gave yellow crystals of the desired compound, 0.80 g (67 % yield). IR (KBr) 3374.4, 3138.5, 2205.1, 1384.6, 1312.8, 1246.2, 1112.8, 825.6, 753.8, 692.3, 615.4 cm^{-1}. ^{1}H NMR (400 MHz, CDCl$_3$) δ 8.27 (s, 1 H), 7.59 (m, 2 H), 7.55 (d, J=8.0 Hz, 2 H), 7.42 (d, J=8.2 Hz, 2 H), 7.38 (m, 3 H), 6.92 (s, 1 H), 4.89 (br s, 2 H), 2.45 (s, 3 H). ^{13}C NMR (400 MHz, CDCl$_3$) δ 193.03, 150.99, 139.53, 134.36, 132.12, 132.08, 130.24, 129.23, 129.19, 128.441, 123.21, 122.55, 121.06, 118.01, 106.88, 97.66, 96.53, 85.98, 84.89, 30.51. HRMS calcd $C_{24}H_{16}N_2O_3S$: 412.0882. Found: 412.0882.

4-Nitro-3-phenylethynyl-6-trimethylsilylethynylaniline. See the general procedure for the Pd/Cu-catalyzed coupling reaction. The compounds used were 2-bromo-4-nitro-5-(phenylethynyl)aniline (0.26 g, 0.83 mmol), TMSA (0.17 mL, 1.25 mmol), CuI (0.02 g, 0.08 mmol), ((C_6H_5)$_3$P)$_2$PdCl$_2$ (0.03 g, 0.04 mmol), diisopropylethylamine (0.58 mL, 3.32 mmol), and THF (10 mL) at 75 °C for 3 d. The desired material was purified by gravity liquid chromatography using silica gel as the stationary phase and a mixture of 3:1 CH_2Cl_2:hexanes as the eluent. R_f = 0.72. The reaction afforded 0.22 g (81 % yield) of the desired compound. IR (KBr) 3465.06, 3350.39, 3214.34, 2958.03, 2360.06, 2341.17, 2146.27, 1625.20, 1539.10, 1507.32, 1305.69, 1247.56, 1199.99, 1091.12, 878.19, 843.71, 756.00, 663.28, 472.37 cm^{-1}. ^{1}H NMR (400 MHz, CDCl$_3$) δ 8.23 (s, 1 H), 7.65-7.60 (m, 2 H), 7.43-7.38 (m, 3 H), 6.91 (s, 1 H), 4.87 (br s, 2 H), 0.31 (s, 9 H). ^{13}C NMR (400 MHz, CDCl$_3$) δ 151.80, 139.68, 132.47, 130.77, 129.62, 128.85, 122.95, 121.37, 118.22, 107.50, 103.95, 99.08, 97.83, 86.32, 0.30. HRMS calcd $C_{19}H_{18}N_2O_2Si$: 334.1138. Found: 334.1135. This material was deprotected using the standard K_2CO_3 protocol described above, and then further coupled with **3** by the Pd/Cu protocol to afford **70** in 82% yield. The spectra were identical to that described above for **70**.

2′-Acetamido-4,4′-diphenylethynyl-5′-nitro-1-thioacetylbenzene (71). See the general procedure for the Pd/Cu-catalyzed coupling reaction. **69** (0.10 g, 0.28 mmol) was coupled to **9**^{54} (0.08 g, 0.45 mmol) as described above using CuI (0.01 g, 0.02 mmol), ((C_6H_5)$_3$P)$_2$PdCl$_2$ (0.01 g, 0.01 mmol), (C_6H_5)$_3$P (0.01 g, 0.04 mmol), diisopropylethylamine (0.19 mL, 1.12 mmol), and THF (10 mL) in a screw capped pressure tube equipped with a magnetic stirbar. The reaction mixture was allowed to stir at 80 °C for 3 d. The resultant mixture was subjected to an aqueous workup as described above. The desired material was purified by gravity liquid chromatography using silica gel as the stationary phase and CH_2Cl_2 as the eluent. R_f = 0.40. The compound was further purified by a hexanes wash to give 0.10 g (82 % yield) of the desired compound as bright yellow crystals. IR (KBr) 3138.5, 2205.1, 1384.6, 1333.3,

1241.0, 1117.9, 953.8, 897.4, 825.6, 753.6, 687.2, 615.4 cm^{-1}. ^{1}H NMR (400 MHz, CDCl$_3$) δ 8.41 (s, 1 H), 8.29 (s, 1 H), 8.06 (br s, 1 H), 7.62 (m, 2 H), 7.57 (d, J=8.4 Hz, 2 H), 7.46 (d, J=8.5 Hz, 2H), 7.38 (m, 3 H), 2.64 (s, 3 H), 2.32 (s, 3 H). ^{13}C NMR (400 MHz, CDCl$_3$) δ 192.77, 168.29, 143.82, 142.02, 134.51, 132.23, 132.17, 130.17, 129.47, 128.61, 128.46, 123.57, 122.27, 122.21, 120.70, 111.15, 99.43, 98.68, 85.55, 83.51, 30.58, 25.33.   HRMS calcd C$_{26}$H$_{18}$N$_2$O$_4$S: 454.0987. Found: 454.0987.

4-Iodophenyl methyl sulfide. 1,4-Diiodobenzene (6.60g, 20.0 mmol) was added to an oven-dried 2-neck round bottom flask equipped with a stir bar. Air was removed and N$_2$ backfilled (3×). THF (2.5 mL) was then added under N$_2$ and the apparatus was cooled in a dry ice/acetone bath to –78°C. *tert*-BuLi (23.4 mL of 1.7 M solution) was then added drop wise over a period of 45 min. The mixture was allowed to stir for 30 min and sulfur (0.769 g, 24 mmol) was then added to the flask. This mixture was allowed to stir for 10 min and subsequently heated to 0°C and stirred for 10 min. The mixture was then cooled to –78°C and CH$_3$I (1.87 mL, 30 mmol) added. The reaction was allowed to warm to room temperature overnight while maintaining stirring. The reaction was then quenched with water and washed with brine and CH$_2$Cl$_2$ (3×). Gravity column chromatography (silica gel with hexanes as eluent) afforded the desired product (3.14 g, 63 % yield). IR (KBr) 3070.5, 2910.5, 2851.5, 1883.0, 1469.0, 1426.3, 1381.1, 1092.3, 1000.2, 801.5, 482.2 cm^{-1}. ^{1}H NMR (400 MHz, CDCl$_3$) δ 7.60 (dt, J=8.6, 2.0 Hz, 2 H), 7.01 (dt, J=8.6, 2.0 Hz, 2 H), 2.48 (s, 3 H). ^{13}C NMR (100 MHz, CDCl$_3$) δ 139, 138.06, 128.68, 90, 16.10. HRMS calc'd for C$_7$,H$_7$,S,I: 249.9313. Found: 249.9307.

4-Thiomethyl-1-(trimethylsilylethynyl)benzene.   See the general procedure for the Pd/Cu-catalyzed coupling reaction. The compounds used were 4-iodophenyl methyl sulfide (2.0 g, 8.0 mmol), ((C$_6$H$_5$)$_3$P)$_2$PdCl$_2$ (0.281 g, 0.40 mmol), CuI (0.15 g, 0.80 mmol), THF (30 mL), diisopropylethylamine (5.57 mL, 32.0 mmol), and TMSA (1.47 mL, 10.4 mmol) at 50 °C for 10 h. Flash column chromatography (hexanes as eluent) afforded the desired product (1.74 g, 99% yield). ^{1}H NMR (400 MHz, CDCl$_3$) δ 7.39 (dt, J=8.6, 2.0 Hz, 2 H), 7.17 (dt, J=8.6, 2.0 Hz, 2 H), 2.50 (s, 3 H), 0.27 (s, 9 H). ^{13}C NMR (100 MHz, CDCl$_3$) δ 139.99, 132.63, 126.05, 119.78, 105.27, 94.56, 15.72, 0.39. IR (KBr) 3740.6, 3645.4, 3070.5, 3026.7, 2956.4, 2920.0, 2157.7, 1898.8, 1590.8, 1488.9, 1438.7, 1320.2, 1250.8, 1092.2, 1014.6. HRMS calculated for C$_{12}$H$_{16}$SSi: 220.0742. Found: 220.0737.

1-Ethynyl-4-thiomethylbenzene (72). See the general procedure for the deprotection of a trimethylsilyl-protected alkyne. The compounds used were 4-thiomethyl-1-(trimethylsilylethynyl)benzene (0.29 g, 1.33 mmol), K$_2$CO$_3$ (0.92 g, 6.63 mmol), MeOH (20 mL), and CH$_2$Cl$_2$ (20 mL) for 2 h. Due to the instability of conjugated terminal alkynes, the material was immediately used in the next step without additional purification.

73. 2-Bromo-4-nitro-5-(phenylethynyl)aniline (317 mg, 1.00 mmol), bis(($(C_6H_5)_3$P)palladiumdichloride (35 mg, 0.05 mmol), CuI (19 mg, 0.1 mmol), diisopropylethylamine (0.70 mL, 4.0 mmol), **72** (178 mg, 1.2 mmol), and THF (25 mL) were coupled according to the general coupling procedure except that **72** was dissolved in THF and transferred via cannula into the reaction. The reaction mixture was heated at 75 °C overnight. The crude product was then separated via flash chromatography (1:1 CH_2Cl_2:hexanes) to afford 143 mg (37%) as a yellow solid. IR (KBr) 3474.1, 3366.0, 2360.1, 2204.9, 1616.3, 1541.0, 1517.0, 1473.0, 1286.3, 1248.4, 1148.4, 1090.1, 814.8, 754.1, 686.4 cm^{-1}. ^{1}H NMR (400 MHz, CDCl$_3$) δ 8.26 (s, 1H), 7.60-7.53 (m, 2 H), 7.42 (d, J = 8.4, 2 H), 7.37-7.35 (m, 3 H), 7.21 (d, J = 8.5, 2 H) 4.85 (br s, 1 H). ^{13}C NMR (100 MHz, CDCl$_3$) δ 151.3, 141.1, 132.8, 132.5, 132.3, 132.1, 130.5, 129.6, 129.0, 128.9, 126.2, 126.1, 123.0, 121.1, 118.5, 118.3, 107.9, 97.7, 15.6. HRMS Calc'd for $C_{23}H_{16}N_2O_2S$: 384.0933. Found: 384.0932.

2-Bromo-4-nitro-5-(trimethylsilylethynyl)acetanilide. See the general procedure for the Pd/Cu-catalyzed coupling reaction. The compounds used were **68** (4.00 g, 11.84 mmol), TMSA (1.30 mL, 11.8 mmol), CuI (0.22 g, 1.18 mmol), (($(C_6H_5)_3$P)$_2$PdCl$_2$ (0.41 g, 0.59 mmol), diisopropylethylamine (8.25 mL, 47.36 mmol), and THF (80 mL) at 70 °C for 2 d. The desired material was purified by gravity liquid chromatography using silica gel as the stationary phase and a mixture of 3:1 diethyl ether:hexanes as the eluent. R_f (product): 0.43. The reaction afforded 1.46 g (35 % yield, 54 % based on a recovered 1.44 g of starting material) of the desired product. IR (KBr) 3384.6, 3107.7, 3056.4, 2964.1, 2143.6, 1717.9, 1559.0, 1523.1, 1492.3, 1446.2, 1379.5, 1333.3, 1246.2, 1225.6, 1097.4, 846.2, 764.1, 712.8 cm^{-1}. ^{1}H NMR (400 MHz, CDCl$_3$) δ 8.84 (s, 1 H), 8.29 (s, 1 H), 7.75 (br s, 1 H), 2.30 (s, 3 H), 0.27 (s, 9 H). ^{13}C NMR (100 MHz, CDCl$_3$) δ 169.38, 145.60, 140.82, 129.90, 126.63, 120.52, 112.46, 106.70, 100.03, 26.45, 0.93. HRMS Calcd $C_{13}H_{15}BrN_2O_3Si$: 354.0035. Found: 354.0034.

2-(Ethynylphenyl)-4-nitro-5-(trimethylsilylethynyl)acetanilide (74). See the general procedure for the Pd/Cu-catalyzed coupling reaction. 2-Bromo-4-nitro-5-(trimethylsilylethynyl)acetanilide (1.20 g, 3.38 mmol) was coupled to phenylacetylene (0.56 mL, 5.07 mmol) as described above using CuI (0.06 g, 0.34 mmol), (($(C_6H_5)_3$P)$_2$PdCl$_2$ (0.12 g, 0.17 mmol), diisopropylethylamine (2.36 mL, 13.52 mmol), and THF (25 mL) at 75 °C for 3 d. The desired material was purified by gravity liquid chromatography using silica gel as the stationary phase and a mixture of 3:1 CH_2Cl_2:hexanes as the eluent. R_f (product): 0.38. The reaction afforded 1.00 g (79 % yield) of the desired product. IR (KBr) 3384.6, 3128.2, 2953.8, 2215.4, 2153.8, 1707.7, 1543.6, 1523.1, 1497.4, 1456.4, 1384.6, 1338.5, 1225.6, 1169.2, 1112.8, 1051.3, 846.2, 748.7, 687.2, 620.5 cm^{-1}. ^{1}H NMR (400 MHz, CDCl$_3$) δ 8.77 (s, 1 H), 8.21 (d, J=0.03 Hz, 1 H), 8.07 (br s, 1 H), 7.57-7.52 (m, 2 H), 7.47-7.39 (m, 3 H), 2.30 (s, 3 H), 0.29 (s, 9 H). ^{13}C NMR (100 MHz, CDCl$_3$) δ 169.35,

145.49, 142.99, 132.82, 131.01, 129.94, 129.34, 125.26, 122.25, 121.06, 112.98, 107.23, 100.90, 100.76, 83.10, 26.45, 0.96. HRMS Calcd $C_{21}H_{20}N_2O_3Si$: 376.1243. Found: 376.1235.

5-Ethynyl-2-(ethynylphenyl)-4-nitroaniline. See the general procedure for the deprotection of trimethylsilyl-protected alkynes. **74** (0.10 g, 0.27 mmol) was deprotected to the terminal alkyne and the free amine using the procedure described above using K_2CO_3 (0.19 g, 1.35 mmol), MeOH (15 mL), and CH_2Cl_2 (15 mL). The reaction mixture was allowed to stir at room temperature for 2 h. The resultant mixture was subjected to an aqueous workup as described above. Due to the instability of conjugated terminal alkynes, the material was immediately used in the next step without additional purification or identification.

5'-Amino-4,4'-diethynylphenyl-2'-nitro-1-thioacetylbenzene   (75). See the general procedure for the Pd/Cu-catalyzed coupling reaction. The compounds used were 5-ethynyl-2-(ethynylphenyl)-4-nitroaniline (0.08 g, 0.27 mmol), **3** (0.09 g, 0.32 mmol), CuI (0.005 g, 0.01 mmol), $((C_6H_5)_3P)_2PdCl_2$ (0.01 g, 0.01 mmol), diisopropylethylamine (0.20 mL, 1.08 mmol), and THF (20 mL) at 70 °C for 12 h. The desired material was purified by gravity liquid chromatography using silica gel as the stationary phase and a mixture of 3:1 CH_2Cl_2:hexanes as the eluent. R_f = 0.32. The reaction afforded 0.09 g (82 % yield over 3 steps) of the desired product as yellow solid which turned yellowish-green upon standing. IR (KBr) 3466.7, 3364.1, 2205.1, 1702.6, 1615.4, 1548.7, 1507.7, 1476.9, 1307.7, 1246.2, 1117.9, 948.7, 912.8, 871.8, 820.5, 748.7, 682.1 cm^{-1}. ^{1}H NMR (400 MHz, CDCl$_3$) δ 8.28 (s, 1 H), 7.61 (½ABq, J = 8.4 Hz, 2H), 7.55 (m, 2 H), 7.41 (½ABq, J = 8.4 Hz, 2 H), 7.41-7.35 (m, 3 H), 6.90 (s, 1 H), 4.93 (br s, 2 H), 2.44 (s, 3 H). ^{13}C NMR (100 MHz, CDCl$_3$) δ 193.09, 150.93, 139.33, 134.98, 134.13, 132.46, 131.55, 130.06, 129.07, 128.49, 123.69, 121.94, 120.25, 117.91, 107.50, 97.51, 96.26, 87.50, 83.16, 30.45. HRMS Calcd $C_{24}H_{16}N_2O_3S$: 412.0882. Found: 412.0883.

1-Bromo-3-nitro-4-(trimethylsilylethynyl)benzene. See the general procedure for the Pd/Cu-catalyzed coupling reaction. The compounds used were 2,5-dibromonitrobenzene (1.37 g, 4.89 mmol), $((C_6H_5)_3P)_2PdCl_2$ (0.17 g, 0.25 mmol), CuI (0.09g, 0.49 mmol), THF (30 mL), Hünig's base (3.41 mL, 19.56 mmol), and TMSA (0.69 mL, 4.9 mmol) at 70 °C for 18 h. Due to difficulty in separation of products, full characterization was not achieved and the resulting mixture was carried on to the next reaction step. ^{1}H NMR (300 MHz, CDCl$_3$) δ 8.14 (d, J=2.0 Hz, 1 H), 7.66 (dd, J=8.3, 2.0 Hz, 1 H), 7.49 (d, J=8.3 Hz, 1 H), 0.26 (s, 9 H).

2-Ethynyl-5-ethynylphenyl-1-nitrobenzene   (77).   2,5-Dibromonitrobenzene **(76)** (4.0 g, 14.24 mmol), $((C_6H_5)_3P)_2PdCl_2$ (0.300 g, 0.427 mmol), CuI (0.163g, 0.854 mmol), THF (30 mL), diisopropylethylamine (9.9 mL, 57.0 mmol), and TMSA (2.21 mL, 15.66 mmol) were used at room temperature for 10 h following the general procedure for couplings. Flash column chromatography (silica gel using 2:1 hexanes:CH$_2$Cl$_2$ as eluent) afforded a mixture of products that was taken onto the next step. The product mixture (3.09 g), ((C$_6$H$_5$)$_3$P)$_2$PdCl$_2$ (0.217 g, 0.31 mmol), CuI (0.118g, 0.62 mmol), THF (30 mL), diisopropylethylamine (7.2 mL, 41.44 mmol), and phenylacetylene (1.7 mL, 15.54 mmol) were used following the general procedure for couplings at 50 °C for 15 h. Flash column chromatography (silica gel using 1:1 hexanes: CH$_2$Cl$_2$ as eluent) afforded a mixture of products that was taken onto the next step. The product mixture (1.95 g), K$_2$CO$_3$ (4.2 g, 30.4 mmol), MeOH (50 mL), and CH$_2$Cl$_2$ (50 mL) were used following the general procedure for deprotection. Flash column chromatography (silica gel using 1:1 hexanes:CH$_2$Cl$_2$ as eluent) afforded the desired product as an orange solid (1.23 g, 37% yield for three steps). IR (KBr) 3267.2, 3250.1, 3079.6, 2208.4, 2102.6, 1541.6, 1522.5, 1496.0, 1347.1, 1275.2, 900.9, 840.5, 825.0, 759.0, 688.0, 528.8 cm^{-1}. ^{1}H NMR (400 MHz, CDCl$_3$) δ 8.16 (d, J = 1.5 Hz, 1 H), 7.67 (dd, J = 8.1, 1.5 Hz, 1 H), 7.64 (d, J = 7.8 Hz, 1 H), 7.53 (m, 2 H), 7.37 (m, 3 H), 3.58 (s, 1 H). ^{13}C NMR (100 MHz, CDCl$_3$) δ 150.62, 135.82, 135.65, 132.24, 129.72, 128.97, 127.80, 125.51, 122.33, 117.01, 94.35, 87.04, 86.97, 78.82. HRMS calc'd for C$_{16}$,H$_9$,N,O$_2$: 247.0633. Found: 247.0632.

4,4'-Di(ethynylphenyl)-2'-nitro-1-thioacetylbenzene (78). See the standard procedure for Pd/Cu couplings. The compounds used were **77** (0.500 g, 2.02 mmol), **3** (0.675 g, 2.43 mmol), bis(dibenzylideneacetone)palladium(0) (0.232 g, 0.404 mmol), CuI (0.077 g, 0.404 mmol), (C$_6$H$_5$)$_3$P (0.212 g, 0.808 mmol), THF (10 mL), and diisopropylethylamine (0.7 mL, 4.04 mmol) at 50 °C oil bath for 2 d. Column chromatography (silica gel using 2:1 CH$_2$Cl$_2$:hexanes as eluent) afforded the desired product as orange solid (0.381 g, 47% yield). IR (KBr) 3100, 2924, 2213.1, 1697.1, 1537.3, 1346.9, 1131.9, 831.9, 751.4, 684.9, 623.0. ^{1}H NMR (400 MHz, CDCl$_3$) δ 8.22 (dd, J = 1.1, 0.3 Hz, 1 H), 7.70 (dd, J = 8.1, 1.5 Hz, 1 H), 7.67 (d, J = 8.0 Hz, 1 H), 7.61 (dt, J = 8.5, 1.9 Hz, 2 H), 7.54 (m, 2 H), 7.42 (dt, J = 8.5, 1.8 Hz, 2 H), 7.37 (m, 3 H), 2.43 (s, 3 H). ^{13}C NMR (75 MHz, CDCl$_3$) δ 193.10, 149.5, 135.30, 134.55, 134.27, 132.57, 131.84, 129.56, 129.26, 128.56, 127.65, 124.47, 123.40, 122.05, 117.5, 97.84, 93.82, 86.86, 86.31, 30.36. HRMS calculated for C$_{24}$,H$_{15}$,N,O$_3$, S: 397.0076. Found: 397.0773.

2-Bromo-5-(ethynylphenyl)acetanilide. See the general procedure for the Pd/Cu-catalyzed coupling reaction. 2,5-Dibromoacetanilide (6.00 g, 17.76 mmol) was coupled to phenylacetylene (1.95 mL, 17.76 mmol) using CuI (0.34 g, 1.78 mmol), ((C$_6$H$_5$)$_3$P)$_2$PdCl$_2$ (0.62 g, 0.89 mmol), diisopropylethylamine (12.37 mL, 71.04 mmol), and THF (75 mL) at 75 °C for 2.5 d. The desired material was purified by gravity liquid chromatography using silica gel as the stationary phase and CH$_2$Cl$_2$ as the eluent. R$_f$ = 0.38. An additional purification was performed using gravity liquid chromatography using silica gel as the stationary phase and a mixture of 3:1 hexanes:EtOAc as the eluent. R$_f$ = 0.50. The reaction afforded 1.79 g (32 % yield, 42 % based on a recovered 0.69 g of starting material) of the desired compound as a white solid.   IR (KBr) 3282.1, 3159.0, 1661.5, 1559.0, 1507.7, 1461.5, 1405.1, 1379.5, 1271.8, 1107.7, 1066.7, 1015.4, 964.1, 892.3, 861.5, 820.5, 748.7, 682.1, 610.3 cm^{-1}. ^{1}H NMR (400 MHz, CDCl$_3$) δ 8.66 (br s, 1 H), 7.92 (br s, 1 H), 7.55-7.49 (m, 2 H), 7.41-7.37 (m, 3 H), 7.32 (½ABq, J=8.3 Hz, 1 H), 7.20 (½ABq d, J= 6.4, J=1.8 Hz, 1 H), 2.25 (s, 3 H). ^{13}C NMR (100 MHz, CDCl$_3$) δ 169.15, 140.81, 133.62, 132.61, 130.32, 129.80, 127.70, 124.93, 123.33, 123.15, 111.69, 98.63, 84.65, 26.32. HRMS Calcd C$_{16}$H$_{12}$BrNO: 313.0102. Found: 313.0107.

3-Ethynylphenyl-6-(trimethylsilylethynyl)acetanilide (80). See the general procedure for the Pd/Cu-catalyzed coupling reaction.   2-Bromo-5-(ethynylphenyl)acetanilide (0.91 g, 2.90 mmol) was coupled to TMSA (0.47 mL, 4.35 mmol) using CuI (0.06 g, 0.29 mmol), ((C$_6$H$_5$)$_3$P)$_2$PdCl$_2$ (0.11 g, 0.15 mmol), diisopropylethylamine (2.02 mL, 11.60 mmol), and THF (20 mL) at 70 °C for 3 d. The desired material was purified by gravity liquid chromatography using silica gel as the stationary phase and CH$_2$Cl$_2$ as the eluent. R$_f$ = 0.33. The reaction afforded 0.81 g (84 % yield) of the desired compound as yellow foam after drying in a vacuum atmosphere. IR (KBr) 3394.9, 3138.5, 2953.8, 2143.6, 1702.6, 1553.85, 1553.8, 1523.1, 1410.3, 1384.6, 1271.8, 1246.2, 1169.2, 1112.8, 1015.4, 846.2, 753.8, 687.2, 620.5 cm^{-1}. ^{1}H NMR (400 MHz, CDCl$_3$) δ 8.53 (br s, 1 H), 7.91 (br s, 1 H), 7.55-7.49 (m, 2 H), 7.43-7.36 (m, 4 H), 7.15 (dd, J= 6.6, 1.5 Hz, 1 H), 2.24 (s, 3 H), 0.25 (s, 9 H). ^{13}C NMR (100 MHz, CDCl$_3$) δ 169.09, 139.72, 132.62, 132.49, 130.27, 129.79, 128.06, 125.58, 123.63, 123.24, 112.98, 105.68, 99.09, 97.67, 85.26, 26.33, 1.28. HRMS Calcd C$_{21}$H$_{21}$BrNOSi: 331.1392. Found: 331.1391.

3-Ethynylphenyl-6-(trimethylsilylethynyl)aniline. A 100 mL round bottom flask equipped with a magnetic stirbar was charged with 3-ethynylphenyl-6-(trimethylsilylethynyl)acetanilide (0.25 g, 0.75 mmol), HCl (15 mL, 1.5 M), and THF (15 mL). The reaction mixture was heated to reflux for 2.5 h. The reaction progress was monitored by TLC. The reaction was quenched and extracted with water (3×) and diluted with CH$_2$Cl$_2$. The organic layers were combined and dried over MgSO$_4$. Volatiles were removed in vacuo. Crude ^{1}H NMR and TLC showed two inseparable products with similar amine and aromatic resonances. Therefore, the crude reaction mixture was reacted further without purification.

2-Ethynyl-5-(ethynylphenyl)aniline. See the general procedure for the deprotection of trimethylsilyl-protected alkynes. The compounds used were 3-ethynylphenyl-6-(trimethylsilylethynyl)aniline (0.22 g, 0.75 mmol) K$_2$CO$_3$ (0.52 g, 3.75 mmol), MeOH (15 mL), and CH$_2$Cl$_2$ (15 mL) for 2 h. Due to the instability of conjugated terminal alkynes, the material was immediately used in the next step without additional purification or identification.

2′-Amino-4,4′-di(phenylethynyl)-1-thioacetylbenzene (81). See the general procedure for the Pd/Cu-catalyzed coupling reaction. The compounds used were 2-ethynyl-5-(ethynylphenyl)aniline (0.16 g, 0.75 mmol), **3** (0.25 g, 0.90 mmol), CuI (0.02 g, 0.08 mmol), $((C_6H_5)_3P)_2PdCl_2$ (0.03 g, 0.04 mmol), diisopropylethylamine (0.53 mL, 3.00 mmol), and THF (15 mL) at 45 °C for 12 h. The desired material was purified by gravity liquid chromatography using silica gel as the stationary phase and a mixture of 1:3 diethyl ether:hexanes as the eluent. R_f (product): 0.40. The reaction afforded 0.28 g (43% yield, over three steps) of the desired compound as bright yellow solid. IR (KBr) 3138.5, 2205.1, 1702.6, 1610.3, 1384.6, 1117.9, 943.6, 825.6, 753.8, 692.3, 615.4 cm^{-1}. ^{1}H NMR (400 MHz, CDCl$_3$) δ 7.56-7.50 (m, 4 H), 7.42-7.31 (m, 6 H), 6.92-6.87 (m, 2 H), 4.32 (br s, 2 H), 4.44 (s, 3 H). ^{13}C NMR (100 MHz, CDCl$_3$) δ 193.66, 148.20, 134.72, 132.40, 131.75, 128.89, 128.86, 128.84, 128383, 124.60, 124.10, 123.32, 121.49, 117.14, 108.58, 96.57, 91.32, 89.72, 85.78, 30.48. HRMS Calcd C$_{24}$H$_{17}$NOS: 367.1031. Found: 367.1032.

    2-Ethynyl-5-(ethynylphenyl)acetanilide. See the general procedure for the deprotection of trimethylsilyl-protected alkynes. The compounds used were **80** (0.20 g, 0.60 mmol) K$_2$CO$_3$ (0.25 g, 1.80 mmol), MeOH (15 mL), and CH$_2$Cl$_2$ (15 mL) for 2 h. Due to the instability of conjugated terminal alkynes, the material was immediately used in the next step without additional purification or identification.

    2'-Acetamido-4,4'-di(phenylethynyl)-1-thioacetylbenzene (82). See the general procedure for the Pd/Cu-catalyzed coupling reaction. The compounds used were 2-ethynyl-5-(ethynylphenyl)acetanilide (0.16 g, 0.60 mmol), **3** (0.20 g, 0.72 mmol) CuI (0.01 g, 0.06 mmol), $((C_6H_5)_3P)_2PdCl_2$ (0.02 g, 0.03 mmol), diisopropylethylamine (0.42 mL, 2.40 mmol), and THF (20 mL) at 70 °C for 12 h. The desired material was purified by gravity liquid chromatography using silica gel as the stationary phase and a mixture of 3:1 EtOAc:hexanes as the eluent. R_f (product): 0.35. The reaction afforded 0.12 g (50 % yield, two steps) of the desired compound as an off-white solid. IR (KBr) 3138.5, 2933.3, 1702.6, 1656.4, 1543.6, 1379.5, 1261.5, 1112.8, 1010.3, 948.7, 882.1, 820.5, 748.7, 682.1, 610.3 cm^{-1}. ^{1}H NMR (400 MHz, CDCl$_3$) δ 8.62 (br s, 1 H), 7.96 (br s, 1 H), 7.58-7.52 (m, 4 H), 7.46 (½ABq, J=7.8 Hz, 1 H), 7.42-7.37 (m, 5 H), 7.23 (½ABq d, J= 8.1, 1.4 Hz, 1 H), 2.43 (s, 3 H), 2.27 (s, 3 H). ^{13}C NMR (100 MHz, CDCl$_3$) δ 193.62, 168.51, 139.39, 134.70, 132.48, 132.02, 131.88, 129.52, 129.09, 129.00, 126.87, 124.41, 124.25, 122.45, 122.27, 112.45, 98.38, 91.06, 90.61, 84.24, 30.48, 25.10. HRMS Calcd C$_{26}$H$_{19}$NO$_2$S: 410.1215. Found: 410.1212.

    2,5-Bis(trimethylsilylethynyl)-4-nitroacetanilide. See the general procedure for the Pd/Cu-catalyzed coupling reaction. The compounds used were **68**[64] (0.60 g, 1.78 mmol), TMSA (0.78 mL, 7.12 mmol), CuI (0.07 g, 0.37 mmol), $((C_6H_5)_3P)_2PdCl_2$ (0.13 g, 0.18 mmol), diisopropylethylamine (2.48 mL, 14.24 mmol), and THF (20 mL) at 75 °C for 3 d. The desired material was purified by gravity liquid chromatography using silica gel as the stationary phase and a mixture of 3:1 diethyl ether:hexanes as the eluent. R_f (product):

0.80. The reaction afforded 0.63 g (95% yield; 0.26 g of material as the product with the deprotected amino moiety instead of the acetamide was also obtained) of the desired product. IR (KBr) 3374.4, 3117.9, 2964.1, 2143.6, 1723.1, 1610.3, 1543.6, 1502.6, 1456.4, 1400.0, 1379.5, 1333.3, 1251.3, 1220.5, 1112.8, 882.1, 846.2, 759.0, 620.5 cm^{-1}. ^{1}H NMR (400 MHz, CDCl$_3$) δ 8.72 (s, 1 H), 8.11 (s, 1 H), 8.07 (br s, 1 H), 2.25 (s, 3 H), 0.31 (s, 9 H), 0.26 (s, 9 H). ^{13}C NMR (100 MHz, CDCl$_3$) δ169.27, 145.23, 143.53, 129.21, 124.94, 121.27, 112.68, 107.86, 107.26, 100.70, 98.57, 26.25, 1.08, 0.94. HRMS Calcd C$_{18}$H$_{24}$N$_2$O$_3$Si$_2$: 372.1325. Found: 372.1332.

2,5-Di(ethynyl)-4-nitroaniline. See the general procedure for the deprotection of trimethylsilyl-protected alkynes. The compounds used were 2,5-bis(trimethylsilylethynyl)-4-nitroacetanilide (0.60 g, 1.61 mmol), K$_2$CO$_3$ (2.22 g, 16.10 mmol), MeOH (40 mL), and CH$_2$Cl$_2$ (40 mL) for 2 h. Due to the instability of conjugated terminal alkynes, the material was immediately used in the next step without additional purification or identification.

2,5-Diphenylethynyl-4',4''-dithioacetyl-4-nitroaniline (83). See the general procedure for the Pd/Cu-catalyzed coupling reaction. The compounds used were 2,5-di(ethynyl)-4-nitroaniline (0.30 g, 1.61 mmol), **3** (1.09 g, 3.86 mmol) CuI (0.06 g, 0.32 mmol), ((C$_6$H$_5$)$_3$P)$_2$PdCl$_2$ (0.11 g, 0.16 mmol), diisopropylethylamine (2.25 mL, 12.88 mmol), and THF (40 mL) at 50 °C for 2 d. The desired material was purified by gravity liquid chromatography using silica gel as the stationary phase and a mixture of 1:1 EtOAc:hexanes as the eluent. R$_f$ = 0.52. The reaction afforded 0.47g (57% over three steps) of the desired compound as bright yellow solid. IR (KBr) 3476.9, 3364.1, 3117.9, 1687.2, 1625.6, 1543.6, 1507.7, 1476.9, 1384.6, 1307.7, 1246.2, 1117.9, 1010.3, 948.7, 825.6, 615.4 cm^{-1}. ^{1}H NMR (400 MHz, CDCl$_3$) δ 8.28 (s, 1 H), 7.62 (½ABq, J=8.2 Hz, 2 H), 7.56 (½ABq, J=8.4 Hz, 2 H), 7.44 (½ABq, J=4.4 Hz, 2 H), 7.42 (½ABq, J=4.2 Hz, 2 H), 6.92 (s, 1 H), 4.90 (br s, 2 H), 2.45 (s, 3 H), 2.44 (s, 3 H). ^{13}C NMR (100 MHz, CDCl$_3$) δ 193.00, 192.91, 150.99, 134.26, 134.25, 134.25, 134.13, 134.11, 132.45, 132.04, 130.16, 123.64, 123.09, 120.52, 118.01, 107.00, 96.59, 96.45, 87.44, 84.79, 30.44, 30.42. HRMS Calcd C$_{24}$H$_{17}$NOS: 487.0786. Found: 487.0792.

4-(Trimethylsilylethynyl)aniline. See the general procedure for the Pd/Cu cross couplings. The compounds used were 4-bromoaniline (6.88 g, 40 mmol), TMSA (11.3 mL, 80 mmol), tetrakis((C$_6$H$_5$)$_3$P)palladium(0) (393 mg, 0.34 mmol), CuI (76 mg, 0.4 mmol) and diisopropylamine (40 mL) at 110-120° C for 12 h. The mixture was concentrated and filtered through a plug of silica get using 1:1 EtOAc:hexane. The filtrate was concentrated and purified on a silica gel column (1:5 EtOAc:hexane). IR (KBr) 3469, 3374, 2958, 2156, 2144, 1624, 1508, 1294, 1249 cm^{-1}. ^{1}H NMR (300 MHz, CDCl$_3$) δ 7.26 (d, J = 8.6 Hz, 2 H), 6.57 (d, J = 8.6 Hz, 2 H), 0.22 (s, 9 H).

4-Ethynylaniline. See the general procedure for the deprotection of trimethylsilyl alkynes. The compounds used were 4-

(trimethylsilylethynyl)aniline (2.48 g, 13 mmol), K_2CO_3 (11 g) and MeOH (100 mL) for 12 h. Hexane was added to a highly concentrated ether solution to give fine crystals. The collected crystals were washed with hexane and dried in vacuo to afford 1.14 g (74 %) of the title compound. ^1H NMR (400 MHz, $CDCl_3$) δ 7.28 (d, J = 8.6 Hz, 2 H), 6.58 (d, J = 8.6 Hz, 2 H), 3.72- 3.88 (br, 2 H), 2.94 (s, 1 H). ^{13}C NMR (100 MHz, $CDCl_3$) δ 147.0, 133.5, 114.6, 111.4, 84.4, 74.9.

4'-Ethynylformanilide (84). A solution of 4-ethynylaniline (0.87 g, 6.0 mmol) in ethyl formate (40 mL) was heated to reflux for 24 h. After removal of the solvents by a rotary evaporation, another portion of ethyl formate (40 mL) was added and the solution was heated to reflux for 24 h. The evaporated residue was chromatographed on silica gel (1:2 EtOAc:hexane) to afford 0.65 g (75 %) of a slightly brown-white solid of the title compound. IR (KBr) 3292, 2107, 1686, 1672, 1601, 1536, 1407, 1314, 842, 668, 606 cm^{-1}. ^1H NMR (400 MHz, $CDCl_3$) δ 8.71 (d, J = 11.3 Hz, a), 8.38 (d, J = 1.5 Hz, b), 7.82-7.92 (br d, J = 10 Hz, c), 7.48 (dt, J = 14.5, 8.7 Hz, 3 H), 7.2 (br, d), 7.02 (d, J = 8.6 Hz, 1 H), 3.07 (s, e), 3.04 (s, f) where (a + b = 1 H, c + d = 1 H, e + f = 1 H).

85. See the general procedure for the Pd/Cu-catalyzed coupling reaction. The compounds used were 2-bromo-4-nitro-5-(phenylethynyl)aniline (0.26 g, 0.83 mmol), 84 (0.15 g, 1.00 mmol), CuI (0.02 g, 0.08 mmol), $((C_6H_5)_3P)_2PdCl_2$ (0.03 g, 0.04 mmol), diisopropylethylamine (0.58 mL, 3.32 mmol), and THF (25 mL) at 70 °C for 3 d. The desired material was purified by gravity liquid chromatography using silica gel as the stationary phase and a mixture of 1:1 EtOAc:hexanes as the eluent. R_f = 0.09. An additional purification was performed using gravity liquid chromatography using silica gel as the stationary phase and a mixture of EtOAc as the eluent. R_f = 0.63. The reaction afforded an impure product of 0.23 g. The crude reaction product was taken on to the next synthetic step.

2'-Amino-4,4'-diphenylethynyl-5'-nitrobenzeneisonitrile (86). To an oven dried 100 mL round bottom flask equipped with a stirbar and a West condenser was added 85 (0.04 g, 0.10 mmol), $(C_6H_5)_3P$ (0.09 g, 0.33 mmol), Et_3N (0.04 mL, 0.39 mmol), CCl_4 (0.03 mL, 0.31 mmol), and CH_2Cl_2 (10 mL).[68] The reaction was heated to 60 °C for 5 h. The reaction mixture was cooled and quenched with water and extracted with CH_2Cl_2 (3×). Organic layers were combined and dried over $MgSO_4$. The volatiles were removed in vacuo. The crude reaction mixture was purified by gravity liquid chromatography using silica gel as the stationary phase and EtOAc as the eluent. R_f = 0.85. An additional purification was performed using gravity liquid chromatography using silica gel as the stationary phase and a mixture of 1:1 CH_2Cl_2:hexanes as the eluent. R_f (product): 0.30. The reaction afforded 0.03 g (83 % yield, two steps) of the desired material. IR (KBr) 3450.62, 3358.15, 2925.78, 2855.52, 2200.00, 2114.03, 1618.06, 1542.38, 1506.39,

1432.51, 1367.16, 1309.39, 1246.34, 1203.57, 1144.72, 1097.07, 995.30, 835.22, 749.25, 470.10 cm^{-1}. ^{1}H NMR (400 MHz, CHCl$_3$) δ 8.32 (s, 1 H), 7.68-7.55 (m, 4 H), 7.45-7.37 (m, 5 H), 6.97 (s, 1 H), 4.89 (br s, 2 H). ^{13}C NMR (100 MHz, CDCl$_3$) δ 151.38, 133.03, 132.52, 132.07, 130.82, 129.74, 129.03, 128.89, 127.11, 126.98, 123.92, 122.86, 121.80, 118.54, 106.73, 98.30, 95.78, 86.36, 86.16. HRMS Calcd C$_{23}$H$_{13}$N$_3$O$_2$: 363.1008. Found: 363.1008.

1-Bromo-4-n-hexylbenzene. The procedure of Ranu et al. was followed.[70] In an 125 mL flask, bromine (0.52 mL, 10 mmol) was absorbed onto neutral, Brockmann grade I, alumina (10 g). 1-Phenylhexane (1.88 mL, 10 mmol) was absorbed onto neutral alumina (10 g) in a second 125 mL flask. The contents of both flasks were combined in a 250 mL flask equipped with a magnetic stirbar. The reaction was complete within 1 min when the dark orange color of the bromine became light yellow. The solid mass was then poured in a column that contained a short plug of silica gel. The desired product was eluted with CH$_2$Cl$_2$ to give 2.58 g of a 80:15:5 mixture (desired product: starting material: ortho-substituted product) as judged by ^{1}H NMR. ^{1}H NMR (400 MHz, CDCl$_3$) δ 7.37 (d, J=8.2 Hz, 2 H), 7.03 (d, J=8.2 Hz, 2 H), 2.54 (t, J=7.5 Hz, 2 H), 1.60 (p, J=7.1 Hz, 2 H), 1.38-1.24 (m, 8 H), 0.92-0.84 (m , 3 H).

1-n-Hexyl-4-(trimethylsilylethynyl)benzene. See the general procedure for the Pd/Cu-catalyzed coupling reaction. The compounds used were 1-bromo-4-n-hexylbenzene (7.23 g, 30.0 mmol) TMSA (5.94 mL, 42.0 mmol), CuI (0.69 g, 3.6 mmol), ((C$_6$H$_5$)$_3$P)$_2$PdCl$_2$ (0.84 g, 1.2 mmol), (C$_6$H$_5$)$_3$P (1.57 g, 6.0 mmol), Et$_3$N (30.36 mL, 300 mmol), and THF (30 mL) at 85 °C for 3 d. The resultant mixture was subjected to an aqueous workup as described above. The desired material was purified by gravity liquid chromatography using silica gel as the stationary phase and hexanes as the eluent. The reaction afforded 5.26 g (68 % yield) of the desired material. IR (KBr) 2923.1, 2851.3, 2158.2, 1923.1, 1507.7, 1461.5, 1405.1, 1246.2, 1220.5, 861.5, 835.9, 753.8, 600.0 cm^{-1}. ^{1}H NMR (400 MHz, CDCl$_3$) δ 7.35 (d, J=8.0 Hz, 2 H), 7.07 (d, J=8.1 Hz, 2 H), 2.56 (t, J=7.7 Hz, 2 H), 1.62-1.50 (m, 2 H), 1.28 (br s, 8 H), 0.86 (br t, 3 H), 0.22 (s, 9 H). ^{13}C NMR (400 MHz, CDCl$_3$) δ 143.48, 131.75, 128.18, 120.17, 105.36, 93.15, 35.95, 31.76, 31.24, 28.95, 22.68, 14.18, 0.17. HRMS calcd C$_{17}$H$_{26}$Si: 258.1804. Found: 258.1793.

1-Ethynyl-4-n-hexylbenzene. See the general procedure for the deprotection of trimethylsilyl-protected alkynes. The compounds used were 1-n-hexyl-4-(trimethylsilylethynyl)benzene (0.18 g, 0.7 mmol), K$_2$CO$_3$ (0.48 g, 3.5 mmol), MeOH (10 mL), and CH$_2$Cl$_2$ (10 mL) for 2 h. The material was immediately reacted in the next step without additional purification or identification.

2-Bromo-5-(4´-n-hexylphenylethynyl)-4-nitroacetanilide (88). See the general procedure for the Pd/Cu-catalyzed coupling reaction. The compounds used were **68** (1.42 g, 4.21 mmol) 1-ethynyl-4-n-hexylbenzene

(0.95 g, 3.83 mmol), CuI (0.02 g, 0.08 mmol), $((C_6H_5)_3P)_2PdCl_2$ (0.07 g, 0.38 mmol), diisopropylethylamine (2.69 mL, 15.38 mmol), and THF (20 mL) at 75 °C for 3 d. The desired material was purified by gravity liquid chromatography using silica gel as the stationary phase and a mixture of CH_2Cl_2 as the eluent. R_f (product): 0.58. The reaction afforded 0.52 g (31% yield, two steps) of the desired material. IR (KBr) 3276.65, 3086.57, 3016.72, 2926.07, 2852.18, 2213.34, 1671.79, 1592.84, 1560.91, 1534.12, 1500.11, 1460.89, 1389.78, 1337.01, 1260.89, 1093.45, 1020.52, 894.33, 813.73, 743.88, 631.04, 464.48, 442.99 cm^{-1}. ^{1}H NMR (400 MHz, CHCl$_3$) δ 8.83 (s, 1 H), 8.39 (s, 1 H), 7.81 (br s, 1 H), 7.35 (ABq, J=8.3 Hz, Δv=110.6 Hz, 4 H), 2.65 (t, J=7.6 Hz, 2 H), 2.33 (s, 3 H), 1.63 (p, J=7.8, 6.1, 2 H), 1.40-1.22 (m, 6 H), 0.93 (t, J= 7.2 Hz, 3 H). ^{13}C NMR (100 MHz, CDCl$_3$) δ 168.83, 145.36, 144.28, 140.22, 132.55, 129.38, 129.03, 125.18, 120.44, 119.69, 111.21, 99.49, 84.76, 36.43, 32.08, 31.53, 29.32, 25.49, 22.99, 14.49. HRMS Calcd $C_{22}H_{23}{}^{79}BrN_2O_3$: 442.0892. Found: 442.0895.

Methyl 4-(trimethylsilylethynyl)benzoate. See the general procedure for the Pd/Cu-catalyzed coupling reaction. The compounds used were methyl 4-iodobenzoate (5.00 g, 19.1 mmol), $((C_6H_5)_3P)_2PdCl_2$ (0.670 g, 0.955 mmol), CuI (0.36 g, 1.91 mmol), THF (50 mL), diisopropylethylamine (13.31 mL, 76.4 mmol) and TMSA (3.51 mL, 24.8 mmol) at 60°C for 18 h. Column chromatography (silica gel, 1:1 hexanes:CH_2Cl_2) afforded the desired product (4.34 g, 98% yield) as orange crystals. IR (KBr) 2958.6, 2159.9, 1720.7, 1603.2, 1443.2, 1404.8, 1278.3, 1243.6, 1171.1, 1110.5, 1017.0, 841.6, 771.1 cm^{-1}. ^{1}H NMR (400 MHz, CDCl$_3$) δ 7.99 (dt, J=8.7 Hz, 1.7 Hz, 2 H), 7.54 (dt, J=8.6, 1.7 Hz, 2 H), 3.94 (s, 3 H), 0.28 (s, 9 H). ^{13}C NMR (100 MHz, CDCl$_3$) δ 166.68, 132.06, 129.89, 129.57, 127.97, 104.27, 97.88, 52.40, 0.029. HRMS calculated for $C_{13}H_{16}O_2Si$: 232.091959. Found: 232.0919

Methyl 4-ethynylbenzoate (89). See the general procedure for the deprotection of trimethylsilyl-protected alkynes. The compounds used were methyl 4-(trimethylsilylethynyl)benzoate (0.75 g, 3.23 mmol), K_2CO_3 (2.23 g, 16.15 mmol), MeOH (50 mL) and CH_2Cl_2 (50 mL) for 2 h. Extraction of the product afforded 0.49 g of the desired product that was immediately reacted in the next step.

Methyl 2′-acetamido-4,4′-diphenylethynyl-4′′-n-hexyl-5′-nitrobenzoate. See the general procedure for the Pd/Cu-catalyzed coupling reaction. The compounds used were 88 (0.23 g, 0.52 mmol), 89 (0.11 g, 0.68 mmol), CuI (0.01 g, 0.05 mmol), $((C_6H_5)_3P)_2PdCl_2$ (0.02 g, 0.03 mmol), diisopropylethylamine (0.36 mL, 2.08 mmol), and THF (15 mL) at 75 °C for 3 d. The desired material was purified by gravity liquid chromatography using silica gel as the stationary phase and CH_2Cl_2 as the eluent. R_f: 0.20. The reaction afforded 0.26 g (96 % yield) of the desired material. IR (KBr) 3426.99, 3286.07, 2926.72, 2844.78, 2361.19, 2334.33, 2194.63, 1722.66, 1671.72, 1602.42, 1546.11, 1494.92, 1426.27, 1407.39, 1339.77, 1276.39,

1173.17, 1105.95, 760.00 cm^{-1}. ^{1}H NMR (400 MHz, CHCl$_3$) δ 8.86 (s, 1 H), 8.33 (s, 1 H), 8.06 (br s, 1 H), 7.85 (ABq, J=6.8 Hz, Δv=188.8 Hz, 4 H), 7.35 (ABq, J=8.2 Hz, Δv=170.20 Hz, 4 H), 3.98 (s, 3 H), 2.66 (t, J=7.6 Hz, 2 H), 2.35 (s, 3 H), 1.65 (p, J=7.8, 6.1 Hz, 2 H), 1.40-1.37 (m, 6 H), 0.93 (t, J=6.8 Hz, 3 H). ^{13}C NMR (100 MHz, CDCl$_3$) δ 168.70, 166.52, 145.43, 144.23, 142.39, 132.65, 132.03, 131.37, 130.30, 129.15, 129.03, 126.09, 123.95, 121.65, 119.74, 111.04, 100.61, 98.76, 85.46, 85.05, 52.86, 36.45, 32.08, 31.54, 29.32, 25.56, 22.99, 14.48. HRMS Calcd C$_{32}$H$_{30}$N$_2$O$_5$: 522.2155. Found: 522.2147.

Methyl   2′-amino-4,-4′-diphenylethynyl-4′′-n-hexyl-5′-nitrobenzoate (90). To a 100 mL round bottom flask equipped with a magnetic stirbar was added methyl 2′-acetamido-4,4′-diphenylethynyl-4′′-n-hexyl-5′-nitrobenzoate (0.10 g, 0.19 mmol), K$_2$CO$_3$ (0.16 g, 1.15 mmol), MeOH (15 mL), and CH$_2$Cl$_2$ (15 mL). The reaction mixture was allowed to react at room temperature for 1 h. The reaction was quenched with water and extracted with CH$_2$Cl$_2$ (3×). Organic layers were combined and dried over MgSO$_4$. Volatiles were removed in vacuo. No further purification was needed. The reaction afforded 0.09 g (99 % yield) of the desired material. IR (KBr) 3475.47, 3362.54, 2914.63, 2850.15, 2205.37, 1706.79, 1629.40, 1596.36, 1543.77, 1519.70, 1426.27, 1316.01, 1290.51, 1279.59, 1173.73, 1141.49, 1114.87, 760.00, 679.40, 614.93, 469.85 cm^{-1}. ^{1}H NMR (400 MHz, CHCl$_3$) δ 8.31 (s, 1 H), 7.85 (ABq, J= 8.6 Hz, Δv=182.9 Hz, 4 H), 7.36 (ABq, J=8.2 Hz, Δv=129.83 Hz, 4 H), 6.95 (s, 1 H), 4.92 (br s, 2 H), 3.96 (s, 3 H), 2.64 (t, J=7.6 Hz, 2 H), 1.65 (p, J=7.7, 6.8 Hz, 2 H), 1.36-1.27 (m, 6 H), 0.91 (t, J=7.1 Hz, 3 H). ^{13}C NMR (100 MHz, CDCl$_3$) δ 166.74, 151.42, 145.13, 139.96, 132.46, 131.92, 130.77, 130.64, 130.10, 129.00, 127.11, 121.96, 120.01, 118.38, 106.80, 98.72, 96.69, 86.51, 85.81, 52.75, 36.42, 32.08, 31.56, 29.32, 22.99, 14.47. HRMS Calcd C$_{30}$H$_{28}$N$_2$O$_4$: 480.2049. Found: 480.2050.

2′-Amino-4,4′-diphenylethynyl-4′′-n-hexyl-5′-nitrobenzoic   acid (91). The procedure by Corey et al. was followed.[71] To a 250 mL round bottom flask equipped with a magnetic stirbar was added **90** (0.07 g, 0.15 mmol), LiOH (0.02, 0.75 mmol), MeOH (9 mL), CH$_2$Cl$_2$ (5 mL), and water (3 mL). The reaction mixture was allowed to stir at room temperature for 2.5 d. The reaction was quenched with water and extracted with CH$_2$Cl$_2$ (3×). The yellow aqueous phases were combined and acidified to pH = 3 whereupon a yellow solid precipitated. The solid material was collected on a fritted funnel. The collected solid reaction mixture was purified by gravity column chromatography using silica gel as the stationary phase and CH$_2$Cl$_2$ as the eluent. R$_f$ (product): 0.10. The reaction afforded 0.065 g (94 % yield) of the desired material. IR (KBr) 3460.77, 3378.60, 2957.49, 2921.54, 2844.51, 2207.7, 1580.98, 1542.74, 1428.19, 1385.56, 1307.71, 1242.23, 1108.70, 774.89, 646.51, 615.69, 456.49 cm^{-1}. ^{1}H NMR (400 MHz, MeOH) δ 8.22 (s, 1 H), 7.72 (ABq, J=8.5 Hz, Δv=142.14 Hz, 4 H), 7.38 (ABq, J=8.2 Hz, Δv=97.07

Hz, 4 H), 6.99 (s, 1 H), 2.61 (t, J=7.6 Hz, 2 H), 1.69-1.59 (m, 2 H), 1.42-1.28 (m, 6 H), 0.96-0.86 (m, 3 H).

4,4´-Dibromo-2,2´-dinitrobiphenyl (92).[72] In a large oven dried screw capped tube equipped with a magnetic stirbar was added 2,2´-dinitrobiphenyl (2.44 g, 10.0 mmol) and silver acetate (4.01 g, 24.0 mmol). Glacial acetic acid (20 mL), sulfuric acid (2.03 mL, 38.0 mmol), and Br_2 (1.54 mL, 30.0 mmol) were sequentially added and the reaction vessel was capped. The reaction vessel was heated to 80 °C for 16 h. The reaction mixture was cooled and was poured into ice water. The solid material was then collected by filtration. The desired material was purified by gravity liquid chromatography using silica gel as the stationary phase and a mixture of 1:1 CH_2Cl_2:hexanes as the eluent. R_f (product): 0.58. The reaction afforded 1.43 (36 % yield) of the desired material as yellow solid. IR (KBr) 3097.4, 2861.5, 1523.1, 1384.6, 1338.5, 1271.8, 1241.0, 1148.7, 1092.3, 1000.0, 892.3, 835.9, 764.1, 723.1, 697.4 cm^{-1}. ^{1}H NMR (400 MHz, CDCl$_3$) δ 8.37 (d, J=2.0 Hz, 2 H), 7.81 (dd, J=2.0, 8.2 Hz, 2 H), 7.15 (d, J=8.0 Hz, 2 H). ^{13}C NMR (100 MHz, CDCl$_3$) δ 147.07, 136.34, 131.76, 131.69, 127.81, 122.66. HRMS Calcd $C_{12}H_6Br_2N_2O_4$: 399.8694. Found: 399.8675.

4,4´-Bis(trimethylsilylethynyl)-2,2´-dinitrobiphenyl (93). See the general procedure for the Pd/Cu-catalyzed coupling reaction. The compounds used were 4,4´-dibromo-2,2´-dinitrobiphenyl (1.50 g, 3.73 mmol), TMSA (1.32 mL, 9.33 mmol), CuI (0.07 g, 0.37 mmol), $((C_6H_5)_3P)_2PdCl_2$ (0.13 g, 0.19 mmol), $(C_6H_5)_3P$ (0.20 g, 0.75 mmol), Et$_3$N (1.62 mL, 14.92 mmol), and THF (25 mL) at 75 °C for 3 d. The desired material was purified by gravity liquid chromatography using silica gel as the stationary phase and a mixture of 1:1 CH_2Cl_2:hexanes as the eluent. R_f (product): 0.55. The reaction afforded 1.44 g (88 % yield) of the desired compound as a very viscous yellow liquid. IR (KBr) 3743.6, 3651.3, 3076.9, 2953.8, 2892.3, 2153.8, 2061.8, 1943.6, 1876.9, 1805.1, 1610.4, 1523.3, 1477.1, 1405.3, 1338.6, 1256.6, 1215.6, 1143.8, 1092.5, 1000.2, 928.4, 851.5, 759.2, 692.5, 641.2 cm^{-1}. ^{1}H NMR (400 MHz, CDCl$_3$) δ 8.28 (d, J=1.6 Hz, 2 H), 7.71 (dd, J=6.2, 0.7 Hz, 2 H), 7.20 (d, J=6.9 Hz, 2 H), 0.19 (s, 18 H). ^{13}C NMR (100 MHz, CDCl$_3$) δ 146.76, 136.22, 133.26, 130.71, 128.04, 124.97, 101.67, 98.74, -0.07. HRMS Calcd $C_{22}H_{24}N_2O_4Si_2$: 436.1275. Found: 436.1281.

2-Amino-4,4´-bis(trimethylsilylethynyl)-2´-nitrobiphenyl (94). 93 (0.70 g, 1.60 mmol), glacial acetic acid (15 mL), and THF (15 mL) were added to a 100 mL round bottom flask equipped with a magnetic stirbar and a West condenser. The reaction mixture was heated to reflux. Iron powder (0.20 g, 3.52 mmol) was carefully added to the refluxing reaction mixture.[73] The reaction mixture was allowed to reflux for 2 h while being monitored by TLC. The reaction mixture was cooled, quenched with water, and filtered through filter paper to remove unreacted iron. The filtrate was extracted with brine (3×) and diluted with CH_2Cl_2. Organic layers were combined and dried over

MgSO$_4$. Volatiles were removed in vacuo. The crude reaction mixture was purified by gravity liquid chromatography using silica gel as the stationary phase and a mixture of 3:1 CH$_2$Cl$_2$:hexanes as the eluent. R$_f$ (product): 0.68. The reaction afforded 0.13 g (21 % yield, 33 % based on a recovered 0.26 g of starting material) of the desired material. IR (KBr) 3469.7, 3382.5, 2953.8, 2154.7, 1617.1, 1529.9, 1479.1, 1413.7, 1346.2, 1242.5, 848.4, 759.5 cm^{-1}. ^{1}H NMR (400 MHz, CDCl$_3$) δ 8.02 (d, J=1.7 Hz, 1 H), 7.68 (dd, J=7.8, 1.6 Hz, 1 H), 7.36 (d J= 7.8 Hz, 1 H), 6.93-6.86 (m, 3 H), 3.49 (s, 2 H), 0.28 (s, 9 H), 0.25 (s, 9 H). ^{13}C NMR (100 MHz, CDCl$_3$) δ 149.08, 143.25, 135.73, 132.67, 132.41, 128.92, 127.58, 124.33, 124.19, 123.13, 122.56, 118.91, 104.72, 101.77, 98.24, 94.43, 0.09, -0.12. HRMS Calcd C$_{22}$H$_{26}$N$_2$O$_2$Si$_2$: 406.1533. Found: 406.1532.

2-Amino-4,4′-diethynyl-2′-nitrobiphenyl. See the general procedure for the deprotection of trimethylsilyl-protected alkynes. The compounds used were **94** (0.13 g, 0.33 mmol), K$_2$CO$_3$ (0.46 g, 3.30 mmol), MeOH (10 mL) and CH$_2$Cl$_2$ (10 mL) for 2 h. Due to the instability of conjugated terminal alkynes, the material was immediately used in the next step without additional purification or identification.

95. See the general procedure for the Pd/Cu-catalyzed coupling reaction. The compounds used were 2-amino-4,4′-diethynyl-2′-nitrobiphenyl (0.09 g, 0.33 mmol), **3** (0.22 g, 0.79 mmol), CuI (0.02 g, 0.10 mmol), ((C$_6$H$_5$)$_3$P)$_2$PdCl$_2$ (0.02 g, 0.03 mmol), diisopropylethylamine (0.46 mL, 2.64 mmol), and THF (10 mL) at 50 °C for 2 d. The desired material was purified by gravity liquid chromatography using silica gel as the stationary phase and CH$_2$Cl$_2$ as the eluent. R$_f$ (product): 0.55. The reaction afforded 0.11 g (61 % yield, two steps) of the desired compound as bright yellow solid. IR (KBr) 3128.2, 2924.8, 2859.4, 1718.8, 1348.6, 1261.1, 1108.5, 948.7, 825.2, 614.5 cm^{-1}. ^{1}H NMR (400 MHz, CDCl$_3$) δ 8.12 (d, J=1.2 Hz, 1 H), 7.78 (dd, J=6.2, 1.6 Hz, 1 H), 7.58 (dd, J=6.6, 1.8 Hz, 2 H), 7.54 (d, J=8.6 Hz, 2 H), 7.46-7.36 (m, 5 H), 7.02-6.94 (m, 3 H), 3.59 (s, 2 H), 2.45 (s, 3 H), 2.43 (s, 3 H). ^{13}C NMR (100 MHz, CDCl$_3$) δ 193.26, 192.96, 149.29, 143.50, 135.56, 134.24, 134.14, 132.74, 132.66, 132.44, 132.36, 132.26, 132.14, 129.17, 128.02, 127.35, 124.42, 124.22, 124.09, 123.21, 122.38, 118.631, 91.64, 90.84, 88.86, 88.12, 30.48, 30.42. HRMS Calcd C$_{32}$H$_{22}$N$_2$O$_4$S$_2$: 563.1099. Found: 563.1094.

3,3′-Dinitro-2,2′-bipyridyl (97).[74] To a 250 mL round bottom flask equipped with a magnetic stirbar and a West condenser was added 2-chloro-3-nitropyridine (15.0 g, 94.61 mmol) and copper bronze (15.03 g, 236.53 mmol). DMF (100 mL) was added and the reaction mixture was heated to reflux for 18 h. The reaction mixture was cooled and filtered through a pad of celite. The filter cake was washed with hot DMF. The filtrate was poured into 1 L of water and the desired material precipitated. The solid material was collected on a fritted funnel to give 3.57 g (35 % yield) of golden-brown solid. ^{1}H NMR (400 MHz, CDCl$_3$) δ 8.91 (dd, J=4.8, 1.5 Hz, 2 H), 8.60 (dd, J=8.3, 1.5 Hz, 2

H), 7.67 (dd, J=8.4, 4.8 Hz, 2 H). ^{13}C NMR (100 MHz, CDCl$_3$) δ 153.52, 151.79, 144.33, 133.44, 124.65.

5,5´-Dibromo-3,3´-dinitro-2,2´-bipyridyl (98). To a 100 mL round bottom flask equipped with a magnetic stirbar was added **97** (1.00 g, 4.06 mmol). The starting material was dissolved in MeOH (50 mL) and CH$_2$Cl$_2$ (50 mL). In a separate 100 mL two necked round bottom flask was added KBr (9.66 g, 81.2 mmol), and then Br$_2$ (4.33 mL, 81.2 mmol) was slowly added.[75] The KBr/Br$_2$ mixture was slowly transferred via cannula over 30 min to the first flask containing the bipyridine. The desired material precipitated and was collected on a fritted funnel. The collected solid was added to an oven dried pressure tube equipped with a magnetic stirbar and capped with a septum. Br$_2$ (0.42 mL, 8.12 mmol) was added, the septum was removed and the reaction vessel was quickly sealed with a screw cap then heated to 180 °C for 3 d. The reaction was cooled and poured into a solution of ice water. 1 M NaHSO$_3$ (aq) was added to react with any unreacted Br$_2$. The solution was made alkaline with NaOH (s). The resulting solution was extracted with CH$_2$Cl$_2$ (4×). The organic layers were combined and dried over MgSO$_4$. Volatiles were removed in vacuo. The reaction mixture was purified by gravity liquid chromatography using silica gel as the stationary phase and 2:3 EtOAc:hexanes as the eluent mixture. R$_f$ = 0.41. The reaction afforded 0.52 g (45 % yield). IR (KBr) 3425.07, 3059.70, 1578.41, 1544.96, 1428.03, 1345.68, 1232.84, 1104.05, 1027.57, 897.37, 879.49, 789.60, 749.49, 649.64, 551.72, 475.22 cm^{-1}. ^{1}H NMR (400 MHz, CDCl$_3$) δ 8.89 (d, J=2.0 Hz, 2 H), 8.67 (d, J= 2.1 Hz, 2 H). ^{13}C NMR (100 MHz, CDCl$_3$) δ154.26, 148.55, 143.76, 135.50, 120.86. HRMS Calcd C$_{10}$H$_4$Br$_2$N$_4$O$_4$: 401.8600. Found: 401.8603.

4-(Trimethylsilylethynyl)benzaldehyde. See the general procedure for the Pd/Cu-catalyzed coupling reaction. The compounds used were 4-iodobenzaldehyde (0.5 g, 2.15 mmol), THF (2.7 mL), TMSA (0.44 mL, 0.31g, 3.18 mmol), diisopropylethylamine (0.6 mL, 3.5 mmol), ((C$_6$H$_5$)$_3$P)$_2$PdCl$_2$ (4 mg, 0.21 mmol) and CuI (0.0020 g, 2.1 mmol) at room temperature for 24 h. After workup, the residue was purified by silica gel column chromatography using hexane:CH$_2$Cl$_2$ (1:1) to provide 0.063 g (73%) of the title compound as a brown solid. MP: 60-66°C. IR (KBr) 2955.6, 2833.4, 2722.2, 2144.5, 1700.0, 1594.5, 1555.6, 1383.3, 1294.5, 1244.5, 1200.0, 1155.6, 861.1, 838.9, 755.6, 661.1 cm^{-1}. ^{1}H NMR (300 MHz, CDCl$_3$) δ 9.98 (s, 1 H), 7.81 (d, J = 8.37 Hz, 2 H), 7.59 (d, J = 8.28 Hz, 2 H), 0.13 (s, 9 H). ^{13}C NMR (100 MHz, CDCl$_3$) δ 191.38, 135.59, 132.47, 129.34, 103.83, 99.02, -0.21. Anal. Calcd for C$_{12}$H$_{14}$OSi: C, 71.00; H, 6.95. Found: C, 71.29; H, 6.96.

4-Ethynylbenzaldehyde. According to the general procedure, the compounds used were 4-(trimethylsilylethynyl)benzaldehyde (0.093 g, 0.45 mmol), CH$_2$Cl$_2$ (5 mL), MeOH (5 mL) and K$_2$CO$_3$ (0.47 g, 3.42 mmol) for 6 h The residue was purified by silica gel column chromatography using CH$_2$Cl$_2$ to provide 0.056 g (95%) of the title compound as pale yellow solid. MP: 84-

86°C. IR (KBr) 3210.3, 1696.9, 1682.0, 1600.0, 1550.0, 1384.6, 1205.1, 1164.1, 825.6, 738.5 cm^{-1}. ^{1}H NMR (300 MHz, CDCl$_3$) δ 10.01 (s, 1 H), 7.81 (d, J = 8.4 Hz, 2 H), 7.63 (d, J = 8.25 Hz, 2 H), 3.27 (s, 1 H). ^{13}C NMR (100 MHz, CDCl$_3$) δ 192.38, 137.06, 133.83, 130.62, 129.43, 83.84, 82.28. FABMS Calcd for C$_9$H$_6$O: 130. Found: 130.

4-Thioacetyldiphenylethynylcarboxaldehyde (110). See the Pd/Cu coupling protocol. The compounds used were 4-ethynylbenzaldehyde (0.049 g, 0.37 mmol), **3** (0.123 g, 0.44 mmol), ((C$_6$H$_5$)$_3$P)$_2$PdCl$_2$ (0.013 g, 0.06 mmol), CuI (0.35 mg, 0.18 mmol) and THF (0.2 mL). The residue was purified by silica gel column chromatography using CH$_2$Cl$_2$/hexane (1:1) as the eluent. The solvent was removed in vacuo to afford 0.078 g (75%) of the title product as a yellow solid. MP: 122-123°C. IR (KBr) 3138.5, 2841.0 1697.4, 1594.9, 1379.5, 1287.2, 1123.1, 959.0, 820.5, 723.1 cm^{-1}. ^{1}H NMR (300 MHz, CDCl$_3$) δ 10.10 (s, 1 H), 7.84 (d, J = 8.4 Hz, 2 H), 7.65 (d, J = 8.25 Hz, 2 H), 7.55 (d, J = 8.4 Hz, 2 H), 7.39 (d, J = 8.4 Hz, 2 H), 2.43 (s, 3 H). ^{13}C NMR (100 MHz, CDCl$_3$) δ 194.14, 192.31, 136.67, 135.34, 134.17, 133.27, 130.67, 130.26, 130.04, 124.76, 93.68, 91.18, 31.66. FABMS Calcd for C$_{17}$H$_{12}$O$_2$S: 280; Found: 280. Anal. Calcd for C$_{17}$H$_{12}$O$_2$S: C, 72.83; H, 4.34. Found: C, 72.21; H, 4.35.

5,15-Bis(4-thioacetyldiphenylethynyl)-10,20-bis(phenyl)porphyrin (111). A solution of **110** (0.10 g, 0.35 mmol) and meso-phenyldipyrrromethane **(102)**[78] (0.079 g, 0.36 mmol), in CHCl$_3$ (36 mL) at room temperature was degassed under N$_2$ for 15 min. This was followed by the addition of two drops of BF$_3$·OEt$_2$. The solution was left stirring under N$_2$ for 1 h after which time DDQ (0.081 g, 0.36 mmol) was added and stirring continued for another 1 h. The solvent was removed in vacuo and the crude sample was purified by silica gel column chromatography using CH$_2$Cl$_2$ as the eluent followed by a second column purification with CH$_2$Cl$_2$:hexane 1:1 to provide 0.047 g (27%) of the title compound in the first major fraction as a purple powder. MP: 200-204°C. IR (KBr) 3435.9, 3128.2, 1625.6, 1384.6, 1123.1, 800 cm^{-1}. ^{1}H NMR (300 MHz, CDCl$_3$) δ 8.85 (m , 8 H), 8.19 (d, J = 7.92 Hz, 8 H), 7.91 (d, J = 7.83 Hz, 4 H), 7.70-7.76 (m, 10 H), 7.45 (d, J = 8.19 Hz, 4 H), 2.46 (s, 6 H), -2.79 (s, 2 H). ^{13}C NMR (100 MHz, CDCl$_3$) δ 193.49, 142.47, 141.99, 134.59, 134.52, 134.32, 132.31, 130.06, 128.24, 127.79, 126.71, 124.52, 122.48, 122.43, 120.53, 120.44, 119.47, 119.37, 119.31, 119.21, 91.0, 89.81, 30.32. UV/Vis (CH$_2$Cl$_2$) λ$_{max}$ (log ε): 450.92 (5.52), 570.12 (3.23), 619.50 (3.91), 670.58 (4.73). FABMS Calcd for C$_{64}$H$_{42}$N$_4$O$_2$S$_2$: 962. Found: 962. Anal. Calcd for C$_{64}$H$_{42}$N$_4$O$_2$S$_2$·CHCl$_3$: C, 72.11; H, 4.00; N, 5.17. Found: C, 73.15; H, 4.33; N, 5.17.

5,15-Bis(4-thioacetyldiphenylethynyl)-10,20-bis(4-methylphenyl)porphyrin (112). See the preparation of **111** for the synthetic protocol. The compounds used were **110** (0.125 g, 0.45 mmol), meso-(4-

methylphenyl)dipyrrromethane (**103**) [78] (0.1 g, 0.45 mmol), $CHCl_3$ (36.66 mL), two drops of $BF_3 \cdot OEt_2$, and DDQ (0.10 g, 0.45 mmol). The solvent was removed in vacuo and the sample was purified by silica gel column chromatography using CH_2Cl_2 as the eluent followed by a second column purification with CH_2Cl_2:hexane 1:1 to provide 0.059 g (27%) of the title compound in the first major fraction as a purple solid. MP: 214-216°C. IR (KBr) 3433.3, 3128.2, 1704.3, 1464.5, 1384.6, 1108.5, 963.2, 796.2, 730.8 cm^{-1}. ^{1}H NMR (300 MHz, $CDCl_3$) δ 8.89-8.84 (m, 8 H), 8.19 (d, J = 8.19 Hz, 4 H), 8.07 (d, J = 7.89 Hz, 4 H), 7.9 (d, J = 7.95 Hz, 4 H), 7.68 (d, J = 8.22 Hz, 4 H), 7.53 (d, J = 7.89 Hz, 4 H), 7.45 (d, J = 8.34 Hz, 4 H), 2.69 (s, 6 H), 2.46 (s, 6 H), -2.75 (s, 2 H). ^{13}C NMR (100 MHz, $CDCl_3$) δ 193.50, 142.58, 139.07, 137.45, 134.58, 134.48, 134.32, 132.31, 130.04, 131.00, 128.22, 127.44, 124.54, 122.37, 120.70, 120.53, 119.16, 119.00, 91.03, 89.77, 30.32, 21.52. UV/Vis (CH_2Cl_2) λ_{max} (log ε): 456.03 (5.20), 617.80 (3.59), 679.10 (4.42). HRFABMS Calcd for $C_{66}H_{46}N_4O_2S_2$: 990.3062; Found: 990.3080. Anal. Calcd for $C_{66}H_{46}N_4O_2S_2$: C, 79.97; H, 4.67; N, 5.65. Found: C, 80.42; H, 4.98; N, 5.97.

5,15-Bis(4-thioacetyldiphenylethynyl)-10,20-bis(4-bromophenyl)porphyrin (113). See the preparation of **111** for the synthetic protocol. The compounds used were **110** (0.061 g, 0.22 mmol), meso-(4-bromophenyl)dipyrromethane (**104**)[78] (0.065 g, 0.22 mmol), $CHCl_3$ (21.87 mL), two drops of $BF_3 \cdot OEt_2$ and DDQ (0.049 g, 0.22 mmol). The solvent was removed in vacuo and the crude sample was purified by silica gel column chromatography using CH_2Cl_2 followed by a second column purification with CH_2Cl_2:hexane (1:1) to provide 0.034 g (28%) of the title compound in the first major fraction as a purple solid. MP: 204-206 °C. IR (KBr) 3435.9, 3138.5, 2923.1, 1625.6, 1461.5, 1384.6, 1117.9, 800 cm^{-1}. ^{1}H NMR (300 MHz, $CDCl_3$) δ 8.87 (m, 8 H), 8.21 (d, J = 8.07 Hz, 4 H), 8.05 (d, J = 8.04 Hz, 4 H), 7.91 (m, 8 H), 7.68 (d, J = 8.22 Hz, 4 H), 7.48 (d, J = 8.19 Hz, 4 H), 2.46 (s, 6 H), -2.82 (s, 2 H). ^{13}C NMR (100 MHz, $CDCl_3$) δ 193.50, 142.25, 142.21, 140.86, 135.82, 134.57, 134.33, 132.43, 132.31, 130.10, 129.96, 128.27, 124.47, 122.59, 122.58, 119.69, 118.90, 90.90, 89.91, 30.34. UV/Vis (CH_2Cl_2) λ_{max} (log ε): 458.68 (5.69), 571.82 (3.32), 675.69 (4.92), 621.20 (4.10). HRFABMS calcd for $C_{64}H_{40}Br_2N_4O_2S_2$: 1119.1038. Found: 1119.1039. Anal. Calcd for $C_{64}H_{40}Br_2N_4O_2S_2$: C, 68.57; H, 3.59; N, 4.99. Found: C, 67.81; H, 3.92; N; 4.86.

5,15-Bis(4-thioacetyldiphenylethynyl)-10,20-bis(4-iodophenyl)porphyrin (114). See the preparation of **111** for the synthetic protocol. The compounds used were **110** (0.060 g, 0.21 mmol), meso-(4-iodophenyl)dipyrrromethane (**105**)[78] (0.075 g, 0.21 mmol), $CHCl_3$ (43 mL), two drops of $BF_3 \cdot OEt_2$ and DDQ (0.049 g, 0.21 mmol). The solvent was removed in vacuo and the crude sample was purified by silica gel column chromatography using CH_2Cl_2 as the eluent followed by a second column purification with CH_2Cl_2:hexanes 1:1 to provide 0.034 g (28%) of the title compound in the first major fraction as purple powder. MP = 216-218 °C. IR (KBr) 3435.9, 3128.2, 1466.7, 1384.6, 1117.9, 800 cm^{-1}. ^{1}H NMR (300 MHz, $CDCl_3$) δ 8.86-8.83 (m, 8 H), 8.17 (d, J = 8.13 Hz, 4 H), 8.05 (d, J = 8.19 Hz, 4 H), 7.89 (d, J = 7.95 Hz, 8 H), 7.67 (d, J = 8.22 Hz, 4 H), 7.45 (d, J = 8.19 Hz, 4 H), 2.46 (s, 6 H), -2.84 (s, 2 H). ^{13}C NMR (100 MHz, $CDCl_3$) δ 194.26, 143.26, 142.51, 137.13, 136.94, 135.59, 135.34, 133.34, 132.10, 131.14, 129.40 125.54, 123.67, 120.76, 120.07, 95.40, 92.07, 91.07, 31.61. HRFABMS Calcd for $C_{64}H_{40}I_2N_4O_2S_2$: 1214.0682. Found: 1214.0759. UV/Vis (CH_2Cl_2) λ_{max} (log ε): 457.88 (5.44), 578.63 (3.35), 614.39 (3.85), 673.99 (4.66). Anal. Calcd for $C_{64}H_{40}I_2N_4O_2S_2$·$CHCl_3$: C, 58.51; H, 3.09; N; 4.19. Found: C, 57.85; H, 3.15; N, 4.54.

Meso-(4-thioacetyldiphenylethynyl)dipyrromethane (115). A solution of pyrrole (6 mL, 87 mmol) and **110** (0.055 g, 0.19 mmol) in MeOH (0.27 mL) was treated with acetic acid (0.82 mL) under N_2 at room temperature for 20 h. The reaction mixture was diluted with CH_2Cl_2 and washed with water. The organic phase was dried over $MgSO_4$ and the solvents were removed in vacuo. The crude sample was purified by silica gel column chromatography using CH_2Cl_2:Et_3N (100:1, v/v) and was isolated as the second light yellow band. The solvent was removed in vacuo to provide 0.067 g (86%) of the title product as tan viscous oil. IR (KBr) 3394.9, 3169.2, 1384.6, 1117.9, 1025.6, 769.2, 717.9 cm^{-1}. ^{1}H NMR (300 MHz, $CDCl_3$) δ 8.02 (br s, 2 H), 7.53 (d, J = 8.13 Hz, 2 H), 7.45 (d, J = 8.25 Hz, 2 H), 7.37 (d, J = 8.04 Hz, 2 H), 7.19 (d, J = 8.31 Hz, 2 H), 6.7 (br s, 2 H), 6.17 (dd, J = 5.4, 2.6 Hz, 2 H), 5.89 (br s, 2 H), 2.41 (s, 3 H), 5.41 (s, 1 H). ^{13}C NMR (100 MHz, $CDCl_3$) δ 193.43, 142.75, 134.17, 132.12, 131.92, 128.44, 127.96, 124.485, 121.405, 117.45, 108.43, 107.39, 90.96, 88.72, 43.94, 30.43. HRFABMS Calcd for $C_{25}H_{20}N_2OS$: 396.1296. Found: 396.1303.

5,10,15,20-tetrakis(4-thioacetyldiphenylethynyl)porphyrin (116). To a stirred solution of **110** (0.062 g, 0.22 mmol) and pyrrole (0.015 g, 0.22 mmol) in $CHCl_3$ (22 mL) that contained 0.75% EtOH was added two drops of BF_3·OEt_2. The reaction mixture was allowed to stir under N_2 for 5 h. After 5 h, p-chloranil (0.05 g, 0.22 mmol) was added and the reaction mixture stirred for another 1 h. The solvent was removed in vacuo and the crude residue was purified by silica gel column chromatography using CH_2Cl_2 as the eluent followed by a second column purification using CH_2Cl_2:hexane (5:1). The solvent was removed in vacuo to provide 0.02 g (29%) of purple solid. MP: 168-170°C. IR (KBr): 3403.2, 3128.2, 1703.3, 1464.5, 1336.4, 1304.7, 11108.5, 956.0, 883.3, 796.2, 738 cm^{-1}. ^{1}H NMR (300 MHz, $CDCl_3$) δ 8.87 (s, 8 H), 8.19 (d, J = 8.13 Hz, 8 H), 7.91 (d, J = 8.13 Hz, 8 H), 7.62 (d, J = 8.67 Hz, 8 H), 7.45 (d, J = 8.13 Hz, 8 H), 2.46 (s, 12 H), -2.78 (s, 2 H). ^{13}C NMR (100 MHz, $CDCl_3$) δ 193.14, 142.13, 134.43, 134.17, 132.16, 129.96, 128.15,

124.37, 122.43, 119.51, 90.91, 89.85, 30.43. UV/Vis (CH_2Cl_2) λ_{max} (log ε): 464.55 (5.69), 628.01 (3.90), 685.91 (4.91). FABMS Calcd for $C_{84}H_{54}N_4O_4S_4$: 1310. Found: 1310.

4-Trimethylsilylethynylbenzonitrile. See the general procedure for the Pd/Cu-catalyzed coupling reaction. The compounds used were 4-bromobenzonitrile (0.50 g, 2.75 mmol), TMSA (0.59 mL, 4.13 mmol), CuI (0.05 g, 0.28 mmol), $((C_6H_5)_3P)_2PdCl_2$ (0.10 g, 0.14 mmol), $(C_6H_5)_3P$ (0.14 g, 0.55 mmol), Et_3N (1.19 mL, 11.00 mmol), and THF (15 mL) at 65 °C for 60 h. The desired material was purified by gravity liquid chromatography using silica gel as the stationary phase and a mixture of 1:1 CH_2Cl_2:hexanes as the eluent. R_f (product): 0.60. The reaction afforded 0.52 g (93 % yield) of the desired compound as off white crystals. IR (KBr) 3128.2, 2953.8, 2225.6, 2143.6, 1600.0, 1492.3, 1384.6, 1246.2, 1174.4, 841.0, 753.8 cm^{-1}. ^{1}H NMR (400 MHz, $CDCl_3$) δ 7.57 (d, J=8.4 Hz, 2 H), 7.52 (d, J=8.3 Hz, 2 H), 0.26 (s, 9 H). ^{13}C NMR (100 MHz, $CDCl_3$) δ132.15, 131.63, 127.73, 118.17, 111.53, 102.74, 99.35, -0.30. HRMS calcd $C_{12}H_{13}NSi$: 199.0817. Found: 199.0816.

4-Ethynylbenzonitrile. See the general procedure for the deprotection of a trimethylsilyl-protected alkyne. The compounds used were 4-trimethylsilylethynylbenzonitrile (0.35 g, 1.72 mmol), K_2CO_3 (1.19 g, 8.60 mmol), MeOH (10 mL), and CH_2Cl_2 (10 mL) for 2 h. The material was immediately reacted in the next step without additional purification or identification.

118. See the general procedure for the Pd/Cu-catalyzed coupling reaction. The compounds used were 4-ethynylbenzonitrile (0.22 g, 1.65 mmol), **3** (0.60 g, 2.15 mmol), CuI (0.03 g, 0.17 mmol), $((C_6H_5)_3P)_2PdCl_2$ (0.06 g, 0.09 mmol), $(C_6H_5)_3P$ (0.09 g, 0.34 mmol), Et_3N (0.96 mL, 6.88 mmol) and THF (20 mL) at 65 °C for 3 d. The desired material was purified by gravity liquid chromatography using silica gel as the stationary phase and a mixture of 3:1 CH_2Cl_2:hexanes as the eluent. R_f = 0.49. The compound was further purified by a hexanes wash to give 0.28 g (76 % yield over two steps) of the desired compound as yellow crystals. IR (KBr) 3117.9, 2225.6, 1692.3, 1379.5, 1266.7, 1164.1, 1112.8, 1010.3, 959.0, 825.6, 615.4 cm^{-1}. ^{1}H NMR (400 MHz, $CDCl_3$) δ 7.63 (d, J=8.6 Hz, 2 H), 7.59 (d, J=8.6 Hz, 2 H), 7.56 (d, J=8.62 Hz, 2 H), 7.42 (d, J=8.6 Hz, 2 H), 2.42 (s, 3 H). ^{13}C NMR (400 MHz, $CDCl_3$) δ 192.94, 134.22, 132.25, 132.10, 132.02, 129.16, 127.78, 132.32, 118.41, 111.77, 92.88, 89.22, 30.48. HRMS calcd $C_{17}H_{11}NOS$: 277.0561. Found: 277.0573.

2-Trimethylsilylethynylbenzonitrile. See the general procedure for the Pd/Cu-catalyzed coupling reaction. The compounds used were 2-bromobenzonitrile (0.50 g, 2.75 mmol), TMSA (0.59 mL, 4.13 mmol), CuI (0.05 g, 0.28 mmol), $((C_6H_5)_3P)_2PdCl_2$ (0.10 g, 0.14 mmol), $(C_6H_5)_3P$ (0.14 g, 0.55 mmol), Et_3N (1.19 mL, 11.00 mmol) and THF (15 mL) at 65 °C for 60 d. The desired material was purified by gravity liquid chromatography using silica gel as the stationary phase and a mixture of 1:1 CH$_2$Cl$_2$:hexanes as the eluent. R$_f$ = 0.60. The reaction afforded 0.52 g (93 % yield) of the desired compound as off white crystals. IR (KBr) 3066.7, 2953.8, 2902.6, 225.6, 2153.8, 1589.7, 1559.0, 1476.9, 1446.2, 1405.1, 1251.3, 1220.5, 1164.1, 1092.3, 1035.9, 953.8, 861.5, 764.1, 733.3, 697.4, 641.0 cm^{-1}. ^{1}H NMR (400 MHz, CDCl$_3$) δ 7.62 (d, J=7.7 Hz, 1 H), 7.53 (t, J=13.7 Hz, 1 H), 7.52 (d, J=11.7 Hz, 1 H), 7.38 (t, J=8.8 Hz, 1 H), 0.30 (s, 9 H). ^{13}C NMR (100 MHz, CDCl$_3$) δ 132.45, 132.35, 132.08, 128.35, 126.87, 117.20, 115.73, 102.16, 100.49, -0.19. HRMS calcd C$_{12}$H$_{13}$NSi: 199.0817. Found: 199.0814.

2-Ethynylbenzonitrile. See the general procedure for the deprotection of a trimethylsilyl-protected alkyne. The compounds used were 2-trimethylsilylethynylbenzonitrile (0.35 g, 1.72 mmol), K$_2$CO$_3$ (1.19 g, 8.60 mmol), MeOH (10 mL) and CH$_2$Cl$_2$ (10 mL) for 2 h. The material was immediately reacted in the next step without additional purification or identification.

120. See the general procedure for the Pd/Cu-catalyzed coupling reaction. The compounds used were 2-ethynylbenzonitrile (0.22 g, 1.72 mmol), **3** (0.61 g, 2.15 mmol) as described above using CuI (0.03 g, 0.17 mmol), ((C$_6$H$_5$)$_3$P)$_2$PdCl$_2$ (0.06 g, 0.09 mmol), (C$_6$H$_5$)$_3$P (0.09 g, 0.34 mmol), Et$_3$N (0.96 mL, 6.88 mmol), and THF (20 mL) at 65 °C for 48 h. The resultant mixture was subjected to an aqueous workup as described above. The desired material was purified by gravity liquid chromatography using silica gel as the stationary phase and a mixture of 1:3 EtOAc:hexanes as the eluent. R$_f$ = 0.38. The compound was further purified by a hexanes wash to give 0.23 g (48 % yield over two steps) of the desired compound as a yellow solid. IR (KBr) 3425.6, 3138.5, 2369.2, 2225.6, 1702.6, 1656.4, 1384.6, 1112.8, 1015.4, 943.6, 825.6, 769.2, 620.5 cm^{-1}. ^{1}H NMR (400 MHz, CDCl$_3$) δ 7.68 (d, J=7.7 Hz, 1 H), 7.65 (d, J=8.4 Hz, 2 H), 7.64 (buried d, 1 H), 7.57 (t, J=7.6, 1 H), 7.44 (buried d, 1 H), 7.41 (d, J=8.7 Hz, 2 H), 2.44 (s, 3 H). ^{13}C NMR (100 MHz, CDCl$_3$) δ 192.96, 134.20, 132.64, 132.46, 132.35, 132.14, 129.27, 128.46, 126.81, 123.14, 117.41, 115.45, 95.08, 87.06, 30.48. HRMS calcd C$_{17}$H$_{11}$NOS: 277.0561. Found: 277.0574.

2-Trimethylsilylethynylpyridine. See the general procedure for the Pd/Cu-catalyzed coupling reaction. The compounds used were 2-bromopyridine (**121**) (0.45 mL, 3.16 mmol), TMSA (0.68 mL, 4.74 mmol), CuI (0.06 g, 0.32 mmol), ((C$_6$H$_5$)$_3$P)$_2$PdCl$_2$ (0.11 g, 0.16 mmol), (C$_6$H$_5$)$_3$P (0.17 g, 0.63 mmol), Et$_3$N (1.38 mL, 12.64 mmol), and THF (15 mL) at 70 °C for 48 h. The desired material was purified by gravity liquid chromatography using silica gel as the stationary phase and a mixture 3:1 CH$_2$Cl$_2$:hexanes as the eluent. R$_f$ (product): 0.15. The reaction afforded 0.50 g (88 % yield) of the desired compound. IR (KBr) 3056.4, 2953.8, 2902.6, 2153.8, 1579.5, 1559.0, 1456.4, 1425.6, 1246.2, 1220.5, 1148.7, 1046.2, 984.6, 866.7, 841.0, 774.4, 759.0, 733.3, 697.4, 651.3 cm^{-1}. ^{1}H NMR (400 MHz, CDCl$_3$) δ 8.55 (d, J = 3.1 Hz, 1

H), 7.63 (t, J=6.1 Hz, 1 H), 7.43 (d, J=7.7 Hz, 1 H), 7.20 (t, J=3.6 Hz, 1 H), 0.27 (s, 9 H). ^{13}C NMR (100 MHz, CDCl$_3$) δ 149.87, 143.03, 135.97, 127.20, 122.95, 103.65, 94.76, -0.07. HRMS calcd C$_{10}$H$_{13}$NSi: 175.0817. Found: 175.0812.

2-Ethynylpyridine. See the general procedure for the deprotection of a trimethylsilyl-protected alkyne. The compounds used were 2-trimethylsilylethynylpyridine (0.35 g, 1.95 mmol), K$_2$CO$_3$ (1.35 g, 9.75 mmol), MeOH (15 mL), and CH$_2$Cl$_2$ (15 mL) for 2 h. The material was immediately reacted in the next step without additional purification or identification.

122. See the general procedure for the Pd/Cu-catalyzed coupling reaction. The compounds used were 2-ethynylpyridine (0.20 g, 1.95 mmol), **3** (0.66 g, 2.34 mmol), CuI (0.02 g, 0.12 mmol), ((C$_6$H$_5$)$_3$P)$_2$PdCl$_2$ (0.04 g, 0.06 mmol), (C$_6$H$_5$)$_3$P (0.06 g, 0.23 mmol), diisopropylethylamine (1.36 mL, 7.80 mmol), and THF (15 mL) at 50 °C for 16 h. The desired material was purified by gravity liquid chromatography using silica gel as the stationary phase and a mixture of 1:1 EtOAc:hexanes as the eluent. R$_f$ (product): 0.38. The reaction afforded 0.26 g (53 % yield over two steps) of the desired compound as a yellow solid. IR (KBr) 3128.2, 2215.4, 1697.4, 1574.4, 1461.5, 1384.6, 1276.9, 1117.9, 1005.1, 948.7, 830.8, 779.5, 733.3, 615.4 cm^{-1}. ^{1}H NMR (400 MHz, CDCl$_3$) δ 8.60 (d, J=4.0 Hz, 1 H), 7.65 (t, J=5.8 Hz, 1 H), 7.59 (d, J=8.0 Hz, 2 H), 7.51 (d, J=4.0 Hz, 1 H), 7.38 (d, J=8.6 Hz, 2 H), 7.22 (t, J=3.7 Hz, 1 H), 2.41 (s, 3 H). ^{13}C NMR (400 MHz, CDCl$_3$) δ 193.01, 150.05, 143.04, 136.14, 134.12, 132.50, 128.94, 127.26, 123.39, 122.96, 90.12, 88.28, 30.46. HRMS calcd C$_{15}$H$_{11}$NOS: 253.0561. Found: 253.0562.

124. See the general procedure for the Pd/Cu-catalyzed coupling reaction. The compounds used were 5-bromopyrimidine (0.18 g, 1.15 mmol), **9**^{54} (0.24 g, 1.38 mmol), CuI (0.02 g, 0.12 mmol), ((C$_6$H$_5$)$_3$P)$_2$PdCl$_2$ (0.04 g, 0.06 mmol), (C$_6$H$_5$)$_3$P (0.06 g, 0.23 mmol), Et$_3$N (0.51 mL, 4.60 mmol), and THF (15 mL) at 75 °C for 4 d. The desired material was purified by gravity liquid chromatography using silica gel as the stationary phase and a mixture of 1:1 EtOAc:hexanes as the eluent. R$_f$ (product): 0.53. The reaction afforded 0.15 g (52%) of the desired compound as bright yellow solid. IR (KBr) 3425.6, 3128.2, 2215.4, 1702.6, 1656.4, 1543.6, 1384.6, 1117.9, 1097.4, 943.6, 820.6, 717.9, 615.4 cm^{-1}. ^{1}H NMR (400 MHz, CDCl$_3$) δ 9.14 (s, 1 H), 8.85 (s, 2 H), 7.76 (d, J=8.1 Hz, 2 H), 7.42 (d, J=8.0 Hz, 2 H), 2.44 (s, 3 H). ^{13}C NMR (400 MHz, CDCl$_3$) δ 192.84, 158.57, 156.79, 134.26, 132.22, 129.49, 122.82, 119.59, 95.46, 83.84, 30.50. HRMS calcd C$_{14}$H$_{10}$N$_2$OS: 254.0514. Found: 254.0513.

3-Bromo-6-(trimethylsilylethynyl)pyridine. See the general coupling procedure. The compounds used were 2,5-dibromopyridine **(125)** (2.37 g, 10.0 mmol), ((C$_6$H$_5$)$_3$P)$_2$PdCl$_2$ (0.35 g, 0.50 mmol), CuI (0.19 g, 1.0 mmol), (C$_6$H$_5$)$_3$P (0.52 g, 2.0 mmol), Et$_3$N (4.35 mL, 40.0 mmol), THF (50 mL), and TMSA (1.4 mL, 10 mmol) at 65 °C for 2 d. The reaction was separated via flash chromatography affording light brown solid (2.130 g, 84% yield), R_f = 0.22 (50% hexanes:CH$_2$Cl$_2$). IR (KBr) 3031.8, 2958.2, 2163.8, 1561.3, 1543.6, 1451.4, 1367.0, 1248.9, 1089.4, 1001.1, 844.6, 760.9, 678.6, 642.87, 534.52 cm^{-1}. ^{1}H NMR (400 MHz, CDCl$_3$) δ 8.61 (dd, J = 2.4, 0.73 Hz, 1 H), 7.76 (dd, J = 8.4, 2.4 Hz, 1 H), 7.32 (dd, J = 8.2, 0.73 Hz, 1 H). ^{13}C NMR δ 151.05, 141.36, 138.72, 128.19, 120.24, 102.58, 96.40, -0.40.

3-Ethynylphenyl-6-(trimethylsilylethynyl)pyridine (126). See the general procedure for the coupling reaction. The compounds used were 5-bromo-2-(trimethylsilylethynyl)pyridine (2.00 g, 7.90 mmol), ((C$_6$H$_5$)$_3$P)$_2$PdCl$_2$ (0.28 g, 0.40 mmol), CuI (0.15 g, 0.8 mmol), THF (20 mL), diisopropylethylamine (5.50 mL, 31.6 mmol), and phenylacetylene (0.87 mL, 7.9 mmol) at 55 °C overnight. The reaction was separated via flash chromatography affording a light brown solid (1.37 g, 63%), R_f = 0.36 (2:1 CH$_2$Cl$_2$ to hexanes). IR (KBr) 2959.5, 2157.9, 1492.3, 1463.6, 1384.0, 1247.7, 1019.9, 844.4, 754.8, 690.3 cm^{-1}. ^{1}H NMR (400 MHz, CDCl$_3$) δ 8.70 (d, J = 1.3 Hz, 1 H), 7.74 (dd, J = 6.0, 2.6 Hz, 1 H), 7.53 (m, 2 H), 7.43 (d, J = 8.0 Hz, 1 H), 7.36 (m, 3 H). ^{13}C NMR (75 MHz, CDCl$_3$) δ 152.32, 141.52, 138.38, 131.73, 129.02, 128.52, 126.59, 122.33, 119.76, 103.51, 96.95, 94.49, 85.93, -0.12. HRMS Calc'd for C$_{19}$H$_{17}$NSi: 275.1130. Found: 275.1126.

2-Ethynyl-5-ethynylphenylpyridine. See the general procedure for the deprotection of a trimethylsilyl-protected alkyne. The compounds used were **126** (272 mg, 1.00 mmol), K$_2$CO$_3$ (690 mg, 5.00 mmol), MeOH (30 mL) and CH$_2$Cl$_2$ (30 mL) for 2.5 h. The product was used without purification.

127. See the general procedure for the Pd/Cu-catalyzed coupling reaction. The compounds used were 2-ethynyl-5-ethynylphenylpyridine (0.167 g, 1.00 mmol), 3 (0.334 g, 1.20 mmol), ((C$_6$H$_5$)$_3$P)$_2$PdCl$_2$ (0.035 g, 0.050 mmol), CuI (0.019 g, 0.10 mmol), (C$_6$H$_5$)$_3$P (0.026 g, 0.10 mmol), THF (30 mL) and diisopropylethylamine (0.70 mL, 4.0 mmol) at 50 °C for 2 d. Column chromatography eluting with 3:1 CH$_2$Cl$_2$ to hexanes yielded 199 mg (56%) of a light brown solid. IR (KBr) 3052.1, 2923.3, 2214.1, 1703.5, 1571.6, 1536.4, 1493.7, 1460.4, 1397.5, 1359.9, 1222.2, 1124.8, 1107.4, 1081.7, 1013.6, 943.1, 824.9, 754.3, 687.7, 618.5, 523.7 cm^{-1}. ^{1}H NMR (300 MHz, CDCl$_3$) δ 8.76 (br s, 1 H), 7.80 (dd, J = 2.0, 8.1 Hz, 1 H), 7.62 (1/2ABq, J = 8.5 Hz, 2 H), 7.54 (m, 3 H), 7.41 (1/2ABq, J = 8.1 Hz, 2 H), 7.37 (m, 3 H). ^{13}C NMR (75 MHz, CDCl$_3$) δ 193.15, 152.51, 141.57, 138.50, 134.23, 132.61, 131.75, 129.28, 129.05, 128.53, 126.63, 123.23, 122.33, 119.73, 94.61, 90.22, 85.97, 30.37. HRMS C$_{23}$H$_{16}$NOS Calc'd: 353.0870. Found: 353.0874.

129. See the general procedure for the Pd/Cu-catalyzed coupling reaction. The compounds used were 2-iodoaniline (**128**) (0.607 g, 2.77 mmol), ((C$_6$H$_5$)$_3$P)$_2$PdCl$_2$ (0.098 g, 0.139 mmol), CuI (0.053 g, 0.277 mmol), diisopropylethylamine (1.93 mL, 11.08 mmol), **89** (0.488 g, 3.05 mmol) and THF (25 mL) at 70 °C for 7 d. Column chromatography (silica gel with CH$_2$Cl$_2$ as eluent) afforded the desired product (0.40 g, 57% yield). IR (KBr) 3468.06,

3375.97, 2941.49, 2210.75, 1711.99, 1602.74, 1485.37, 1453.52, 1308.06, 1280.46, 1099.68, 770.84, 753.54, 695.52 cm^{-1}. ^{1}H NMR (400 MHz, CDCl$_3$) δ 8.04 (dt, J = 8.5 Hz, 1.8 Hz, 2 H), 7.59 (dt, J = 8.5, 1.7 Hz, 2 H), 7.40 (dd, J = 7.8, 1.5 Hz, 1 H), 7.18 (td, J = 7.6, 1.5 Hz, 1 H), 6.75 (m, 2 H), 4.33 (br s, 2 H), 3.94 (s, 3 H). ^{13}C NMR (100 MHz, CDCl$_3$) δ 166.95, 148.44, 132.74, 131.70, 130.70, 129.98, 129.76, 128.46, 118.44, 114.87, 107.64, 94.43, 89.54, 52.65. HRMS calculated for C$_{16}$H$_{13}$NO$_2$,: 251.094629. Found: 251.0940.

4-(2'-Aminoethynylphenyl)benzoic acid (130). 129 (0.300 g, 1.194 mmol), lithium hydroxide (0.250 g, 5.97 mmol), MeOH (30 mL), water (10 mL), CH$_2$Cl$_2$ (20 mL) and a stir bar were added to a 100 mL round bottom flask.[71] The mixture was stirred at room temperature for 2 d. The mixture was washed with CH$_2$Cl$_2$ and the layers separated. The aqueous portion was adjusted to pH = 4 and washed with CH$_2$Cl$_2$ to afford 0.277 g of product (98% yield). IR (KBr) 3468.1, 3376.3, 3054.3, 2957.6, 2656.7, 2538.5, 2205.4, 1681.3, 1604.8, 1488.4, 1422.2, 1318.8, 1281.9, 860.4, 758.7 cm^{-1}. ^{1}H NMR (400 MHz, d-DMSO) δ 7.95 (dt, J = 8.5, 1.8 Hz, 2 H), 7.72 (dt, J = 8.5, 1.7 Hz, 2 H), 7.26 (dd, J = 7.7, 1.5 Hz, 1 H), 7.11 (td, J = 7.7, 1.6 Hz, 1 H), 6.75 (dd, J = 8.3, 0.6 Hz, 1 H), 6.55 (td, J = 7.6, 1.0 Hz, 1 H), 5.59 (br s, 2 H). ^{13}C NMR (100 MHz, d-DMSO) δ 167.65, 150.85, 132.88, 132.12, 131.19, 130.72, 130.24, 128.32, 116.66, 114.94, 105.64, 94.14, 90.90. HRMS calculated for C$_{15}$H$_{11}$NO$_2$,: 237.0790. Found: 237.0792.

Methyl 4-(2'-methoxyethynylphenyl)benzoate (132). See the general procedure for the Pd/Cu-catalyzed coupling reaction. The compounds used were 2-iodoanisole **(131)** (0.49 mL, 3.74 mmol), **89** (0.50 g, 3.12 mmol), CuI (0.06 g, 0.31 mmol), ((C$_6$H$_5$)$_3$P)$_2$PdCl$_2$ (0.11 g, 0.16 mmol), diisopropylethylamine (2.17 mL, 12.48 mmol) and THF (15 mL) at 75 °C for 2.5 d. The desired material was purified by gravity liquid chromatography using silica gel as the stationary phase and CH$_2$Cl$_2$ as the eluent. R$_f$ (product): 0.59. An additional purification was performed using gravity liquid chromatography using silica gel as the stationary phase and a mixture of 1:1 diethyl ether:hexanes as the eluent. R$_f$ = 0.54. The reaction afforded 0.47 g (57 % yield) of the desired compound as a white solid. IR (KBr) 3426.87, 2941.49, 2828.66, 2200.00, 1720.89, 1597.73, 1487.07, 1463.09, 1433.24, 1275.68, 1245.54, 1167.90, 1102.30, 1018.13, 853.62, 753.68, 691.05, 474.58 cm^{-1}. ^{1}H NMR (400 MHz, CDCl$_3$) δ 7.80 (ABq, J = 8.7 Hz, Δν = 159.91 Hz, 4 H), 7.52 (dd, J = 7.6, 1.8, 1 H), 7.36 (td J = 7.4, 1.7 Hz, 1 H), 6.98 (td, J = 7.5, 1.0 Hz, 1 H), 6.94 (dd, J = 8.4, 0.7 Hz, 2 H), 3.95 (s, 6 H). ^{13}C NMR (100 MHz, CDCl$_3$) δ 167.05, 160.50, 134.09, 132.88, 131.95, 130.73, 129.99, 129.84, 129.69, 128.76, 120.95, 112.31, 111.14, 93.06, 89.29, 56.26, 52.61. HRMS Calcd C$_{17}$H$_{14}$O$_3$: 266.0943. Found: 266.0945.

4-(2'-Methoxyphenylethynyl)benzoic acid (133). To a 100 mL round bottom flask equipped with a magnetic stirbar was added **132** (0.30 g, 1.16 mmol), LiOH (0.14, 5.82 mmol), MeOH (18 mL), CH$_2$Cl$_2$ (10 mL), and water (6 mL). [71] The reaction mixture was allowed to stir at room temperature for 2 d. The reaction was quenched with water and extracted with CH_2Cl_2 (3×). The yellow aqueous phases were combined and acidified to pH = 3 whereupon a white solid precipitated. The solid material was collected on a fritted funnel. No further purification was needed. The reaction afforded 0.28 g (97 % yield) of the desired material. IR (KBr) 3445.36, 2962.62, 2829.10, 2659.63, 2536.38, 2212.84, 1681.14, 1604.93, 1488.82, 1457.92, 1425.90, 1317.19, 1297.57, 1278.77, 1244.42, 1178.84, 1098.43, 1016.26, 954.64, 858.43, 757.58, 697.86, 554.07 cm^{-1}. ^{1}H NMR (400 MHz, d-DMSO) δ 13.00 (br s, 1 H), 7.80 (ABq, J=8.2 Hz, Δv=135.77 Hz, 4 H), 7.52 (dd, J=7.5, 1.7 Hz, 1 H), 7.42 (td, J=7.7, 1.7 Hz, 1 H), 7.12 (d, J=8.4 Hz, 1 H), 7.00 (td, J=7.4, 0.6 Hz, 1 H), 3.33 (s, 3 H). ^{13}C NMR (100 MHz, d-DMSO) δ 167.59, 160.67, 134.09, 132.20, 131.78, 131.27, 130.43, 127.88, 121.43, 112.32, 111.60, 93.08, 89.81, 56.61. HRMS Calcd $C_{16}H_{12}O_3$: 252.0786. Found: 252.0782.

3.4 Highly Functional Molecular Wires and Devices with Diverse Alligator Clips and Experimental Details

3.4.1. *Introduction*

Recent advances in the field of molecular electronics have shown that oligo(phenylene ethynylene)s containing nitro groups are good candidates for electronic switching and storage devices.[5,82] Since the mechanism of switching and electron storage involves a reduction upon applying a potential,[56] the search continues for a second generation of storage and switching devices containing reversibly reducible functional groups. To this end, two polynitro compounds as well as several quinone derivatives have been synthesized. Quinones are frequently found in nature as electron acceptors and can be easily reduced and re-oxidized, thus making them ideal for study in molecular electronic devices.[83,84]

In order to make molecular electronic components an alternative to the presently utilized silicon analogs will demand reliable low energy contacts between molecules and conductive leads. Typically, this is accomplished via a thiol group, which connects the active molecule to a metal contact forming a well-ordered self-assembled monolayer (SAM).[85,23a] In an attempt to improve on the presently used thiol alligator clip, we utilized the well-established reactivity of aromatic diazonium salts toward transition metals[86] to produce a direct connection of the organic molecules to metal surfaces through a carbon-metal bond.[87] We have also built on early work showing that pyridine can assemble on a metal surface[88] by synthesizing molecules containing pyridine alligator clips.[89]

3.4.2. *Switches and Memory Components*

In an effort to improve the electron storage time by adding more nitro groups, synthetic targets **1** and **2** were chosen. The synthesis of compound **1** is outlined in Scheme 3.50.

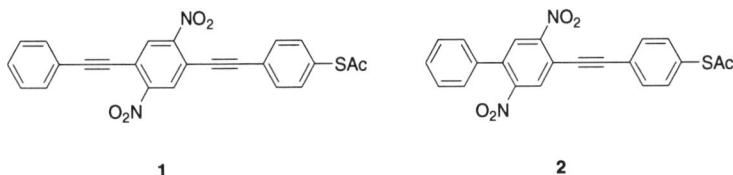

Scheme 3.50 Dinitro-containing synthesis targets.

The synthesis of **1** began by Sonogashira coupling[41] 2,5-dibromo-4-nitroaniline[40] (**3**) to phenylacetylene affording **4** that was subjected to an HOF oxidation[90] forming **5**. A final coupling produced desired compound **1** in 24% yield. The low yield in this coupling may be indicative of the easily deprotected thiol or a stable palladacycle intermediate that formed during coupling (Scheme 3.51).

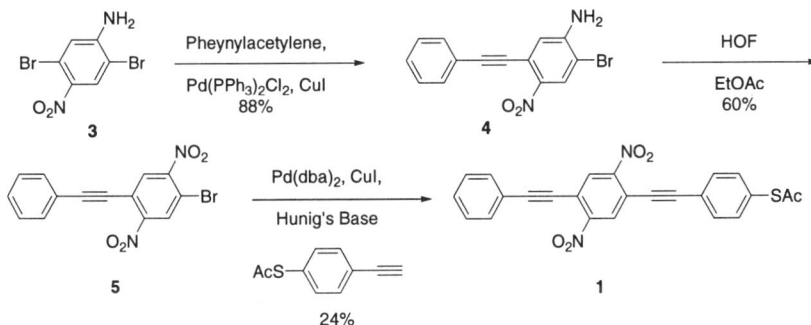

Scheme 3.51 Synthesis of Target **1**.

In order to conduct electrons all the phenyl rings in the conjugated molecule should be preferentially planar to each other.[56] If a phenyl group replaces the terminal phenylethynyl group, the system cannot attain planarity. In an effort to determine the effect of a rotational barrier (i.e. conduction barrier), the synthesis of compound **2** was initiated via a Suzuki[91] coupling of 2,5-dibromo-4-nitroacetanilide (**6**)[40] to phenyl boronic acid to form compound **7** (Scheme 3.52). The acetyl group was removed to provide the aniline (**8**) functionality that would subsequently undergo an HOF oxidation to afford **9** in nearly quantitative yield. A final Sonogashira coupling provided **2**.

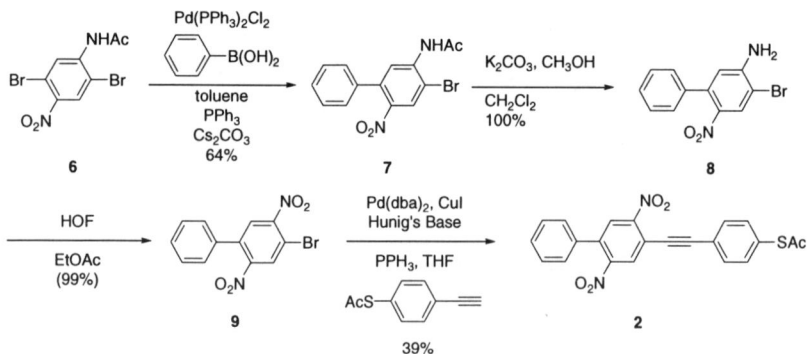

Scheme 3.52 Synthesis of Target **2**.

13 was synthesized for the purpose of studying the electrochemical properties of the quinone-containing molecular system.[92]   Scheme 3.53 shows the synthesis of **13** from 1,4-dimethoxybenzene (**10**). **10** was converted to **11** using bromine and glacial acetic acid in good yield.[93] Compound **11** was then cross-coupled with an excess of phenylacetylene to afford compound **12** which was then oxidized to the quinone affording desired compound **13**. This synthetic route had to be used because quinones generally cannot be used in the palladium-catalyzed couplings since quinones are known to oxidize Pd(0) to Pd(II), terminating the catalytic cycle.[94]   Ceric ammonium nitrate (CAN) is a mild and neutral oxidizing agent known to generate quinones from dimethoxybenzenes and therefore was a logical choice for this procedure.[95]   This oxidation afforded the desired quinone compound in 47 % yield. The optimum conditions for the oxidation have not yet been obtained for these systems.

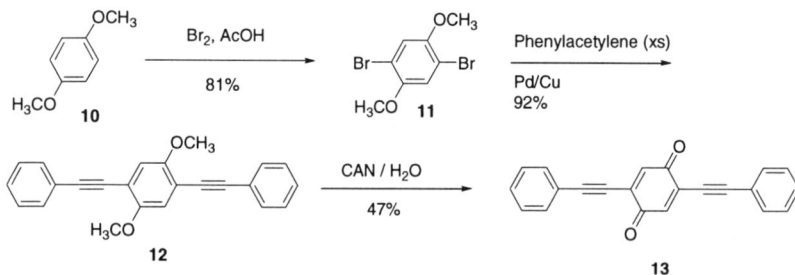

Scheme 3.53 Synthesis of quinone-containing target **13**.

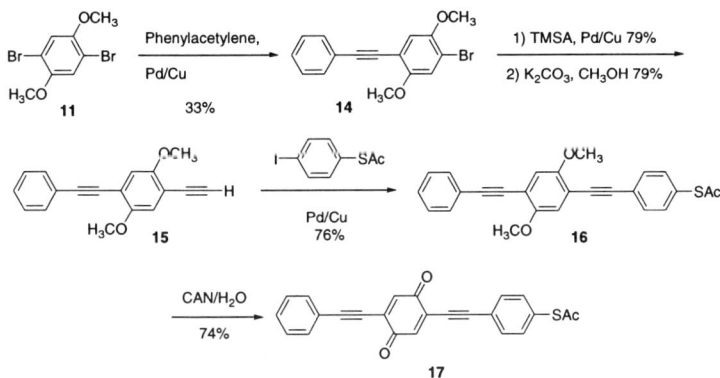

Scheme 3.54 Synthesis of quinone target **17** containing an alligater clip.

Scheme 3.54 shows the synthesis of the quinone-containing molecular system with one thioacetate group serving as a protected alligator clip. Cross-coupling of **11** with phenylacetylene afforded **14** in a modest yet statistically expected yield of 33% due to the equal reactivity of both aryl bromides of **11** under Sonogashira coupling conditions. **15** was prepared by the cross-coupling[41] of TMSA with **14** followed by deprotection of the alkyne to afford **15**. Further palladium-catalyzed cross-coupling with 4-iodobenzenethioacetate afforded compound **16**. The final compound **17** was obtained in 74% yield via the CAN oxidation.[95] However, this yield was an isolated incident. Other attempts resulted in much lower yields (~ 20 %).

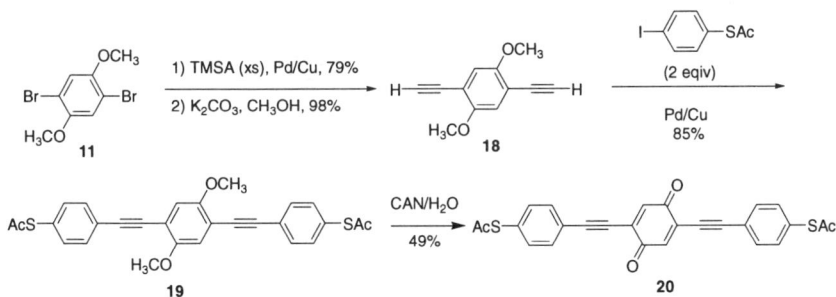

Scheme 3.55 Synthesis of quinone-containing molecule **20** with double alligator clips.

Scheme 3.55 shows the synthesis of the quinone-containing molecular system with alligator clips on both ends (**5**). This compound can be used to crosslink metallic nanoparticles for bridging connections in future molecular electronic devices. **11** was cross-coupled with an excess of TMSA followed by a subsequent deprotection to cleanly afforded the diyne **18**. This was subsequently cross-coupled with 2 equivalents of 4-iodobenzenethioacetate to afford compound **19**. Finally, **19** was oxidized using the CAN procedure to generate **20** in modest yield.

3.4.3. *Alligator Clips*

The synthesis of several compounds containing a pyridine alligator clip for incorporation into a molecular electronic device began with compound **21**. The synthesis of **22** was accomplished by coupling pyridine **21**[29,96] with 2,5-dibromonitrobenzene as shown in Scheme 3.56. The low yield may be due to a stable copper acetylide formed after the TMS group is cleaved. If an in situ deprotection was not used, the pyridine alkyne proved to be unstable.

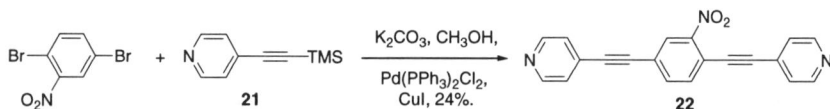

Scheme 3.56 Synthesis of **22** containing two pyridine alligator clips.

24 was synthesized according to Scheme 3.57. The synthesis began by coupling one equivalent of **21** to 2,5-dibromonitrobenzene selectively to the position ortho to the nitro group affording **23**. Coupling **23** to phenylacetylene to produce **24** completed the synthesis.

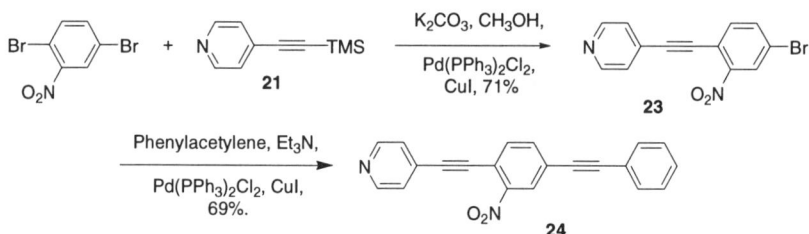

Scheme 3.57 Synthesis of molecule **24** containing pyridine alligator clip.

The synthesis of compound **26** was initiated to study the effect of the nitro group in relation to the chemisorbed pyridine alligator clip. To this end, compound **24** was synthesized in a manor analogous to the synthesis of **23** as shown in Scheme 3.58. Coupling one equivalent of phenylacetylene selectively to 2,5-dibromonitrobenzene to produce **25** then coupling to **21** to afford **26** in good yield completed the synthesis.

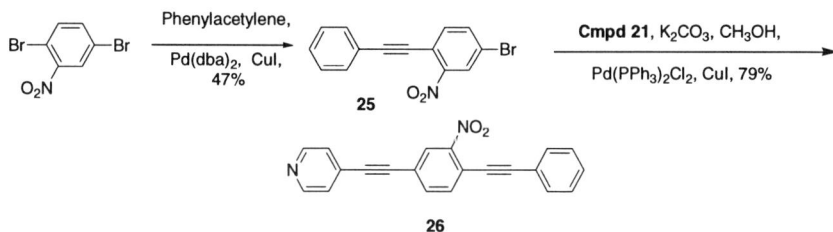

Scheme 3.58 Synthesis of molecule with pyridine alligator clip and alternative nitro group position.

Linker **28** was synthesized according to Scheme 3.59. The synthesis commenced with the coupling of 2,5-dibromo-4-nitroacetanilide[40] with excess TMSA to give **27**, which was then deprotected in-situ and coupled with 4-iodopyridine to produce **28** in poor yield. The low yield of the coupling reactions could be due to the cyclization process reported by Rosen et al.[97] between the nitro and the alkyne unit.

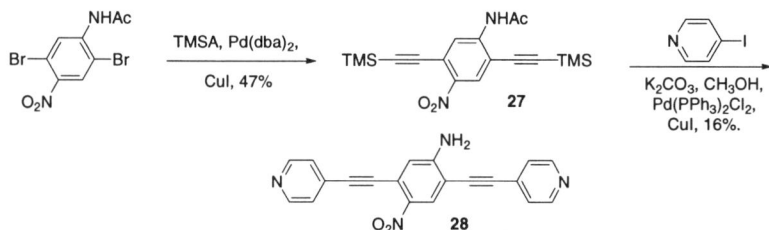

Scheme 3.59 Synthesis of nitro aniline **28** with two pyridine-based alligator clips.

Compound **31** was synthesized in an effort to form a SAM via the protected benzenethiol terminal group enabling the pyridyl end of the molecule to serve as a better top contact with metal than a phenyl when incorporated into a device. **31** was produced by coupling the 2,5-dibromo-4-nitroacetanilide[40] with **21** in a low yield to afford compound **29**. **29** was then coupled with TMSA, followed by deprotection with K_2CO_3 to yield **30**. Finally, **30** was coupled with 4-iodobenzenethioacetate, which afforded the molecular device **31** in good yield (75 %, Scheme 3.60).

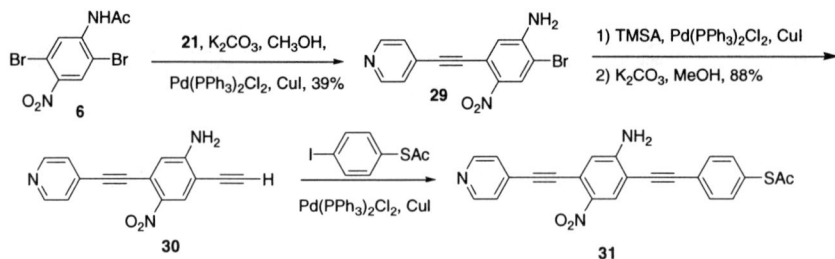

Scheme 3.60 Synthesis of molecule **31** with mixed alligator clips.

32 was synthesized to study the effect of a rotational barrier analogous to that described for **2**. The synthesis of **32** began with previously synthesized **7** and coupling to **21** in good yield as shown in Scheme 3.61.

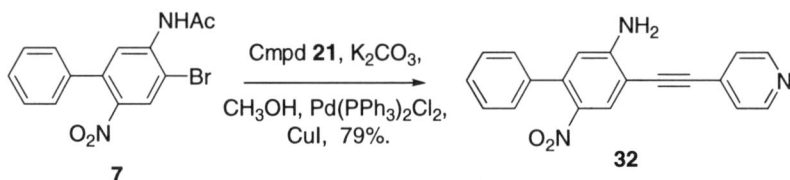

Scheme 3.61 Synthesis of rotationally-challenged molecule **32** containing pyridine-based alligator clip.

Compound **34** was synthesized according to Scheme 3.62 using the previously described **33**. Compound **34** is analogous to a thiol-terminated nitroaniline that previously exhibited negative differential resistance (NDR) in a device embodiment.[82]

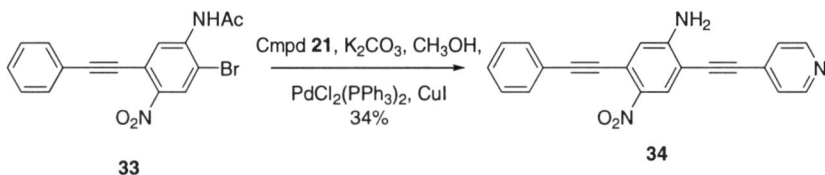

Scheme 3.62   Synthesis of nitroaniline core-containing molecule **34** with pyridine alligator clip.

In addition to the pyridine containing systems, three potential memory and switching components terminated by diazonium salts were synthesized. **38** is analogous to the thioacetyl terminated NDR and memory component[82] and the pyridyl terminated **24**. The synthesis of **38** began by coupling **35**[87] to 2,5-

dibromonitrobenzene in moderate yield to afford **36** that was then coupled to phenylacetylene to produce compound **37**. Diazotization of **37** produced the completed molecule **38** in good yield (Scheme 3.63).

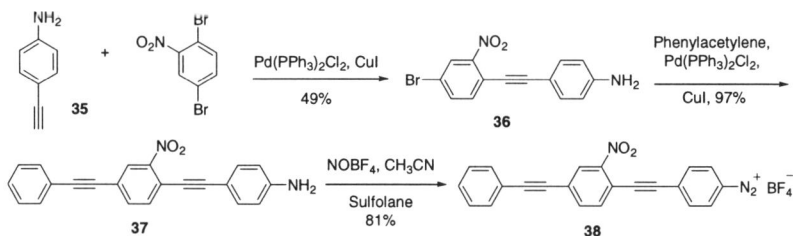

Scheme 3.63 Synthesis of diazonium alligator clip-containing molecule 38.

40 is similar in structure to **26** except the pyridyl group has been replaced with the aryl diazonium salt. The synthesis of **40** is shown is Scheme 3.64. Coupling aniline **35** to nitrocompound **25** produced diazonium precursor **39** in moderate yield. Diazotization of aniline **42** afforded desired product **37**.

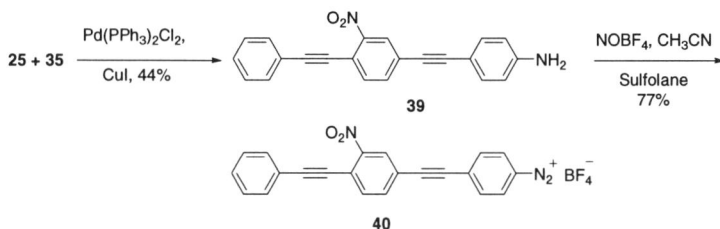

Scheme 3.64 Synthesis of diazonium alligator clip-containing molecule 40 with "away" configuration of nitro group.

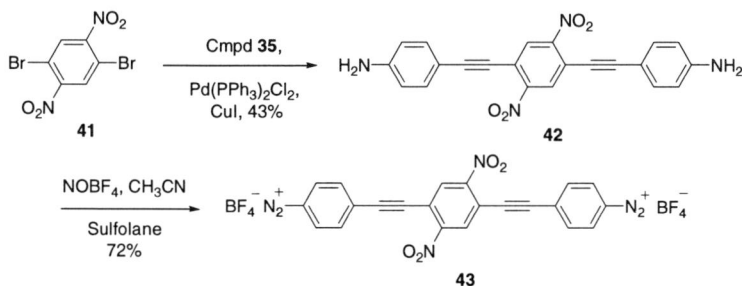

Scheme 3.65 Synthesis of molecule 43 containing two diazonium alligator clips to serve as a nanoparticle linker.

Nanoparticle linker **43** was synthesized according to Scheme 3.65. Starting from dinitro **41**[90] and coupling aniline **35** afforded dinitrodianiline **42** that was subsequently diazotized to produce **43** in good yield.

3.4.4. Conclusions

Many oligo(phenylene ethynylene)s containing reversibly reducible functionalities based on quinone and nitro cores have been synthesized. These molecules have methods of attachment to a metal surface ranging from the standard protected thiol groups to the novel diazonium and pyridyl linkages. Work is currently underway to examine the assembly of these various compounds on metal surfaces as well as their efficiency as switches and memory devices.

3.4.5. Experimental

General Pd/Cu Coupling Reaction Procedures.[41] To an oven dried glass screw capped tube were added all solids including the aryl halide (bromide or iodide), alkyne, CuI, $(C_6H_5)_3P$ and palladium catalyst. The atmosphere was removed via vacuum and replaced with dry N_2 (3×). THF, remaining liquids, and Hünig's base or Et_3N were added and the reaction was heated in an oil bath while stirring. Upon cooling the reaction mixture was filtered via gravity filtration to remove solids and diluted with CH_2Cl_2. The reaction mixture was extracted with an aqueous solution of ammonium chloride (NH_4Cl) (3×). The organic layer was dried with $MgSO_4$ and filtered. The solvent was then removed in vacuo.

General Procedure for the Deprotection of Trimethylsilyl-Protected Alkynes. To a round bottom flask equipped with a stir bar were added the protected alkyne, K_2CO_3 (5 equiv per protected alkyne), MeOH, and CH_2Cl_2. The reaction was heated, and upon completion the reaction mixture was diluted with CH_2Cl_2 and washed with brine (3×). The organic layer was dried over $MgSO_4$, and the solvent removed in vacuo.

General HOF Oxidation Procedure.[90] To a 125 mL polyethylene bottle were added H_2O (2 mL) and CH_3CN (60 mL) and cooled to –20 °C. F_2 (20% in He) was then bubbled through the solution at a rate of 50 sccm for 2 h. The resulting HOF/CH_3CN solution was purged with He for 15 min. The species to be oxidized was added in acetone or EtOAc (10 mL) and mixed at –20 °C for 5 min before being neutralized by pouring into a saturated $NaHCO_3$ solution. The organic phase was then separated, dried over $MgSO_4$ and the solvent were removed in vacuo.

General Procedure for the Diazotization of Anilines with Nitrosonium Tetrafluoroborate in the Acetonitrile - Sulfolane System.[87] $NOBF_4$ was weighed out in a N_2 filled dry box and placed in a round bottom flask equipped with a magnetic stirring bar and sealed with a septum. Acetonitrile and sulfolane were injected in a 5 to 1 volume ratio and the resulting suspension was cooled in a dry ice/acetone bath to – 40 °C. The solution of the aniline was prepared by adding warm sulfolane (45-50 °C) to the amine under a N_2 blanket, sonication for 1 min and subsequent addition of acetonitrile (10-20% by volume). The aniline solution was then added to the nitrosonium salt suspension over a period of 10 min. The reaction mixture was kept at – 40 °C for 30 min and was then allowed to warm to the room temperature. At this point, the diazonium salt was precipitated by the addition of ether or CH_2Cl_2, collected by filtration, washed with ether or CH_2Cl_2 and dried. Additional purification of the salt was accomplished by re-precipitation from DMSO by CH_2Cl_2 and/or ether.

4-Ethynlphenyl-2,4-dinitrobromobenzene (5).[90] 2-Bromo-4-nitro-5-ethynlphenylaniline (490 mg, 1.48 mmol) in EtOAc (10 mL) was oxidized according to the general HOF oxidation procedure to yield 320 mg (60 %) of a yellow solid. IR (KBr) 3442.7, 3101.4, 2216.8, 1610.6, 1540.9, 1461.3, 1384.8, 1358.7, 1337.1, 1264.4, 906.2, 849.6, 824.4, 760.2, 689.8 cm^{-1}. ^{1}H NMR (400 MHz, CDCl$_3$) δ 8.41 (s, 1 H), 8.09 (s, 1 H), 7.60-7.58 (m, 2 H), 7.41-7.39 (m, 3 H). ^{13}C NMR (100 MHz, CDCl$_3$) δ 152.1, 150.4, 132.7, 131.7, 131.0, 130.7, 129.1, 121.5, 119.8, 113.9, 102.0. HRMS Calc'd for 345.9589. Found: 345.9585.

2',5'-Dinitro-4,4'-diethynylphenyl-1-thioacetylbenzene (1). 4 (300 mg, 0.86 mmol), 4-ethynyl(thioacetyl)benzene (183 mg, 1.04 mmol), bis(dibenzylideneacetone)palladium (12 mg, 0.02 mmol), CuI (4 mg, 0.02 mmol), $(C_6H_5)_3P$ (13 mg, 0.05 mmol), Hunig's base (0.60 mL) and THF (20 mL) were reacted according to the general coupling procedure. The reaction mixture was heated at 60 °C overnight and worked up according to the procedure above. The crude compound was purified via flash chromatography (silica, 3:1 CH_2Cl_2:hexane) to yield 90 mg (24%) of a bright yellow solid. IR (KBr) 2220.2, 1705.2, 1545.5, 1499.81, 1396.8, 1337.5, 1286.1, 1252.1, 1108.6, 1087.2, 953.2, 926.0, 868.3, 827.2, 756.7, 684.1, 618.3 cm^{-1}. ^{1}H NMR (400 MHz, CDCl$_3$) δ 8.34 (d, J = 0.4, 1 H), 8.35 (d, J = 0.4, 1 H), 7.63-7.59 (m, 4 H), 7.46-7.40 (m, 5 H), 2.49 (s, 3 H). ^{13}C NMR (100 MHz, CDCl$_3$) δ 193.2, 151.1, 134.7, 133.1, 132.7, 131.0, 130.7, 129.1, 122.8, 121.7, 119.4, 118.6, 102.4, 100.9, 84.8, 83.5, 30.8. HRMS Calc'd for 442.0623. Found: 442.0634.

2-Bromo-4-nitro-5-phenylacetanilide (7). 6 (676 mg, 2 mmol), $(C_6H_5)_3P$ (52 mg, 0.2 mmol), phenylboronic acid (293 mg, 2.4 mmol), $((C_6H_5)_3P)_2PdCl_2$ (70 mg, 0.1 mmol), and $CsCO_3$ (977 mg, 3 mmol) were placed in a 100 mL round bottom flask and the atmosphere was removed and replaced with N_2. Toluene (30 mL) was added and the reaction was heated at 60 °C for 2 d. The reaction was worked up by diluting with ether, washing with aqueous ammonium chloride (2×), drying over MgSO$_4$, and removing the solvents in vacuo. The crude product was purified via flash chromatography

(CH$_2$Cl$_2$) to yield 430 mg (64%) of a white solid. IR (KBr) 3373.6, 3322.4, 3086.5, 1774.0, 1681.7, 1568.9, 1528.8, 1445.8, 1389.4, 1358.6, 1245.8, 1179.1, 1112.5, 1056.1, 1030.4, 999.6, 872.0, 850.9, 768.9, 697.1 cm^{-1}. ^{1}H NMR (400 MHz, CDCl$_3$) δ 8.54 (s, 1H), 8.15 (s, 1H), 7.80 (br s, 1H), 7.40-7.38 (m, 3H), 7.29-7.27 (m, 2H) 2.26 (s, 3H). ^{13}C NMR (100 MHz, CDCl$_3$) δ 168.44, 143.77, 139.34, 137.74, 136.81, 128.67, 128.51, 128.47, 127.85, 123.31, 110.59, 25.05. HRMS Calc'd for C$_{14}$H$_{11}$BrN$_2$O$_3$: 333.9953. Found: 333.9952.

2-Bromo-4-nitro-5-phenylaniline   (8).   7 (500 mg, 1.49 mmol), K$_2$CO$_3$ (1.031 g, 7.46 mmol), MeOH (30 mL), and CH$_2$Cl$_2$ (30 mL) were added to a 100 mL round bottom flask and stirred at room temperature under a N$_2$ blanket for 2 h. The reaction was worked up by filtering off the K$_2$CO$_3$ and washing with CH$_2$Cl$_2$ to yield 437 mg (100%) of the title compound. IR (KBr) 3463.7, 3349.2, 3221.3, 1623.9, 1584.6, 1555.4, 1495.5, 1443.6, 1406.9, 1305.6, 1259.4, 1123.9, 1051.7, 896.7, 846.5, 760.1, 701.3, 632.1, 563.8 cm^{-1}. ^{1}H NMR (400 MHz, CDCl$_3$) δ 8.21 (s, 1H), 7.39-7.36 (m, 3H), 7.23-7.21 (m (overlapping), 2H), 6.61 (s, 1H). ^{13}C NMR (100 MHz, CDCl$_3$) δ 148.5, 139.4, 138.5, 130.7, 128.8, 128.4, 128.2, 128.1, 117.2, 106.3. HRMS Calc'd for 291.9848. Found: 291.9846.

2,5-Dinitro-4-phenylbromobenzene (9). 8 (373 mg, 1.28 mmol) in EtOAc (10 mL) was oxidized according to the general HOF oxidation procedure to yield 407 mg (99 %) of orange solid IR (KBr) 3446.7, 3090.4, 1542.8, 1461.1, 1443.1, 1347.3, 1257.7, 1114.6, 1076.2, 1051.8, 1021.0, 904.5, 842.5, 768.8, 743.7, 699.9, 551.0, 485.16 cm^{-1}. ^{1}H NMR (400 MHz, CDCl$_3$) δ 8.16 (s, 1 H), 7.89 (s, 1 H), 7.47-7.45 (m, 3 H), 7.31-7.29 (m, 2 H). ^{13}C NMR (100 MHz, CDCl$_3$) δ 151.5, 150.6, 137.2, 134.4, 130.8, 130.1, 129.7, 128.9, 128.1, 114.1. HRMS Calc'd for 321.9589. Found: 321.9592.

2',4'-Dinitro-5'-phenyl-4-ethynylphenyl-1-thioacetylbenzene (2). 9 (147 mg, 0.46 mmol), 4-ethynyl(thioacetyl)benzene (106 mg, 0.60 mmol), bis(dibenzylideneacetone)palladium (26 mg, 0.05 mmol), CuI (9 mg, 0.05 mmol), (C$_6$H$_5$)$_3$P (12 mg, 0.05 mmol), Hunig's base (0.16 mL) and THF (20 mL) were coupled according to the general coupling procedure. The reaction mixture was stirred and heated overnight at 45 °C. Crude product was purified via column chromatography (silica, 3:1 CH$_2$Cl$_2$:hexanes) to yield 75 mg of an orange solid (39%). IR (KBr) 2922.7, 2214.3, 1702.7, 1542.8, 1488.1, 1357.1, 1271.1, 1115.1, 1088.6, 956.0, 908.6, 829.9, 770.5, 707.0, 623.4 cm^{-1}. ^{1}H NMR (400 MHz, CDCl$_3$) δ 8.16 (s, 1 H), 8.10 (s, 1 H), 7.63 (d, J = 8.4, 2 H), 7.48-7.44 (m, 5 H), 7.36-7.33 (m, 2 H), 2.44 (s, 3 H). ^{13}C NMR (100 MHz, CDCl$_3$) δ 193.3, 151.2, 150.6, 136.8, 134.8, 134.7, 133.1, 130.8, 130.2, 130.1, 129.6, 128.6, 128.1, 122.9, 119.0, 99.8, 84.5, 30.8. HRMS Calc'd for 418.0623. Found: 418.0619.

2,5-Dibromo-1,4-dimethoxybenzene (11). In a 100 mL round bottom flask, 1,4-dimethoxybenzene (10.0 g, 72.4 mmol) was dissolved in glacial acetic acid (20 mL). A solution of bromine (7.42 mL, 145.0 mmol) in glacial acetic acid (7.5 mL) was added dropwise to the first solution at room temperature over 40 min. The resulting mixture was allowed to stir for 2 h. The crude product was washed with ice-cold water and ice-cold MeOH to afford fine white crystals. The mother liquor was concentrated and cooled to afford more white crystals (15.9 g, 74% yield). Mp 136-138 °C (lit[98] mp 144-145 °C). IR (KBr) 3091.9, 3022.1, 2968.8, 2944.4, 2842.8, 1694.9, 1494.2, 1475.6, 1436.5, 1358.2, 1275.0, 1211.8, 1185.0, 1065.4, 1021.9, 860.5, 760.4, 441.8 cm^{-1}. ^{1}H NMR (400 MHz, CDCl$_3$) δ 7.13 (s, 2 H), 3.87 (s, 6 H). ^{13}C NMR (100 MHz, CDCl$_3$) δ 150.93, 117.53, 110.90, 57.43.

2,5-Di(ethynylphenyl)-1,4-dimethoxybenzene (12). **11** (8.745 g, 29.55 mmol), ((C$_6$H$_5$)$_3$P)$_2$PdCl$_2$ (0.415 g, 0.591 mmol), CuI (0.225 g, 1.182 mmol), (C$_6$H$_5$)$_3$P (0.310 g, 1.182 mmol), THF (35 mL), Hünig's base (20.5 mL, 118 mmol), and phenylacetylene (7.8 mL, 70.92 mmol) were used following the general procedure for couplings. The solution was heated in a 65 °C oil bath for 3 d. Recrystallization from benzene afforded the desired product mp 175-177 °C (lit.[16] 176-177 °C) (9.22 g, 92 %). ^{1}H NMR (400 MHz, CDCl$_3$) δ 7.57 (m, 4 H), 7.34 (m, 6H), 7.03 (s, 2H), 3.89 (s, 6 H). ^{13}C NMR (100 MHz, CDCl$_3$) δ 154.10, 131.89, 128.60, 128.50, 123.39, 115.86, 113.57, 95.23, 85.86, 56.66.

2,5-Di(ethynylphenyl)benzoquinone (13). **12** (0.300 g, 0.886 mmol) and THF (6 mL) were added to a 25 mL round bottom flask containing a stir bar. A solution of ceric ammonium nitrate (1.46 g, 2.658 mmol) in water (3 mL) was slowly added to the flask and allowed to stir for 15 min. Water was added and the organic materials were extracted with CH$_2$Cl$_2$. Flash column chromatography (silica gel using 1:1 hexane:CH$_2$Cl$_2$ as eluent) afforded the desired product (0.129 g, 47 %). IR (KBr) 3047.5, 2203.0, 1716.2, 1655.3, 1568.3, 1215.4, 1100.6, 902.1, 757.6, 686.4 cm^{-1}. ^{1}H NMR (400 MHz, CDCl$_3$) δ 7.58 (dd, J = 7.9, 1.5 Hz, 4 H), 7.38 (m, 6 H), 6.99 (s, 2 H). ^{13}C NMR (100 MHz, CDCl$_3$) δ 182.87, 136.55, 133.34, 132.83, 130.57, 128.97, 121.83, 105.26, 82.90. HRMS calc'd for C$_{22}$H$_{12}$O$_2$: 308.0837 Found: 308.0834.

2-Bromo-5-ethynylphenyl-1,4-dimethoxybenzene (14). **11** (2.96 g, 10.0 mmol), bis(dibenzylideneacetone)palladium (0.115 g, 0.20 mmol), CuI (0.038 g, 0.20 mmol), (C$_6$H$_5$)$_3$P (0.131 g, 0.50 mmol), THF (15 mL), Hünig's base (6.97 mL, 40.0 mmol) and phenylacetylene (1.21 mL, 11.0 mmol) were used following the general procedure for coupling. The tube was heated in a 50 °C oil bath for 18 h. Column chromatography (silica gel using 19:1 hexanes:diethyl ether as eluent) afforded the desired product, somewhat impure (approximately 15% impurities by NMR) in moderate yield (1.02 g, 32% yield). This was taken onto the next step in this impure form. ^{1}H NMR (400 MHz, CDCl$_3$) δ 7.54 (m, 2 H), 7.33 (m, 3 H), 7.09 (s, 1 H), 7.02 (s, 1 H), 3.86 (s, 6 H).

1,4-Dimethoxy-2-ethynylphenyl-5-(trimethylsilylethynyl)benzene.
14 (1.0 g, 3.15 mmol), bis(dibenzylideneacetone)palladium (0.036 g, 0.063 mmol), CuI (0.012 g, 0.063 mmol), $(C_6H_5)_3P$ (0.042 g, 0.16 mmol), THF (20 mL), Hünig's base (2.2 mL, 12.6 mmol), and TMSA (0.89 mL, 6.3 mmol) were used following the general procedure for couplings. The tube was capped and heated in a 60 °C oil bath for 1 d. Flash column chromatography (silica gel using 24:1 hexanes:EtOAc as eluent) afforded the desired product slightly impure (0.83 g, 79% yield). ^1H NMR (400 MHz, CDCl$_3$) δ 7.55 (m, 2 H), 7.32 (m, 3 H), 6.98 (s, 1 H), 6.95 (s, 1 H), 3.84 (s, 3 H), 3.83 (s, 3 H), 0.27 (s, 9 H).

1,4-Dimethoxy-2-ethynyl-5-(ethynylphenyl)benzene **(15).** 1,4-dimethoxy-2-ethynylphenyl-5-(trimethylsilylethynyl)benzene (0.830 g, 2.48 mmol), K$_2$CO$_3$ (1.71 g, 12.4 mmol), MeOH (50 mL), and CH$_2$Cl$_2$ (50 mL) were used following the general procedure for deprotection to afford the desired product (0.513 g, 79% yield). ^1H NMR (400 MHz, CDCl$_3$) δ 7.55 (m, 2 H), 7.33 (m, 3 H), 7.00 (s, 1 H), 6.98 (s, 1 H), 3.87 (s, 3 H), 3.86 (s, 3 H), 3.39 (s, 1 H).

4,4'-Di(ethynylphenyl)-2',5'-dimethoxy-1-benzenethioacetate **(16).**
15 (0.513 g, 1.96 mmol), bis(dibenzylideneacetone)palladium(0) (0.058 g, 0.10 mmol), CuI (0.019 g, 0.10 mmol), $(C_6H_5)_3P$ (0.066 g, 0.25 mmol), THF (20 mL), Hünig's base (1.37 mL, 7.84 mmol), and 4-(thioacetyl)iodobenzene (0.608 g, 2.16 mmol) were used following the general procedure for couplings. The tube was capped and heated in a 55 °C oil bath for 3 d. Flash column chromatography (silica gel using CH$_2$Cl$_2$ as eluent) afforded the desired product slightly impure (0.621 g, 76% yield). ^1H NMR (400 MHz, CDCl$_3$) δ 7.57 (m, 4 H), 7.38 (d, J = 8.1 Hz, 2 H), 7.33 (m, 3 H), 7.03 (s, 1 H), 7.02 (s, 1 H), 3.874 (s, 3 H), 3.870 (s, 3 H), 2.40 (s, 3 H).

2-Ethynylphenyl-5-((4'-thioacetyl)ethynylphenyl)benzoquinone
(17). **16** (0.050 g, 0.12 mmol), acetonitrile (5 mL), and THF (5 mL) were added to a 25 mL round bottom flask containing a stir bar. A solution of ceric ammonium nitrate (0.13 g, 0.24 mmol) in water (1 mL) was added in one portion. After stirring at room temperature for 30 min, another equivalent solution of ceric ammonium nitrate (0.13 g, 0.24 mmol) was added. After 20 additional min, the reaction was quenched by adding water (30 mL) to effect precipitation of an orange solid. Flash column chromatography (silica gel using CH$_2$Cl$_2$ as eluent) afforded the desired product (0.034 g, 74% yield). IR (KBr) 3053.0, 2924.3, 2852.6, 2205.4, 1703.4, 1652.7, 1568.8, 1483.7, 1442.2, 1354.8, 1221.3, 1105.4, 1089.4, 949.6, 920.1, 830.9, 758.2, 688.2, 620.6 cm^{-1}. ^1H NMR (400 MHz, CDCl$_3$) δ 7.58 (m, 4 H), 7.42 (m, 2 H), 7.38 (m, 3 H), 6.98 (s, 1 H), 6.97 (s, 1 H), 2.42 (s, 3 H). ^{13}C NMR (100 MHz, CDCl$_3$) δ 193.22, 182.74, 182.67, 136.88, 136.51, 134.63, 133.34, 133.24, 132.99, 132.84, 130.94, 130.63, 128.99, 122.81, 121.80, 105.38, 103.99, 84.17, 82.92, 30.80. HRMS calc'd for C$_{24}$H$_{14}$O$_3$S: 382.0664. Found: 382.0663.

1,4-Dimethoxy-2,5-bis(trimethylsilylethynyl)benzene. 11 (1.75 g, 5.91 mmol), $((C_6H_5)_3P)_2PdCl_2$ (0.207 g, 0.296 mmol), CuI (0.113 g, 0.591 mmol), $(C_6H_5)_3P$ (0.155 g, 0.591 mmol), THF (20 mL), Hünig's base (4.1 mL, 23.64 mmol), and TMSA (2.51 mL, 17.73 mmol) were used following the general procedure for couplings. The tube was capped and heated in a 55 °C oil bath for 2 d. Flash column chromatography (silica gel using 1:1 hexanes:CH_2Cl_2 as eluent) afforded the desired product (1.54 g, 79 % yield). IR (KBr) 2957.0, 2898.2, 2851.2, 2829.0, 2149.1, 1496.8, 1464.1, 1449.1, 1388.2, 1283.7, 1249.0, 1223.6, 1203.1, 1172.4, 1039.6, 883.2, 841.3, 757.4, 714.9, 696.2, 626.4 cm^{-1}. ^1H NMR (400 MHz, CDCl$_3$) δ 6.89 (s, 2 H), 3.81 (s, 6 H), 0.25 (s, 18 H) ^{13}C NMR (100 MHz, CDCl$_3$) δ 154.56, 116.59, 113.81, 101.22, 100.84, 56.83, 0.40. HRMS calc'd for $C_{18}H_{26}O_2Si_2$: 330.1471, Found: 330.1468.

1,4-Dimethoxy-2,5-diethynylbenzene (18). 1,4-Dimethoxy-2,5-bis(trimethylsilylethynyl)benzene (1.50 g, 4.54 mmol), K_2CO_3 (6.27 g, 45.4 mmol), MeOH (50 mL), and CH_2Cl_2 (50 mL) were used following the general procedure for deprotection to give the desired product (0.829 g, 98 %). ^1H NMR (400 MHz, CDCl$_3$) δ 6.96 (s, 2 H), 3.84 (s, 6 H), 3.37 (s, 2 H).

2,5-Bis(4'-(thioacetyl)ethynylphenyl)-1,4-dimethoxybenzene (19). **18** (0.810 g, 4.35 mmol), bis(dibenzylideneacetone)palladium (0.253 g, 0.44 mmol), CuI (0.084 g, 0.44 mmol), $(C_6H_5)_3P$ (0.115 g, 0.44 mmol), THF (30 mL), Hünig's base (4.5 mL, 26.1 mmol), and 4-(thioacetyl)iodobenzene (2.54 g, 9.14 mmol) were used following the general procedure for couplings. The solution was stirred in a 60 °C oil bath for 16 h. Crystallization from CH_2Cl_2:hexanes afforded the desired product (1.81 g, 85 %). IR (KBr) 3129.1, 3057.4, 3006.2, 2975.5, 2940.0, 2847.4, 2207.2, 1697.7, 1506.8, 1483.1, 1463.1, 1396.2, 1279.2, 1223.5, 1122.2, 1034.2, 949.5, 898.8, 825.5, 765.6, 616.8 cm^{-1}. ^1H NMR (400 MHz, CDCl$_3$) δ 7.57 (dt, J = 8.5, 1.9 Hz, 4 H), 7.39 (dt, J = 8.5, 2.0 Hz, 4 H), 7.01 (s, 2 H), 3.89 (s, 6 H), 2.42 (s, 6 H). ^{13}C NMR (100 MHz, CDCl$_3$) δ 193.85, 154.43, 134.58, 132.65, 128.64, 124.84, 116.08, 113.75, 94.76, 87.73, 56.91, 30.70. HRMS calc'd for $C_{28}H_{22}O_4S_2$,: 486.0960 Found: 486.0956.

2,5-Bis(4'-(thioacetyl)ethynylphenyl)benzoquinone (20). **19** (0.050 g, 0.103 mmol), acetonitrile (5 mL), and THF (3 mL) were added to a 25 mL round bottom flask containing a stir bar. A solution of ceric ammonium nitrate (0.339 g, 0.618 mmol) in water (2 mL) was added in two portions at 30 min intervals. After stirring at room temperature for 3 h, the reaction was quenched by adding water to effect precipitation of an orange solid. Flash column chromatography (silica gel using CH_2Cl_2 as eluent) afforded the desired product (0.023 g, 49 % yield). IR (KBr) 2922.2, 2847.4, 2203.4, 1694.9, 1660.1, 1569.9, 1351.8, 1212.3, 1119.7, 1084.6, 1013.2, 960.3, 826.8, 620.6 cm^{-1}. ^1H NMR (400 MHz, CDCl$_3$) δ 7.60 (dt, J = 8.3, 1.6 Hz, 4 H), 7.42 (dt, J = 8.3, 1.6 Hz, 4 H), 7.00 (s, 2 H), 2.43 (s, 6 H). ^{13}C NMR (100 MHz, CDCl$_3$) δ 193.23,

182.61, 136.86, 134.64, 133.25, 133.07, 130.97, 122.78, 104.14, 84.08, 30.79. HRMS calc'd for $C_{26}H_{16}O_4S_2$: 456.0500, Found: 456.0490.

2,5-Bis(4'-ethynylpyridyl)-1-nitrobenzene (22). To a solution of 2,5-dibromonitrobenzene (0.28 g, 0.997 mmol), $((C_6H_5)_3P)_2PdCl_2$ (0.07 g, 0.098 mmol), CuI (0.019 g, 0.098 mmol), $(C_6H_5)_3P$ (0.106 g, 0.40 mmol) and K_2CO_3 (1.1 g, 7.96 mmol) in THF (4 mL) were added via a cannula **21** (0.377 g, 2.15 mmol) in THF (4 mL) and MeOH (2 mL). The mixture was heated at 64 °C for 20 h. The solvent was removed by rotary evaporation and the black residue was washed with aqueous K_2CO_3 and extracted with Et_2O. The combined organic layers were dried over Na_2SO_4, filtered, and the solvent evaporated in vacuo. Purification by flash chromatography (silica gel, hexane:EtOAc 70/30, 50/50, 20/80, 0/100) afforded 60 mg (24% yield) of the title compound as a yellow solid. Mp: 178-180 °C. IR (KBr) 3414.0, 3036.7, 1616.0, 1589.4, 1538.1, 1519.9, 1407.9, 1345.7, 1271.1, 1214.1, 828.3 cm^{-1}. ^{1}H NMR (400 MHz, DMSO-d) δ 8.69 (br s, 4 H), 8.44 (d, J=1.4 Hz, 1 H), 8.04 (1/2 ABqd, J=8.0, 1.4 Hz, 1 H), 7.99 (1/2 ABq, J=8.0 Hz, 1 H), 7.60 (d, J=5.8 Hz, 2 H), 7.57 (d, J=5.8 Hz, 2 H). ^{13}C NMR (100 MHz, DMSO-d) δ 150.21, 150.13, 149.42, 136.27, 135.36, 129.16, 129.11, 127.96, 125.50, 125.39, 123.25, 116.55, 94.98, 90.63, 90.59, 88.13. HRMS calc'd for $C_{20}H_{11}N_3O_2$: 325.0851, found: 325.0847.

1-Bromo-4-(4'-ethynylpyridyl)-3-nitrobenzene (23). To a solution of 2,5-dibromonitrobenzene (0.43 g, 1.53 mmol), $((C_6H_5)_3P)_2PdCl_2$ (0.052 g, 0.074 mmol), CuI (0.015 g, 0.078 mmol), $(C_6H_5)_3P$ (0.079 g, 0.30 mmol) and K_2CO_3 (0.83 g, 6.0 mmol) in THF (2 mL) were added via a cannula **21** (0.342 g, 1.95 mmol) in THF (4 mL) and MeOH (1.5 mL). The mixture was heated at 23 °C for 2 d. The solvent was removed by rotary evaporation and the residue was diluted with water and extracted with Et_2O. The combined organic layers were dried over Na_2SO_4, filtered, and the solvent evaporated in vacuo. Purification by flash chromatography (silica gel, hexane:EtOAc 90/10, 70/30, 50/50) afforded 330 mg (71% yield) of the title compound as an off-white solid. Mp: 166-171 °C. IR (KBr) 3424.4, 3093.3, 1592.3, 1521.4, 1409.3, 1341.4, 1272.6 cm^{-1}. ^{1}H NMR (400 MHz, CDCl$_3$) δ 8.68 (br s, 2 H), 8.29 (d, J=1.9 Hz, 1 H), 7.79 (dd, J=8.3, 2.0 Hz, 1 H), 7.62 (d, J=8.3 Hz, 1 H), 7.44 (d, J=4.7 Hz, 2 H). ^{13}C NMR (100 MHz, CDCl$_3$) δ 149.96, 136.22, 135.69, 130.14, 128.08, 126.67, 125.65, 123.19, 116.48, 94.80, 87.81. HRMS calc'd for $C_{13}H_7BrN_2O_2$: 303.9672, found: 303.9682 .

5-Ethynylphenyl-2-(4'-ethynylpyridyl)-1-nitrobenzene (24). To a solution of **23** (88.8 mg, 0.293 mmol), $((C_6H_5)_3P)_2PdCl_2$ (0.011 g, 0.016 mmol), CuI (0.004 g, 0.021 mmol) and $(C_6H_5)_3P$ (0.008 g, 0.029 mmol) in THF (4 mL) were added Et_3N (0.25 mL, 1.76 mmol) and phenylacetylene (0.1 mL, 9.1 mmol). The mixture was stirred at 56 °C for 36 h. The solvent was evaporated in vacuo. The residue was diluted with water and extracted with Et_2O. The combined organic layers were dried over $MgSO_4$, filtered, and the solvent evaporated in vacuo. Purification by flash chromatography (silica gel, EtOAc:hexane 20:80) afforded 65 mg (69% yield) of the title compound as a yellow solid. Mp: 130-132 °C. IR (KBr) 3445.3, 3046.3, 2203.5, 1548.5, 1529.1, 1399.9, 1341.6 cm^{-1}. ^{1}H NMR (400 MHz, CDCl$_3$) δ 8.67 (br d, J = 4.9 Hz, 2 H), 8.27 (d, J = 1.5 Hz, 1 H), 7.76 (1/2 ABqd, J = 8.0, 1.6 Hz, 1 H), 7.72 (1/2 ABqd, J = 8.0, 0.5 Hz, 1 H), 7.56 (m, 2 H), 7.45 (dd, J = 5.9, 1.7 Hz, 2 H), 7.40 (m, 3 H). ^{13}C NMR (100 MHz, CDCl$_3$) δ 149.58, 135.39, 134.65, 131.81, 129.34, 128.54, 127.67, 125.32, 121.85, 116.66, 95.30, 94.30, 88.52, 86.63. HRMS calc'd for C$_{21}$H$_{12}$N$_2$O$_2$: 324.0899, found: 324.0895.

1-Bromo-4-ethynylphenyl-3-nitrobenzene (25). To a solution of 2,5-dibromonitrobenzene (0.937 g, 3.34 mmol), bis(dibenzylideneacetone)palladium (0.095 g, 0.166 mmol), CuI (0.032 g, 0.168 mmol) and (C$_6$H$_5$)$_3$P (0.173 g, 0.66 mmol) in THF (4 mL) were added Et$_3$N (1 mL, 7.2 mmol) and phenylacetylene (0.5 mL, 4.56 mmol). The mixture was stirred at 23 °C for 48 h. The mixture was washed with a saturated solution of NH$_4$Cl and then extracted with Et$_2$O. The combined organic layers were dried over Na$_2$SO$_4$, filtered, and the solvent evaporated in vacuo. Purification by flash chromatography (silica gel, CH$_2$Cl$_2$:hexane 1/8) afforded 0.48 g (47% yield) of the title compound as a yellow solid. Mp: 58-74 °C. IR (KBr) 3421.9, 3085.5, 2213.4, 1595.7, 1545.9, 1521.3, 1336.5, 1269.2 cm^{-1}. ^{1}H NMR (400 MHz, CDCl$_3$) δ 8.23 (d, J=1.9 Hz, 1 H), 7.72 (dd, J=8.3 Hz, 2.0, 1 H), 7.59 (m, 3 H), 7.40 (m, 3 H). ^{13}C NMR (100 MHz, CDCl$_3$) δ 149.71, 135.91, 135.45, 131.99, 129.44, 128.46, 127.78, 122.03, 121.75, 117.69, 98.43, 84.00. HRMS calc'd for C$_{14}$H$_8$NO$_2$Br: 302.9720, found: 302.9725.

2-Ethynylphenyl-5-(4'-ethynylpyridyl)-1-nitrobenzene (26). To a solution of **25** (0.306 g, 1.01 mmol), K$_2$CO$_3$ (0.713 g, 5.16 mmol), ((C$_6$H$_5$)$_3$P)$_2$PdCl$_2$ (0.035 g, 0.05 mmol), CuI (0.009 g, 0.047 mmol) and (C$_6$H$_5$)$_3$P (0.052 g, 0.198 mmol) in THF (2 mL) were added via a cannula **21** (0.217 g, 1.24 mmol) in THF (2 mL) and MeOH (1 mL). The mixture was heated at 60 °C for 18 h. The solvent was removed by rotary evaporation and the brown residue was diluted with water and extracted with Et$_2$O. The combined organic layers were dried over Na$_2$SO$_4$, filtered, and the solvent evaporated in vacuo. Purification by flash chromatography (silica gel, EtOAc:hexane 20:80, 40:60) afforded 260 mg (79% yield) of the title compound as a yellow solid. Mp: 144-146 °C. IR (KBr) 3442.3, 3053.0, 2209.4, 1631.3, 1584.8, 1524.7, 1404.3, 1344.7, 1269.0, 826.4, 755.2, 686.6 cm^{-1}. ^{1}H NMR (400 MHz, CDCl$_3$) δ 8.67 (dd, J=4.4, 1.6 Hz, 2 H), 8.27 (br s, 1 H), 7.74 (m, 2 H), 7.63 (d, J=1.8 Hz, 1 H), 7.60 (m, 1 H), 7.42 (m, 5 H). ^{13}C NMR (100 MHz, CDCl$_3$) δ 149.99, 135.41, 134.65, 132.14, 130.19, 129.61, 128.54, 127.95, 125.50, 122.68, 122.06, 119.15, 99.67, 90.83, 90.27, 84.62. HRMS calc'd for C$_{21}$H$_{12}$N$_2$O$_2$: 324.0899, found: 324.0897.

2,5-Bis(trimethylsilylethynyl)-4-nitroacetanilide (27). To a solution of **6** (0.78 g, 2.3 mmol), bis(dibenzylideneacetone)palladium (0.068 g, 0.118

mmol), CuI (0.023 g, 0.012 mmol), $(C_6H_5)_3P$ (0.123 g, 0.47 mmol) in THF (8 mL) were added Et_3N (1 mL, 7.2 mmol) and TMSA (1 mL, 7.0 mmol). The mixture was heated at 67 °C for 48 h. The solvent was removed by rotary evaporation and the brown residue was diluted with water and extracted with Et_2O. The combined organic phases were dried over Na_2SO_4, filtered, and the solvent evaporated in vacuo. Purification by flash chromatography (silica gel, CH_2Cl_2:hexane 35:65) afforded 410 mg (47% yield) of the title compound as an off-white solid. Mp: 162-164°C. IR (KBr) 3372.9, 2962.9, 2146.0, 1727.2, 1611.2, 1544.9, 1501.5, 1457.1, 1404.3, 1338.2, 1250.6, 1222.3, 881.9 cm⁻¹. ¹H NMR (400 MHz, $CDCl_3$) δ 8.75 (s, 1 H), 8.15 (s, 1 H), 8.10 (br s, 1 H), 2.27 (s, 3 H), 0.33 (s, 9 H), 0.28 (s, 9 H). ¹³C NMR (100 MHz, $CDCl_3$) δ 168.21, 144.19, 142.41, 128.11, 123.82, 120.18, 111.52, 106.66, 106.16, 99.50, 97.44, 24.90, -0.31, -0.46. HRMS calc'd for $C_{18}H_{24}N_2O_3Si_2$: 372.1326, found: 372.1326.

2,5-Bis(4'-ethynylpyridyl)-4-nitroaniline (28). To a solution of **27** (0.056 g, 0.15 mmol), 4-iodopyridine (0.08 g, 0.39 mmol), K_2CO_3 (0.17 g, 1.2 mmol), $((C_6H_5)_3P)_2PdCl_2$ (0.01 g, 0.015 mmol), CuI (0.004 g, 0.021 mmol) and $(C_6H_5)_3P$ (0.016 g, 0.061 mmol) in THF (4 mL) was added MeOH (1 mL). The mixture was heated at 60 °C for 50 h. The solvent was removed by rotary evaporation and the brown residue was diluted with water and extracted with EtOAc. The combined organic phases were dried over Na_2SO_4, filtered, and the solvent evaporated in vacuo. Purification by flash chromatography (silica gel, EtOAc) afforded 8 mg (16% yield) of the title compound as a yellow solid. Mp: 154-160 °C. IR (KBr) 3730.2, 3438.6, 2204.8, 1592.4, 1541.1, 1409.8, 1308.5, 1249.9, 818.8 cm⁻¹. ¹H NMR (400 MHz, $CDCl_3$) δ 8.67 (dd, J=4.4, 1.7 Hz, 2 H), 8.65 (dd, J= 4.5, 1.7 Hz, 2 H), 8.34 (s, 1 H), 7.44 (dd, J=4.5, 1.7 Hz, 2 H), 7.40 (dd, J=4,4, 1.6 Hz, 2 H), 6.99 (s, 1 H), 5.03 (br s, 2 H). ¹³C NMR (100 MHz, $CDCl_3$) δ 151.26, 150.03, 149.90, 139.56, 130.71, 130.52, 130.00, 125.65, 125.33, 120.33, 118.52, 106.57, 94.67, 94.19, 89.55, 87.27. HRMS calc'd for $C_{20}H_{12}N_4O_2$: 340.0960, found: 340.0958.

2-Amino-4-(4'-ethynylpyridyl)-5-nitrobromobenzene (29). To a solution of **6** (0.877 g, 8.84 mmol), K_2CO_3 (1.08 g, 7.81 mmol), $((C_6H_5)_3P)_2PdCl_2$ (0.054 g, 0.077 mmol), CuI (0.025 g, 0.13 mmol) and $(C_6H_5)_3P$ (0.068 g, 0.26 mmol) in THF (4 mL) were added via a cannula **21** (0.404 g, 2.30 mmol) in THF (8 mL) and MeOH (3 mL). The mixture was stirred at 23 °C for 1 d. The solvent was evaporated in vacuo. The residue was diluted with water and extracted with EtOAc. The combined organic phases were dried over $MgSO_4$, filtered and the solvent evaporated in vacuo. Purification by flash chromatography (silica gel, EtOAc:hexane 40:60 50:50) afforded 290 mg (39% yield) of the title compound as a yellow solid. Mp: 226-228 °C. IR (KBr) 3385.4, 3297.7, 3171.3, 1646.8, 1591.7, 1556.9, 1471.3, 1297.8 cm⁻¹. ¹H NMR (400 MHz, DMSO-d) δ 8.66 (br d, J = 3.8 Hz, 2 H), 8.32 (d, J = 1.3 Hz, 1 H), 7.53 (br d, J = 4.5 Hz, 2 H), 7.06 (d, J= .3 Hz, 1 H), 6.94 (br s, 2 H). ^{13}C NMR (100 MHz, DMSO-d) δ 151.33, 150.12, 136.44, 130.70, 129.64, 125.32, 118.13, 117.73, 106.02, 91.85, 89.72. HRMS calc'd for $C_{13}H_8BrN_3O_2$: 316.9800, found: 316.9801.

4-Amino-2-(4'-ethynylpyridyl)-1-nitro-5-(trimethylsilylethynyl)benzene. To a solution of **29** (0.310 g, 0.975 mmol), $((C_6H_5)_3P)_2PdCl_2$ (0.035 g, 0.05 mmol), CuI (0.011 g, 0.05 mmol) and $(C_6H_5)_3P$ (0.026 g, 0.10 mmol) in THF (10 mL) were added Et$_3$N (0.9 mL, 6.5 mmol) and TMSA (0.2 mL, 1.4 mmol). The mixture was stirred at 60 °C for 2 d. The solvent was evaporated in vacuo. The residue was diluted with water and extracted with EtOAc. The combined organic phases were dried over MgSO$_4$, filtered, and the solvent evaporated in vacuo. Purification by flash chromatography (silica gel, Et$_2$O) afforded 160 mg (49% yield) of the title compound as a yellow solid. Mp: 145-150 °C. IR (KBr) 3451.9, 3379.1, 2960.5, 2149.5, 1620.4, 1597.9, 1545.5, 1512.2, 1317.0 cm^{-1}. ^{1}H NMR (400 MHz, CDCl$_3$) δ 8.65 (dd, J=4.6, 1.5 Hz, 2 H), 8.25 (s, 1 H), 7.44 (dd, J=4.3, 1.5 Hz, 2 H), 6.93 (s, 1 H), 4.90 (s, 2 H), 0.30 (s, 9 H). ^{13}C NMR (100 MHz, CDCl$_3$) δ 151.44, 149.90, 139.35, 130.65, 130.43, 125.65, 119.56, 118.06, 107.93, 104.28, 98.37, 93.70, 89.79, -0.15. HRMS calc'd for $C_{18}H_{17}N_3O_2Si$: 335.1090, found: 335.1089.

4-Amino-5-ethynyl-2-(4'-ethynylpyridyl)-1-nitrobenzene. (30). To a solution of 4-Amino-2-(4'-ethynylpyridyl)-1-nitro-5-(trimethylsilylethynyl)benzene (160 mg, 0.477 mmol) in MeOH (15 mL) and CH$_2$Cl$_2$ (15 mL) was added K$_2$CO$_3$ (0.66 g, 4.77 mmol). The solution was stirred at 23 °C for 2 h. The reaction mixture was diluted with water and extracted with EtOAc. The combined organic layers were dried over MgSO$_4$, filtered, and the solvent evaporated in vacuo. The reaction afforded 0.11 g (88% yield) of the title compound as a yellow solid. The product was too unstable to attain its complete characterization data. ^{1}H NMR (400 MHz, DMSO-d) δ 8.67 (dd, J=4.5, 1.6 Hz, 2 H), 8.12 (s, 1 H), 7.53 (dd, J=4.5, 1.6 Hz, 2 H), 7.03 (s, 1 H), 6.97 (br s, 2 H), 4.70 (s, 1 H).

4-Amino-2-(4'-ethynylpyridyl)-5-(4'-thioacetylphenylethynyl)-1-nitrobenzene (31). To a solution of **30** (0.110 g, 0.418 mmol), 4-thioacetyliodobenzene[41] (0.124 g, 0.446 mmol), $((C_6H_5)_3P)_2PdCl_2$ (0.015 g, 0.021 mmol), CuI (0.004 g, 0.021 mmol) and $(C_6H_5)_3P$ (0.014 g, 0.053 mmol) in THF (13 mL) was added Et$_3$N (0.4 mL, 2.9 mmol). The mixture was stirred at 50 °C for 2 d. The reaction was checked by TLC (EtOAc/hex 75/25). More $((C_6H_5)_3P)_2PdCl_2$ (0.014 g, 0.020 mmol), CuI (0.035 g, 0.018 mmol) and $(C_6H_5)_3P$ (0.085 g, 0.324 mmol) were added and the reaction was stirred at 60 °C for 1 d. The solvent was evaporated in vacuo. The residue was diluted with water and extracted with EtOAc. The combined organic layers were dried over MgSO$_4$, filtered, and the solvent evaporated in vacuo. Purification by flash chromatography (silica gel, EtOAc:hexane 66:33) afforded 130 mg (75% yield) of the title compound as a yellow solid. Mp: 185-188 °C. IR (KBr) 3438.2,

3195.9, 2922.4, 1695.4, 1627.7, 1596.5, 1545.1, 1514.8, 1477.2, 1402.8, 1316.4, 1249.9 cm^{-1}. ^{1}H NMR (400 MHz, DMSO-d) δ 8.68 (br d, J=4.0 Hz, 2 H), 8.23 (s, 1 H), 7.79 (d, J=8.1 Hz, 2 H), 7.54 (d, J=5.0 Hz, 2 H), 7.49 (d, J= 8.0 Hz, 2 H), 7.13 (br s, 2 H), 7.06 (s, 1 H), 2.46 (s, 3 H). ^{13}C NMR (100 MHz, DMSO-d) δ 192.98, 153.79, 150.13, 136.28, 134.31, 132.32, 130.69, 129.67, 128.66, 125.34, 123.05, 118.70, 118.26, 105.43, 95.72, 92.51, 90.12, 85.54, 30.32. HRMS calc'd for $C_{23}H_{15}N_3O_3S$: 413.0834, found: 413.0940.

2-(4'-Ethynylpyridyl)-4-nitro-5-phenylaniline (32). To a solution of **7** (80.5 mg, 0.241 mmol), K_2CO_3 (0.151 g, 1.09 mmol), $((C_6H_5)_3P)_2PdCl_2$ (0.009 g, 0.014 mmol), CuI (0.003 g, 0.014 mmol) and $(C_6H_5)_3P$ (0.014 g, 0.053 mmol) in THF (2 mL) were added via a cannula **1** (0.053 g, 0.3 mmol) in THF (2 mL) and MeOH (1 mL). The mixture was heated to 70 °C for 3 d. The solvent was removed by rotary evaporation and the brown residue was diluted with water and extracted with Et$_2$O. The combined organic layers were dried over Na$_2$SO$_4$, filtered and the solvent evaporated in vacuo. Purification by flash chromatography (silica gel, EtOAc:hexane 30:70) afforded 60 mg (79% yield) of the title compound as yellow solid. Mp: 187-190 °C. IR (KBr) 3410.2, 3323.4, 3212.1, 2215.1, 1627.6, 1592.4, 1548.4, 1511.7, 1410.5, 1331.9 cm^{-1}. ^{1}H NMR (400 MHz, CDCl$_3$) δ 8.64 (br d, J=4.8, 2 H), 8.16 (s, 1 H), 7.39 (m, 5 H), 7.27 (m, 2 H), 6.62 (s, 1 H), 5.03 (br s, 2 H). ^{13}C NMR (100 MHz, CDCl$_3$) δ 151.23, 149.82, 140.65, 138.82, 138.19, 130.49, 128.36, 128.06, 127.52, 125.34, 116.41, 104.85, 93.24, 87.89.   HRMS calc'd for $C_{19}H_{13}N_3O_2$: 315.1008, found: 315.1011.

1-Bromo-4-(4'-ethynyl)pyridine-3-nitrobenzene (34). To a solution of **33** (0.84 g, 2.34 mmol), $((C_6H_5)_3P)_2PdCl_2$ (0.083 g, 0.117 mmol), CuI (0.022 g, 0.117 mmol), K_2CO_3 (1.94 g, 14.04 mmol) in THF (4 mL) were added **21** (0.451 g, 2.57 mmol) in THF (12 mL) via a cannula and CH$_3$OH (4 mL). The mixture was heated to 55 °C for 14 h. The solvent was removed by rotary evaporation and the residue was diluted with water, washed with brine and extracted with EtOAc. The combined organic phases were dried over MgSO$_4$, filtered and the solvent evaporated in vacuo.   Purification by flash chromatography (silica gel, EtOAc) afforded 271 mg (34% yield) of the title compound as a yellow solid.   Mp: 224-229 °C.   IR (KBr) 3451.7, 3351.1, 3202.6, 2206.4, 1622.9, 1588.4, 1539.0, 1474.4, 1306.7, 1249.8 cm^{-1}. ^{1}H NMR (400 MHz, DMSO-d) δ 8.64 (d, J= 5.7 Hz, 2 H), 8.25 (s, 1 H), 7.67 (dd, J=4.5, 1.5 Hz, 2 H), 7.59 (m, 2 H), 7.47 (m, 3 H), 7.15 (br s, 1 H), 7.03 (s, 1 H). ^{13}C NMR (100 MHz, DMSO-d) δ 153.97, 149.83, 136.31, 131.67, 131.17, 130.01, 129.69, 128.99, 125.45, 121.78, 120.40, 118.06, 103.92, 96.13, 93.41, 88.37, 86.25.   HRMS calc'd for $C_{21}H_{13}N_3O_2$: 339.1008, found: 339.1004.

1-Bromo-3-nitro-4-(4-aminophenylethynyl)benzene   (36).   1,4-Dibromo-2-nitrobenzene (5.62 g, 20.0 mmol), $((C_6H_5)_3P)_2PdCl_2$ (0.140 g, 0.20 mmol), CuI (0.038 g, 0.20 mmol), Et$_3$N (10.0 mL), THF (10 mL) and **35** (1.170 g, 10.0 mmol) were used following the general procedure for couplings.

The reaction mixture was stirred at room temperature for 4 h. After solvent removal in vacuo, the residue was chromatographed on a column of silica (CH_2Cl_2 as eluent) to give a mixture of the desired product along with its regioisomer as a red solid. The desired product was isolated from the mixture by a two-fold recrystallization from CH_2Cl_2:hexanes as fine bright red needles (1.561 g, 49% yield). Mp 147-149 °C. IR (KBr) 3457, 3367, 2194, 1623, 1593, 1513, 1550, 1334, 1273, 1136, 834, 817, 528 cm^{-1}. ^{1}H NMR (400 MHz, CDCl$_3$) δ 8.21 (d, J=2.0 Hz), 7.67 (dd, J=8.4, 2.0 Hz), 7.51 (d, J=8.4 Hz), 7.96 (m, AA' part of AA'XX' pattern, J=8.2, 2.7, 1.9, 0.4 Hz, 2 H), 7.93 (m, XX' part of AA'XX' pattern, J=8.2, 2.7, 1.9, 0.4 Hz, 2 H), 3.39 (s, 2 H). ^{13}C NMR (100 MHz, CDCl$_3$) δ 149.27, 147.85, 135.82, 135.12, 133.71, 127.73, 120.62, 118.59, 114.63, 111.09, 100.24, 82.86. HRMS calc'd for $C_{14}H_9N_2BrO_2$: 315.9848, found: 315.9845.

4-(2-Nitro-4-phenylethynylphenylethynyl)aniline (37).   **36** (0.697 g, 2.20 mmol), $((C_6H_5)_3P)_2PdCl_2$ (0.062 g, 0.088 mmol), CuI (0.0084 g, 0.044 mmol), Et$_3$N (10.0 mL) and ethynylbenzene (0.306 g, 3.00 mmol) were used following the general procedure for couplings. The reaction mixture was stirred at 80 °C for 2 h. After solvent removal in vacuo, the residue was chromatographed on a column of silica with CH_2Cl_2 to give red needles of the desired product (0.72 g, 97% yield) Mp 166-168 °C. IR (KBr) 3454, 3381, 3360, 2177, 2197, 1594, 1623, 1539, 1520, 1299, 1342, 1133, 829, 758, 690, 527 cm^{-1}. ^{1}H NMR (400 MHz, CDCl$_3$) δ8.20 (dd, J=1.6, 0.3 Hz), 7.66 (dd, J=8.2, 1.6, Hz), 7.61 (d, J=8.1 Hz), 7.52-7.57 (m, 2 H), 7.36-7.43 (m, 5 H), 3.94 (s, 2 H). ^{13}C NMR (100 MHz, CDCl$_3$) δ 148.93, 147.81, 135.12, 134.04, 133.76, 131.74, 129.04, 128.49, 127.59, 122.97, 122.18, 118.95, 114.64, 111.29, 100.75, 93.03, 87.05, 83.71. HRMS calc'd for $C_{22}H_{14}N_2O_2$: 338.1055, found: 338.1058.

4-(2-Nitro-4-phenylethynylphenylethynyl)benzenediazonium tetrafluoroborate (38). Following the general diazotization procedure **37** (0.0845 g, 0.250 mmol) was treated with NOBF$_4$ (0.0322 g, 0.275 mmol) in acetonitrile (2 mL)/sulfolane (2 mL). The product was precipitated with ether (12 mL) as dark orange scales. The salt was washed with ether and reprecipitated from DMSO (0.5 mL) and CH_2Cl_2 (20 mL) as lustrous dark orange plates (0.0885 g, 81% yield). IR (KBr) 3103, 2279, 2209, 1576, 1345, 1540, 1084, 841, 764 cm^{-1}. ^{1}H NMR (400 MHz, CDCl$_3$/DMSO-d$_6$, line width of about 1.9 Hz was observed) δ 8.78 (d, J=8.9 Hz, 2 H), 8.30 (s, 1 H), 8.03 (d, J=8.9 Hz, 2 H), 7.85-7.92 (m, 2 H), 7.57-7.60 (m, 2 H), 7.42-7.44 (m, 3H). ^{13}C NMR (100 MHz, CDCl$_3$/DMSO-d$_6$) δ 149.00, 135.46, 134.85, 134.15, 133.31, 132.84, 1.34, 129.13, 128.21, 127.15, 125.66, 121.06, 114.81, 114.25, 94.57, 94.42, 94.11, 86.29.

4-(3-Nitro-4-phenylethynylphenylethynyl)aniline (39).   **25** (1.208 g, 4.0 mmol), $((C_6H_5)_3P)_2PdCl_2$ (0.070 g, 0.10 mmol), CuI (0.019 g, 0.10 mmol), Et$_3$N (6.0 mL), THF (6.0 mL) and **35** (0.479 g, 4.10 mmol) were used following the general procedure for couplings. The reaction mixture was stirred at room temperature for 15 h. After solvent removal in vacuo, the residue was chromatographed on a short column of silica with CH_2Cl_2:hexanes (1:1) to afford the desired product as an orange solid (0.560 g, 44% yield): mp 175-177 °C. IR (KBr) 3303, 2985, 1696, 1587, 1522, 1406, 1314, 1243, 1153, 1060, 839, 757, 692 cm^{-1}. ^1H NMR (400 MHz, CDCl$_3$) δ 8.16 (t, J=1.0 Hz, 1H), 7.64 (d, J=1.0 Hz, 2H), 7.58-7.61 (m, 2H), 7.34-7.40 (m, 3H), 7.35 (m, AA' part of AA'XX' pattern, J=8.0, 2.5, 2.0, 0.4 Hz, 2 H), 6.65 (m, XX' part of AA'XX' pattern, J=8.0, 2.5, 2.0, 0.4 Hz, 2 H), 3.91 (s, 2H). ^{13}C NMR (100 MHz, CDCl$_3$) 149.4, 147.5, 134.9, 134.3, 133.3, 132.0, 129.3, 128.5, 127.1, 124.9, 122.3, 117.1, 114.7, 11.1, 98.4, 94.9, 85.3, 85.0. HRMS calc'd for $C_{22}H_{14}N_2O_2$: 338.1055, found: 338.1059.

4-(3-Nitro-4-phenylethynylphenylethynyl)benzenediazonium tetrafluoroborate (40).

Following the general diazotization procedure, **39** (0.0676 g, 0.200 mmol) was treated with NOBF$_4$ (0.025 g, 0.210 mmol) in acetonitrile (2 mL)/sulfolane (2 mL). The product was precipitated with ether (20 mL) as fine orange-red crystals. The salt was washed with ether and reprecipitated from DMSO (0.6 mL) and CH_2Cl_2 (10 mL) as heavy lustrous red plates (0.0676 g, 77% yield). IR (KBr) 3101, 2279, 2209, 1576, 1540, 1346, 1083, 1034, 840, 764 cm^{-1}. ^1H NMR (400 MHz, CDCl$_3$/DMSO-d$_6$) δ 7.94 (m, AA' part of AA'XX' pattern, J=8.7, 2.4, 1.7, 0.4 Hz, 2 H), 7.82 (dd, J=1.7, 0.4 Hz, 1 H), 7.49 (m, XX' part of AA'XX' pattern, J=8.7, 2.4, 1.7, 0.4 Hz, 2 H), 7.62 (dd, J=8.1, 1.7 Hz, 1 H), 7.56 (dd, J=8.1, 0.4 Hz, 1 H), 7.07 (m, AA' part of AA'XX'Y pattern, J=7.8, 7.6, 1.8, 1.3, 1.3, 0.6 Hz, 2 H), 6.94 (tt, J= 7.6, 1.3 Hz, 1 H), 6.91 (m, YY' part of AA'XX'Y pattern, J=7.8, 7.6, 1.8, 1.3, 1.3, 0.6 Hz, 2 H). ^{13}C NMR (100 MHz, CDCl$_3$/DMSO-d$_6$) δ 137.24, 136.97, 136.23, 135.40, 133.72, 133.00, 131.08, 129.96, 129.48, 122.81, 122.75, 120.68, 114.12, 100.47, 98.81, 91.04, 85.57.

4-(2,5-Dinitro-4-(4-aminophenylethynyl)phenylethynyl)aniline (42).

1,4-Dibromo-2,5-dinitrobenzene[90] (0.977 g, 3.0 mmol), ((C$_6$H$_5$)$_3$P)$_2$PdCl$_2$ (0.042 g, 0.06 mmol), CuI (0.011 g, 0.06 mmol), Et$_3$N (5.0 mL), THF (5.0 mL) and 4-ethynylaniline (0.468 g, 4.00 mmol) were used following the general procedure for couplings. The reaction mixture was stirred at room temperature for 12 h. After solvent removal in vacuo, the residue was sonicated with CH_2Cl_2 (10 mL) and filtered. The filter cake was washed 5X with CH_2Cl_2 (10 mL) and dried in vacuo to afford dark purple crystals of the diamine **42** (0.432 g, 36% yield). Mp >270 °C. IR (KBr) 3494, 3387, 2184, 1600, 1400, 1523, 1537, 1308, 1337, 1251, 1136 cm^{-1}. ^1H NMR (400 MHz, DMSO-d$_6$) δ 8.37 (s, 2 H), 7.27-7.29 (m, 2 H), 6.59-6.61 (m, 2 H), 5.93 (br s, 4 H). ^{13}C NMR (100 MHz, DMSO-d$_6$) δ 151.18, 149.89, 133.67, 129.43, 116.95, 113.66, 106.10, 103.45, 82.23. HRMS calc'd for $C_{22}H_{14}N_4O_4$: 398.1015, found 398.1018.

4-(2,5-Dinitro-4-(4-diazoniophenylethynyl)phenylethynyl)benzenediazonium tetrafluoroborate

(43). Following the general diazotization procedure **42** (0.199 g, 0.500 mmol) was treated with $NOBF_4$ (0.128 g, 1.10 mmol) in acetonitrile (5.0 mL) /sulfolane (5.0 mL). The product was precipitated with ether (20 mL). The salt was washed with ether and reprecipitated from DMSO and CH_2Cl_2 as light-sensitive yellow crystals (0.215 g, 72% yield). IR (KBr) 3107, 2291, 1579, 1546, 1342, 1078, 830 cm^{-1}. ^1H NMR (400 MHz, CDCl$_3$/DMSO-d$_6$) δ 8.85 (s, 2H), 8.79 (d, J = 9 Hz, 2 H), 8.20 (d, J = 9 Hz, 2 H). ^{13}C NMR (100 MHz, CDCl$_3$/DMSO-d$_6$) δ 150.60, 133.93, 133.83, 133.14, 132.40, 131.75, 117.62, 116.32, 96.91, 91.51.

3.5 Combinatorial Routes to Molecular Wires and Devices and Experimental Details

3.5.1. *Introduction*

Combinatorial chemistry has become one of the fastest growing fields in chemistry because of its ability to rapidly generate small molecule libraries in a short period of time.[99] The application of combinatorial methodology has been extensively utilized in the past decade[100] for the synthesis of libraries of small organic molecules as potential drug candidates. Lately, the combinatorial approach has been applied to materials research,[101] mainly on inorganic solid-state materials, polymeric materials, and catalytic systems, in order to speed discovery and development. By utilizing the combinatorial method with an effective screening technique, a wide variety of substrates can be tested for a desired property or activity, producing structure-activity relationships and leading to optimization of the molecular properties.

To date, most well-examined oligo(phenylene ethynylene)s bearing functional ends for adhesion to metal surfaces are trimers.[5] Thus, in an attempt to broaden our library of candidate compounds, we have chosen to expand our synthetic efforts to tetrameric systems in which two central phenyl units can be functionalized. We report here the development of combinatorial strategies for the preparation of tetrameric phenylene ethynylenes in an effort to find better candidate molecules for applications in molecular electronics.

3.5.2. *Results and Discussion*

3.5.2.1. Monomer Synthesis for the Combinatorial Approach

In order to adequately explain and predict the relationship between molecular structure of the tested molecules and desired results, i.e. structure activity relationship (SAR), monomers with electron-donating groups (EDGs) such as alkyl, or electron-withdrawing groups (EWGs) such as F, CN, and CF_3 were designed and synthesized. The syntheses of the five scaffold monomers were performed as shown in Scheme 3.66. Sonogashira coupling[29,41] of aniline derivatives **1a-e**[102] with TMSA in Et_3N gave high yields of corresponding products **2a-e**. Without further purification, **2a-e** were converted to the corresponding diazonium salts using the method of Doyle et al,[39] followed by quenching with NaI and I_2[86] to give desired monomers **3a-e** in good yields.

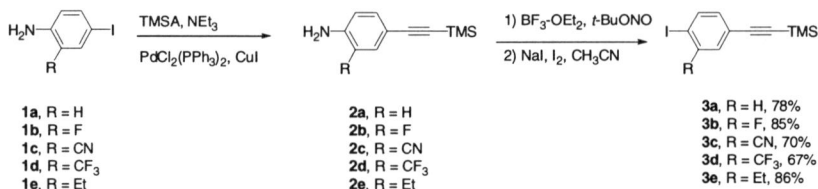

H_2N—◯—I →(TMSA, NEt$_3$ / PdCl$_2$(PPh$_3$)$_2$, CuI)→ H_2N—◯—≡—TMS →(1) BF$_3$-OEt$_2$, *t*-BuONO / 2) NaI, I$_2$, CH$_3$CN)→ I—◯—≡—TMS

1a, R = H	**2a**, R = H	**3a**, R = H, 78%
1b, R = F	**2b**, R = F	**3b**, R = F, 85%
1c, R = CN	**2c**, R = CN	**3c**, R = CN, 70%
1d, R = CF$_3$	**2d**, R = CF$_3$	**3d**, R = CF$_3$, 67%
1e, R = Et	**2e**, R = Et	**3e**, R = Et, 86%

Scheme 3.66 Synthesis of scaffold monomers.

3.5.2.2. Oligomer Syntheses in Solution

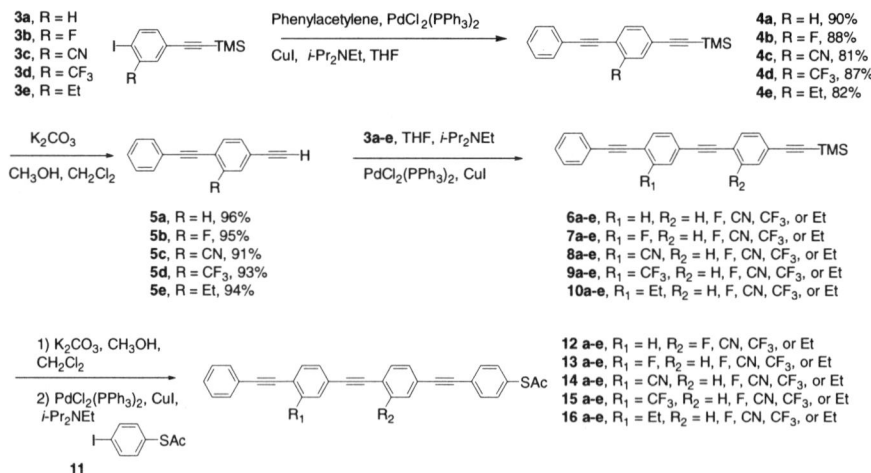

3a, R = H	
3b, R = F	
3c, R = CN	
3d, R = CF$_3$	
3e, R = Et	

I—◯—≡—TMS →(Phenylacetylene, PdCl$_2$(PPh$_3$)$_2$ / CuI, *i*-Pr$_2$NEt, THF)→ ◯—≡—◯—≡—TMS

4a, R = H, 90%
4b, R = F, 88%
4c, R = CN, 81%
4d, R = CF$_3$, 87%
4e, R = Et, 82%

→(K$_2$CO$_3$ / CH$_3$OH, CH$_2$Cl$_2$)→ ◯—≡—◯—≡—H

5a, R = H, 96%
5b, R = F, 95%
5c, R = CN, 91%
5d, R = CF$_3$, 93%
5e, R = Et, 94%

→(3a-e, THF, *i*-Pr$_2$NEt / PdCl$_2$(PPh$_3$)$_2$, CuI)→ ◯—≡—◯—≡—◯—≡—TMS

6a-e, R$_1$ = H, R$_2$ = H, F, CN, CF$_3$, or Et
7a-e, R$_1$ = F, R$_2$ = H, F, CN, CF$_3$, or Et
8a-e, R$_1$ = CN, R$_2$ = H, F, CN, CF$_3$, or Et
9a-e, R$_1$ = CF$_3$, R$_2$ = H, F, CN, CF$_3$, or Et
10a-e, R$_1$ = Et, R$_2$ = H, F, CN, CF$_3$, or Et

→(1) K$_2$CO$_3$, CH$_3$OH, CH$_2$Cl$_2$ / 2) PdCl$_2$(PPh$_3$)$_2$, CuI, *i*-Pr$_2$NEt / I—◯—SAc **11**)→ ◯—≡—◯—≡—◯—≡—◯—SAc

12 a-e, R$_1$ = H, R$_2$ = F, CN, CF$_3$, or Et
13 a-e, R$_1$ = F, R$_2$ = H, F, CN, CF$_3$, or Et
14 a-e, R$_1$ = CN, R$_2$ = H, F, CN, CF$_3$, or Et
15 a-e, R$_1$ = CF$_3$, R$_2$ = H, F, CN, CF$_3$, or Et
16 a-e, R$_1$ = Et, R$_2$ = H, F, CN, CF$_3$, or Et

Scheme 3.67 Synthesis of oligomers in solution.

The synthesis of a library of 25 tetramers of oligo(phenylene ethynylene)s was carried out by a combinatorial approach in conjunction with the use of five different monomers **3a-e**. The solution-phase synthetic approach of tetramers is outlined in Scheme 3.67. Sonogashira coupling five monomers **3a-e** to phenylacetylene afforded five corresponding dimers **4a-e** in excellent yields. Protodesilylation[54,58] of the terminal alkynes **4a-e** with K_2CO_3 in CH_3OH and CH_2Cl_2 gave **5a-e** in which each of the five alkynes **5a-e** was subjected to coupling with the five monomers **3a-e** to generate a library of 25 trimers. The coupling reaction of monomers bearing CF_3 or Et required higher temperatures and longer reaction times, and they gave lower yields. The relatively lower yields in this coupling may be caused by steric effects. The final step is the attachment of 1-iodo-4-thioacetylbenzene **11**,[54] acting as an adhesive group for metal surface attachment (molecular alligator clip). A library of 24 potential molecular devices **12b-e** and **13a-e** – **16a-e** were obtained in moderate yields (Table 3.2). Unsubstituted tetramer **12a** was not prepared due to the low solubility of trimer **6a**. Tetramers bearing only one functional group in two central phenyl units exhibited low solubility in organic solvents.

Table 3.2 Isolated yields of solution-phase synthesized trimers (6-10) and tetramers (12-16). 12a was not prepared because of low solubility of 6a.

	$R_1 = H$	$R_1 = F$	$R_1 = CN$	$R_1 = CF_3$	$R_1 = Et$
$R_2 = H$	6a (44%)	7a (86%)	8a (72%)	9a (75%)	10a (52%)
	12a[a]	13a (38%)	14a (58%)	15a (40%)	16a (46%)
$R_2 = F$	6b (66%)	7b (87%)	8b (88%)	9b (77%)	10b (70%)
	12b (43%)	13b (41%)	14b (42%)	15b (36%)	16b (53%)
$R_2 = CN$	6c (84%)	7c (97%)	8c (77%)	9c (70%)	10c (69%)
	12c (41%)	13c (48%)	14c (42%)	15c (26%)	16c (46%)
$R_2 = CF_3$	6d (71%)	7d (62%)	8d (72%)	9d (51%)	10d (64%)
	12d (47%)	13d (41%)	14d (30%)	15d (56%)	16d (51%)
$R_2 = Et$	6e (55%)	7e (54%)	8e (71%)	9e (68%)	10e (51%)
	12e (47%)	13e (48%)	14e (43%)	15e (21%)	16e (37%)

3.5.2.3. Oligomer Syntheses on a Solid Support

The advantages of employing solid-phase synthesis for the preparation of conjugated oligomers over solution-phase synthesis have been demonstrated by many research groups.[58,103] The most substantial advantages include higher overall yields as well as rapid purification processes by avoiding time-consuming chromatography, which is one of the bottlenecks in conventional solution-phase chemistry. By adapting the sequence of oligomer growth in the solution-phase synthesis, we explored the utilization of solid-phase synthesis toward oligo(phenylene ethynylene)s synthesis. An important issue in the design of a solid-phase combinatorial synthesis is to choose the proper linker to attach starting monomer to the resin. The triazene linkage **17** has been used by us[58] and other research groups[28c,104,105] as there are various cleavage methods for further manipulation of the cleavage products. The linker is also stable under the various reaction conditions encountered in the synthesis. The triazene linker can be cleaved with trifluoroacetic acid (TFA)[104] to give the diazonium salt **18** or reacted with CH_3I[28c,58,105] to transform it into the corresponding aryl iodide **19**. Both functionalized cleavage products can be further coupled with aromatic alkynes to afford oligo(phenylene ethylene)s. The third method, a "traceless cleavage"[106] converts the triazene linker into a C-H bond[104a] to give the corresponding arenes **20** (Scheme 3.68).

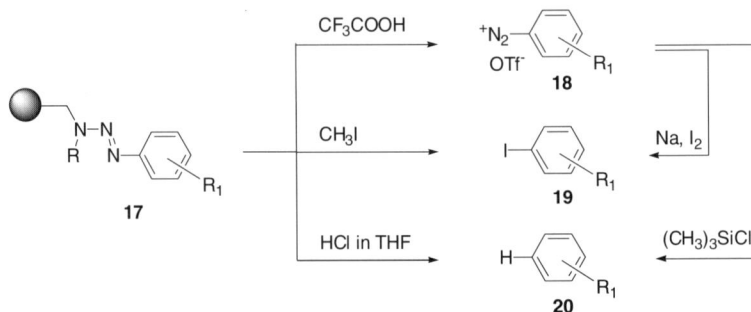

Scheme 3.68 Methods of cleavage of the product from the resin.

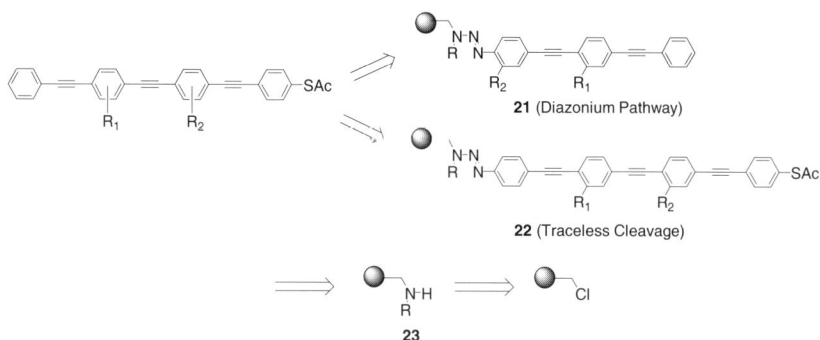

Scheme 3.69 Retrosynthetic analysis.

The retrosynthetic approach to the solid-phase combinatorial syntheses of our target oligo(phenylene ethylene)s is outlined in Scheme 3.69. The product molecules could be prepared from the diazonium pathway **21** by cleavage of the triazene with TFA followed by coupling with 4-ethynyl-1-thioacetylbenzene or directly by traceless cleavage of the corresponding resin-bound oligomers **22**. It is worth mentioning that these two different pathways could produce two different sets of target molecules from the same monomers in which the substituent groups are ortho or meta in the orientation to the alligator clip.

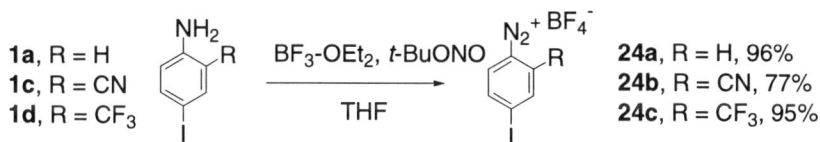

Scheme 3.70 Preparation of diazonium starting materials.

The initial step in the solid-phase synthesis of oligo(phenylene ethynylene)s was connection of the starting monomers to the resin. The functionalized diazonium salts **24a-c** were prepared[39] in high yields from the corresponding aniline derivatives **1a, c**, and **d**, respectively (Scheme 3.70).

Propylaminomethylate polystyrene **23** was prepared by heating a suspension of Merrifield's resin with n-propylamine in a screw cap tube at 70 °C for 3-4 days.[28c] Functionalized diazonium salts **24a-c** were reacted with resin **23** in DMF-THF solution in the presence of K_2CO_3 to form the corresponding triazene resin **25a-c**, respectively (Scheme 3.71). The attachment procedure was repeated to maximize the loading and a total of 2-3 equiv of diazonium salts were used. Small portions of resin were taken out and cleaved to give the corresponding diiodobenzene derivatives to monitor the loading. This method was found to be more accurate than elemental analysis or weight changes in calculating yields and loading.[28c,107] Treatment of resin-bound monomer **25a** with CH_3I in a screw cap tube at 120 °C for 24 h effected the liberation of 1,4-diiodobenzene in 70% yield over three steps. This result indicated that the first two steps were achieved in high loadings. Unfortunately, cleavage of resin-bound monomer **25b** in CH_3I at 120 °C only gave 7–16% yields over three steps; varying reaction time from 12–60 h were screened. Slightly higher yields (29–35%) were obtained by adding co-solvent such as CH_3CN. Similar results were obtained from the liberation of resin-bound monomer **25c**. We suspected that the ortho-functional group adjacent to a triazene moiety may produce low loading of the triazene resin or cause low yielding cleavage of the desired products because of steric hindrance.

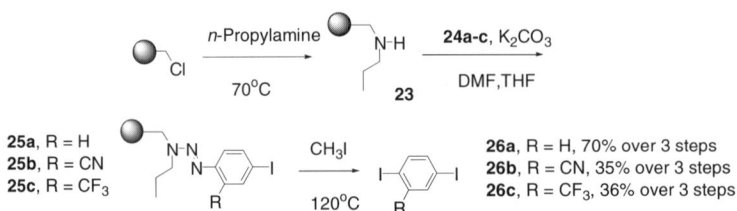

Scheme 3.71 Cleavage of products 26a-c from resins 25a-c using CH_3I.

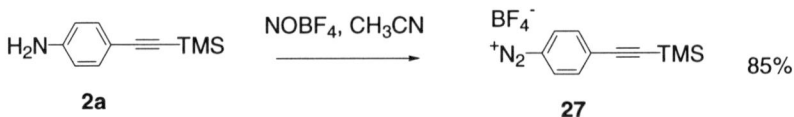

Scheme 3.72 Diazotization of aniline 2a.

As a result of these low isolated yields of products, we altered our synthetic pathway through the route of traceless cleavage in order to obtain satisfactory results. The linker diazonium salt **27** was prepared in high yield from diazotization of 4-(trimethylsilylethynyl)aniline (**2a**) with NOBF$_4$[87] (Scheme 3.72).

Diazonium salt **27** was then reacted with resin **23** in DMF–THF solution in the presence of K$_2$CO$_3$ to give the corresponding triazene monomer **28** (Scheme 3.73). Resin **28** was deprotected by treatment of the resin-bound TMSA group with a solution of tetrabutylammonium fluoride (TBAF) in THF to give the corresponding resin-bound alkynes **29**.[58,28c] Infrared analysis was used as a simple, non-destructive method to monitor oligomer synthesis. The IR spectra were assigned as follows: 2153 cm^{-1} (medium) carbon-carbon stretch of the trimethylsilylethynyl group; 3317 cm^{-1} (strong) alkynyl carbon-hydrogen stretch; and 2106 cm^{-1} (weak) alkyne carbon-carbon stretch.

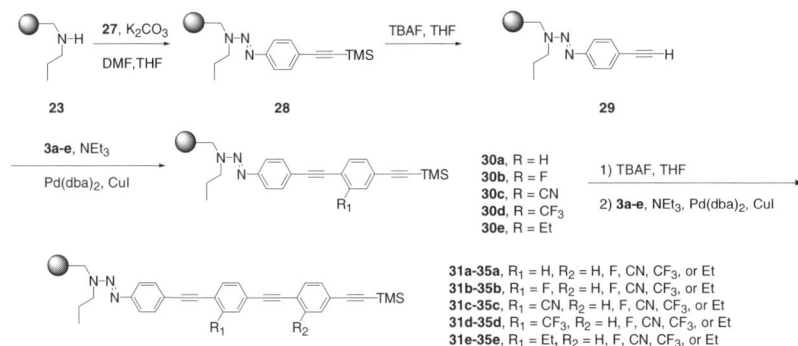

Scheme 3.73 Synthesis of trimers 31a-e and 35a-e.

The preparation of resin-bound alkyne **29** followed a parallel synthesis. The subsequent "split-and-pool" synthesis of individual oligomers was performed in micro-reactors, specifically MacroKans,[108] with unique identification system.[109] Resin-bound monomer **29** sealed in MacroKans were subjected to a Pd/Cu-catalyzed cross-coupling with monomers **30a-e** in Et$_3$N. This catalyst (Pd(dba)$_2$/PPh$_3$/CuI) solution[58,28c] was pre-mixed with stirring at 70 °C for 2 h prior to use. The suspended MacroKans were heated in a screw cap tube at 65 °C for 44-48 h. At the completion of the coupling reaction, the MacroKans were collected and washed according to a literature procedure.[58,28c] The next step was to pool one (or more) of each resin-bound dimers **30a-e** for another split combinatorial approach with monomers **3a-e**. Resin-bound dimmers **30a-e** were deprotected with a solution TBAF in THF to give the corresponding resin-bound alkynes, following a subsequent Pd/Cu-catalyzed cross-coupling with monomers **3a-e** to afford a library of 25 different polymer-supported trimers **31a-e - 35a-e** (Scheme 3.73).

In order to monitor resin loading, **30a** and **30e** were liberated into corresponding dimers via the traceless cleavage method[104a] by sonication at 50 °C using 10% HCl in THF solution (Scheme 3.74). The isolated yields of **4a** and **4e** (over five steps from Merrifield's resin) were 37% and 43%, respectively. This result indicate that high loadings and cleavage (> 82%, on average) were achieved in each step.

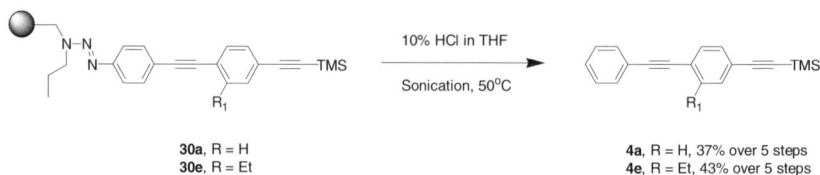

30a, R = H
30e, R = Et

4a, R = H, 37% over 5 steps
4e, R = Et, 43% over 5 steps

Scheme 3.74 Traceless cleavage of dimer products..

The final step before the cleavage of the triazene linker to release desired product is to attach "alligator clip" **11** (Scheme 3.75). Unfortunately, the ^{1}H NMR spectra of crude liberation product **37**, from using 10% HCl in THF solution, indicated a plethora of products. One of the possible side-reactions may be hydrolysis of thioester to form disulfides, which under these conditions would exhibit extremely low solubility.

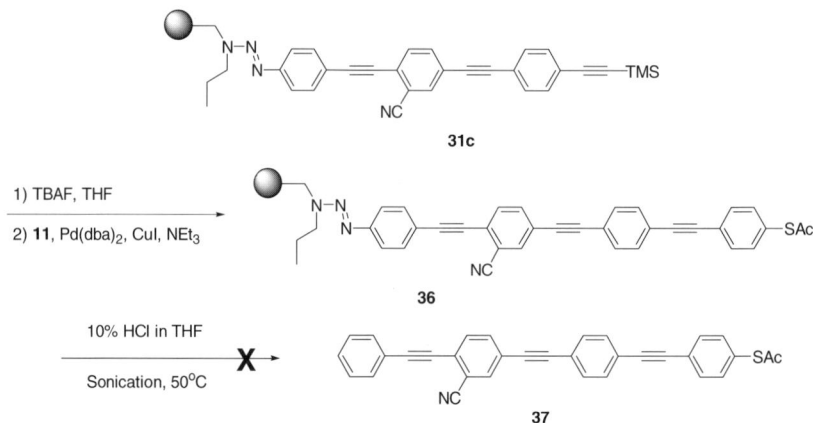

Scheme 3.75 Cleavage of alligator clip-containing products.

Due to these unsuccessful attempts, we decided to liberate the trimers prior to the coupling with the alligator clip. The traceless cleavage method was applied to cleave the polymer-supported trimers **31a-e – 35a-e** to afford desired products (Scheme 3.76). The final isolated yields of trimers over seven steps from commercially available Merrifield's resin ranged from 18 to 34%, an average of 78 to 86% yield over each step (Table 3.3).

31a-35a, R$_1$ = H, R$_2$ = H, F, CN, CF$_3$, or Et
31b-35b, R$_1$ = F, R$_2$ = H, F, CN, CF$_3$, or Et
31c-35c, R$_1$ = CN, R$_2$ = H, F, CN, CF$_3$, or Et
31d-35d, R$_1$ = CF$_3$, R$_2$ = H, F, CN, CF$_3$, or Et
31e-35e, R$_1$ = Et, R$_2$ = H, F, CN, CF$_3$, or Et

6a-e, R$_1$ = H, R$_2$ = H, F, CN, CF$_3$, or Et
7a-e, R$_1$ = F, R$_2$ = H, F, CN, CF$_3$, or Et
8a-e, R$_1$ = CN, R$_2$ = H, F, CN, CF$_3$, or Et
9a-e, R$_1$ = CF$_3$, R$_2$ = H, F, CN, CF$_3$, or Et
10a-e, R$_1$ = Et, R$_2$ = H, F, CN, CF$_3$, or Et

10% HCl in THF
Sonication, 50°C

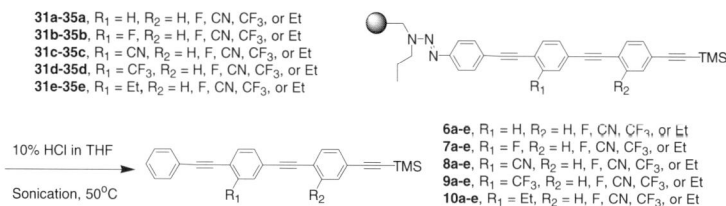

Scheme 3.76 Traceless cleavage of 31a-e through 35a-e to yield trimer products 6a-e through 10a-e.

Table 3.3 Overall yields of solid-phase synthesized trimers. Yields are based on isolated products (amount shown in parentheses) and calculated according to the loading of original resin.

	R$_1$ = H	R$_1$ = F	R$_1$ = CN	R$_1$ = CF$_3$	R$_1$ = Et
R$_2$ = H	6a, 30% (33 mg)	7a, 23% (26 mg)	8a, 32% (37 mg)	9a, 24% (41 mg)	10a, 27% (32 mg)
R$_2$ = F	6b, 31% (36 mg)	7b, 33% (40 mg)	8b, 34% (42 mg)	9b, 33% (45 mg)	10b, 24% (30 mg)
R$_2$ = CN	6c, 33% (39 mg)	7c, 29% (35 mg)	8c, 25% (30 mg)	9c, 28% (39 mg)	10c, 18% (25 mg)
R$_2$ =CF$_3$	6d, 34% (44 mg)	7d, 31% (42 mg)	8d, 29% (40 mg)	9d, 32% (48 mg)	10d, 23% (28 mg)
R$_2$ = Et	6e, 27% (32 mg)	7e, 23% (28 mg)	8e, 26% (32 mg)	9e, 22% (30 mg)	10e, 18% (23 mg)

3.5.3. Summary

We demonstrated the combinatorial synthesis of oligo(phenylene ethynylene)s both in solution phase and on solid supports. The construction of oligomers is accomplished by following an iterative approach starting from Sonogashira coupling of aryl halides with alkynes and desilylation. Five monomers with electron-donating group (Et) or electron-withdrawing groups (F, CN, or CF$_3$) were synthesized and used to generate a library of 25 trimers by "split-and-pool" synthesis. The coupling reaction with the monomers bearing CF$_3$ or Et functional groups required higher temperatures and longer reaction times, and gave lower yields. A library of 24 tetramers bearing a thioacetyl end group was synthesized from those 25 trimers. The solid-phase combinatorial synthesis offers advantages over the conventional solution-phase approach by maximizing overall yields and facilitating purification process by avoiding time-consuming chromatography. Traceless cleavage from the resin was effectively achieved by sonication of resin-bound oligomer in 10% HCl in THF solution to liberate oligomer. An average yield of 78 to 86% (for each step over seven steps from solid-phase combinatorial synthesis) was obtained.

This study presents a fast combinatorial synthesis of oligo(phenylene ethynylene)s incorporating diversity of substituent group, which could be used to deduce structure-activity relationships.

3.5.4. Experimental Section

General Procedure. Merrifield's resin (1.1 mmol Cl⁻/g, 1% cross-linker divinylbenzene copolymer, 70-90 mesh) was obtained from Aldrich Chemical Co. MacroKans were obtained from IRORI.[108] FT-IR characterizations of the polymer-supported reactions were carried out by placing ca. 10 mg of the polymer-supported material on a NaCl plate. After the beads were swollen with 2-3 drops of CCl_4, a second NaCl plate was pressed onto the beads, and an FTIR spectrum was recorded.

General Procedure for the Coupling of Aniline Derivatives with TMSA. To a stirring solution in a septum-capped tube of the aniline derivative was added at room temperature $((C_6H_5)_3P)_2PdCl_2$ (0.5 mol %), CuI (1 mol %) in Et_3N, and TMSA (1.07 equiv). The tube was capped and the solution was stirred overnight. The reaction mixture was filtered and the filtrate was washed with ether. The combined organic phase was then washed with NH_4Cl (2 M) (2x), brine and dried over Na_2SO_4 to give the desired product.

4-Trimethylsilylethynylaniline (2a).[102b] According to the general procedure, 4-iodo-aniline **1a** (25 g, 114 mmol), $((C_6H_5)_3P)_2PdCl_2$ (400 mg, 0.57 mmol, 0.5 mol %), CuI (217 mg, 1.14 mmol, 1 mol %), and TMSA (17.3 mL, 122 mmol, 1.07 equiv) in Et_3N (200 mL) gave the desired product (21.33 g, 113 mmol, 99%).

2-Fluoro-4-trimethylsilylethynylaniline (2b). According to the general procedure, **1b** (12 g, 50.6 mmol), $((C_6H_5)_3P)_2PdCl_2$ (178 mg, 0.253 mmol, 0.5 mol %), CuI (96 mg, 0.506 mmol, 1 mol %) and TMSA (7.7 mL, 54.5 mmol, 1.07 equiv) in Et_3N (120 mL) gave the desired product (10.4 g, 50.1 mmol, 99%). IR (KBr) 3439, 3393, 2148 cm⁻¹; ¹H NMR (400 MHz, CDCl₃) δ7.08-7.14 (m, 2H), 6.68 (dd, J = 1.0 Hz, J = 8.2 Hz, 1H), 3.95 (br s, 2H), 0.25 (s, 9H); ¹³C NMR (100 MHz, CDCl₃) δ150.9 (d, J_{C-F} = 239 Hz), 135.8 (d, J_{C-F} = 12 Hz), 129.2 (d, J_{C-F} = 3 Hz), 119.1 (d, J_{C-F} = 20 Hz), 116.5 (d, J_{C-F} = 4 Hz), 113.1 (d, J_{C-F} = 8 Hz), 105.1 (d, J_{C-F} = 26 Hz), 92.6, 0.4; HRMS calcd for $C_{11}H_{14}FSi$: 207.0880. Found: 207.0877.

2-Amino-5-trimethylsilylethynylbenzonitrile (2c). According to the general procedure, **1c** (23.5 g, 104.5 mmol), $((C_6H_5)_3P)_2PdCl_2$ (338 mg, 0.48 mmol, 0.5 mol %), CuI (183 mg, 0.96 mmol, 1 mol %) and TMSA (14.6 mL, 103 mmol, 1.07 equiv) in Et_3N (200 mL) gave the desired product (20.5 g, 95.6 mmol, 91%). IR (KBr) 3447, 3352, 2219, 2145 cm⁻¹; ¹H NMR (400 MHz, CDCl₃) δ7.53 (d, J = 1.9 Hz, 1H), 7.42 (dd, J = 1.9 Hz, J = 8.6 Hz, 1H), 6.67 (d, J = 8.6 Hz, 1H), 4.59 (br s, 2H), 0.25 (s, 9H); ¹³C NMR (100 MHz, CDCl₃) δ149.8, 137.8, 136.4, 117.2, 115.5, 113.1, 103.8, 96.1, 93.7, 0.4; HRMS calcd for $C_{12}H_{14}N_2Si$: 214.0926. Found: 214.0924.

2-Trifluoromethyl-4-trimethylsilylethynylaniline (2d). According to the general procedure, **1d** (30 g, 104.5 mmol), $((C_6H_5)_3P)_2PdCl_2$ (367 mg, 0.52 mmol, 0.5 mol %), CuI (199 mg, 1.04 mmol, 1 mol %) and TMSA (15.8 mL, 112 mmol, 1.07 equiv) in Et_3N (200 mL) gave the desired product (26.8 g, 104 mmol, 99%). IR (KBr) 3486, 3396, 2150 cm^{-1}; ^{1}H NMR (400 MHz, CDCl$_3$) δ7.58 (d, J = 1.8 Hz, 1H), 7.40 (dd, J = 1.8 Hz, J = 8.5 Hz, 1H), 6.66 (d, J = 8.5 Hz, 1H), 4.33 (br s, 2H), 0.25 (s, 9H); ^{13}C NMR (100 MHz, CDCl$_3$) δ144.9, 136.6, 131.0 (q, J_{C-F} = 5 Hz), 124.8 (q, J_{C-F} = 272 Hz), 117.2, 113.7 (q, J_{C-F} = 30 Hz), 112.4, 104.8, 93.2, 0.4; HRMS calcd for $C_{12}H_{14}F_3NSi$: 257.0848. Found: 257.0845.

2-Ethyl-4-trimethylsilylethynylaniline (2e). According to the general procedure, **1e** (7.4 g, 30 mmol), $((C_6H_5)_3P)_2PdCl_2$ (105 mg, 0.15 mmol, 0.5 mol %), CuI (57 mg, 0.3 mmol, 1 mol %) and TMSA (4.5 mL, 31.8 mmol, 1.07 equiv) in Et_3N (80 mL) gave the desired product (6.5 g, 29.9 mmol, 99%). IR (neat) 3480, 2140 cm^{-1}; ^{1}H NMR (400 MHz, CDCl$_3$) δ7.19 (d, J = 1.8 Hz, 1H), 7.14 (dd, J = 1.8 Hz, J = 8.1 Hz, 1H), 6.53 (d, J = 8.1 Hz, 1H), 3.74 (br s, 2H), 2.43 (q, J = 7.5 Hz, 2H), 1.21 (t, J = 7.5 Hz, 3H), 0.23 (s, 9H); ^{13}C NMR (100 MHz, CDCl$_3$) δ145.0, 132.6, 131.3, 127.9, 115.2, 113.0, 106.9, 91.5, 24.1, 13.1, 0.6; HRMS calcd for $C_{13}H_{19}NSi$: 217.1287. Found: 217.1285.

General Procedure for the Conversion of Aniline Derivatives into Iodoarenes Through the Diazonium Salt. To a round-bottom flask fitted with an addition funnel and N_2 inlet was added boron trifluoride etherate (4 equiv) that was then chilled in a dry ice-acetone bath (-20 °C). To the reaction flask was added dropwise over 5 min a solution of the aniline derivative (1 equiv) in dry ether, followed by a solution of tert-butylnitrite (3.5 equiv) in dry ether over 30 min. The chilled mixture was stirred an additional 10 min, and the cold bath was allowed to warm to 5 °C over 20 min. To the mixture was added diethyl ether, and the mixture was chilled in an ice-bath for 15 min. The solid was collected by filtration, washed with chilled (0-5 °C) diethyl ether, and dried to give diazonium salt. After briefly air-drying, the diazonium salt was dissolved in CH_3CN and then added dropwise via cannula to a solution of NaI (1.1 equiv) and I_2 (0.1 equiv) in CH_3CN. The mixture was stirred at room temperature for 1 h, then $Na_2S_2SO_3$(aq) (2 M) was added to the mixture. The mixture was extracted with CH_2Cl_2 (3x), and the organic phase was washed with brine and dried over Na_2SO_4. The crude product was purified by column chromatography (silica gel) with hexane or a hexane:EtOAc mixture to give the desired product.

(4-Iodo-phenylethynyl)trimethylsilane (3a).[110] According to the general procedure, 13.2 g (44 mmol, 78%) of **3a** was produced from **2a** (10.7 g, 56.5 mmol).

(3-Fluoro-4-iodophenylethynyl)trimethylsilane (3b). According to the general procedure, 8.97 g (28.2 mmol, 85%) of **3b** was produced from **2b** (6.9 g, 33.3 mmol). IR (neat) 2160 cm^{-1}; ^{1}H NMR (400 MHz, CDCl$_3$) δ7.69 (dd, J = 6.6 Hz, J = 8.1 Hz, 1H), 7.16 (dd, J = 1.7 Hz, J = 8.5 Hz, 1H), 7.00 (dd, J = 1.7 Hz, J = 8.6 Hz, 1H), 0.27 (s, 9H); ^{13}C NMR (100 MHz, CDCl$_3$) δ161.2 (d, J$_{C\text{-}F}$ = 246 Hz), 139.1 (d, J$_{C\text{-}F}$ = 2 Hz), 129.1 (d, J$_{C\text{-}F}$ = 3 Hz), 125.2 (d, J$_{C\text{-}F}$ = 8 Hz), 118.6 (d, J$_{C\text{-}F}$ = 25 Hz), 102.6 (d, J$_{C\text{-}F}$ = 3 Hz), 96.9, 81.8 (d, J$_{C\text{-}F}$ = 26 Hz), -0.2; HRMS calcd for C$_{11}$H$_{12}$FISi: 317.9737. Found: 317.9737.

2-Iodo-5-trimethylsilanylethynyl-benzonitrile (3c). According to the general procedure, 9.6 g (29.5 mmol, 70%) of **3c** was produced from **2c** (9.0 g, 42 mmol). IR (neat) 2233, 2151 cm^{-1}; ^{1}H NMR (400 MHz, CDCl$_3$) δ7.86 (d, J = 8.3 Hz, 1H), 7.66 (d, J = 2.0 Hz, 1H), 7.31 (dd, J = 2.0 Hz, J = 8.3 Hz, 1H), 0.26 (s, 9H); ^{13}C NMR (100 MHz, CDCl$_3$) δ139.7, 137.1, 136.5, 124.2, 121.0, 118.7, 101.6, 99.1, 98.1, -0.1; HRMS calcd for C$_{12}$H$_{12}$INSi: 324.9784. Found: 324.9784.

(4-Iodo-3-trifluoromethylphenylethynyl)trimethylsilane   (3d). According to the general procedure, 8.6 g (23.4 mmol, 67%) of **3d** was produced from **2d** (9.0 g, 35 mmol). IR (neat) 2164 cm^{-1}; ^{1}H NMR (400 MHz, CDCl$_3$) δ7.97 (d, J = 8.2 Hz, 1H), 7.74 (d, J = 1.8 Hz, 1H), 7.26 (dd, J = 1.8 Hz, J = 8.2 Hz, 1H), 0.28 (s, 9H); ^{13}C NMR (100 MHz, CDCl$_3$) δ142.2, 135.7, 134.0 (q, J$_{C\text{-}F}$ = 31 Hz), 130.5 (q, J$_{C\text{-}F}$ = 6 Hz), 123.8, 122.9 (q, J$_{C\text{-}F}$ = 274 Hz), 102.6, 98.1, 91.0 (q, J$_{C\text{-}F}$ = 2 Hz), -0.4; HRMS calcd for C$_{12}$H$_{12}$F$_3$INSi: 367.9705. Found: 367.9702.

(3-Ethyl-4-iodophenylethynyl)trimethylsilane (3e). According to the general procedure, 9.72 g (29.6 mmol, 86%) of **3e** was produced from **2e** (7.5 g, 34.5 mmol). IR (neat) 2160 cm^{-1}; ^{1}H NMR (400 MHz, CDCl$_3$) δ7.75 (d, J = 8.1 Hz, 1H), 7.33 (d, J = 8.1 Hz, 1H), 6.98 (dd, J = 2.0 Hz, J = 8.1 Hz, 1H), 2.71 (q, J = 7.5 Hz, 2H), 1.22 (t, J = 7.5 Hz, 3H), 0.27 (s, 9H); ^{13}C NMR (100 MHz, CDCl$_3$) δ147.1, 139.8, 132.2, 131.2, 123.9, 104.8, 101.3, 95.8, 34.5, 14.9, 0.5; HRMS calcd for C$_{13}$H$_{17}$ISi: 328.0144. Found: 328.0142.

General Procedure for the Coupling of a Terminal Alkyne with an Aryl Halide Utilizing a Palladium-Copper Cross-Coupling (Sonogashira Protocol). To a stirring solution of the aryl halide (1 equiv), ((C$_6$H$_5$)$_3$P)$_2$PdCl$_2$ (2 mol %), CuI (4 mol %), and additional (C$_6$H$_5$)$_3$P (only used at temperature > 25 °C, 2 equiv based on Pd) in THF was added the terminal alkyne (1-2 equiv) followed by the amine (4 equiv based on the aryl halide) at room temperature (unless otherwise stated) under N$_2$ in a screw cap tube. The tube was flushed with N$_2$, capped, and allowed to stir 24 h. The reaction mixture was then subjected to an aqueous workup and the aqueous layer extracted with CH$_2$Cl$_2$. After drying the combined organic layers over Na$_2$SO$_4$, the solvent was removed in vacuo. The crude product was purified by column chromatography (silica gel) with hexanes or hexane:EtOAc mixtures to give the desired product.

Trimethyl(4-phenylethynylphenylethynyl)silane (4a).[5b] **3a** (1.2 g, 4 mmol), phenylacetylene (0.48 mL, 4.4 mmol), ((C_6H_5)$_3$P)$_2$PdCl$_2$ (56 mg, 0.08 mmol), CuI (30 mg, 0.16 mmol), and N,N-diisopropylethylamine (2.8 mL, 16.0 mmol) in THF (40 mL) for 1 d gave the desired product (0.99 g, 90%).

(3-Fluoro-4-phenylethynylphenylethynyl)trimethylsilane (4b). **3b** (2.65 g, 8.33 mmol), phenylacetylene (0.87 mL, 7.92 mmol), ((C_6H_5)$_3$P)$_2$PdCl$_2$ (56 mg, 0.08 mmol), CuI (30 mg, 0.16 mmol), and N,N-diisopropylethylamine (5.5 mL, 31.6 mmol) in THF (40 mL) for 1 d gave the desired product (2.05 g, 88%). IR (KBr) 2216, 2154 cm^{-1};^{1}H NMR (400 MHz, CDCl$_3$) δ7.56-7.59 (m, 2H), 7.46 (dd, J = 8 Hz, J = 8 Hz, 1H), 7.38-7.40 (m, 3H), 7.21-7.26 (m, 2H), 0.29 (s, 9H); ^{13}C NMR (100 MHz, CDCl$_3$) δ161.2 (d, J_{C-F} = 252 Hz), 133.5 (d, J_{C-F} = 2 Hz), 132.1, 129.8, 128.8, 128.1 (d, J_{C-F} = 3 Hz), 125.1 (d, J_{C-F} = 6 Hz), 123.1, 119.2 (d, J_{C-F} = 23 Hz), 112.8 (d, J_{C-F} = 16 Hz), 103.7 (d, J_{C-F} = 3 Hz), 97.8, 96.5 (d, J_{C-F} = 4 Hz), 82.8, 0.2; HRMS calcd for $C_{19}H_{17}FSi$: 292.1084. Found: 292.1082.

2-Phenylethynyl-5-trimethylsilanylethynylbenzonitrile (4c). **3c** (2.14 g, 6.58 mmol), phenylacetylene (0.69 mL, 6.28 mmol), ((C_6H_5)$_3$P)$_2$PdCl$_2$ (44 mg, 0.06 mmol), CuI (24 mg, 0.12 mmol), and N,N-diisopropylethylamine (4.4 mL, 25.2 mmol) in THF (30 mL) for 1 d gave the desired product (1.52 g, 81%). IR (KBr) 2230, 2211, 2152 cm^{-1}; ^{1}H NMR (400 MHz, CDCl$_3$) δ7.77 (d, J = 1.6 Hz, 1H), 7.62-7.65 (m, 5H), 7.57 (d, J = 8.2 Hz, 1H), 7.36-7.43 (m, 3H), 0.29 (s, 9H); ^{13}C NMR (100 MHz, CDCl$_3$) δ136.1, 135.8, 132.5, 132.4, 129.9, 128.9, 127.0, 124.0, 122.2, 117.2, 115.9, 102.6, 99.5, 98.3, 86.0, 0.2; HRMS calcd for $C_{20}H_{17}NSi$: 299.1130. Found: 299.1134.

Trimethyl-(4-phenylethynyl-3-trifluoromethylphenylethynyl)silane (4d). **3d** (3.23 g, 8.77 mmol), phenylacetylene (1.06 mL, 9.65 mmol), ((C_6H_5)$_3$P)$_2$PdCl$_2$ (123 mg, 0.17 mmol), CuI (67 mg, 0.35 mmol), (C_6H_5)$_3$P (92 mg, 0.35 mmol), and N,N-diisopropylethylamine (6.1 mL, 35.0 mmol) in THF (45 mL) at 40 °C for 3 d gave the desired product (2.61 g, 87%). IR (KBr) 2210, 2159 cm^{-1}; ^{1}H NMR (400 MHz, CDCl$_3$) δ7.77 (m, 1H), 7.56-7.60 (m, 2H), 7.54-7.56 (m, 2H), 7.36-7.40 (m, 3H), 0.28 (s, 9H); ^{13}C NMR (100 MHz, CDCl$_3$) δ134.8, 134.0, 132.1, 132.0 (q, J_{C-F} = 31 Hz), 129.9 (q, J_{C-F} = 5 Hz), 129.4, 128.8, 123.6 (q, J_{C-F} = 274 Hz), 123.4, 122.9, 121.7 (q, J_{C-F} = 2 Hz), 103.6, 98.5, 97.2 (q, J_{C-F} = 2 Hz), 85.6, 0.2; HRMS calcd for $C_{20}H_{17}F_3Si$: 342.1051. Found: 342.1054.

(3-Ethyl-4-phenylethynylphenylethynyl)trimethylsilane (4e). **3e** (2.65 g, 8.07 mmol), phenylacetylene (0.93 mL, 8.47 mmol), ((C_6H_5)$_3$P)$_2$PdCl$_2$ (93 mg, 0.16 mmol), CuI (61 mg, 0.32 mmol), (C_6H_5)$_3$P (212 mg, 0.81 mmol), and N,N-diisopropylethylamine (5.6 mL, 32.1 mmol) in THF (40 mL) at 40 °C for 3 d gave the desired product (2.01 g, 82%). IR (KBr) 2210, 2157 cm^{-1}; ^{1}H NMR (400 MHz, CDCl$_3$) δ7.53-7.56 (m, 2H), 7.45 (d, J = 7.9 Hz, 1H), 7.35-7.40 (m, 4H), 7.30 (dd, J = 1.7 Hz, J = 7.9 Hz, 1H), 2.86 (q, J = 7.6 Hz, 2H), 1.32 (t, J = 7.6 Hz, 3H), 0.28 (s, 9H); ^{13}C NMR (100 MHz, CDCl$_3$) δ146.5,

132.4, 131.9, 131.8, 129.6, 128.8, 128.7, 123.8, 123.4, 105.5, 96.0, 95.0, 88.2, 28.0, 15.0, 0.4; HRMS calcd for $C_{21}H_{22}Si$: 302.1491. Found: 302.1492.

General Procedures for the Desilylation of Alkynes. The silylated alkyne (1 equiv) was dissolved in CH_3OH or CH_2Cl_2/CH_3OH mixtures. K_2CO_3 (2 equiv) was added, and the reaction was stirred overnight. The reaction mixture was then subjected to an aqueous workup and the aqueous layer extracted with CH_2Cl_2. After drying the combined organic layers over Na_2SO_4, the solvent was removed in vacuo. The crude product was purified by column chromatography (silica gel) with hexanes or hexane:EtOAc mixtures to give the desired product.

1-Ethynyl-4-phenylethynylbenzene (5a). **4a** (1.3 g, 4.7 mmol), CH_3OH (20 mL), CH_2Cl_2 (20 mL), and K_2CO_3 (1.3 g, 9.4 mmol) afforded 0.92 g (96%) of the title compound. IR (KBr) 3276 cm^{-1}; ^1H NMR (400 MHz, CDCl$_3$) δ7.55-7.58 (m, 2H), 7.43-7.52 (m, 4H), 7.37-7.40 (m, 3H), 3.20 (s, 1H); ^{13}C NMR (100 MHz, CDCl$_3$) δ132.6, 132.2, 132.0, 129.0, 128.9, 124.3, 123.4, 122.4, 91.9, 89.4, 83.8, 79.5; HRMS calcd for $C_{16}H_{10}$: 202.0782. Found: 202.0784.

4-Ethynyl-2-fluoro-1-phenylethynylbenzene (5b). **4b** (1.2 g, 4.1 mmol), CH_3OH (15 mL), CH_2Cl_2 (15 mL), and K_2CO_3 (1.13 g, 8.2 mmol) afforded 0.748 g (83%) of the title compound. IR (KBr) 3267, 2216 cm^{-1}; ^1H NMR (500 MHz, CDCl$_3$) δ7.57-7.60 (m, 2H), 7.48 (dd, J = 7.6 Hz, J = 7.6 Hz, 1H), 7.37-7.41 (m, 3H), 7.24-7.28 (m, 2H), 0.29 (s, 9H); ^{13}C NMR (125 MHz, CDCl$_3$) δ162.4 (d, J_{C-F} = 253 Hz), 133.6, 132.2, 129.3, 128.8, 128.2 (d, J_{C-F} = 4 Hz), 124.0 (d, J_{C-F} = 9 Hz), 123.0, 119.4 (d, J_{C-F} = 23 Hz), 113.3 (d, J_{C-F} = 16 Hz), 96.7 (d, J_{C-F} = 4 Hz), 82.7, 82.5 (d, J_{C-F} = 3 Hz), 80.2; HRMS calcd for $C_{16}H_9F$: 220.0688. Found: 220.0690.

5-Ethynyl-2-phenylethynylbenzonitrile (5c). **4c** (425 mg, 1.4 mmol), CH_3OH (10 mL), CH_2Cl_2 (10 mL), and K_2CO_3 (392 mg, 2.8 mmol) afforded 295 mg (91%) of the title compound. IR (KBr) 3262, 2234, 2207 cm^{-1}; ^1H NMR (400 MHz, CDCl$_3$) δ7.79 (d, J = 1.6 Hz, 1H), 7.66 (dd, J = 1.6 Hz, J = 8.2 Hz, 1H), 7.63-7.65 (m, 2H), 7.60 (d, J = 8.2 Hz, 1H), 7.39-7.45 (m, 3H), 3.30 (s, 1H); ^{13}C NMR (100 MHz, CDCl$_3$) δ136.3, 136.0, 132.5, 132.4, 129.9, 128.9, 127.6, 123.0, 122.2, 117.1, 116.1, 85.8, 81.5; HRMS calcd for $C_{17}H_9N$: 227.0735. Found: 227.0737.

4-Ethynyl-1-phenylethynyl-2-trifluoromethylbenzene (5d). **4d** (1.23 g, 3.59 mmol), CH_3OH (35 mL), and K_2CO_3 (0.99 g, 7.16 mmol) afforded 0.9 g (93%) of the title compound. IR (KBr) 2218, 2197 cm^{-1}; ^1H NMR (400 MHz, CDCl$_3$) δ 7.82 (s, 1H), 7.60 (br s, 2H), 7.55-7.59 (m, 2H), 7.38-7.42 (m, 3H), 3.27 (s, 1H); ^{13}C NMR (100 MHz, CDCl$_3$) δ 135.1, 134.1, 132.2, 132.0 (q, J_{C-F} = 31 Hz), 130.0 (q, J_{C-F} = 5 Hz), 122.8, 122.5 (q, J_{C-F} = 274 Hz), 122.4, 122.2 (q, J_{C-F} = 2 Hz), 97.5 (q, J_{C-F} = 2 Hz), 85.5, 82.4, 80.8; HRMS calcd for $C_{17}H_9F_3$: 270.0656. Found: 270.0652.

2-Ethyl-4-ethynyl-1-phenylethynylbenzene (5e). **4e** (1.43 g, 4.7 mmol), CH_3OH (25 mL), and K_2CO_3 (1.3 g, 9.4 mmol) afforded 1.02 g (94%) of the title compound. IR (neat) 3292, 2202 cm^{-1}; ^1H NMR (400 MHz, CDCl$_3$) δ7.54-7.58 (m, 2H), 7.48 (d, J = 7.9 Hz, 1H), 7.37-7.42 (m, 4H), 7.34 (dd, J = 1.6 Hz, J = 7.9 Hz, 1H), 3.18 (s, 1H), 2.90 (q, J = 7.6 Hz, 2H), 1.33 (t, J = 7.6 Hz, 3H); ^{13}C NMR (100 MHz, CDCl$_3$) δ146.7, 132.5, 132.1, 132.0, 129.9, 128.9, 128.8, 123.7, 123.5, 122.4, 95.2, 88.2, 84.1, 78.9, 28.1, 15.0; HRMS calcd for $C_{18}H_{14}$: 230.1096. Found: 230.1093.

Trimethyl[4-(4-phenylethynylphenylethynyl)phenylethynyl]silane (6a). **5a** (250 mg, 1.24 mmol), **3a** (742 mg, 2.47 mmol), $((C_6H_5)_3P)_2PdCl_2$ (17 mg, 0.024 mmol), CuI (9 mg, 0.047 mmol), and N,N-diisopropylethylamine (860 μL, 4.94 mmol) in THF (8 mL) for 1 d gave the desired product (205 mg, 44%). IR (KBr) 2152 cm^{-1}; ^1H NMR (400 MHz, CDCl$_3$) δ7.52 (m, 7H), 7.45-7.51 (m, 3H), 7.37-7.41 (m, 3H), 0.28 (s, 9H); ^{13}C NMR (125 MHz, CDCl$_3$) □132.3, 132.1, 132.0, 131.9, 131.8, 128.9, 128.8, 123.8, 123.6, 123.5, 123.4, 123.2, 105.0, 96.9, 91.8, 91.4, 89.4, 0.3; HRMS calcd for $C_{27}H_{22}Si$: 374.1491. Found: 374.1488.

[3-Fluoro-4-(4-phenylethynylphenylethynyl)phenylethynyl]trimethylsilane (6b). **5a** (240 mg, 1.19 mmol), **3b** (756 mg, 2.38 mmol), $((C_6H_5)_3P)_2PdCl_2$ (17 mg, 0.024 mmol), CuI (9 mg, 0.047 mmol), and N,N-diisopropylethylamine (830 μL, 4.76 mmol) in THF (8 mL) for 1 d gave the desired product (310 mg, 66%). IR (KBr) 2211, 2152 cm^{-1}; ^1H NMR (500 MHz, CDCl$_3$) δ7.53-7.58 (m, 6H), 7.46 (dd, J = 7.6 Hz, J = 7.6 Hz, 1H), 7.37-7.40 (m, 3H), 7.21-7.26 (m, 2H), 0.28 (s, 9H); ^{13}C NMR (125 MHz, CDCl$_3$) δ162.4 (d, J_{C-F} = 252 Hz), 133.5, 132.1, 132.0, 129.9, 128.9, 128.8, 128.1 (d, J_{C-F} = 3 Hz), 125.4, 125.3, 124.1, 123.4, 122.8, 119.2 (d, J_{C-F} = 23 Hz), 112.5 (d, J_{C-F} = 16 Hz), 103.6 (d, J_{C-F} = 3 Hz), 98.0, 96.2 (d, J_{C-F} = 4 Hz), 91.9, 89.4, 84.6, 0.2; HRMS calcd for $C_{27}H_{21}FSi$: 392.1396. Found: 392.1394.

2-(4-Phenylethynylphenylethynyl)-5-trimethylsilanylethynylbenzonitrile (6c). **5a** (150 mg, 0.74 mmol), **3c** (241 mg, 0.74 mmol), $((C_6H_5)_3P)_2PdCl_2$ (10 mg, 0.014 mmol), CuI (6 mg, 0.032 mmol), and N,N-diisopropylethylamine (517 μL, 2.96 mmol) in THF (8 mL) for 1 d gave the desired product (248 mg, 84%). IR (KBr) 2230, 2210, 2149 cm^{-1}; ^1H NMR (400 MHz, CDCl$_3$) δ7.77 (d, J = 1.2 Hz, 1H), 7.63 (dd, J = 1.7 Hz, J = 8.2 Hz, 1H), 7.58-7.61 (m, 2H), 7.52-7.57 (m, 5H), 7.37-7.40 (m, 3H), 0.29 (s, 9H); ^{13}C NMR (100 MHz, CDCl$_3$) δ136.2, 135.9, 132.1, 132.0, 131.9, 129.0, 128.8, 126.8, 124.8, 124.2, 123.3, 121.9, 117.2, 115.9, 102.5, 99.8, 97.8, 92.4, 89.4, 87.6, 0.1; HRMS calcd for $C_{28}H_{21}NSi$: 399.1443. Found: 399.1436.

Trimethyl[4-(4-phenylethynylphenylethynyl)-3-trifluoromethylphenyl-ethynyl]silane (6d). **5a** (150 mg, 0.74 mmol), **3d** (410 mg, 1.11 mmol), $((C_6H_5)_3P)_2PdCl_2$ (10 mg, 0.014 mmol), CuI (6 mg, 0.032 mmol), $(C_6H_5)_3P$ (8 mg, 0.03 mmol), and N,N-diisopropylethylamine (517 μL,

2.96 mmol) in THF (8 mL) at 50 °C for 7 d gave the desired product (233 mg, 71%). IR (KBr) 2210, 2157 cm^{-1}; ^{1}H NMR (400 MHz, CDCl$_3$) δ7.82 (s, 1H), 7.51-7.60 (m, 8H), 7.36-7.41 (m, 3H), 0.33 (s, 9H); ^{13}C NMR (100 MHz, CDCl$_3$) δ134.9, 134.0, 132.1, 132.0, 131.9 (q, J_{C-F} = 31 Hz), 129.9 (q, J_{C-F} = 5 Hz), 129.0, 128.8, 124.4, 123.7, 123.4, 123.3 (q, J_{C-F} = 274 Hz), 122.6, 121.4 (q, J_{C-F} = 2 Hz), 103.6, 98.7, 96.9, 92.2, 89.5, 87.4, 0.2; HRMS calcd for C$_{28}$H$_{21}$F$_3$Si: 442.1365. Found: 442.1366.

[3-Ethyl-4-(4-phenylethynylphenylethynyl)phenylethynyl]trimethylsilane (6e). **5a** (155 mg, 0.77 mmol), **3e** (330 mg, 1.0 mmol), ((C$_6$H$_5$)$_3$P)$_2$PdCl$_2$ (11 mg, 0.016 mmol), CuI (6 mg, 0.032 mmol), (C$_6$H$_5$)$_3$P (8 mg, 0.03 mmol), and N,N-diisopropylethylamine (517 μL, 2.96 mmol) in THF (8 mL) at 50 °C for 5 d gave the desired product (171 mg, 55%). IR (KBr) 2156 cm^{-1}; ^{1}H NMR (500 MHz, CDCl$_3$) δ7.51-7.58 (m, 6H), 7.46 (d, J = 7.9 Hz, 1H), 7.37-7.41 (m, 4H), 7.31 (dd, J = 1.4 Hz, J = 7.9 Hz, 1H), 2.88 (q, J = 7.6 Hz, 2H), 1.33 (t, J = 7.6 Hz, 3H), 0.29 (s, 9H); ^{13}C NMR (125 MHz, CDCl$_3$) δ146.5, 132.4, 132.1, 132.0, 131.9, 131.8, 129.7, 128.9, 128.8, 123.7, 123.6, 123.5, 123.4, 122.8, 105.4, 96.2, 94.7, 91.8, 90.1, 89.5, 28.0, 15.0, 0.4; HRMS calcd for C$_{29}$H$_{26}$Si: 402.1804. Found: 402.1798.

[4-(3-Fluoro-4-phenylethynylphenylethynyl)phenylethynyl]trimethylsilane (7a). **5b** (185 mg, 0.84 mmol), **3a** (252 mg, 0.84 mmol), ((C$_6$H$_5$)$_3$P)$_2$PdCl$_2$ (18 mg, 0.026 mmol), CuI (10 mg, 0.052 mmol), and N,N-diisopropylethylamine (586 μL, 3.36 mmol) in THF (8 mL) for 1 d gave the desired product (285 mg, 86%). IR (KBr) 2210, 2154 cm^{-1}; ^{1}H NMR (500 MHz, CDCl$_3$) δ7.56-7.59 (m, 2H), 7.48 (dd, J = 7.6 Hz, J = 7.6 Hz, 1H), 7.38-7.40 (m, 3H), 7.24-7.28 (m, 2H), 0.26 (s, 9H); ^{13}C NMR (125 MHz, CDCl$_3$) δ161.5 (d, J_{C-F} = 252 Hz), 133.7, 132.4, 132.1, 131.9, 129.2, 128.8, 127.7 (d, J_{C-F} = 3 Hz), 125.0 (d, J_{C-F} = 9 Hz), 123.9, 123.0 (d, J_{C-F} = 11 Hz), 118.8 (d, J_{C-F} = 23 Hz),112.8 (d, J_{C-F} = 16 Hz), 104.9, 97.1, 96.6 (d, J_{C-F} = 3 Hz), 92.0, 90.2 (d, J_{C-F} = 4 Hz), 0.3; HRMS calcd for C$_{27}$H$_{21}$FSi: 392.1396. Found: 396.1397.

[3-Fluoro-4-(3-fluoro-4-phenylethynylphenylethynyl)phenylethynyl]-trimethylsilane (7b). **5b** (150 mg, 0.68 mmol), **3b** (217 mg, 0.68 mmol), ((C$_6$H$_5$)$_3$P)$_2$PdCl$_2$ (10 mg, 0.014 mmol), CuI (5 mg, 0.026 mmol), and N,N-diisopropylethylamine (474 μL, 2.72 mmol) in THF (8 mL) for 1 d gave the desired product (231 mg, 82%). IR (KBr) 2209, 2153 cm^{-1}; ^{1}H NMR (400 MHz, CDCl$_3$) δ7.56-7.60 (m, 2H), 7.52 (dd, J = 7.6 Hz, J = 7.6 Hz, 1H), 7.45 (dd, J = 7.6 Hz, J = 7.6 Hz, 1H), 7.22-7.27 (m, 2H), 0.28 (s, 9H); ^{13}C NMR (100 MHz, CDCl$_3$) δ162.5 (d, J_{C-F} = 252 Hz), 162.4 (d, J_{C-F} = 252 Hz), 133.7, 133.6, 133.5, 133.4, 132.2, 129.2, 128.8, 128.2 (d, J_{C-F} = 4 Hz), 128.8 (d, J_{C-F} = 4 Hz), 127.7 (d, J_{C-F} = 9 Hz), 124.6 (d, J_{C-F} = 9 Hz), 123.0, 119.2 (d, J_{C-F} = 22 Hz), 118.9 (d, J_{C-F} = 22 Hz), 113.1 (d, J_{C-F} = 16 Hz), 112.1 (d, J_{C-F} = 16 Hz), 103.5 (d, J_{C-F} = 3 Hz), 98.3, 96.8 (d, J_{C-F} = 3

Hz), 95.0 (d, J_{C-F} = 3 Hz), 94.9 (d, J_{C-F} = 3 Hz), 85.4, 82.8, 0.2; HRMS calcd for $C_{26}H_{17}F_2Si$: 410.1302. Found: 410.1302.

2-(3-Fluoro-4-phenylethynylphenylethynyl)-5-trimethylsilanylethynyl-benzonitrile (7c). **5b** (150 mg, 0.68 mmol), **3c** (221 mg, 0.68 mmol), $((C_6H_5)_3P)_2PdCl_2$ (10 mg, 0.014 mmol), CuI (5 mg, 0.026 mmol), and N,N-diisopropylethylamine (474 µL, 2.72 mmol) in THF (8 mL) for 1 d gave the desired product (275 mg, 97%). IR (KBr) 2231, 2217, 2152 cm^{-1}; ^{1}H NMR (400 MHz, CDCl$_3$) δ7.77 (d, J = 1.3 Hz, 1H), 7.62 (dd, J = 1.6 Hz, J = 8.2 Hz, 1H), 7.57-7.61 (m, 3H), 7.53 (dd, J = 7.4 Hz, J = 7.4 Hz, 1H), 7.37-7.41 (m, 4H), 7.34 (dd, J = 1.4 Hz, J = 9.5 Hz, 1H), 0.28 (s, 9H); ^{13}C NMR (100 MHz, CDCl$_3$) δ161.8 (d, J_{C-F} = 253 Hz), 135.6, 135.3, 133.2, 133.1, 131.9, 131.6, 128.7, 128.2, 127.6 (d, J_{C-F} = 3 Hz), 125.6, 124.0, 123.0 (d, J_{C-F} = 9 Hz), 122.3, 118.5 (d, J_{C-F} = 22 Hz), 116.4, 115.4, 113.2 (d, J_{C-F} = 16 Hz), 101.8, 99.4, 96.6 (d, J_{C-F} = 3 Hz), 95.8 (d, J_{C-F} = 3 Hz), 87.6, 82.2, -0.4; HRMS calcd for $C_{28}H_{20}FNSi$: 417.1349. Found: 417.1346.

[4-(3-Fluoro-4-phenylethynylphenylethynyl)-3-trifluoromethyl-phenylethynyl]trimethylsilane (7d). **5b** (150 mg, 0.68 mmol), **3d** (251 mg, 0.68 mmol), $((C_6H_5)_3P)_2PdCl_2$ (10 mg, 0.014 mmol), CuI (5 mg, 0.026 mmol), $(C_6H_5)_3P$ (8 mg, 0.03 mmol), and N,N-diisopropylethylamine (474 µL, 2.72 mmol) in THF (8 mL) at 50 °C for 7 d gave the desired product (195 mg, 62%). IR (KBr) 2207, 2159 cm^{-1}; ^{1}H NMR (500 MHz, CDCl$_3$) δ7.80 (d, J = 0.4 Hz, 1H), 7.60-7.62 (m, 2H), 7.56-7.60 (m, 2H), 7.53 (dd, J = 7.5 Hz, J = 7.5 Hz, 1H), 7.38-7.42 (m, 3H), 7.30-7.34 (m, 2H), 0.30 (s, 9H); ^{13}C NMR (125 MHz, CDCl$_3$) δ162.5 (d, J_{C-F} = 252 Hz), 134.9, 134.1, 133.8, 133.7, 132.2, 132.1 (q, J_{C-F} = 31 Hz), 129.9 (q, J_{C-F} = 5 Hz), 129.3, 128.8, 127.8 (d, J_{C-F} = 3 Hz), 124.4, 124.3, 124.1, 123.5 (q, J_{C-F} = 274 Hz), 123.1, 120.9 (q, J_{C-F} = 2 Hz), 118.9 (d, J_{C-F} = 22 Hz), 113.4 (d, J_{C-F} = 16 Hz), 103.4, 99.0, 97.0 (d, J_{C-F} = 3 Hz), 95.5 (d, J_{C-F} = 3 Hz), 88.0, 82.9, 0.2; HRMS calcd for $C_{28}H_{20}F4Si$: 460.1270. Found: 460.1273.

[3-Ethyl-4-(3-fluoro-4-phenylethynylphenylethynyl)phenylethynyl]-trimethylsilane (7e). **5b** (150 mg, 0.68 mmol), **3e** (223 mg, 0.68 mmol), $((C_6H_5)_3P)_2PdCl_2$ (10 mg, 0.014 mmol), CuI (5 mg, 0.026 mmol), $(C_6H_5)_3P$ (8 mg, 0.03 mmol), and N,N-diisopropylethylamine (474 µL, 2.72 mmol) in THF (8 mL) at 50 °C for 4 d gave the desired product (150 mg, 54%). IR (KBr) 2219, 2152, 2131 cm^{-1}; ^{1}H NMR (400 MHz, CDCl$_3$) δ7.57-7.61 (m, 2H), 7.52 (dd, J = 7.5 Hz, J = 7.5 Hz, 1H), 7.46 (d, J = 7.9 Hz, 1H), 7.36-7.42 (m, 4H), 7.26-7.34 (m, 3H), 2.89 (q, J = 7.6 Hz, 2H), 1.33 (t, J = 7.6 Hz, 3H), 0.30 (s, 9H); ^{13}C NMR (100 MHz, CDCl$_3$) δ162.5 (d, J_{C-F} = 252 Hz), 133.7, 133.6, 132.5, 132.0, 129.7, 129.2, 128.8, 127.6 (d, J_{C-F} = 3 Hz), 124.0, 123.1, 122.3, 118.6 (d, J_{C-F} = 22 Hz), 112.6 (d, J_{C-F} = 16 Hz), 105.3, 96.6 (d, J_{C-F} = 3 Hz), 96.5, 93.5 (d, J_{C-F} = 3 Hz), 90.9, 82.9, 28.0, 15.0, 0.4; HRMS calcd for $C_{29}H_{25}FSi$: 420.1710. Found: 420.1711.

2-Phenylethynyl-5-(4-trimethylsilanylethynylphenylethynyl)benzonitrile (8a). **5c** (150 mg, 0.66 mmol), **3a** (208 mg, 0.69 mmol), $((C_6H_5)_3P)_2PdCl_2$ (9 mg, 0.013 mmol), CuI (5 mg, 0.026 mmol), and N,N-diisopropylethylamine (460 µL, 2.64 mmol) in THF (4 mL) for 22 h gave the desired product (190 mg, 72%). IR (KBr) 2233, 2211, 2152 cm^{-1}; ^{1}H NMR (500 MHz, CDCl$_3$) δ7.82 (d, J = 1.6 Hz, 1H), 7.69 (dd, J = 1.6 Hz, J = 8.2 Hz, 1H), 7.61-7.65 (m, 3H), 7.49 (br, 4H), 7.39-7.45 (m, 3H), 0.29 (s, 9H); ^{13}C NMR (125 MHz, CDCl$_3$) δ135.8, 135.5, 132.6, 132.5, 132.4, 132.0, 129.9, 128.9, 127.0, 124.3, 123.9, 122.5, 122.3, 117.2, 116.2, 104.8, 98.4, 97.4, 93.3, 89.1, 86.0, 0.3; HRMS calcd for $C_{28}H_{21}NSi$: 399.1443. Found: 399.1446.

5-(2-Fluoro-4-trimethylsilanylethynylphenylethynyl)-2-phenylethynyl-benzonitrile (8b). **5c** (180 mg, 0.79 mmol), **3b** (252 mg, 0.79 mmol), $((C_6H_5)_3P)_2PdCl_2$ (11 mg, 0.016 mmol), CuI (6 mg, 0.032 mmol), and N,N-diisopropylethylamine (550 µL, 3.16 mmol) in THF (8 mL) for 1 d gave the desired product (293 mg, 88%). IR (KBr) 2231, 2210, 2149 cm^{-1}; ^{1}H NMR (400 MHz, CDCl$_3$) δ7.84 (dd, J = 1.7 Hz, J = 0.3 Hz, 1H), 7.72 (dd, J = 1.7 Hz, J = 8.2 Hz, 1H), 7.62-7.66 (m, 3H), 7.46 (dd, J = 7.6 Hz, J = 7.6 Hz, 1H), 7.39-7.46 (m, 3H), 7.23-7.28 (m, 2H), 0.29 (s, 9H); ^{13}C NMR (100 MHz, CDCl$_3$) δ162.5 (d, J_{C-F} = 252 Hz), 135.8, 135.5, 133.6, 133.5, 132.6, 132.5, 129.9, 128.9, 128.2 (d, J_{C-F} = 3 Hz), 127.3, 126.1 (d, J_{C-F} = 9 Hz), 123.5, 122.2, 119.3 (d, J_{C-F} = 22 Hz), 117.2, 116.2, 111.6 (d, J_{C-F} = 16 Hz), 103.4 (d, J_{C-F} = 3 Hz), 98.7, 98.6, 93.8 (d, J_{C-F} = 3 Hz), 86.7, 86.0, 0.2; HRMS calcd for $C_{28}H_{20}FNSi$: 417.1349. Found: 417.1349.

2-(3-Cyano-4-phenylethynylphenylethynyl)-5-trimethylsilanylethynyl-benzonitrile (8c). **5c** (126 mg, 0.55 mmol), **3c** (189 mg, 0.58 mmol), $((C_6H_5)_3P)_2PdCl_2$ (12 mg, 0.017 mmol), CuI (6 mg, 0.032 mmol), $(C_6H_5)_3P$ (9 mg, 0.034 mmol), and N,N-diisopropylethylamine (380 µL, 2.18 mmol) in THF (4 mL) at 40 °C for 3 d gave the desired product (182 mg, 77%). IR (KBr) 2196, 1704 cm^{-1}; ^{1}H NMR (400 MHz, CDCl$_3$) δ7.87 (dd, J = 0.5 Hz, J = 1.7 Hz, 1H), 7.78 (dd, J = 1.7 Hz, J = 8.3 Hz, 1H), 7.76 (dd, J = 0.5 Hz, J = 1.7 Hz, 1H), 7.62-7.67 (m, 4H), 7.57 (dd, J = 0.5 Hz, J = 8.1 Hz, 1H), 7.37-7.42 (m, 3H), 0.27 (s, 9H); ^{13}C NMR (100 MHz, CDCl$_3$) δ136.2, 136.0, 135.9, 132.6, 132.5, 130.0, 128.9, 128.0, 125.7, 125.0, 122.6, 122.1, 117.0, 116.9, 116.3, 116.2, 102.3, 100.4, 99.0, 95.1, 89.3, 85.9, 0.1; HRMS calcd for $C_{29}H_{20}N_2Si$: 424.1396. Found: 424.1391.

2-Phenylethynyl-5-(2-trifluoromethyl-4-trimethylsilanylethynyl-phenylethynyl)benzonitrile (8d). **5c** (130 mg, 0.57 mmol), **3d** (220 mg, 0.6 mmol), $((C_6H_5)_3P)_2PdCl_2$ (12 mg, 0.017 mmol), CuI (7 mg, 0.037 mmol), $(C_6H_5)_3P$ (9 mg, 0.034 mmol), and N,N-diisopropylethylamine (400 µL, 2.30 mmol) in THF (4 mL) at 40 °C for 2 d gave the desired product (192 mg, 72%). IR (KBr) 2231, 2209, 2156 cm^{-1}; ^{1}H NMR (400 MHz, CDCl$_3$) δ7.81

(dd, J = 0.5 Hz, J = 1.7 Hz, 1H), 7.80 (br s, 1H), 7.69 (dd, J = 1.7 Hz, J = 8.2 Hz, 1H), 7.62-7.65 (m, 5H), 7.37-7.42 (m, 3H), 0.28 (s, 9H); ^{13}C NMR (100 MHz, CDCl$_3$) δ135.8, 135.5, 135.0, 134.1, 132.6, 132.5, 132.2 (q, J$_{C-F}$ = 31 Hz), 130.0 (q, J$_{C-F}$ = 5 Hz), 129.9, 127.6, 124.5, 123.4 (q, J$_{C-F}$ = 274 Hz), 123.3, 122.2, 120.4 (q, J$_{C-F}$ = 2 Hz), 117.1, 116.2, 103.3, 99.4, 98.8, 94.2, 89.2, 86.0, 0.2; HRMS calcd for C$_{29}$H$_{20}$F$_3$NSi: 467.1317. Found: 467.1313.

5-(2-Ethyl-4-trimethylsilanylethynylphenylethynyl)-2-phenylethynyl-benzonitrile (8e). **5c** (180 mg, 0.79 mmol), **3e** (260 mg, 0.79 mmol), ((C$_6$H$_5$)$_3$P)$_2$PdCl$_2$ (11 mg, 0.016 mmol), CuI (6 mg, 0.032 mmol), (C$_6$H$_5$)$_3$P (8 mg, 0.03 mmol), and N,N-diisopropylethylamine (550 µL, 3.16 mmol) in THF (7 mL) at 50 °C for 5 d gave the desired product (242 mg, 71%). IR (KBr) 2230, 2215, 2152 cm^{-1}; ^{1}H NMR (400 MHz, CDCl$_3$) δ7.81 (d, J = 1.6 Hz, 1H), 7.68 (dd, J = 1.6 Hz, J = 1.78.1 Hz, 1H), 7.61-7.69 (m, 3H), 7.43 (d, J = 8.1 Hz, 2H), 7.40-7.43 (m, 4H), 7.32 (dd, J = 1.6 Hz, J = 7.9 Hz, 1H), 2.86 (q, J = 7.6 Hz, 2H), 1.32 (t, J = 7.6 Hz, 3H), 0.29 (s, 9H); ^{13}C NMR (100 MHz, CDCl$_3$) δ146.8, 135.6, 135.3, 132.6, 132.5, 132.4, 132.0, 129.9, 129.8, 128.9, 126.8, 124.4, 124.2, 122.3, 121.9, 117.3, 116.2, 105.2, 98.4, 96.8, 92.4, 92.2, 86.0, 28.0, 15.0, 0.4; HRMS calcd for C$_{30}$H$_{25}$NSi: 427.1756. Found: 427.1747.

Trimethyl[4-(4-phenylethynyl-3-trifluoromethylphenylethynyl)-phenylethynyl]silane (9a). **5d** (150 mg, 0.55 mmol), **3a** (167 mg, 0.55 mmol), ((C$_6$H$_5$)$_3$P)$_2$PdCl$_2$ (8 mg, 0.011 mmol), CuI (4 mg, 0.021 mmol), and N,N-diisopropylethylamine (387 µL, 2.22 mmol) in THF (6 mL) for 1 d gave the desired product (185 mg, 75%). IR (KBr) 2211, 2149 cm^{-1}; ^{1}H NMR (500 MHz, CDCl$_3$) δ7.86 (s, 1H), 7.64-7.68 (m, 2H), 7.57-7.60 (m, 2H), 7.48-7.52 (m, 4H), 7.39-7.42 (m, 3H), 0.29 (s, 9H); ^{13}C NMR (125 MHz, CDCl$_3$) δ134.5, 134.2, 132.4, 132.1 (q, J$_{C-F}$ = 31 Hz), 132.0, 129.5 (q, J$_{C-F}$ = 5 Hz), 129.4, 128.9, 124.1, 123.5 (q, J$_{C-F}$ = 274 Hz), 123.4, 123.0, 122.9, 121.6 (q, J$_{C-F}$ = 2 Hz), 104.9, 97.4, 97.2, 92.6, 90.1, 85.7, 0.3; HRMS calcd for C$_{28}$H$_{21}$F$_3$Si: 442.1364. Found: 442.1364.

[3-Fluoro-4-(4-phenylethynyl-3-trifluoromethylphenylethynyl)-phenylethynyl]-trimethyl-silane (9b). **5d** (150 mg, 0.55 mmol), **3b** (177 mg, 0.55 mmol), ((C$_6$H$_5$)$_3$P)$_2$PdCl$_2$ (8 mg, 0.011 mmol), CuI (4 mg, 0.021 mmol), and N,N-diisopropylethylamine (387 µL, 2.22 mmol) in THF (8 mL) for 1 d gave the desired product (198 mg, 77%). IR (KBr) 2204, 2151 cm^{-1}; ^{1}H NMR (400 MHz, CDCl$_3$) δ7.86 (d, J = 0.5 Hz, 1H), 7.64-7.69 (m, 2H), 7.55-7.59 (m, 2H), 7.46 (dd, J = 7.5 Hz, J = 7.5 Hz, 1H), 7.38-7.41 (m, 3H), 7.23-7.28 (m, 2H), 0.28 (s, 9H); ^{13}C NMR (125 MHz, CDCl$_3$) δ162.5 (d, J$_{C-F}$ = 252 Hz), 134.6, 134.2, 133.6, 133.5, 132.1 (q, J$_{C-F}$ = 31 Hz), 129.5 (q, J$_{C-F}$ = 5 Hz), 129.5, 128.5, 128.1 (d, J$_{C-F}$ = 3 Hz), 126.0, 125.9, 123.5 (q, J$_{C-F}$ = 274 Hz), 123.0, 122.9, 122.1 (q, J$_{C-F}$ = 2 Hz), 119.3 (d, J$_{C-F}$ = 23 Hz), 112.0 (d, J$_{C-F}$ = 16 Hz), 103.5 (d, J$_{C-F}$ = 3 Hz), 98.4, 97.6, 94.7 (d, J$_{C-F}$ = 3 Hz), 86.0, 85.7, 0.2; HRMS calcd for C$_{28}$H$_{20}$F$_4$Si: 460.1270. Found: 460.1266.

2-(4-Phenylethynyl-3-trifluoromethylphenylethynyl)-5-trimethylsilanylethynylbenzonitrile (9c). **5d** (150 mg, 0.55 mmol), **3c** (180 mg, 0.55 mmol), $((C_6H_5)_3P)_2PdCl_2$ (8 mg, 0.011 mmol), CuI (4 mg, 0.021 mmol), and N,N-diisopropylethylamine (387 µL, 2.22 mmol) in THF (6 mL) for 1 d gave the desired product (182 mg, 70%). IR (KBr) 2231, 2210, 2154 cm^{-1}; ^{1}H NMR (500 MHz, CDCl$_3$) δ7.90 (s, 1H), 7.78 (d, J = 1.3 Hz, 1H), 7.75 (dd, J = 1.3 Hz, J = 8.1 Hz, 1H), 7.68 (d, J = 8.1 Hz, 1H), 7.64 (dd, J = 1.6 Hz, J = 8.1 Hz, 1H), 7.56-7.60 (m, 3H), 7.39-7.42 (m, 3H), 0.30 (s, 9H); ^{13}C NMR (125 MHz, CDCl$_3$) δ136.2, 135.9, 135.0, 134.3, 132.6, 132.3 (q, J_{C-F} = 31 Hz), 132.2, 129.7 (q, J_{C-F} = 5 Hz), 129.6, 128.9, 126.1, 124.8, 123.4 (q, J_{C-F} = 274 Hz), 122.8, 122.7 (q, J_{C-F} = 2 Hz), 122.0, 117.0, 116.1, 102.4, 100.2, 98.2, 96.2, 88.6, 85.6, 0.1; HRMS calcd for $C_{29}H_{20}F_3NSi$: 467.1317. Found: 467.1314.

Trimethyl[4-(4-phenylethynyl-3-trifluoromethylphenylethynyl)-3-trifluoromethylphenylethynyl]silane (9d). **5d** (225 mg, 0.83 mmol), **3d** (260 mg, 0.79 mmol), $((C_6H_5)_3P)_2PdCl_2$ (17 mg, 0.024 mmol), CuI (9 mg, 0.047 mmol), $(C_6H_5)_3P$ (13 mg, 0.049 mmol), and N,N-diisopropylethylamine (580 µL, 3.32 mmol) in THF (5 mL) at 50 °C for 3 d gave the desired product (230 mg, 51%). IR (KBr) 2202, 2158 cm^{-1}; ^{1}H NMR (400 MHz, CDCl$_3$) δ7.84 (br s, 1H), 7.79 (d, J = 0.7 Hz, 1H), 7.66 (d, J = 1.1 Hz, 2H), 7.61 (d, J = 1.1 Hz, 2H), 7.55-7.59 (m, 2H), 7.35-7.40 (m, 3H), 0.28 (s, 9H); ^{13}C NMR (100 MHz, CDCl$_3$) δ134.9, 134.6, 134.2, 134.1, 132.3 (q, J_{C-F} = 31 Hz), 132.2 (q, J_{C-F} = 31 Hz), 132.1, 129.9 (q, J_{C-F} = 5 Hz), 129.4 (q, J_{C-F} = 5 Hz), 128.9, 124.9, 124.8, 124.2, 123.6 (q, J_{C-F} = 274 Hz), 123.5 (q, J_{C-F} = 274 Hz), 122.8, 122.7, 122.3 (q, J_{C-F} = 2 Hz), 120.7 (q, J_{C-F} = 2 Hz), 103.4, 99.1, 97.7, 95.2, 88.5, 85.6, 0.2; HRMS calcd for $C_{29}H_{20}F_6Si$: 510.1238. Found: 510.1234.

[3-Ethyl-4-(4-phenylethynyl-3-trifluoromethylphenylethynyl)-phenylethynyl]trimethylsilane (9e). **5d** (180 mg, 0.67 mmol), **3e** (219 mg, 0.67 mmol), $((C_6H_5)_3P)_2PdCl_2$ (9 mg, 0.013 mmol), CuI (5 mg, 0.026 mmol), $(C_6H_5)_3P$ (7 mg, 0.026 mmol), and N,N-diisopropylethylamine (550 µL, 3.16 mmol) in THF (7 mL) at 50 °C for 4 d gave the desired product (213 mg, 68%). IR (KBr) 2216, 2154 cm^{-1}; ^{1}H NMR (400 MHz, CDCl$_3$) δ 7.79 (s, 1H), 7.56-7.64 (m, 2H), 7.52-7.55 (m, 2H), 7.42 (d, J = 7.9 Hz, 1H), 7.34-7.38 (m, 4H), 7.27 (dd, J = 1.6 Hz, J = 7.9 Hz, 1H), 2.83 (q, J = 7.6 Hz, 2H), 1.28 (t, J = 7.6 Hz, 3H), 0.25 (s, 9H); ^{13}C NMR (125 MHz, CDCl$_3$) δ146.7, 134.4, 134.2, 132.6, 132.3 (q, J_{C-F} = 31 Hz), 132.2, 132.0, 129.7, 129.5, 129.3 (q, J_{C-F} = 5 Hz), 128.9, 125.0, 123.6, 123.5 (q, J_{C-F} = 274 Hz), 122.9, 122.2, 121.5 (q, J_{C-F} = 2 Hz), 105.2, 97.3, 96.6, 93.3, 91.4, 85.7, 28.0, 15.0, 0.4; HRMS calcd for $C_{30}H_{25}F_3Si$: 470.1678. Found: 470.1678.

[4-(3-Ethyl-4-phenylethynylphenylethynyl)phenylethynyl]trimethylsilane (10a). **5e** (150 mg, 0.65 mmol), **3a** (196 mg, 0.65 mmol), $((C_6H_5)_3P)_2PdCl_2$ (9 mg, 0.013 mmol), CuI (5 mg, 0.026 mmol), and N,N-diisopropylethylamine (454 µL, 2.60 mmol) in THF (8 mL) for 1 d gave the desired product (136 mg, 52%). IR

(KBr) 2201, 2154 cm^{-1}; ^{1}H NMR (500 MHz, CDCl$_3$) δ7.55-7.58 (m, 2H), 7.51 (d, J = 7.9 Hz, 1H), 7.43-7.50 (m, 4H), 7.44 (d, J = 1.3 Hz, 1H), 7.35-7.42 (m, 4H), 2.92 (q, J = 7.6 Hz, 2H), 1.35 (t, J = 7.6 Hz, 3H), 0.29 (s, 9H); ^{13}C NMR (125 MHz, CDCl$_3$) δ146.7, 132.5, 132.3, 131.9, 131.8, 131.5, 129.3, 128.8, 123.7, 123.6, 123.4, 123.3, 123.1, 105.0, 96.8, 95.1, 91.8, 90.8, 88.2, 28.0, 15.0, 0.3; HRMS calcd for C$_{29}$H$_{26}$Si: 402.1804. Found: 402.1800.

[4-(3-Ethyl-4-phenylethynylphenylethynyl)-3-fluorophenylethynyl]trimethyl-silane (10b). **5e** (160 mg, 0.69 mmol), **3b** (221 mg, 0.69 mmol), ((C$_6$H$_5$)$_3$P)$_2$PdCl$_2$ (10 mg, 0.014 mmol), CuI (4 mg, 0.026 mmol), and N,N-diisopropylethylamine (484 μL, 2.78 mmol) in THF (8 mL) for 1 d gave the desired product (206 mg, 70%). IR (KBr) 2205, 2150 cm^{-1}; ^{1}H NMR (400 MHz, CDCl$_3$) δ7.59-7.59 (m, 2H), 7.53 (d, J = 7.9 Hz, 1H), 7.46-7.49 (m, 2H), 7.37-7.42 (m, 4H), 7.23-7.27 (m, 2H), 2.93 (q, J = 7.6 Hz, 2H), 1.37 (t, J = 7.6 Hz, 3H), 0.31 (s, 9H); ^{13}C NMR (100 MHz, CDCl$_3$) δ162.4 (d, J$_{C-F}$ = 252 Hz), 146.7, 133.5, 132.5, 132.0, 131.6, 129.4, 128.8, 128.1 (d, J$_{C-F}$ = 3 Hz), 125.3, 125.2, 123.7, 123.5, 122.9, 119.2 (d, J$_{C-F}$ = 23 Hz), 112.7 (d, J$_{C-F}$ = 16 Hz), 103.7 (d, J$_{C-F}$ = 3 Hz), 98.0, 96.6 (d, J$_{C-F}$ = 3 Hz), 88.2, 84.2, 28.0, 15.0, 0.2; HRMS calcd for C$_{29}$H$_{25}$FSi: 420.1710. Found: 420.1706.

2-(3-Ethyl-4-phenylethynylphenylethynyl)-5-trimethylsilanylethynyl-benzonitrile (10c). **5e** (175 mg, 0.76 mmol), **3c** (247 mg, 0.76 mmol), ((C$_6$H$_5$)$_3$P)$_2$PdCl$_2$ (11 mg, 0.016 mmol), CuI (6 mg, 0.032 mmol), and N,N-diisopropylethylamine (530 μL, 3.04 mmol) in THF (8 mL) for 1 d gave the desired product (225 mg, 69%). IR (KBr) 2230, 2201, 2149 cm^{-1}; ^{1}H NMR (400 MHz, CDCl$_3$) δ7.78 (dd, J = 0.6 Hz, J = 1.6 Hz, 1H), 7.62 (dd, J = 1.6 Hz, J = 8.2 Hz, 1H), 7.51-7.58 (m, 5H), 7.43 (dd, J = 1.6 Hz, J = 7.9 Hz, 1H), 7.37-7.41 (m, 3H), 2.92 (q, J = 7.6 Hz, 2H), 1.36 (t, J = 7.6 Hz, 3H), 0.28 (s, 9H); ^{13}C NMR (125 MHz, CDCl$_3$) δ146.8, 136.2, 135.8, 132.6, 132.4, 132.0, 131.9, 129.7, 128.9, 128.8, 126.9, 124.2, 124.1, 123.6, 122.0, 117.2, 115.9, 102.6, 99.7, 98.3, 95.7, 88.1, 87.1, 28.0, 15.0, 0.1; HRMS calcd for C$_{30}$H$_{25}$NSi: 427.1756. Found: 427.1755.

[4-(3-Ethyl-4-phenylethynylphenylethynyl)-3-trifluoromethylphenylethynyl]-trimethylsilane (10d). **5e** (180 mg, 0.78 mmol), **3d** (432 mg, 1.17 mmol), ((C$_6$H$_5$)$_3$P)$_2$PdCl$_2$ 2 (11 mg, 0.016 mmol), CuI (6 mg, 0.032 mmol), (C$_6$H$_5$)$_3$P (8 mg, 0.032 mmol), and N,N-diisopropylethylamine (545 μL, 3.13 mmol) in THF (8 mL) at 50 °C for 7 d gave the desired product (235 mg, 64%). IR (KBr) 2200, 2157 cm^{-1}; ^{1}H NMR (500 MHz, CDCl$_3$) δ7.79 (s, 1H), 7.58-7.62 (m, 2H), 7.55-7.57 (m, 2H), 7.51 (d, J = 7.9 Hz, 1H), 7.44 (dd, J = 0.5 Hz, J = 1.5 Hz, 1H), 7.37-7.40 (m, 4H), 2.93 (q, J = 7.6 Hz, 2H), 1.35 (t, J = 7.6 Hz, 3H), 0.30 (s, 9H); ^{13}C NMR (125 MHz, CDCl$_3$) δ146.8, 134.9, 134.0, 132.6, 132.1 (q, J$_{C-F}$ = 31 Hz), 132.0, 131.6, 129.6 (q, J$_{C-F}$ = 5 Hz), 129.5, 128.9, 128.8, 123.8, 123.7, 123.6 (q, J$_{C-F}$ = 274 Hz), 123.5, 122.8, 121.6 (q, J$_{C-F}$ = 2 Hz), 103.6, 98.6, 97.3, 95.5, 88.2, 86.9, 28.1, 15.0, 0.2; HRMS calcd for C$_{30}$H$_{25}$F$_3$Si: 470.1678. Found: 470.1681.

[3-Ethyl-4-(3-ethyl-4-
phenylethynylphenylethynyl)phenylethynyl]trimethyl-silane (10e). **5e** (180
mg, 0.78 mmol), **3e** (385 mg, 1.17 mmol), $((C_6H_5)_3P)_2PdCl_2$ (11 mg, 0.016
mmol), CuI (6 mg, 0.032 mmol), $(C_6H_5)_3P$ (8 mg, 0.032 mmol), and N,N-
diisopropylethylamine (545 μL, 3.13 mmol) in THF (8 mL) at 50 °C for 4 d
gave the desired product (170 mg, 51%). IR (KBr) 2200, 2155 cm^{-1}; ^{1}H NMR
(400 MHz, CDCl$_3$) δ7.54-7.58 (m, 2H), 7.51 (d, J = 8.0 Hz, 1H), 7.45 (d, J = 8.4
Hz, 1H), 7.41-7.43 (m, 2H), 7.35-7.40 (m, 4H), 7.35 (dd, J = 1.7 Hz, J = 8.0 Hz,
1H), 7.30 (dd, J = 1.7 Hz, J = 8.0 Hz, 1H), 2.91 (q, J = 7.6 Hz, 2H), 2.90 (q,
J = 7.6 Hz, 2H), 1.35 (t, J = 7.6 Hz, 3H), 1.33 (t, J = 7.6 Hz, 3H), 0.29 (s, 9H);
^{13}C NMR (125 MHz, CDCl$_3$) δ146.7, 146.5, 132.5, 132.4, 131.9, 131.8, 131.4,
129.7, 129.2, 128.8, 128.7, 123.8, 123.6, 123.5, 123.0, 122.9, 105.5, 96.2, 95.1,
95.0, 89.6, 89.3, 28.1, 28.0, 15.0, 14.9, 0.4; HRMS calcd for $C_{31}H_{30}Si$: 430.2117.
Found: 430.2120.

General Procedure for the Preparation of Tetramer Thioacetate.
The silylated alkyne was deprotected as described previously. Without further
purification, the trimer alkyne was subjected to couple with the alligator clipas
follows: To a stirring solution of the 1-iodo-4-thioacetylbenzene (**11**) (1-2
equiv), $((C_6H_5)_3P)_2PdCl_2$ (3 mol %), and CuI (6 mol %) in THF was added the
terminal alkyne followed by the amine (4 equiv based on the alkyne) at room
temperature under N_2 in a screw cap tube. The tube was flushed with N_2,
capped, and allowed to stir 24 h. The reaction mixture was then subjected to an
aqueous workup and the aqueous layer extracted with CH_2Cl_2. After drying the
combined organic layers over Na_2SO_4, the solvent was removed in vacuo. The
crude product was purified by column chromatography (silica gel) with
hexane:EtOAc mixtures to give desired product.

Thioacetic acid S-{4-[3-fluoro-4-(4-phenylethynylphenylethynyl-
phenylethynyl]phenyl} ester (12b). 75 mg (43%) of **12b** was produced
from **6b** (145 mg, 0.37 mmol) and **11** (205 mg, 0.74 mmol). IR (KBr) 2204,
1700 cm^{-1}; ^{1}H NMR (500 MHz, CDCl$_3$) δ7.53-7.59 (m, 8H), 7.52 (dd, J = 7.6
Hz, J = 7.6 Hz, 1H) 7.43 (d, J = 8.4 Hz, 2H), 7.37-7.41 (m, 3H), 7.28-7.34 (m,
2H), 2.47 (s, 3H); ^{13}C NMR (125 MHz, CDCl$_3$) δ193.7, 162.5 (d, J_{C-F} = 252 Hz),
134.7, 133.6, 132.6, 132.1, 132.0, 129.2, 128.9, 128.8, 127.8 (d, J_{C-F} = 3 Hz),
125.1, 125.0, 124.2, 124.1, 123.4, 122.8, 118.9 (d, J_{C-F} = 23 Hz), 112.6 (d,
J_{C-F} = 16 Hz), 96.5, 96.3 (d, J_{C-F} = 3 Hz), 92.0, 91.8, 89.9 (d, J_{C-F} = 3 Hz), 89.4,
84.7, 30.7; HRMS calcd for $C_{32}H_{19}FOS$: 470.1140. Found: 470.1141.

Thioacetic acid S-{4-[3-cyano-4-(4-phenylethynylphenylethynyl)-
phenylethynyl]phenyl} ester (12c). 72 mg (41%) of **12c** was produced from
6c (148 mg, 0.37 mmol) and **11** (122 mg, 0.44 mmol). IR (KBr) 2223, 2209,
1706 cm^{-1}; ^{1}H NMR (400 MHz, CDCl$_3$) δ7.85 (d, J = 1.5 Hz, 1H), 7.71 (dd, J =
1.7 Hz, J = 8.2 Hz, 1H), 7.55-7.65 (m, 9H), 7.45 (d, J = 8.4 Hz, 1H), 7.37-7.41
(m, 3H), 2.48 (s, 3H); ^{13}C NMR (100 MHz, CDCl$_3$) δ193.6, 135.8, 135.6,
134.7, 132.7, 132.5, 132.4, 132.1, 132.0, 129.6, 129.0, 128.8, 126.8, 124.9,

124.0, 123.7, 123.3, 121.9, 117.2, 116.1, 98.0, 93.0, 92.4, 89.4, 88.8, 87.7, 30.8; HRMS calcd for $C_{33}H_{19}NOS$: 477.1187. Found: 477.1193.

Thioacetic acid S-{4-[4-(4-phenylethynylphenylethynyl)-3-trifluoromethyl-phenylethynyl]phenyl} ester (12d). 88 mg (47%) of **12d** was produced from **6d** (158 mg, 0.36 mmol) and **11** (142 mg, 0.51 mmol) IR (KBr) 2214, 1703 cm^{-1}; ^{1}H NMR (400 MHz, CDCl$_3$) δ7.87 (br s, 1H), 7.68 (br s, 2H), 7.56-7.61 (m, 8H), 7.44 (d, J = 8.4 Hz, 2H), 7.36-7.41 (m, 3H), 2.48 (s, 3H); ^{13}C NMR (125 MHz, CDCl$_3$) δ193.7, 134.7, 134.6, 134.2, 132.7, 132.2 (q, J_{C-F} = 31 Hz), 132.21, 132.0, 129.6 (q, J_{C-F} = 5 Hz), 129.3, 129.0, 128.8, 124.4, 124.0, 123.5 (q, J_{C-F} = 274 Hz), 123.4, 123.3, 122.6, 121.5 (q, J_{C-F} = 2 Hz), 97.0, 92.3, 92.1, 89.8, 89.4, 87.4, 30.7; HRMS calcd for $C_{33}H_{19}F_3OS$: 520.1109. Found: 520.1109.

Thioacetic acid S-{4-[3-ethyl-4-(4-phenylethynylphenylethynyl)-phenylethynyl]phenyl} ester (12e). 72 mg (47%) of **12e** was produced from **6e** (128 mg, 0.32 mmol) and **11** (131 mg, 0.47 mmol). IR (KBr) 2200, 1694 cm^{-1}; ^{1}H NMR (400 MHz, CDCl$_3$) δ7.50-7.60 (m, 9H), 7.45-7.46 (m, 1H), 7.43 (d, J = 8.4 Hz, 2H), 7.37-7.41 (m, 4H), 2.91 (q, J = 7.6 Hz, 2H), 2.47 (s, 3H), 1.34 (t, J = 7.6 Hz, 3H); ^{13}C NMR (100 MHz, CDCl$_3$) δ193.8, 146.7, 134.7, 132.6, 132.5, 132.1, 132.0, 131.9, 131.6, 129.4, 128.9, 128.8, 128.6, 124.8, 123.7, 123.5, 123.4, 122.9, 94.9, 91.8, 91.5, 90.6, 90.2, 89.6, 30.7, 28.1, 15.0; HRMS calcd for $C_{34}H_{24}OS$: 480.1548. Found: 480.1457.

Thioacetic acid S-{4-[4-(3-fluoro-4-phenylethynylphenylethynyl)-phenylethynyl]phenyl} ester (13a). 82 mg (38%) of **13a** was produced from **7a** (180 mg, 0.46 mmol) and **11** (120 mg, 0.43 mmol). IR (KBr) 2212, 1692 cm^{-1}; ^{1}H NMR (500 MHz, CDCl$_3$) δ7.56-7.60 (m, 4H), 7.50-7.56 (m, 4H), 7.52 (dd, J = 7.6 Hz, J = 7.6 Hz, 1H) 7.44 (d, J = 8.4 Hz, 2H), 7.35-7.40 (m, 3H), 7.31-7.35 (m, 2H), 2.47 (s, 3H); ^{13}C NMR (125 MHz, CDCl$_3$) δ193.8, 162.5 (d, J_{C-F} = 252 Hz), 133.7, 133.6, 132.6, 132.2, 132.1, 132.0, 129.2, 128.8, 127.7 (d, J_{C-F} = 4 Hz), 125.0 (d, J_{C-F} = 9 Hz), 124.6, 123.7, 123.1, 118.8 (d, J_{C-F} = 23 Hz), 112.8 (d, J_{C-F} = 16 Hz), 96.6 (d, J_{C-F} = 3 Hz), 92.0, 91.2, 91.0, 90.4 (d, J_{C-F} = 4 Hz), 82.9, 30.7; HRMS ; HRMS calcd for $C_{32}H_{19}FOS$: 470.1140. Found: 470.1143.

Thioacetic acid S-{4-[3-fluoro-4-(3-fluoro-4-phenylethynylphenylethynyl)-phenylethynyl]phenyl} ester (13b). 30 mg (23%) of **13b** was produced from **7b** (112 mg, 0.28 mmol) and **11** (75 mg, 0.27 mmol). IR (KBr) 2208, 1700 cm^{-1}; ^{1}H NMR (500 MHz, CDCl$_3$) δ7.57-7.60 (m, 4H), 7.52 (dd, J = 7.3 Hz, J = 14.8 Hz, 2H) 7.44 (d, J = 8.4 Hz, 2H), 7.37-7.41 (m, 3H), 7.30-7.37 (m, 4H), 2.47 (s, 3H); ^{13}C NMR (125 MHz, CDCl$_3$) δ193.7, 162.6 (d, J_{C-F} = 252 Hz), 162.5 (d, J_{C-F} = 252 Hz), 134.7, 133.7, 132.7, 132.2, 129.2 (d, J_{C-F} = 4 Hz), 128.9, 127.9 (d, J_{C-F} = 4 Hz), 127.8, 125.5 (d, J_{C-F} = 9 Hz), 124.4 (d, J_{C-F} = 9 Hz), 124.1, 123.0, 119.0 (d, J_{C-F} = 22 Hz), 118.9 (d, J_{C-F} = 22 Hz), 113.1 (d, J_{C-F} = 16 Hz), 112.1 (d, J_{C-F} = 16 Hz), 96.8 (d, J_{C-F} = 3 Hz),

96.5, 95.1 (d, J_{C-F} = 3 Hz), 92.0, 89.8 (d, J_{C-F} = 3 Hz), 85.4, 82.8, 30.7; HRMS calcd for $C_{32}H_{18}F_2OS$: 488.1046. Found: 488.1034.

Thioacetic acid S-{4-[3-cyano-4-(3-fluoro-4-phenylethynylphenylethynyl)-phenylethynyl]phenyl} ester (13c). 65 mg (48%) of **13c** was produced from **7c** (115 mg, 0.28 mmol) and **11** (76 mg, 0.27 mmol). IR (KBr) 2228, 2208, 1700 cm⁻¹; ¹H NMR (400 MHz, CDCl₃) δ7.80 (d, J = 1.7 Hz, 1H), 7.68 (dd, J = 1.6 Hz, J = 8.2 Hz, 1H), 7.59 (d, J = 8.2 Hz, 1H), 7.52-7.56 (m, 4H), 7.50 (dd, J = 7.7 Hz, J = 7.7 Hz, 1H), 7.41 (d, J = 8.4 Hz, 2H), 7.31-7.37 (m, 5H), 2.43 (s, 3H); ¹³C NMR (100 MHz, CDCl₃) δ193.5, 162.5 (d, J_{C-F} = 252 Hz), 135.9, 135.6, 134.7, 133.9, 133.8, 132.7, 132.6, 132.2, 129.7, 129.3, 128.8, 128.2 (d, J_{C-F} = 3 Hz), 126.3, 124.4, 123.6 (d, J_{C-F} = 9 Hz), 123.0, 119.1 (d, J_{C-F} = 23 Hz), 117.0, 116.3, 113.9 (d, J_{C-F} = 16 Hz), 97.3 (d, J_{C-F} = 3 Hz), 96.6 (d, J_{C-F} = 3 Hz), 93.3, 88.7, 88.2, 82.8, 30.8; HRMS calcd for $C_{33}H_{18}FNOS$: 495.1093. Found: 495.1096.

Thioacetic acid S-{4-[4-(3-fluoro-4-phenylethynylphenylethynyl)-3-trifluoromethylphenylethynyl]phenyl} ester (13d). 80 mg (41%) of **13d** was produced from **7d** (166 mg, 0.36 mmol) and **11** (97 mg, 0.35 mmol). IR (KBr) 2210, 1700 cm⁻¹; ¹H NMR (500 MHz, CDCl₃) δ7.86 (s, 1H), 7.63-7.67 (m, 2H), 7.59-7.61 (m, 4H), 7.53 (dd, J = 7.6 Hz, J = 7.6 Hz, 1H), 7.44 (d, J = 8.3 Hz, 2H), 7.38-7.41 (m, 3H), 7.29-7.34 (m, 2H), 2.47 (s, 3H); ¹³C NMR (125 MHz, CDCl₃) δ193.6, 162.5 (d, J_{C-F} = 252 Hz), 134.7, 134.6, 134.2, 133.7, 133.6, 123.7, 132.3 (q, J_{C-F} = 31 Hz), 132.2, 129.6 (q, J_{C-F} = 5 Hz), 129.4, 128.8, 127.9 (d, J_{C-F} = 3 Hz), 124.4, 124.3, 123.9, 123.8, 123.4 (q, J_{C-F} = 274 Hz), 120.9 (q, J_{C-F} = 2 Hz), 118.9 (d, J_{C-F} = 22 Hz), 113.4 (d, J_{C-F} = 16 Hz), 97.0 (d, J_{C-F} = 3 Hz), 96.6 (d, J_{C-F} = 3 Hz), 92.5, 89.7, 88.1, 82.8, 30.7; HRMS calcd for $C_{33}H_{18}F_4OS$: 538.1014. Found: 538.1014.

Thioacetic acid S-{4-[3-ethyl-4-(3-fluoro-4-phenylethynylphenylethynyl)-phenylethynyl]phenyl} ester (13e). 62 mg (48%) of **13e** was produced from **7e** (110 mg, 0.26 mmol) and **11** (68 mg, 0.24 mmol). IR (KBr) 2218, 1698 cm⁻¹; ¹H NMR (400 MHz, CDCl₃) δ7.56-7.61 (m, 4H), 7.50-7.54 (m, 2H), 7.46 (br, 1H), 7.43 (d, J = 8.2 Hz, 2H), 7.36-7.42 (m, 4H), 2.91 (q, J = 7.6 Hz, 2H), 2.47 (s, 3H), 1.33 (t, J = 7.6 Hz, 3H); ¹³C NMR (100 MHz, CDCl₃) δ193.8, 162.5 (d, J_{C-F} = 252 Hz), 146.9, 134.6, 133.7, 133.6, 132.7, 132.6, 132.2, 131.6, 129.4, 129.2, 128.8, 128.7, 127.6 (d, J_{C-F} = 3 Hz), 125.3, 125.2, 124.7, 123.8, 123.1, 122.4, 118.6 (d, J_{C-F} = 23 Hz), 112.6 (d, J_{C-F} = 16 Hz), 96.6 (d, J_{C-F} = 3 Hz), 93.7 (d, J_{C-F} = 3 Hz), 91.4, 90.9, 90.7, 83.0, 30.7, 28.0, 15.0; HRMS calcd for $C_{34}H_{23}FOS$: 498.1454. Found: 498.1453.

Thioacetic acid S-{4-[4-(3-cyano-4-phenylethynylphenylethynyl)-phenylethynyl]phenyl} ester (14a). 105 mg (58%) of **14a** was produced from **8a** (125 mg, 0.38 mmol) and **11** (111 mg, 0.4 mmol). IR (KBr) 2228, 2213, 1700 cm⁻¹; ¹H NMR (400 MHz, CDCl₃) δ7.82 (d, J = 1.8 Hz, 1H), 7.70 (dd, J = 1.7 Hz, J = 8.2 Hz, 1H), 7.61-7.65 (m, 3H), 7.57 (d, J = 8.4 Hz, 2H), 7.53-7.56 (m, 4H), 7.41 (d, J = 8.4 Hz, 2H), 7.40-7.42 (m, 3H), 0.27 (s, 9H); ¹³C NMR

(100 MHz, CDCl$_3$) δ193.8, 135.8, 135.5, 134.7, 132.6, 132.5, 132.4, 132.2, 132.1, 129.9, 128.8, 127.0, 124.5, 124.1, 123.9, 122.6, 122.2, 117.2, 116.2, 98.4, 93.3, 91.4, 90.9, 89.2, 86.0, 30.7; HRMS calcd for C$_{33}$H$_{19}$NOS: 477.1187. Found: 477.1184.

Thioacetic acid S-{4-[4-(3-cyano-4-phenylethynylphenylethynyl)-3-fluoro-phenylethynyl]phenyl} ester (14b). 62 mg (42%) of **14b** was produced from **8b** (125 mg, 0.3 mmol) and **11** (78 mg, 0.28 mmol). IR (KBr) 2231, 2210, 1695 cm^{-1}; ^{1}H NMR (400 MHz, CDCl$_3$) δ7.87 (d, J = 1.6 Hz, 1H), 7.74 (dd, J = 1.6 Hz, J = 8.2 Hz, 1H), 7.62-7.66 (m, 3H), 7.57 (dd, J = 7.6 Hz, J = 7.6 Hz, 1H), 7.43 (d, J = 8.4 Hz, 2H), 7.41-7.43 (m, 3H), 7.30-7.35 (m, 2H), 2.47 (s, 3H); ^{13}C NMR (100 MHz, CDCl$_3$) δ193.2, 162.2 (d, J$_{C-F}$ = 252 Hz), 135.1, 134.2, 133.3, 133.2, 132.2, 132.1, 132.0, 129.5, 128.9, 128.5, 127.5 (d, J$_{C-F}$ = 3 Hz), 126.9, 125.5 (d, J$_{C-F}$ = 9 Hz), 123.6, 123.1, 121.8, 118.5 (d, J$_{C-F}$ = 22 Hz), 116.7, 115.7, 111.3 (d, J$_{C-F}$ = 16 Hz), 98.2, 93.5 (d, J$_{C-F}$ = 3 Hz), 91.8, 89.3 (d, J$_{C-F}$ = 3 Hz), 86.3, 85.5, 30.3; HRMS calcd for C$_{33}$H$_{18}$FNOS: 495.1093. Found: 495.1095.

Thioacetic acid S-{4-[3-cyano-4-(3-cyano-4-phenylethynylphenylethynyl)-phenylethynyl]phenyl} ester (14c). 49 mg (42%) of **14c** was produced from **8c** (100 mg, 0.24 mmol) and **11** (65 mg, 0.23 mmol). IR (KBr) 2231, 2209, 1706 cm^{-1}; ^{1}H NMR (500 MHz, CDCl$_3$) δ7.90 (d, J = 1.4 Hz, 1H), 7.86 (d, J = 1.4 Hz, 1H), 7.80 (dd, J = 1.6 Hz, J = 8.2 Hz, 1H), 7.74 (dd, J = 1.6 Hz, J = 8.2 Hz, 1H), 7.69 (d, J = 1.4 Hz, 1H), 7.65-7.69 (m, 4H), 7.59 (d, J = 8.4 Hz, 2H), 7.45 (d, J = 8.4 Hz, 2H), 7.40-7.44 (m, 3H), 2.48 (s, 3H); ^{13}C NMR (125 MHz, CDCl$_3$) δ193.4, 136.0, 135.9, 135.8, 135.7, 134.7, 132.7, 132.6, 132.5, 130.0, 129.8, 128.9, 128.1, 125.8, 124.8, 123.5, 122.6, 122.1, 117.0, 116.9, 116.4, 116.3, 99.1, 95.3, 93.6, 89.3, 88.6, 86.0, 30.7; HRMS calcd for C$_{34}$H$_{18}$N$_2$OS: 502.1140. Found: 502.1140.

Thioacetic acid S-{4-[4-(3-cyano-4-phenylethynylphenylethynyl)-3-trifluoromethylphenylethynyl]phenyl} ester (14d). 32 mg (30%) of **14d** was produced from **8d** (90 mg, 0.19 mmol) and **11** (53 mg, 0.19 mmol). IR (KBr) 2230, 2210, 1703 cm^{-1}; ^{1}H NMR (500 MHz, CDCl$_3$) δ7.88 (d, J = 0.6 Hz, 1H), 7.84 (d, J = 1.5 Hz, 1H), 7.72 (dd, J = 1.7 Hz, J = 13.1 Hz, 1H), 7.69 (d, J = 1.4 Hz, 1H), 7.68 (s, 1H) 7.63-7.66 (m, 3H), 7.60 (d, J 8.4 Hz, 2H), 7.45 (d, J = 8.4 Hz, 2H), 7.38-7.44 (m, 3H), 2.48 (s, 3H); ^{13}C NMR (125 MHz, CDCl$_3$) δ193.6, 135.9, 135.6, 134.7, 134.3, 132.7, 132.6, 132.5, 132.4 (q, J$_{C-F}$ = 31 Hz), 130.0, 129.7 (q, J$_{C-F}$ = 5 Hz), 129.5, 128.9, 127.6, 124.3, 123.9, 123.4 (q, J$_{C-F}$ = 274 Hz), 123.3, 122.2, 120.5, 120.4 (q, J$_{C-F}$ = 2 Hz), 98.8, 94.4, 92.8, 89.5, 89.2, 86.0, 30.8; HRMS calcd for C$_{34}$H$_{18}$F$_3$NOS: 545.1061. Found: 545.1060.

Thioacetic acid S-{4-[4-(3-cyano-4-phenylethynylphenylethynyl)-3-ethyl-phenylethynyl]phenyl} ester (14e). 61 mg (43%) of **14e** was produced from **8e** (120 mg, 0.28 mmol) and **11** (75 mg, 0.27 mmol). IR (KBr) 2229, 2212, 1706 cm^{-1}; ^{1}H NMR (400 MHz, CDCl$_3$) δ7.84 (dd, J = 1.6 Hz, J =

0.4 Hz, 1H), 7.69 (dd, J = 1.6 Hz, J = 8.1 Hz, 1H), 7.64-7.68 (m, 3H), 7.60 (d, J = 8.4 Hz, 2H), 7.53 (d, J = 8.0 Hz, 1H), 7.39 (m, 7H), 2.91 (q, J = 7.6 Hz, 2H), 2.49 (s, 3H), 1.37 (t, J = 7.6 Hz, 3H); ^{13}C NMR (100 MHz, CDCl$_3$) δ193.8, 147.0, 135.6, 135.3, 134.6, 132.8, 132.6, 132.5, 132.4, 132.3, 131.7, 129.9, 129.5, 128.9, 128.8, 126.9, 124.6, 124.2, 124.1, 122.3, 121.9, 117.3, 116.2, 98.4, 92.6, 92.3, 91.3, 91.0, 86.0, 30.7, 28.0, 15.0; HRMS calcd for C$_{35}$H$_{23}$NOS: 505.1500. Found: 505.1501.

Thioacetic   acid   S-{4-[4-(4-phenylethynyl-3-trifluoromethylphenylethynyl)-phenylethynyl]phenyl} ester (15a). 60 mg (40%) of **15a** was produced from **9a** (128 mg, 0.29 mmol) and **11** (80 mg, 0.29 mmol). IR (KBr) 2212, 1697 cm^{-1}; ^{1}H NMR (500 MHz, CDCl$_3$) δ7.87 (s, 1H), 7.67 (d, J = 1.1 Hz, 2H), 7.58-7.60 (m, 4H), 7.56 (br, 2H), 7.43 (d, J = 8.4 Hz, 2H), 7.39-7.5422 (m, 3H), 2.47 (s, 3H); ^{13}C NMR (125 MHz, CDCl$_3$) δ193.8, 134.7, 134.5, 134.2, 133.5, 132.6, 132.3 (q, J$_{C-F}$ = 31 Hz), 132.2, 132.1, 132.0, 129.6 (q, J$_{C-F}$ = 5 Hz), 129.5, 128.9, 124.6, 123.8, 123.5 (q, J$_{C-F}$ = 274 Hz), 123.4, 122.9, 121.7 (q, J$_{C-F}$ = 2 Hz), 97.4, 92.6, 91.3, 91.0, 90.2, 85.7, 30.7; HRMS calcd for C$_{33}$H$_{19}$F$_3$OS: 520.1109. Found: 520.1107.

Thioacetic   acid   S-{4-[3-fluoro-4-(4-phenylethynyl-3-trifluoromethyl-phenylethynyl)phenylethynyl]phenyl} ester (15b). 56 mg (36%) of **15b** was produced from **9b** (132 mg, 0.29 mmol) and **11** (77 mg, 0.28 mmol). IR (KBr) 2205, 1699 cm^{-1}; ^{1}H NMR (500 MHz, CDCl$_3$) δ7.89 (s, 1H), 7.69-7.72 (m, 2H), 7.58-7.60 (m, 4H), 7.53 (dd, J = 7.7 Hz, J = 7.7 Hz, 1H), 7.44 (d, J = 8.4 Hz, 2H), 7.40-7.42 (m, 3H), 7.34-7.35 (m, 1H), 7.30-7.33 (m, 1H), 2.47 (s, 3H); ^{13}C NMR (125 MHz, CDCl$_3$) δ193.6, 162.6 (d, J$_{C-F}$ = 252 Hz), 134.7, 134.6, 134.2, 133.7, 132.7, 132.3 (q, J$_{C-F}$ = 31 Hz), 132.2, 129.6 (q, J$_{C-F}$ = 5 Hz), 129.5, 129.3, 128.9, 127.9 (d, J$_{C-F}$ = 3 Hz), 125.7, 125.6, 124.1, 123.5 (q, J$_{C-F}$ = 274 Hz), 122.9, 122.8, 122.0, (q, J$_{C-F}$ = 2 Hz), 119.0 (d, J$_{C-F}$ = 23 Hz), 112.0 (d, J$_{C-F}$ = 16 Hz), 97.6, 94.8 (d, J$_{C-F}$ = 3 Hz), 92.1, 89.9 (d, J$_{C-F}$ = 3 Hz), 85.9, 85.6, 30.7; HRMS calcd for C$_{33}$H$_{18}$F$_4$OS: 538.1014. Found: 538.1013.

Thioacetic   acid   S-{4-[3-cyano-4-(4-phenylethynyl-3-trifluoromethyl-phenylethynyl)phenylethynyl]phenyl} ester (15c). 35 mg (26%) of **15c** was produced from **9c** (116 mg, 0.25 mmol) and **11** (68 mg, 0.24 mmol). IR (KBr) 2229, 2229, 1696 cm^{-1}; ^{1}H NMR (400 MHz, CDCl$_3$) δ7.90 (br s, 1H), 7.84 (d, J = 1.1 Hz, 1H), 7.75 (dd, J = 1.5 Hz, J = 8.1 Hz, 1H), 7.71 (dd, J = 1.6 Hz, J = 8.1 Hz, 1H), 7.68 (d, J = 8.1 Hz, 1H), 7.63 (d, J = 8.2 Hz, 1H), 7.56-7.58 (m, 4H), 7.43 (d, J = 8.4 Hz, 1H), 7.38-7.41 (m, 3H), 2.46 (s, 3H); ^{13}C NMR (125 MHz, CDCl$_3$) δ193.5, 135.9, 135.6, 135.0, 134.7, 134.3, 132.7, 132.3 (q, J$_{C-F}$ = 31 Hz), 132.2, 129.7 (q, J$_{C-F}$ = 5 Hz), 129.6, 128.9, 126.1, 124.6, 123.6, 123.5 (q, J$_{C-F}$ = 274 Hz), 122.8, 122.8, 122.0, 117.0, 116.3, 98.0, 96.3, 93.4, 88.7, 88.6, 85.6, 30.8; HRMS calcd for C$_{34}$H$_{18}$F$_3$NOS: 545.1061. Found: 545.1062.

Thioacetic acid S-{4-[4-(4-phenylethynyl-3-trifluoromethylphenylethynyl)-3-trifluoromethylphenylethynyl]phenyl} ester (15d). 125 mg (56%) of **15d** was produced from **9d** (195 mg, 0.38 mmol) and **11** (106 mg, 0.38 mmol). IR (KBr) 2206, 1699 cm^{-1}; ^{1}H NMR (400 MHz, CDCl$_3$) δ7.88 (d, J = 0.6 Hz, 1H), 7.86 (br s, 1H), 7.68-7.72 (m, 4H), 7.56-7.60 (m, 4H), 7.44 (d, J = 8.4 Hz, 2H), 7.39-7.42 (m, 3H), 2.48 (s, 3H); ^{13}C NMR (100 MHz, CDCl$_3$) δ193.6, 134.7, 134.6, 134.3, 134.2, 132.7, 132.4 (q, J$_{C-F}$ = 31 Hz), 132.3 (q, J$_{C-F}$ = 31 Hz), 132.2, 129.6 (q, J$_{C-F}$ = 5 Hz), 129.5, 129.4 (q, J$_{C-F}$ = 5 Hz), 128.9, 124.0, 123.9, 122.8, 122.7, 123.5 (q, J$_{C-F}$ = 274 Hz), 123.4 (q, J$_{C-F}$ = 274 Hz), 122.4 (q, J$_{C-F}$ = 2 Hz), 120.8 (q, J$_{C-F}$ = 2 Hz), 97.7, 95.3, 92.6, 89.6, 88.5, 85.6, 30.7; HRMS calcd for C$_{34}$H$_{18}$F$_6$OS: 588.0982. Found: 588.0984.

Thioacetic acid S-{4-[3-ethyl-4-(4-phenylethynyl-3-trifluoromethyl-phenylethynyl)phenylethynyl]phenyl} ester (15e). 34 mg (21%) of **15e** was produced from **9e** (142 mg, 0.30 mmol) and **11** (122 mg, 0.3 mmol). IR (KBr) 2219, 1696 cm^{-1}; ^{1}H NMR (500 MHz, CDCl$_3$) δ7.85 (s, 1H), 7.69-7.71 (m, 2H), 7.58-7.61 (m, 4H), 7.53 (d, J = 7.9 Hz, 1H), 7.47 (d, J = 1.1 Hz, 1H), 7.44 (d, J = 8.4 Hz, 1H), 7.39-7.43 (m, 4H), 2.91 (q, J = 7.6 Hz, 2H), 2.47 (s, 3H), 1.36 (t, J = 7.6 Hz, 3H); ^{13}C NMR (125 MHz, CDCl$_3$) δ193.7, 146.9, 134.6, 134.4, 134.2, 132.7, 132.6, 132.4 (q, J$_{C-F}$ = 31 Hz), 132.2, 131.6, 129.4, 129.3 (q, J$_{C-F}$ = 5 Hz), 128.8, 128.7, 124.7, 123.9, 123.6, 123.5 (q, J$_{C-F}$ = 274 Hz), 123.0, 122.3, 121.6 (q, J$_{C-F}$ F = 2 Hz), 97.3, 93.4, 91.4, 91.3, 90.8, 85.7, 30.7, 28.0, 15.0; HRMS calcd for C$_{35}$H$_{23}$F$_3$OS: 548.1422. Found: 548.1423. '

Thioacetic acid S-{4-[4-(3-ethyl-4-phenylethynylphenylethynyl)-phenylethynyl]phenyl} ester (16a). 14 mg (12%) of **16a** was produced from **10a** (98 mg, 0.24 mmol) and **11** (67 mg, 0.24 mmol). IR (KBr) 2203, 1708 cm^{-1}; ^{1}H NMR (400 MHz, CDCl$_3$) δ 7.50-7.60 (m, 9H), 7.43-7.46 (m, 3H), 7.36-7.41 (m, 4H), 2.92 (q, J = 7.6 Hz, 2H), 2.47 (s, 3H), 1.36 (t, J = 7.6 Hz, 3H); ^{13}C NMR (125 MHz, CDCl$_3$) δ193.4, 146.2, 134.2, 132.4, 132.2, 132.1, 132.0, 131.6, 131.58, 131.55, 131.5, 131.1, 128.9, 128.4, 128.3, 127.2, 124.3, 123.3, 122.8, 122.7, 122.6, 94.7, 91.5, 90.7, 90.5, 90.4, 87.8, 30.3, 27.6, 14.5; HRMS calcd for C$_{34}$H$_{24}$OS: 480.1548. Found: 480.1544.

Thioacetic acid S-{4-[4-(3-ethyl-4-phenylethynylphenylethynyl)-3-fluoro-phenylethynyl]phenyl} ester (16b). 88 mg (53%) of **16b** was produced from **10b** (140 mg, 0.33 mmol) and **11** (88 mg, 0.32 mmol). IR (KBr) 2205, 1700 cm^{-1}; ^{1}H NMR (400 MHz, CDCl$_3$) δ7.54-7.60 (m, 4H), 7.50-7.54 (m, 2H), 7.47 (d, J = 1.0 Hz, 1H), 7.44 (d, J = 8.4 Hz, 1H), 7.35-7.40 (m, 4H), 7.29-7.33 (m, 2H), 2.91 (q, J = 7.6 Hz, 2H), 2.47 (s, 3H), 1.36 (t, J = 7.6 Hz, 3H); ^{13}C NMR (125 MHz, CDCl$_3$) δ193.6, 162.5 (d, J$_{C-F}$ = 252 Hz), 146.7, 134.7, 133.7, 133.6, 132.7, 132.5, 131.9, 131.6, 129.4, 129.2, 128.8, 127.8 (d, J$_{C-F}$ = 3 Hz), 125.0, 124.9, 124.2, 123.7, 123.5, 122.9, 119.0 (d, J$_{C-F}$ = 23 Hz), 112.7 (d, J$_{C-F}$ = 16 Hz), 96.8 (d, J$_{C-F}$ = 3 Hz), 95.3, 91.7, 90.0 (d, J$_{C-F}$ = 3 Hz),

88.2, 84.2, 30.7, 28.1, 15.0; HRMS calcd for $C_{34}H_{23}OSF$: 498.1454.   Found: 498.1456.

Thioacetic   acid   S-{4-[3-cyano-4-(3-ethyl-4-phenylethynylphenylethynyl)-phenylethynyl]phenyl} ester (16c).   75 mg (46%) of **16c** was produced from **10c** (138 mg, 0.32 mmol) and **11** (103 mg, 0.37 mmol).   IR (KBr) 2228, 2197, 1696 cm^{-1}; ^{1}H NMR (400 MHz, CDCl$_3$) δ7.85 (d, J = 1.6 Hz, 1H), 7.71 (dd, J = 1.6 Hz, J = 8.2 Hz, 1H), 7.63 (d, J = 9.7 Hz, 1H), 7.52-7.60 (m, 6H), 7.43-7.47 (m, 3H), 2.94 (q, J = 7.6 Hz, 2H), 2.48 (s, 3H), 1.36 (t, J = 7.6 Hz, 3H); ^{13}C NMR (100 MHz, CDCl$_3$) δ193.6, 146.8, 135.9, 135.6, 134.7, 132.7, 132.5, 132.0, 131.9, 129.8, 129.6, 128.9, 128.8, 126.9, 124.2, 123.9, 123.7, 123.6, 122.0, 117.2, 116.1, 98.4, 95.7, 93.0, 88.9, 88.1, 87.2, 30.8, 28.1, 15.0; HRMS calcd for $C_{35}H_{23}NOS$: 505.1500.   Found: 505.1497.

Thioacetic acid S-{4-[4-(3-ethyl-4-phenylethynyl-phenylethynyl)-3-trifluoromethylphenylethynyl]phenyl} ester (16d).   98 mg (51%) of **16d** was produced from **10d** (165 mg, 0.35 mmol) and **11** (80 mg, 0.29 mmol).   IR (KBr) 2192, 1694 cm^{-1}; ^{1}H NMR (500 MHz, CDCl$_3$) δ7.87 (s, 1H), 7.68 (d, J = 1.0 Hz, 1H), 7.60 (d, J = 8.4 Hz, 2H), 7.55-7.59 (m, 2H), 7.52 (d, J = 7.9 Hz, 1H), 7.42-7.46 (m, 3H), 7.37-7.41 (m, 4H), 2.92 (q, J = 7.6 Hz, 2H), 2.48 (s, 3H), 1.36 (t, J = 7.6 Hz, 3H); ^{13}C NMR (125 MHz, CDCl$_3$) δ193.6, 146.8, 134.7, 134.6, 134.2, 132.7, 132.5, 132.2 (q, J_{C-F} = 31 Hz), 131.9, 131.6, 129.6 (q, J_{C-F} = 5 Hz), 129.4, 129.3, 128.9, 128.8, 124.0, 123.7, 123.6, 123.5 (q, J_{C-F} = 274 Hz), 123.3, 122.7, 121.6 (q, J_{C-F} = 2 Hz), 97.4, 95.4, 92.2, 89.9, 88.2, 86.9, 30.7, 28.1, 15.0; HRMS calcd for $C_{35}H_{23}OSF_3$: 548.1422.   Found: 548.1423.

Thioacetic   acid   S-{4-[3-ethyl-4-(3-ethyl-4-phenylethynylphenylethynyl)-phenylethynyl]phenyl} ester (16e).   48 mg (37%) of **16e** was produced from **10e** (110 mg, 0.25 mmol) and **11** (70 mg, 0.25 mmol).   IR (KBr) 2199, 1692 cm^{-1}; ^{1}H NMR (500 MHz, CDCl$_3$) δ7.53-7.58 (m, 4H), 7.49 (dd, J = 9.9 Hz, J = 1.8 Hz, 2H), 7.38-7.44 (m, 4H), 7.34-7.37 (m, 5H), 2.90 (q, J = 7.6 Hz, 4H), 2.44 (s, 3H), 1.34 (t, J = 7.6 Hz, 6H); ^{13}C NMR (125 MHz, CDCl$_3$) δ193.8, 146.8, 146.7, 134.6, 132.6, 132.5, 132.4, 131.9, 131.6, 131.4, 129.4, 129.2, 128.8, 128.6, 124.8, 123.8, 123.6, 123.3, 123.1, 123.0, 95.2, 95.1, 91.5, 90.5, 89.6, 88.3, 30.7, 28.1, 28.0, 15.1, 15.0; HRMS calcd for $C_{36}H_{28}OS$: 508.1861.   Found: 508.1860.

General Procedure for the Diazonium Formation.   To a round-bottom flask fitted with an addition funnel and N_2 inlet was added boron trifluoride etherate (4 equiv) that was then chilled in a dry ice-acetone bath (-20 °C).   To the reaction flask was added dropwise over 5 min a solution of the aniline derivative (1 equiv) in dry THF, followed by a solution of tert-butylnitrite (3.5 equiv) in dry THF over 30 min.   The chilled mixture was stirred an additional 10 min, and the cold bath was allowed to warm to 5 °C over 20 min.   To the mixture was added diethyl ether, and the mixture was chilled in an ice-bath for 15 min.   The solid was collected by filtration, washed with chilled (0-5 °C) diethyl ether, and dried in vacuo to give the desired product.

4-Iodobenzenediazonium Tetrafluoroborate (24a).[110] According to the general procedure, 7.0 g (96%) of **24a** was produced from **1a** (5.0 g, 22.8 mmol).

2-Cyano-4-Iodobenzenediazonium Tetrafluoroborate (24b). According to the general procedure, 6.5 g (77%) of **24b** was produced from **1c** (6.0 g, 25.7 mmol). IR (KBr) 2284, 2244 cm^{-1}; ^{1}H NMR (400 MHz, CD$_3$CN) δ8.76 (d, J = 1.7 Hz, 1H), 8.65 (dd, J = 1.7 Hz, J = 8.8 Hz, 1H), 8.42 (d, J = 8.8 Hz, 1H); ^{13}C NMR (100 MHz, CD$_3$CN) δ146.8, 146.2, 135.2, 116.6, 115.3, 114.7, 111.3; FAB HRMS calcd for (M - BF$_4^-$) C$_7$H$_3$N$_3$I: 255.9372. Found: 255.9379.

2-Trifluoromethyl-4-iodobenzenediazonium Tetrafluoroborate (24c). According to the general procedure, 3.83 g (95%) of **24c** was produced from **1d** (3.0 g, 10.4 mmol). IR (KBr) 2290 cm^{-1}; ^{1}H NMR (400 MHz, CD$_3$CN) δ8.71 (d, J = 1.6 Hz, 1H), 8.66 (dd, J = 1.6 Hz, J = 8.7 Hz, 1H), 8.46 (d, J = 8.7 Hz, 1H); ^{13}C NMR (125 MHz, CD$_3$CN) δ146.3, 141.1 (q, J$_{C-F}$ = 4 Hz), 136.2, 130.5 (q, J$_{C-F}$ = 36 Hz), 120.9 (q, J$_{C-F}$ = 275 Hz), 115.7, 111.8; FAB HRMS calcd for (M - BF$_4^-$) C$_7$H$_3$N$_2$IF$_3$: 298.9293. Found: 298.9292.

Propylaminomethylated Polystyrene (1% cross-linked divinylbenzene copolymer, 7-90 mesh) (23).[28c] A suspension of chloromethyl polystyrene: 1% divinylbenzene copolymer beads (54 g, 1.1 mequiv/g of chlorine, 70-90 mesh) and n-propylamine (350 mL) were degassed and heated at 70 °C with stirring for 4 days in a screw cap tube. The polymer was transferred to a coarse sintered glass filter using CH$_2$Cl$_2$ and washed with CH$_2$Cl$_2$ (500 mL). The resin was thoroughly washed with the following protocol to remove noncovalently-bound material: the resin was stirred slowly with dioxane/2N NaOH (1/1,v/v, 500 mL) at 70 °C for 30 min, and the solvent was removed by aspiration through a coarse sintered glass filter. This was repeated once more with dioxane/2N NaOH (1/1,v/v, 500 mL), twice each with dioxane/H$_2$O (1/1,v/v, 500 mL), DMF (500 mL), CH$_3$OH (500 mL), and finally benzene (500 mL). The resin was then washed with hot CH$_3$OH (800 mL), hot benzene (800 mL), hot CH$_3$OH (800 mL), hot CH$_2$Cl$_2$ (800 mL) and 400 mL of CH$_3$OH and dried in vauco to a constant mass to give 53.97 g.

General Procedure for the Polymer-Supported Triazene Formation. To a chilled (0 °C) suspension of propylaminomethyl polystyrene resin (**23**) and finely ground K$_2$CO$_3$ (2 equiv) in DMF-THF was added the diazonium salt with stirring at 0 °C for 5 min and rt for 1 h. The suspension was transferred to a fritted filter using DMF and washed sequentially with 120 mL of the following solvents: CH$_3$OH, H$_2$O, CH$_3$OH, THF, CH$_3$OH. The loading procedure was repeated twice and the resin was dried in vacuo to a constant mass to give desired functional polymer beads.

1-(4-Iod-phenyl)-3-propyl-3-(benzyl-supported) Triazene (25a).
According to the general procedure, resin 23 (3.00 g), K$_2$CO$_3$ (0.83 g, 6 mmol), and 4-iodo-benzenediazonium tetrafluoroborate (24a) (0.954 g, 3 mmol) in DMF-THF (25 mL–25 mL) gave desired polymer beads 3.72 g.

1-(2-Cyano-4-iodophenyl)-3-propyl-3-(benzyl-supported) Triazene (25b). According to the general procedure, resin 23 (4.00 g), K$_2$CO$_3$ (1.1g, 8 mmol), and 2-cyano-4-iodo-benzenediazonium tetrafluoroborate (24b) (1.37 g, 4 mmol) in DMF-THF (25 mL–25 mL) gave desired polymer beads 4.886 g.

1-(2-Trifluoromethyl-4-iodophenyl)-3-propyl-3-(benzyl-supported) Triazene (25c). According to the general procedure, resin 23 (3.00 g), K$_2$CO$_3$ (0.83 g, 6 mmol), and 2-trifluoromethyl-4-iodo-benzenediazonium tetrafluoroborate (24c)(1.16 g, 3 mmol) in DMF-THF (25 mL–25 mL) gave desired polymer beads 3.60 g.

General Procedure for the Cleavage of Polymer-Supported Triazene using CH$_3$I. A thick-walled oven-dried screw cap tube was charged with a suspension of the polymer-supported triazene and CH$_3$I. The tube was flushed with N$_2$, capped, and heated to 120 °C for 24 h without stirring. The reaction mixture was cooled and passed through a fritted filter before the resin was introduced to hot CH$_2$Cl$_2$ (3x) to extract any residual product trapped in the polymer matrix. The product was purified by chromatography with hexane or a hexane:EtOAc mixture to give the product.

1,4-Diiodobenzene (26a). According to the general procedure, 1-(4-iodophenyl)-3-propyl-3-(benzyl-supported) triazene (25a) (300 mg) in CH$_3$I (4.5 mL) gave the product (62 mg, 70 % yield over 3 steps from Merrifield's resin 1.1 mequiv/g).

2-Cyano-1,4-diiodobenzene (26b). According to the general procedure, 1-(2-cyano-4-iodophenyl)-3-propyl-3-(benzyl-supported) triazene (25b) (300 mg) in CH$_3$I-CH$_3$CN (2.5 mL-2.5 mL) at 120 °C for 18 h gave the product (33 mg, 35 % yield over 3 steps from Merrifield's resin 1.1 mequiv/g). IR (KBr) 2230 cm^{-1}; ^{1}H NMR (400 MHz, CDCl$_3$) δ7.89 (d, J = 2.0 Hz, 1H), 7.62 (d, J = 8.4 Hz, 1H), 7.57 (dd, J = 2.0 Hz, J = 8.4 Hz, 1H), ^{13}C NMR (100 MHz, CDCl$_3$) δ143.2, 142.8, 141.2, 123.0, 118.2, 98.0, 93.1; HRMS calcd for C$_7$H$_3$NI$_2$: 354.8355. Found: 354.8352.

2-Trifluoromethyl-1,4-diiodobenzene (26c). According to the general procedure, 1-(2-trifluoromethyl-4-iodophenyl)-3-propyl-3-(benzyl-supported) triazene (25c) (300 mg) in CH$_3$I-CH$_3$CN (2.5 mL-2.5 mL) at 110 °C for 24 h gave the product (37 mg, 36 % yield over 3 steps from Merrifield's resin 1.1 mequiv/g). ^{1}H NMR (400 MHz, CDCl$_3$) δ7.95 (d, J = 2.0 Hz, 1H), 7.74 (d, J = 8.2 Hz, 1H), 7.53 (dd, J = 2.0 Hz, J = 8.2 Hz, 1H); ^{13}C NMR (125 MHz, CDCl$_3$) δ143.8, 142.3, 136.8 (q, J$_{C-F}$ = 6 Hz), 135.8 (q, J$_{C-F}$ = 31 Hz), 122.0 (q, J$_{C-F}$ = 274 Hz), 93.4, 90.7; HRMS calcd for C$_7$H$_3$F$_3$I$_2$: 397.8276. Found: 397.8272.

4-(Trimethylsilyl)ethynylbenzenediazonium tetrafluoroborate (27).
To a cooled solution of NOBF$_4$ (5.75 g, 49 mmol, 1.05 equiv) in 50 mL CH$_3$CN at –40 °C was added dropwise over 15 min **2a** (8.86 g, 46.8 mmol) in 60 mL CH$_3$CN, and the mixture was stirred at –40 °C for another 50 min. Ether was added to this mixture to effect precipitation of the product. The precipitate was collected by filtration, washed with chilled ether, and dried in vacuo to give the product (11.5 g, 85%). IR (KBr) 2290 cm^{-1}; ^{1}H NMR (400 MHz, CD$_3$CN) δ8.46 (d, J = 9.2 Hz, 2H), 7.90 (d, J = 9.2 Hz, 2H), 0.32 (s, 9H); ^{13}C NMR (100 MHz, CD$_3$CN) δ137.3, 135.4, 133.8, 114.2, 109.6, 102.6, -0.5; FAB HRMS calcd for (M - BF$_4^-$) C$_{11}$H$_{13}$N$_2$SiI: 201.0848. Found: 201.0850.

1-(4-(2-(Trimethylsilyl)ethynyl)phenyl)-3-propyl-3-(benzyl-supported) Triazene. (28). According to the general procedure, 4-(trimethylsilyl)ethynyl-benzenediazonium tetrafluoroborate **27** (5.65 g, 19.6 mmol), K$_2$CO$_3$ (5.4 g, 39.1 mmol), and resin **23** (19.6 g) in DMF-THF (80 mL-80 mL) were allowed to react to give desired polymer beads 23.96 g. The loading procedure was repeated twice (2.8 g and 1.6 g of **27** were used, respectively) and dried in vacuo to a constant mass to give desired polymer beads 23.96 g. IR 2153 cm^{-1}.

1-(4-Ethynyl)phenyl)-3-propyl-3-(benzyl-supported) Triazene (29).
To a suspension of polymer-supported aryl(trimethylsilyl)alkyne **28** (23.96 g) and THF (9 mL/g of polymer) in an Erlenmeyer flask was added a solution of TBAF (50 mL, 50 mmol, 1.0 M in THF). The suspension was swirled periodically for 25 min. The polymer was then transferred to a preweighed fritted filter using THF, washed sequentially (ca. 30 mL/g polymer) with THF followed by CH$_3$OH, and dried to constant mass in a vacuum oven at 60 °C for 36 h to give desired polymer beads 22.11 g. IR 3317, 2106 cm^{-1}.

Macrokans Loading Resin **29** was distributed into 50 Macrokans (300 mg each). The above Macrokans were split into 5 groups each containing 10 Macrokans. Each group was subjected to coupling with the corresponding aryl halides (**3a-e**) (R$_1$, Scheme 3.5.2.2.1) under the following procedure.

General Procedure for Cross-Coupling of Resin-Bound Terminal Alkyne with Aryl Halide Monomers. To a heavy-walled flask equipped with a N$_2$ inlet was added 10 Macrokans. The flask was evacuated and back-filled with dry N$_2$ 3x. In a separate flask, a catalyst solution consisting of tris(dibenzylideneacetone)dipalladium(0) (138 mg, 0.24 mmol), CuI (46 mg, 0.24 mmol), and (C$_6$H$_5$)$_3$P (252 mg, 0.96 mmol) in dry Et$_3$N (50 mL) was degassed and stirred at 70 °C for 2 h. The supernatant from this catalyst solution was transferred via cannula to the reaction flask containing the Macrokans. To the mixture was added aryl halides (**3a-e**) (15 mmol). The flask was sealed and kept at 65 °C for 44 h with stirring. The Macrokans were collected and washed sequentially with 100 mL of the following solvents: CH$_2$Cl$_2$, DMF, 0.05 M solution of sodium diethyl dithiocarbamate in 99/1 DMF/diisopropylethylamine, DMF, CH$_2$Cl$_2$, CH$_3$OH, and dried in vacuo.

General Procedures for the Desilylation of Polymer-Supported Dimers. To a suspension of 10 Macrokans containing polymer-supported aryl(trimethylsilyl)alkyne **30** and THF (50 mL) in an Erlenmeyer flask was added a solution of TBAF (20 mL, 20 mmol, 1.0 M in THF). The suspension was stirred for 30 min. The Macrokans were collected and washed sequentially (100 mL) with THF followed by CH$_3$OH, and dried in vacuo.

Polymer-Supported Trimer Formation. The above Macrokans were re-split into 5 new groups by combining two Macrokans from each of the previous groups. Each of the 5 new groups was subjected to coupling with the previous 5 aryl halides (**3a-e**) by following previous procedure.

Traceless Cleavage of Resin. To a suspension of polymer-supported trimer in 4 mL of 10% conc. HCl/THF solution was treated with ultrasound in a sonicator bath at 50 °C for 10 min. The resulting slurry was filtered and the resin residue was washed with EtOAc and THF. The organic phase was then washed with H$_2$O (2x), brine, and dried over Na$_2$SO$_4$. The crude product was purified by column chromatography (silica gel) with hexanes or a hexane:EtOAc mixture to give desired product.

Trimethyl-(4-phenylethynylphenylethynyl)silane (4a).[5b] 30 mg (37 % yield over five steps from Merrifield's resin) of product was obtained. The yield was calculated according to the loading of original resin.

(3-Ethyl-4-phenylethynylphenylethynyl)trimethylsilane (4e). 36 mg (41 % yield over five steps from Merrifield's resin) of product was obtained. The yield was calculated according to the loading of original resin.

Chapter 4

Molecular Self-Assembly, Device Construction, and Testing

4.1 Self-Assembly and Molecular Ordering

Scores of sophisticated looking molecules are now accessible to researchers wherein function is built into the molecule, 10^{23} units at a time. The question then becomes how to achieve the orienting of 10^{14} molecules in a pre-defined addressable array on a 1-cm^2 chip? Some headway has been made.[5] The atomic order of a surface coupled with molecular packing requirements can give rise to thermodynamically driven self-assembly over large surfaces. The gold-thiolate (R-S-Au) system is the most commonly studied self-assembled monolayer (SAM).[31] The assembly takes place within seconds or minutes, depending on the concentration and molecular structure. Sometimes 24-hour periods are used for the assembly to permit dense packing of otherwise slow-to-assemble molecules. Crystalline ordering of the system can occur in domains ranging to hundreds of square nanometers. The gold-thiolate bond has a strength of ~1.8 eV or ~45 kcal/mole, hence it is quite robust relative to typical metal-molecule bonds that can be formed at ambient temperature. A SAM also slows the inherent surface reconstruction, a phenomenon that often causes rapid randomization of surface atoms. Notice in Figure 1 the protocol. Molecules are constructed that bear some functional component, such as the nitroaromatic as a memory storage unit. These are randomized in solution, but after dipping in a gold substrate, for example, in a matter of a few seconds to minutes, the molecules arrange on the surface as soldiers standing at attention. The aromatic thiolates have a tilt angle of approximately 20° relative to the surface normal while alkanethiolate are tilted approximately 30° to the normal. Few molecules stand perpendicular to the surface and the tilt is a function of the hybridization state of the sulfur and the intermolecular packing requirements.

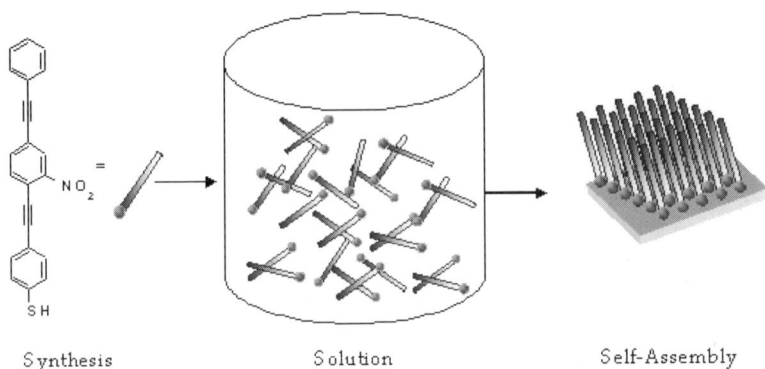

Synthesis   Solution   Self-Assembly

Figure 4.1   The self-assembly process wherein molecules that have programmed functionality built into their structure then self-order on a surface upon dipping the surface into a solution of the molecules. Once the surface is removed and rinsed, it will bear a monolayer of molecules that possess long-range order.

There are numerous other surfaces that have been employed in SAM construction including other conductors such as copper, silver and palladium; semiconductors such as silicon gallium arsenide and cadmium sulfide; and insulators such as silicon oxide. Other alligator clips which have been studied include selenols (-SeH), phosphines (R_3P), and carboxylates (-CO_2^-).[31] We recently developed a method for direct assembly of aryl groups on silicon (n-doped, p-doped, undoped, single crystal or polycrystalline) and gallium arsenide using aryl diazonium salts. There is spontaneous ejection of N_2 and direct carbon-silicon or carbon-As formation.[87,111] But this is more of a grafting, or kinetic assembly, rather than a thermodynamic assembly since the C-Si or C-As bonds are so strong that facile equilibration is unlikely. This grafting method holds great promise since there is no need to make metal lines atop the silicon prior to molecular assembly. And this procedure is quite different from the work of others using alcohol or thiol connections to hydrogen passivated surfaces.[112] Those procedures undoubtedly permit surface silicon oxides or sulfides to form that insulate the molecule from the silicon surface.

One can also construct molecules having a preference for a given surface by judicious choice of the alligator clip. This surface-selective SAM formation could be a critical factor in determining the usefulness of SAMs for device placement, as heterogeneity in device patterns must be attained. In other words, an array of all AND logic gates would be useless. The heterogeneity is best if it is programmable in its design pattern. Using crystalline substrates that are composed of different atom types and yet are regular in their periodicity could provide a method for the predictable arrangement of surface molecules, but programmability to the array architecture would then be limited to one's ability to form tailored substrate crystals.

We recently developed a method to use a voltage promoted assembly of molecules on contact pads using an electrochemical technique. By applying a voltage to a specific contact pad, using it as the working electrode in a solution of the thiolacetate, thiolate assemble on the voltage-regulated pad occurs within seconds.[113] Hence selective assembly of different molecule types can ensue on arrays of contact pads using a voltage-promoted assembly.

Conversely, one could consider a random SAM formation of different molecule types, and hence logic types, followed by testing of the chip properties once the SAM formation is complete. If this random approach were used, it would mean that none of the chips would have a known function a priori; however, the technique could be efficacious since some logic applications could be programmed into the chip post-fabrication. This will be discussed later in the context of the NanoCell approach.

Therefore, self-assembly, or the process of permitting the molecules to attach in place, is quite attractive for molecular ordering which can also be done in three-dimensional arrays. However, in three dimensions, heat dissipation and addressing issues become extreme challenges. Despite the problems associated with using self-assembly, it promises to be far superior to tedious single-molecule manipulations.

Since any potential molecular wire must bridge two electrodes, the question arises as to how these rigid rod difunctional oligomers will order themselves on metallic surfaces. For example, will the oligomers bridge the gold-gold gap as in Figure 4.2a, or will they reside nearly parallel to the surface of the gold by either dithiol or aromatic adsorption to the gold surfaces as in Figures 4.2b and 4.2c, respectively? By making SAMs on gold surfaces, we have demonstrated using ellipsometry, XPS, and grazing angle IR measurements, that the rigid rod systems stand nearly perpendicular to the surface; the thiol groups dominating the adsorption sites on the gold. Even when the oligomers were α,ω-dithiol-substituted, the rigid molecules tended to stand on end as judged by the ellipsometric thicknesses of the adsorbate layers. This trend holds true for molecules up to ~5 nm in length; however, beyond that length, it became difficult to obtain reasonably packed monolayers.[23a]

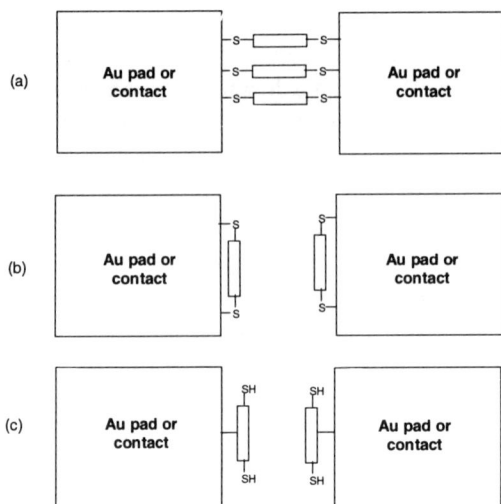

Figure 4.2 Modes for molecular wire assembly (a) as desired between proximal gold probes, (b) adsorbed via the thiols without bridging the probes, or (c) having the adsorption sites dominated by the aromatic units.

Note that we used thioacetate (sometimes referred to as thiolacetate) end groups since these could be selectively deprotected in solution, to afford the free thiol, using NH_4OH or acid during the deposition process.[23a,113] Alkali metal salts can be avoided since they tend to disrupt electronics measurements. Use of the free thiols, rather than the thioacetates, proved to be somewhat problematic since they were prone to very rapid oxidative disulfide formation. But if a quality inert atmosphere box is used, use of the free thiol can be effective; a N_2-flush bag is not adequate to prevent the aromatic disulfide formation. Furthermore, the disulfides formed can self-assemble on gold, but the assembly is approximately 1000 times slower than with the thiols. When using the α,ω-dithiols, oxidative polymerization ensues which rapidly results in insoluble material. Hence, in situ removal of the acetates has proven to be quite effective although not essential if strict exclusion of air is maintained.

4.2 Probe Addressing of Molecules

Our initial efforts, a decade ago, were directed toward straddling groups of longer molecular wires across lithographically patterned proximal gold-coated probes separated by approximately 10 nm (Figure 4.2a). Unfortunately, we were never able to record current across these gap regions, although we conducted many experiments aimed at doing this over a period of two years. The reasons for the failure could have been the inability to bridge these patterned gaps with molecular systems of that length (Figures 4.2b and

4.2c), or that the molecules did assemble across the gaps, but were too electrically resistive over those length scales. Efforts to make smaller, lithographically patterned gaps, were hampered by technical patterning limitations and by our inability to perform scanning probe microscopy within the crevice formed by the proximal probe gap. Hence, definitive gap separation determinations were not achievable at the time that we were conducting the experiments in the early 1990s. Since that time, there are some impressive results where single or small groups of molecules have been assembled between proximal probes including those that have gates constructed below them to modulate the current through the molecules to make them transistor-like.[114]

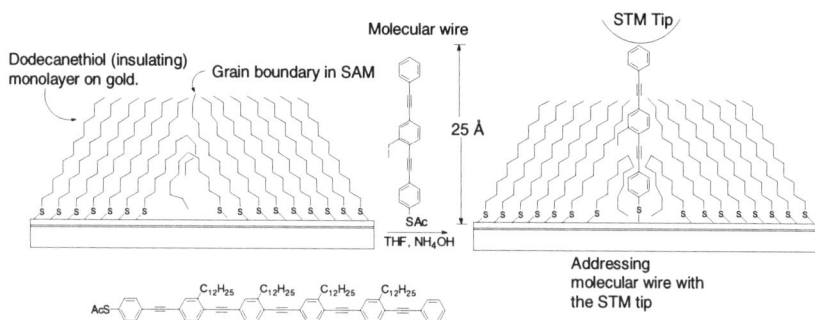

Figure 4.3 Protocol for inserting molecular wires into dodecanethiolate SAMs at grain boundaries and step edges. Relative conductance recording was done with a STM tip. The molecule at the bottom has also been used in this study.

In collaboration with Paul Weiss and David Allara, we have been able to address single molecular wires (oligophenylene ethynylenes = OPEs) that had been inserted at grain boundaries within a self-assembled monolayer of dodecanethiolate on gold (Figure 4.3). This technique has permitted the isolation of single molecular wires from their neighbors, and it has also allowed the addressing of the vertically arranged systems. Using scanning tunneling microscopy (STM), the molecules could then be individually imaged and addressed. Qualitative results of the conductance levels showed that the molecular wires, although topographically higher above the gold surface, were more highly conducting than the surrounding alkanethiolate structures. Such a result is intuitive; however, it had never before been demonstrated in a single conjugated molecule, projecting on end, that was isolated from all its neighbors.[24,59,115]

Further enjoying the collaboration with the Weiss and Allara groups in scanning probe microscopic (SPM) analyses, the next experiments involved the observance of discrete switching states within molecular devices. We have tracked over time the conductance of single and bundled molecular switches isolated in matrices of alkanethiolate monolayers. The persistence times for isolated and bundled molecules in either the "on" or "off" switch state ranged from seconds to tens of hours. The order in the surrounding matrix played a critical role in determining the rate at which the inserted molecules switch; the tighter the packing in the SAM, the slower the switching. We concluded that the switching was a result of conformational changes in the molecules or bundles. See Figure 4.4 where a single molecular device could be resolved and switched.[116]

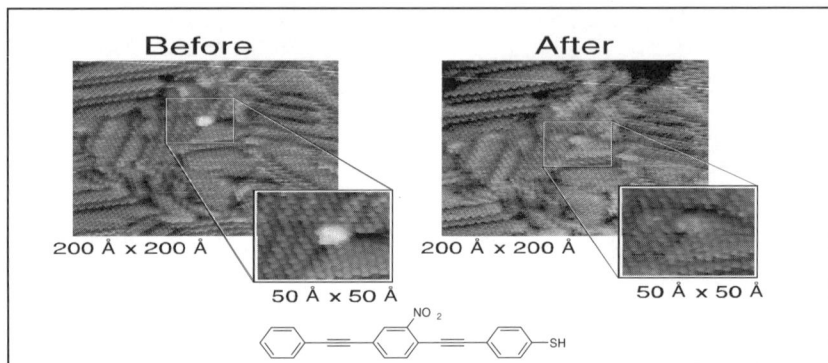

Figure 4.4 A single molecule (nitro OPE structure at bottom) in an alkanethiolate matrix SAM on gold. The dark spheres are the terminal methyl groups of the alkanethiolates. "Before" shows the OPE in an "on" state, and after a voltage pulse from the tip, the OPE is in an "off" state shown under "After". Note that the OPE appears to have too large a diameter; however, when an asperity is sharper than the STM tip, one obtains an image of the tip rather than an image of the asperity.

In collaboration with Al Bard, the interfacial electron transfer processes were probed, particularly those involving unoccupied states, of SAMs of thiolates or arylates on gold by using shear force-based SPM combined with current-voltage and current-distance measurements.[117] The current-voltage curves of hexadecanethiolate on gold in the low bias regime were symmetric around 0 V and the current increased exponentially with voltage at high bias voltage. Different than hexadecanethiolate, reversible peak-shaped current-voltage characteristics were obtained for most of the nitro-based OPE SAMs studied, indicating that part of the conduction mechanism of these junctions involved resonance tunneling. These reversible peaked I(V) curves, often described as a negative differential resistance (NDR) effect of the junction, can be used to define a threshold tip bias, V_{TH}, for resonant conduction. NDR is an inflection in the I(V) curve. We also found that for all of the SAMs studied by this technique, the current decreased with increasing distance, d, between tip and substrate. The attenuation factor β of hexadecanethiolate on gold was high, ranging from 1.3 to 1.4 Å^{-1} and was nearly independent of the tip bias. The β-values for nitro-based molecules were low and depended strongly on the tip bias. Both the V_{TH} and β values of these nitro-based SAMs were also strongly dependent on the structures of the molecules, e.g., the number of the electroactive substituent groups on the central benzene ring of the molecular wire backbone, the anchoring linkage, and the head group. We also observed charge storage on nitro-based OPEs. For one of the SAMs, 25% of charge collected in the negative scan is stored in the molecules and can be collected at positive voltages, thereby suggesting a possible memory effect. A possible mechanism involving lateral electron hopping was proposed to explain this phenomenon that correlates well with findings of the Allara group where they doped SAMs with potassium metal.[118]

A key to the success of Bard's studies was the use of a SPM wherein the probe tip was mounted on a tuning fork, not on a cantilever that can be pulled into the surface, thereby limiting the removal of molecules from the surface to the tip (Figure 4.5).[117] Molecule migration from the SAM to the probe tip is a common occurrence that can exacerbate the problems in data analysis.

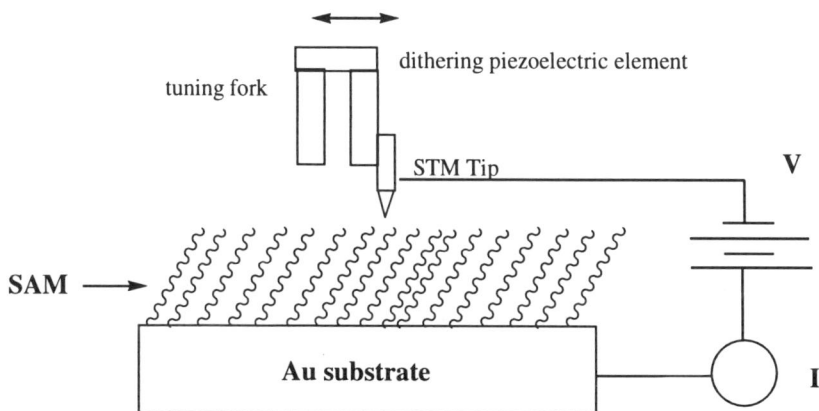

Figure 4.5 Schematic representation of the measurement and formation of the metal-molecule-metal junction with Bard's tuning fork SPM tip contacting the SAM on gold (not to scale).

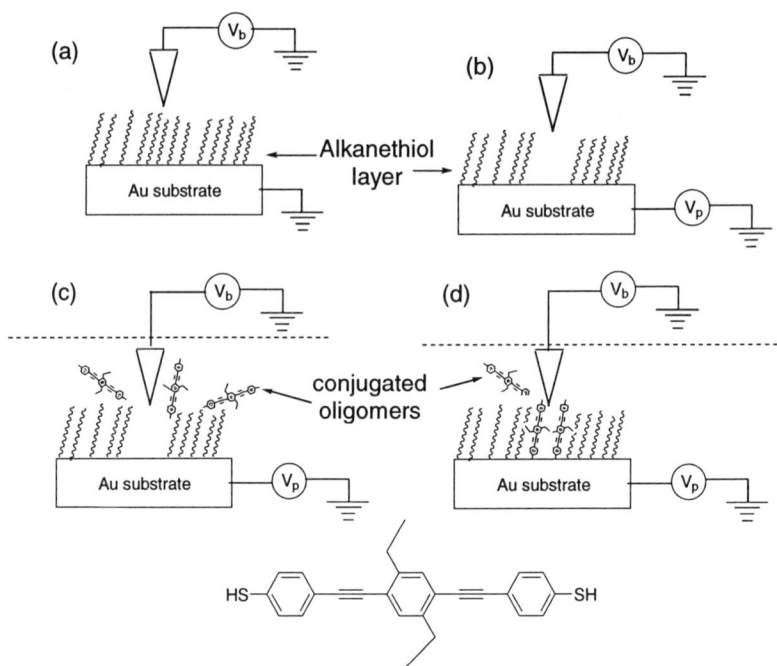

Figure 4.6 Schematic representation of the lithographic patterning and replacement of conjugated molecules in an alkanethiolate matrix. (a) Normal STM imaging of an alkanethiolate SAM with tip bias V_b. (b) SAM removal by applying a voltage pulse V_p to the substrate. (c) Carrying out the same voltage pulse as in (b), but under a solution of molecular wires (expanded structure at bottom) causes (d) insertion of the wires into the newly vacated site.

     With Reed, we further developed methods to insert the molecular wires at controlled rather than at random locations. By applying controlled voltage pulses to an alkanethiolate SAM under a solution of molecular wires (Figure 4.6), we were able to achieve the precise placement of molecular wire bundles (< 10 molecules/bundle) at programmed positions (Figure 4.7).[119]

     In order to quantify the degree of current that could be passed through a single molecule, we again worked with Reed's group who had develop a mechanically controllable break junction method for addressing single molecules. Using this device, two gold tips could be generated and moved in picometer increments with respect to each other by use of an underlying piezo element (Figure 4.8). Conductance quantization (changes in the conductance in discrete steps of $2e^2/h$) observed in the probe system ensured that the movements of the tips were controllable in subatomic-length increments. Benzene-1,4-dithiol (generated in situ from the dithioacetate) was permitted to self-assemble on the two tips that were, initially, widely separated. The two tips were then moved together, in picometer steps during the close contact point, until one molecule bridged the gap (Figure 4.9).[120]

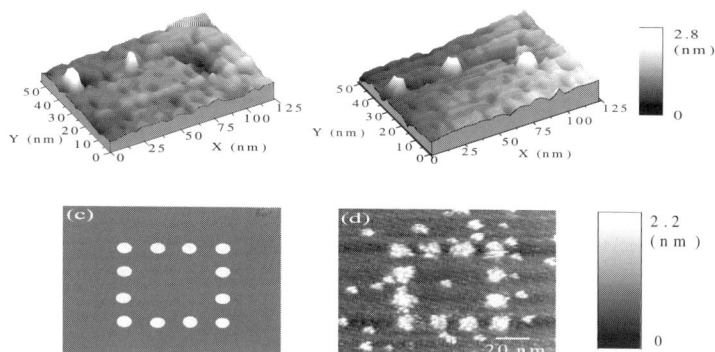

Figure 4.7 (a) A dodecanethiolate SAM surface after three consecutive voltage pulsing events. The first two pulsed locations have molecular wires inserted while the third location remains to be filled. (b) The image taken a few minutes later shows that wire insertion at the third pulse location is now complete. (c) A programmed rectangular pattern for controlled voltage pulses. (d) The image of the patterned SAM after pulsing and molecular wire insertion. Some random insertions at grain boundaries or other defect sites are also evident.

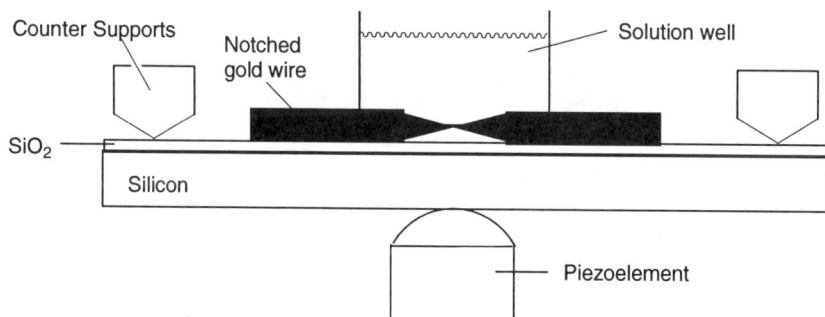

Figure 4.8 A schematic of the mechanically controllable break junction showing the bending beam formed from a silicon wafer, the counter supports, the notched gold wire which is glued to the surface, the piezo element for controlling the tip-to-tip distance through bending of the silicon platform, and the glass capillary tube containing the solution of molecular wires.

Current/voltage responses were recorded for a single molecule bridging the gap. Remarkably, 0.1 microamps current could be recorded through a single molecule. However, few or none of those 10^{12} electrons per second were colliding with the nuclei of the molecule, hence all the heat was dissipated in the contact. Note that the mean free path of an electron in metal is hundreds of angstroms; hence, it is not surprising that collisions did not take place within the small molecule. Most importantly, since most computing instruments operate on microamps of current, the viability of molecular electronics became all the more tangible.

4.3 Switching and Memory in Molecular Bundles

In order to make reliable and measurements on groups of molecules as might ultimately be used in initial device structures, a testbed method was developed by Reed which he call the nanopore. The nanopore system consists of a small (30-50 nm diameter) surface of evaporated metal (which can vary, but most often Au or Pd) on which a SAM of the molecular wires or devices is permitted to form. An upper metal (usually Au or Ti) contact is then evaporated onto the top of the SAM layer making a sandwich of metal-SAM-metal through which I(V) measurements are recorded (Figure 4.10).[121] Using such a small area for the SAM (~1000 molecules), we can probably form SAMs that are defect-free since the entire areas are smaller than the typical defect density of a SAM. This would eliminate electrical shorts that can occur if one had evaporated metal atop a SAM that is larger, for example, micron-sized.

Conductivity of these OPE molecular scale wires and devices is hypothesized to arise from transfer of electrons through the π-orbital backbone that extends over the entire molecule. When the phenyl rings of the OPEs are planar, the π-orbital overlap within the molecules is continuous. Thus transfer over the entire molecule is achieved, electrons can freely flow between the two metal contacts, and conductivity is maximized. But if the phenyl rings become perpendicular with respect to each other, the π-orbitals between the phenylene rings likewise become perpendicular. The discontinuity of the π-orbital network in the perpendicular arrangement minimizes electron transport through the molecular system, thus conductivity is greatly decreased. [122]

In collaboration with Seminario, we used a combination of density functional theory (DFT) and molecular dynamics to simulate five tolane molecules in Figure 4.11a, a model of the nanopore arrangement.[56,122] Notice that there is restricted molecular rotation about the long axis of the molecules at 10 K (Figure 4.11c) which are frozen in the orthogonal positions. At these lower temperatures, phenylene ethynylenes have the tendency to fishbone pack on crystallization in the SAM. [56,123] The aryl rings are therefore in perpendicular arrangements with respect to each other along each molecule, causing a decrease in the π-orbital overlap. This results in the sudden decline in current at lower temperatures since crystallization in the SAM restricts conformational rotation. However, at 30 K, Figure 4.11b, there is free rotation, hence some conformation where there is electron transport through the molecule.

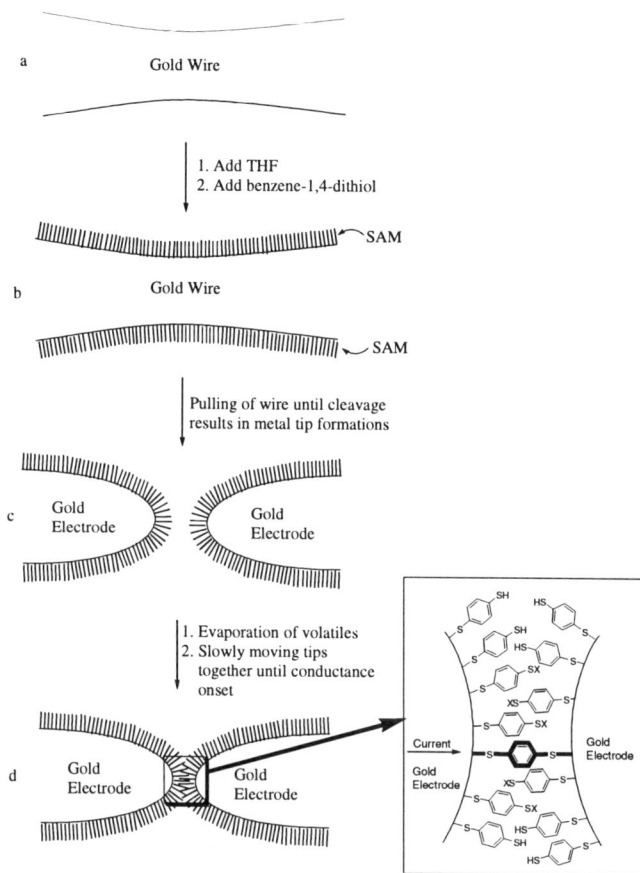

Figure 4.9 A representation of the technique used in the mechanically controllable break junction for recording the current through a single molecule. (a) The Au wire was coated with a SAM of the molecular wires (b) and then broken, under solution (c), via extension of the piezo element under the silicon surface (see Figure 4.8). Evaporation of the volatile components and slow movement of the piezo downward (see Figure 4.8) permits one molecule to bridge the gap (d) that is shown, in expanded view, in the insert. The insert shows a benzene-1,4-dithiolate molecule between proximal Au electrodes. The thiolate is normally H-terminated after deposition; end groups denoted as X can be H or Au, the Au potentially arising from a previous contact/retraction event.

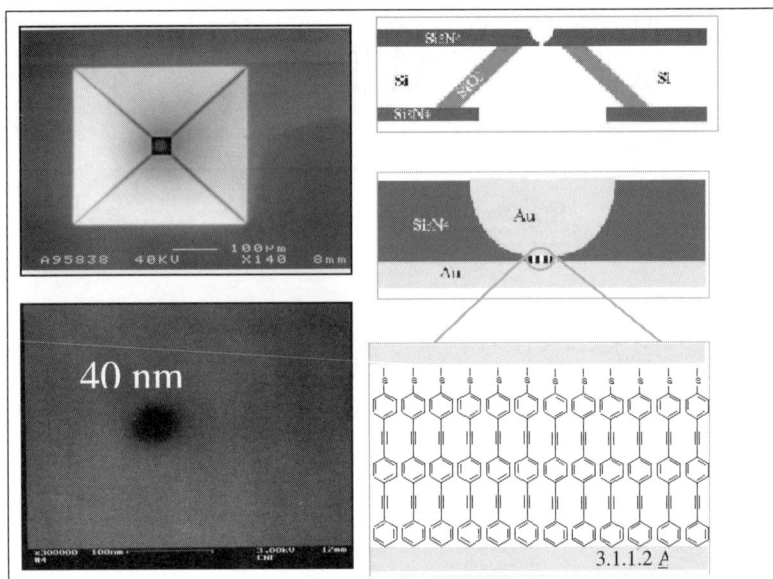

Figure 4.10 (upper right) The starting substrate for the device fabrication is a 250 μm-thick double side polished Si (100) wafer, upon which 50 nm of low stress Si_3N_4 was deposited by low-pressure chemical vapor deposition (LPCVD). On the back surface, the nitride was removed in a (400 μm by 400 μm) square by optical lithography and reactive ion etching (RIE). The exposed Si was etched in an orientation-dependent anisotropic etchant (85°C, 35% KOH solution) through to the top surface to leave a suspended (40 μm by 40 μm) silicon nitride membrane. Reed then grew 100 nm of SiO_2 thermally on the Si sidewalls to improve electrical insulation. A single hole 30 to 50 nm in diameter was made through the membrane by electron beam lithography and RIE. Because of the constrained geometry, the RIE rates are substantially reduced so that the far side opening is much smaller than the actual pattern, thereby rendering the cross section bowl-shaped geometry. (middle and lower right) A Au contact of 200 nm thickness was evaporated onto the topside of the membrane, which filled the pore with Au. The sample was then immediately transferred into a solution to self-assemble the active electronic component, illustrated here with an unfunctionalized sulfur-tipped OPE. The sample was then rinsed, quickly loaded into a vacuum chamber, and mounted onto a liquid N_2 cooling stage for the bottom Au electrode evaporation, where 200 nm Au was evaporated at 77 K at a rate of less than 1 Å/s. The devices were then diced into individual chips, bonded onto packaging sockets and loaded into a variable temperature cryostat and measured with a Semiconductor Parameter Analyzer. (upper left) Scanning electron micrograph (SEM) of the top-side of the nanopore. (lower left) SEM of the bottom-side of the nanopore showing the 40 nm diameter hole in which the SAM formation occurs.

There is experimental evidence for this result as well. As seen in Figure 4.12, two different OPE SAMs in the nanopore show a sharp decrease in conductance between 20 and 40 K in the temperature-current plots.[121,124] As the SAM is permitted to warm above 40 K, the system has enough energy to permit conformational rotation in the molecules. Presumably rotational movement permits the aryl subunits to attain some conformations with near planarity, and conduction thus occurs.

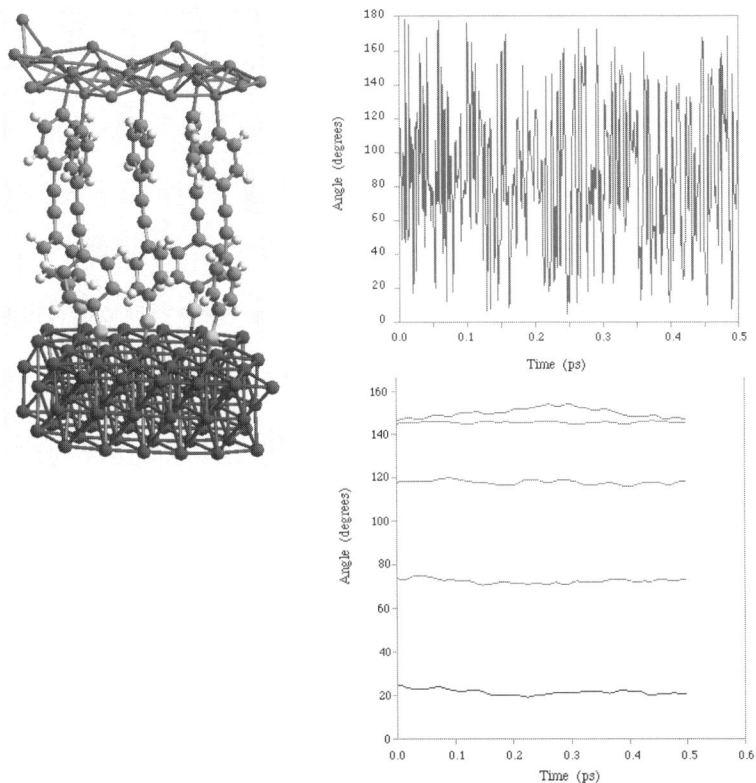

Figure 4.11 (a, top left) A combination of DFT and molecular dynamics was used to simulate five tolanethiols in a SAM with a self-assembled contact to gold on the bottom and a contact to evaporated titanium on top. (b, top right) A plot of the angle in degrees (0-180°) vs. the time (picoseconds) for the adjacent ring conformational changes in one of the five tolanes at 30 K, therefore it is undergoing free rotation about aryl rings. (c, bottom right) All five tolanes are then frozen at 10 K with the adjacent phenylene rings nearly orthogonal relative to each other, i.e. they do not rotate.

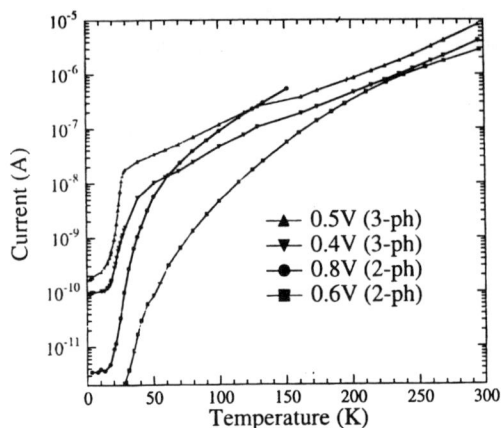

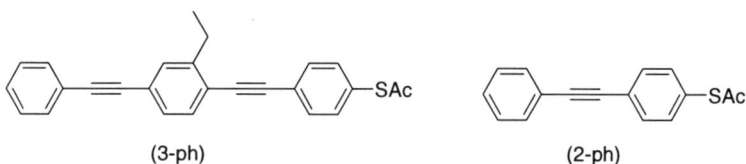

Figure 4.12 Plot of current versus temperature of two different OPEs in nanopores, both with a bottom contact (SAM-contact side) of Au and a top (evaporated) contact of Ti. Each nanopore contains ~1000 molecules.

Since modulation of temperature is an inefficient and impractical way to modulate a structure's conformation and hence conductance, we sought another structural element that would permit altering the degree of a molecule's π-orbital overlap through the use of a third electrode (gate). Thus molecules that have net dipoles that are orthogonal (or simply out of plane from the long molecular axis), could be controlled by use of a third electrode in the nanopore to modulate the conformation, and hence the current through the system. However, since nanopore devices with an electrode perpendicular to the SAM axis had not yet been fabricated, we simply began with the control experiments. Namely, to study the two-electrode nanopore made with molecules bearing dipolar groups.

Accordingly, the nitroaniline OPE was tested in the nanopore, in the absence of an orthogonal external electric field, to determine its electronic characteristics. A series of control experiments were first performed with alkanethiol-derived SAMs and systems containing no molecules. Both the Au–alkanethiolate–Au junctions and the Au-silicon nitride membrane–Au junctions showed current levels at the noise limit of the apparatus (< 1 pA) for both bias polarities at both room and low temperatures. The Au-Au junctions gave ohmic I(V) characteristics with very low resistances. A device containing a SAM of the conjugated OPEs but not bearing the nitroaniline functionalities, namely 3-ph in Figure 4.12, was fabricated and measured in nearly identical conditions[124] and it exhibited essentially linear I(V) behavior within its non-crystalline temperature range (vide supra).

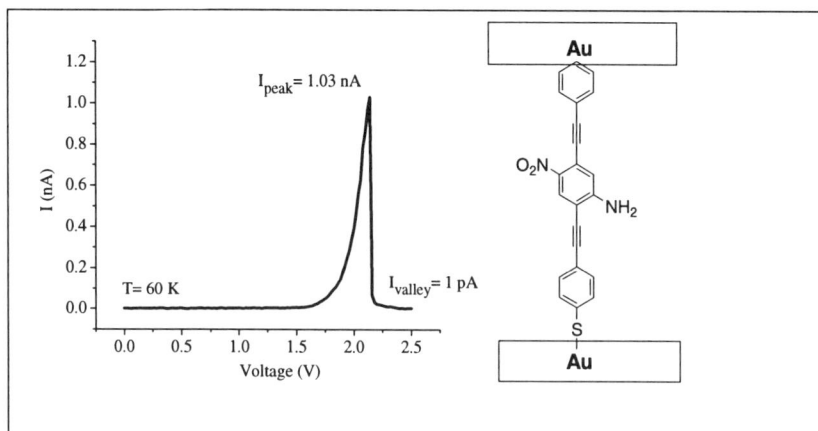

Figure 4.13 I(V) characteristics of an Au-(nitroaniline OPE)-Au device at 60 K in the nanopore.

Remarkably, typical I(V) characteristics of an Au-(nitroaniline OPE)-Au device at 60 K are shown in Figure 4.13.[82a,b] Positive bias corresponds to hole injection from the chemisorbed thiol-Au contact and electron injection from the evaporated contact. Unlike previous devices that also used molecules to form the active region, this device exhibits a robust and large NDR with a valley-to-peak ratio (PVR) of 1030:1.[82a,b] The NDR effect from the system was observed up to 260 K. Beyond that temperature, however, no NDR was observed. More recently room temperature NDR has been seen in the nanopores containing the SAM of the nitro OPEs (Figure 4.14a),[82a,b] but even larger PVRs of 7:1 have been observed at room temperature with the dinitrobiphenylene compound (Figure 4.14b).[125]

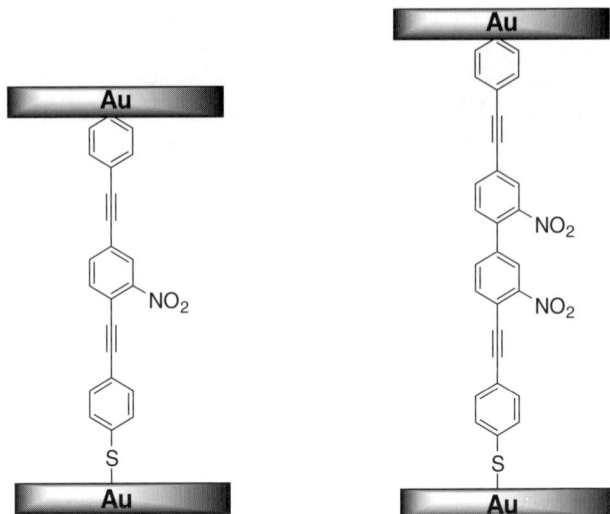

Figure 4.14 Two OPEs, shown as a portion of their SAMs sandwiched between gold contacts, exhibited room temperature NDR with PVRs of (left) 4:1 and (right) 7:1.

Additionally, we demonstrated charge storage in a self-assembled nanoscale molecular device that operated as a molecular dynamic random access memory (mDRAM) with practical thresholds and output under ambient operation.[82b] The memory device operates by the storage of a high or low conductivity state. Hence, we need not address the nanopore and attempt detection of a small number of additional electrons; a problematic feature of typical solid state single electron devices. Conversely, the added electrons dramatically affect the conductivity of the molecular system thus a conductivity check notes the presence of the information state. Figure 4.15 shows the write, read, and erase sequence. An initially low conductivity state (low σ) is changed (written) into a high conductivity state (high σ) upon application of a voltage pulse at approximately 2 V. The direction of current that flows during this "write" pulse is diagrammed. The high σ state persists as a stored "bit", which is read in the low voltage region, approximately 0.3 V, as a nondestructive read. Again, this effect persisted up to 260 K. [82b]

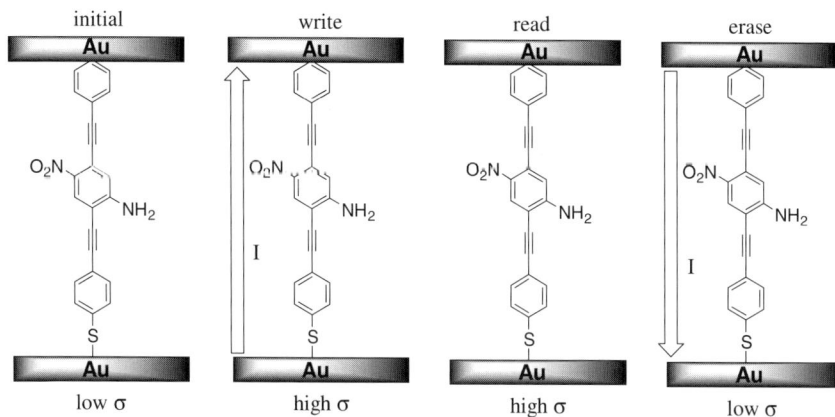

Figure 4.15 Write, read, and erase sequences for a nitroaniline OPE in the nanopore and its use as a one-bit random access memory. There were approximately 1000 molecules in the nanopore.

To further explore the mechanism of this molecular NDR (mNDR) and molecular DRAM (mDRAM) phenomenon, several related compounds have been synthesized such as a nitroacetamide rather than a free nitroaniline moiety. After testing in the nanopore, nitroacetamide exhibited the NDR effect, however, with a smaller PVR of 200:1 at 60 K.

Figure 4.16 is a measured logic diagram demonstrating the mDRAM cell using the mononitro OPV in the nanopore at room temperature. To convert the stored conductivity to standard voltage conventions, the output of the device was dropped across a resistor, sent to a comparator and inverted and gated with the "read" pulse. The upper trace shown in Figure 4.16 is an input waveform applied to the device, and the lower is the mDRAM cell output. The first positive pulse configures the state of the cell by writing a bit, and the second and third positive pulses read the cell. The third pulse (and subsequent read pulses, not shown here for simplicity) demonstrates that the cell is robust and continues to hold the state (up to the limit of the bit retention time). The negative pulse erases the bit, resetting the cell. The second set of four pulses repeats this pattern, and many months of continuous operation have been observed with no degradation in performance.[82b]

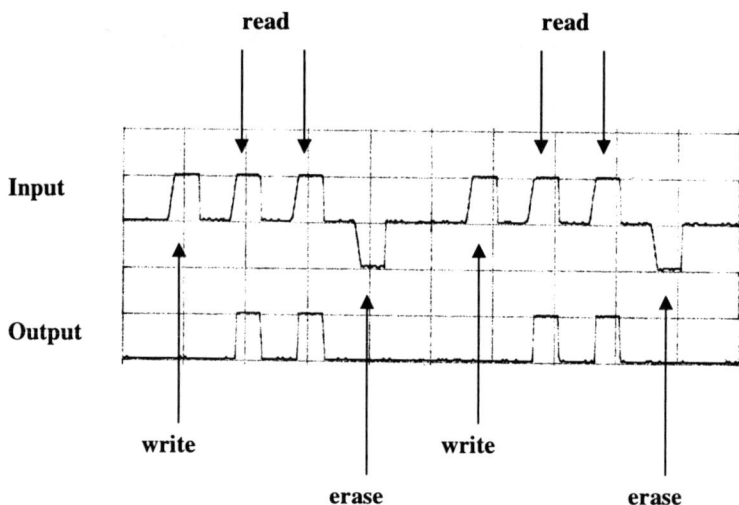

Figure 4.16 The mDRAM cell input and output that is constructed from the mononitro OPV in the nanopore. The mDRAM was built into a circuit that had a transistor and a comparator (as do most commercial solid state DRAMs) and operation was at 300 K.

This memory can be rationalized based upon conduction channels that change upon charge injection as studied by density functional theory (DFT).[126] These DFT studies further corroborate with the experimental results in that the unsubstituted OPE and the amine-substituted OPE would be inactive as devices (having linear I(V) curves) while nitroaniline and nitro OPE would both have switching states (exhibited by sharp nonlinear I(V) characteristics) due to the accepting of electrons during voltage application.   Furthermore, the DFT calculations showed that nitroaniline would need to receive one electron in order to become conductive whereas the nitro OPE would be initially conductive ("on" in the mDRAM) and then become less conductive, "off", upon receipt of one electron.[126] This is precisely the effect observed in the experiment.[82b] Four compounds on which we have mNDR and mDRAM experimental results are shown in Figure 4.17 along with the data from their testing.

Inactive	Inactive	Active	Active
high σ at Q = 0	high σ at Q = 0	low σ at Q = 0	high σ at Q = 0
No switching	No switching	high σ at Q = -1	low σ at Q = -1

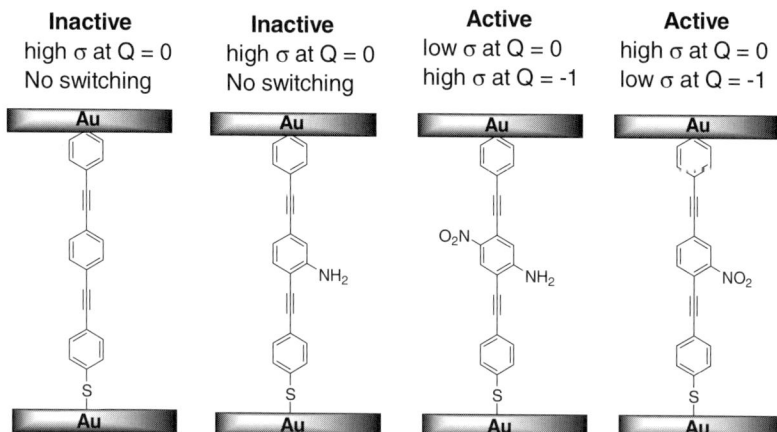

Figure 4.17 Some of the mNDR and mDRAM results in the nanopore where σ is the conductance and Q is charge. "Inactive" and "Active" refer to the device properties wherein a large nonlinearity in the I(V) curve results upon application of a voltage, analogous to observing an NDR peak.

A problem that persists in molecular electronics is the impedance mismatch between the molecule and the metal contact and we have been investing much time and effort to understand and minimize the resistance barrier.[5,123] To reduce this impedance mismatch, the sulfur atoms in our alligator clips have been replaced with more metallic Se and Te termini to allow for greater overlap of the compounds' LUMO and the gold's Fermi levels.[127] Nonetheless, it was determined that neither the selenium or tellurium alligator clip significantly reduced the barrier height.[123]

Recently it has been discovered that the use of an isonitrile as the contact between the organic molecular scale wire and a palladium probe would significantly reduce the conduction barrier, and would allow an increase in the conductivity of the molecular scale wires. Therefore molecular scale devices with an isonitrile attachment moiety was synthesized.[5b,128]

We envisioned a nanopore containing nitroaniline biphenylene could act as a molecular controller wherein the molecular system would have greater contiguous overlap in the presence of an applied orthogonal (gate) field as described in the Figure 4.18. In the ground state, the biphenyl ring system will be non-planar due to steric interactions. This will cause the π-overlap of the molecular device to be non-contiguous thus decreasing the electrical conductivity. In an applied electric field that is perpendicular to the molecular axis gate, the more planar zwitterionic resonance form will be a greater contributor to the overall structure. Hence, gated control of the current through the system might be permitted. It is not essential that the molecule be entirely planar when the gate electrode is activated. It is simply necessary that the applied field lessens the twist angle between the two central rings; hence, current modulation between the top and bottom electrodes could be maintained. The increased conductivity in the perturbed state (gate voltage applied), compared to the ground state, would allow this material to function as a molecular scale switch. Building the requisite three-terminal system has not been completed, although a claim to such a design method for the silicon-based portion has recently been made.[129]

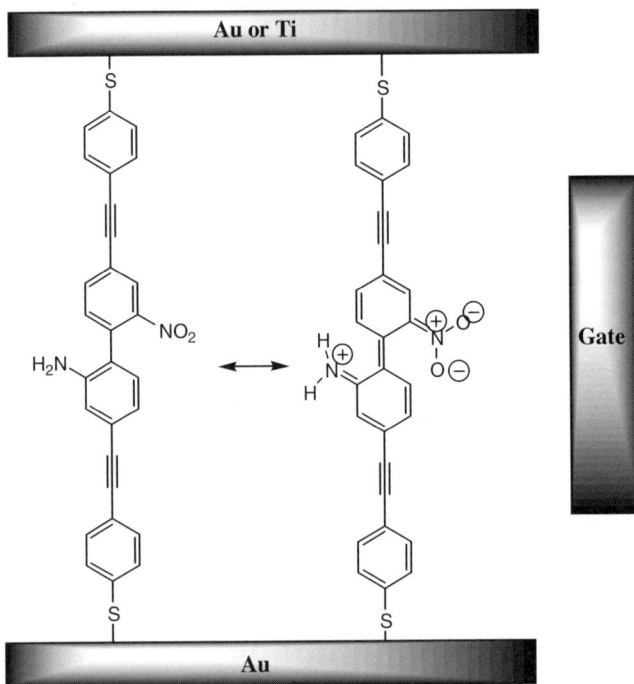

Figure 4.18 Schematic of a molecular device controller where a gate electrode could modulate the overlap in a molecule by preferring the more planar zwitterionic form.

Several of the porphyrin-containing systems bearing alligator clips did not possess significantly non-linear I(V) characteristics in both the forward and reverse bias modes when studied in the nanopores. But we have yet to test the metal-containing porphyrins. Although our device studies on the porphyrins have not afforded positive results, these observations were specifically found in the nanopore using a selected set of symmetric structures and should not be used to exclude the search for other porphyrin-based molecular electronic devices. Related structures have found efficacy in the studies by Lindsey and

Bocian and are the basis of the molecular-FET-like structures of ZettaCore Corp.[112,130]

Aside from the gate below the oxide, which does not directly contact or exploit the molecules' smallness in size,[114] three-terminal molecules have yet to be addressed either as single molecules or as nanometer-sized bundles. This stems from the fact that it is far easier to bring two probes into close proximity than to bring three probes into near contact. Macroscopic leads are akin to basketballs; touching two basketballs together at the same junction point is simple; however, trying to touch three together at a common contact point is impossible. It is the same when addressing single molecules. Unless the three probes have very large aspect ratios (extreme sharpness at the molecular level), or unless one cleverly devises another method, testing of three- or four terminal systems will remain a challenge.

4.4 Large Area Molecular Electronic Devices: The Large Area Contact Problem

The so-called "large area contact problem" has slowed development of easier to access testbeds as well as the development of commercial devices. You recall that Reed smartly devised the nanopore test, thereby making a small area SAM, only over about 1000 molecules, which was smaller than the defect density of typical SAMs. When he then evaporated top metal, there were no pinhole defects through which the metal could punch. Typically, however, this is a critical problem when using a larger area. Reed fashioned his nanopore holes using e-beam patterning. Albeit perfectly acceptable for a laboratory demonstration,[120] the nanopore is impractical for a commercial embodiment or when large numbers of test structures are needed with a high yield. The clean top contact formation over a "large area" or area that is conventionally lithographically attainable by a planar process, such as 0.1-0.2 μm^2, is much harder to achieve. Over such areas, there are significant numbers of grain boundaries and step edges in the SAM. This causes top metal evaporation to easily punch through the 2-3 nm thick SAM and subsequent device failure (Figure 4.19). Note that metals have been deposited by evaporation atop micron-sized LB monolayers when the lower metal was an oxide, specifically aluminum oxide. The oxide inhibits the short circuits of the system, but these devices have not been shown to be active over extended cycling.[69] In general, therefore, this problem has not been solved so as to generate devices in sufficiently high yields to make attractive prototypes for industry. Molecular Electronics Corp. has also been plagued with the difficulties of the large area top contact problem, however, there are some promising candidate methods beginning to emerge to self-heal or cover the naturally occurring pinhole defects.

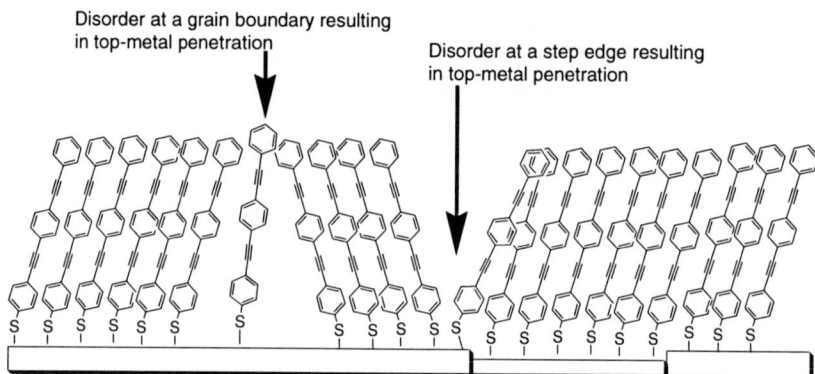

Figure 4.19   Illustration of a possible mechanism for top-metal penetration upon evaporation of the top metal contact onto the SAM. The SAM exhibits "pinhole" defects that can occur at grain boundaries where the natural tilt of the molecules (~20° off the surface normal) are dislocated at differing domains and at step edges where there are single atomic step defects in the underlying metal substrate. These "pinhole" defects could be the source of the metal shorts upon evaporation or continued usage of the devices. The larger the area, the more defects that will be present, thereby exacerbating the top metal punch through problem.

4.5 Summary

Enormous progress has been made in constructing devices for molecule tests. The SPM has provided proof for the ability of single molecules and packets of molecules to be switched and to hold electrons as a memory. The break junction addressing methodology demonstrated that single molecules could carry large amounts of current relative to their size. The nanopore testbed structure provided a method to address bundles of molecules, which are the likely methods of addressing systems in the near future. The nanopore cleverly solved the SAM defect problem by using a surface area that was smaller than the defect density of the SAM, namely 30–40 nm in diameter. However, the nanopore can not be used in a commercially viable design wherein many such memory units need to be integrated with planar processing procedures with SAM areas that are on the order of 0.1 μm^2 or larger. Once this large area contact problem is solved, useful prototypes await that demonstrate the long-awaited silicon/molecule hybrid devices.

Chapter 5

Architectures in Molecular Electronics[131]

5.1 Introduction

Molecular electronics seeks to build computational systems, both memory and logic, wherein individual or small collections of molecules serve as discrete device components. As discussed previously, potential advantages of molecular electronic systems could be many-fold including reducing the complexity and cost of current integrated circuit fabrication technologies, reducing heat generation by using only a few electrons per bit of information, and providing a route to meet the ever-continuing demand for miniaturization. While molecules are approximately one million times smaller in area than their present-day solid-state counterparts, this small size brings with it a new set of problems. In order to take advantage of the ultra-small size of molecules, one ideally needs an interconnect technology that (a) scales from the molecular dimensions, (b) can be structured to permit the formation of the molecular equivalent of large-scale diverse modular logic blocks as found in VLSI (very large-scale interconnect) architectures, and (c) can be selectively connected to mesoscopically (100 nm scale) defined input/output.

The first approach to molecular computing is based on quantum cellular automata (QCA) and related electrostatic information transfers.[132] This method relies on electrostatic field repulsions to transport information throughout the circuitry. One major benefit of the QCA or electrostatics approach is that heat dissipation is less of an issue because only a few or fractions of an electron are used for each bit of information in classical solid-state devices.

The second approach is based on the massively parallel solid-state Teramac computer developed at Hewlett-Packard (HP)[133] and involves building a similarly massively parallel computing device using molecular electronics-

251

based crossbar technologies that are proposed to be defect tolerant. The "NanoFabric" architecture is again a Teramac-like crossbar-based approach that deals more effectively with the numbers of wires emanating from the array.[134] When applied to molecular systems, the crossbar approaches propose to use single-walled carbon nanotubes (SWNTs)[135] or synthetic nanowires (NWs)[136] for the crossbars. Logic functions are performed either by sets of crossed and specially doped nanowires or by molecular switches placed at each crossbar junction.

The third approach involves using molecular scale switches as part of a NanoCell, a method we proposed four years ago.[137] The NanoCell relies on disordered arrays of molecular switches to perform logic functions. It does not require that each switching molecule be individually addressed and furthermore utilizes the principles of chemical self-assembly in construction of the logic circuitry, thereby reducing complexity. While fabrication constraints are greatly eased, programming difficulties increase dramatically. The programming methods will be outlined in Chapter 6.

5.1.1. *Quantum Cellular Automata (QCA) and Electrostatics Architectures*

Quantum dots have been called "artificial atoms" or "boxes for electrons"[138] since they have discrete charge states and energy level structures that are similar to atomic systems and can contain from a few thousand to one electron. They are typically small electrically conducting regions, 1 μm or less in size, with a variety of geometries and dimensions. Because of the small volume, the electron energies are quantized. No shell structure exists, instead the generic energy spectrum has universal statistical properties associated with quantum chaos.[139] Several groups have studied the production of quantum dots.[140] For example, Leifeld and coworkers studied the growth of Ge quantum dots on silicon surfaces that had been pre-covered with 0.05-0.11 monolayer of carbon,[141] i.e. carbon atoms replaced about five to ten of every 100 silicon atoms at the surface of the wafer. It was found that the Ge dots grew directly over the areas of the silicon surface where the carbon atoms had been inserted.

Heath discovered that hexane solutions of Ag nanoparticles, passivated with octanethiol, formed spontaneous patterns on the surface of water when the hexane was evaporated,[142] and has prepared superlattices of quantum dots.[143] Lieber has investigated the energy gaps in "metallic" single-walled carbon nanotubes[135] and has used an atomic-force microscope to mechanically bend SWNT in order to create quantum dots less than 100 nm in length.[135] He found that most metallic SWNT are not true metals, and that by bending the SWNT, a defect was produced that had a resistance of 10 to 100 kΩ. Placing two defects less than 100 nm apart produced the quantum dots.

In the QCA approach toward molecular electronic computing systems, four quantum dots in a square array are placed in a cell such that electrons are able to tunnel between the dots but are unable to leave the cell.[144] Coulomb repulsion will force the electrons to occupy dots on opposite corners. The two ground state polarizations are both energetically equivalent and can be labeled logic "0" or "1." Flipping the logic state of one cell, for instance by applying a negative potential to a lead near the quantum dot occupied by an electron, will result in the next door cell flipping ground states in order to reduce Coulomb repulsion. In this way, a line of QCA cells can be used to do computation.[145] A QCA fan-out structure has been proposed but when the ground state of the input cell is flipped, the energy put into the system may not be enough to flip all the cells of both branches of the structure, producing long-lived metastable states and erroneous calculations. Switching the cells using a quasi-adiabatic approach prevents the production of these metastable states.[146]

Amlani and co-workers have demonstrated experimental switching of six-dot QCA cells.[147] The polarization switching was accomplished by applying biases to the gates of the input double-dot of a cell fabricated on an oxidized Si surface using standard Al tunnel junction technology, with Al islands and leads patterned by e-beam lithography, followed by a shadow evaporation process and an *in situ* oxidation step. The switching was experimentally verified in a dilution refrigerator using the electrometers capacitively coupled to the output double-dot. A functioning majority gate was also demonstrated by Amlani and co-workers,[148] with logic AND and OR operations verified using electrometer outputs after applying inputs to the gates of the cell.

While the use of quantum dots in the demonstration of QCA is a good first step in reduction to practice, the ultimate goal is to use individual molecules to hold the electrons and pass electrostatic potentials down QCA wires. We have synthesized molecules that have been shown by ab initio computational methods to have the capability of transferring information from one molecule to another through electrostatic potentials (Figure 5.1). The potentials use a millionth of an electron per bit of information. This is quite attractive since a major consideration in molecular devices is the energy consumption/dissipation needs.[132c] Considering the fact that there are 10^8 gates/cm^2 (in presently-sized silicon-based systems) functioning at the rate of 10^{-9} sec (present speeds), those gates afford 10^{17} electrons/sec ($\sim$0.02 amperes/cm^2) if only one electron per gate is used to transport, indicate, fetch, or represent a binary digit. At this point heat considerations are already extreme: if the average resistance of the circuit is 30 Ω, this represents 20 watts/cm^2. If an increase of several orders of magnitude in performance is expected with molecular circuits, this would imply a proportional increase in power dissipation. Such levels of power dissipation rule out most conventional current or electron transfer methods for practical molecular devices wherein large numbers of devices are densely configured.

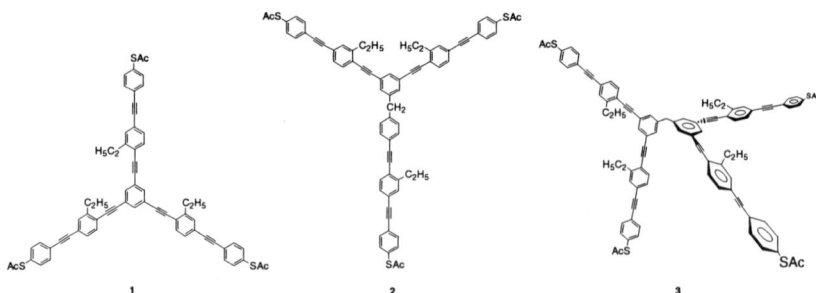

Figure 5.1   Three compounds that were synthesized for studies in computing using electrostatic potentials. 1 is a molecular three-terminal junction that could be used as a molecular interconnect. 2 is a molecular-sized switch for which there is a corresponding equivalent of a source, drain and gate terminals of a bulk solid state FET. 3 is a can be an active OR or a passive NOR gate if positive logic is used or an AND or NAND gate if negative logic is used.[132c]

Electrostatic interactions are produced by small reshapes of the electron density due to the input signals.   In turn, electrostatic potential interactions between molecules could transport the information throughout the central processing unit (CPU).   Using this approach, there is no need for electron currents or electron transfers as in present devices; a small change in the electrostatic potential of one molecule could be enough to send the information to another molecule.   These perturbations of the electrostatic potential imply a very small amount of charge transfer, far less than one electron.   External fields or excitations are able to change the boundary conditions of the molecule producing a change in the electrostatic potential generated by these sources. As an example, a charge or field on the left side of a molecule would reshape the electron density providing a different potential at the other end, the output side, which can be detected.   This is an efficient way to transmit a signal.   There is no need for electron transfer, just a charge reshape. If the excitation ceases, the shape comes back to its original form. Is this method of information transfer really detectable since the change in the electrostatic potential is minimal (~10 kcal/mole/electron where the formal units of electrostatic potential are energy per unit of charge)?   The change observed in the electrostatic potentials is in the range of values of nonbonded interactions, such as van der Waals interactions, which are easily detected by neighboring molecules.   In fact, these are precisely the ranges of signal energies that would be attractive if we will ultimately utilize large-scale integration in very small areas of materials.   If the signals were large, problems of energy consumption and dissipation, as discussed above, rule out the use of molecular-based electronics.   Thus the use of electrostatics directly addresses energy concerns in downsizing.

Distinguishing between two types of gates in electrostatics-based molecular circuits is also in order. There are those gates at the interface of the logical circuits or CPU. The gates at the input interface would have to be able to take signals from standard electronic circuits, lasers, scanning probe microscopy (SPM) tips, or in general, from well localized electrical fields. Molecular gates at the output would have to drive or inject a current onto a minuscule tip or metal cluster, or be able to excite the input of standard bulk devices. The vast majority of molecular gates, however, would be within the molecular CPU and would thus perform the information processing by controllably affecting the electrostatic potential of neighboring molecules.

Present technologies are also beset with problems from working at very high frequencies. The use of electrostatics also addresses this difficulty. Small spurious capacitances (C) usually appear between portions of devices and circuits that otherwise should be totally electrically disconnected, or at infinite impedance (Z). If the operational frequency of the device increases, the impedance would decrease proportionally with the frequency (f) of operation (Z = $1/2\pi fC$) rendering the device useless after a characteristic cutoff frequency. This cutoff frequency constitutes an upper limit to the operational frequency, which in present technology approaches 1 GHz (which corresponds to a 1 ns operation time). Since the capacitance is also inversely dependent on the distance, making smaller circuits yields larger spurious capacitances, and therefore smaller impedances, exacerbating the move toward smaller devices. Since electrostatic potentials are only limited by the time in which the electron density rearranges in a molecule (<1 fs) and they do not imply a net transfer of current, they can lead to the high inter-device impedance levels required.

Although we synthesized molecules that included three-terminal molecular junctions, switches, and molecular logic gates to demonstrate the electrostatics methodology, [132c] none of the molecules were incorporated into an actual assembly. All results were based on computation only because the QCA and electrostatics method have major obstacles to overcome before even simple laboratory tests can be attempted. While relatively large quantum dot arrays can be fabricated using existing methods, a major problem is that placement of molecules in precisely aligned arrays at the nanoscopic level is very difficult to achieve with accuracy and precision. Another problem is that degradation of only one molecule in the array can cause failure of the entire circuit. There has also been some debate about the unidirectionality (or lack thereof) of QCA designs.[149] Even small examples of two-dots have yet to be demonstrated using molecules because addressing of the molecular-sized inputs and the recording of a signal based on fractions of an electron make the hurdles enormous.

Therefore, even though this author has been a vocal proponent of the methodology due to its minimal heat dissipation features, it is his opinion that the QCA and electrostatics computing methods will not be commercially viable for at least 20 years. Nonetheless, these methods are worth studying because electrostatics are the information transfer methods of nature. For example, when a substrate approaches an enzymatic cleft, it is either accepted or rejected based upon van der Waals repulsions or attractive interactions that are all electrostatics-based. Hence, as demonstrated by nature wherein an enormous amount of information is processed, the electrostatics method is altogether fascinating and it might pave the way for distant synthetic computing methodologies.

5.1.2. *Cross-Bar Arrays*

Heath, Kuekes, Snider, and Williams reported on a massively parallel experimental computer that contained 220,000 hardware defects yet operated 100 times faster than a high-end single processor workstation for some configurations.[133] The solid-state-based (not molecular electronic) Teramac computer, built at HP, relied on its fat-tree architecture for its logical configuration. The minimum communication bandwidth needed in the fat-tree architecture was determined by utilizing Rent's rule, which states that the number of wires coming out of a region of a circuit should scale as the power of the number of devices (n) in that region, ranging from $n^{1/2}$ in two dimensions to $n^{2/3}$ in three dimensions. The HP workers built in excess bandwidth, putting in many more wires that needed. The reason for the large number of wires can be understood by considering a simple city map. To get from point A to point B, one can take local streets, main thoroughfares, freeways, interstate highways, or any combination thereof. If there is a roadblock at any point C between A and B, then re-navigation can be assessed using the map to get to point B. In the Teramac computer, "street blockages" are stored in a defect database; when one device needs to communicate with another device, it uses the database and the map to determine how to get there. The Teramac design can therefore tolerate defects.

In the Teramac computer (or a proposed molecular computer based on the Teramac design) the wires that comprise the address lines controlling the settings of the configuration switches and the data lines that link the logic devices are the most important and plentiful part of the computer. It is logical that a large amount of research has been done to develop NWs that could be used in the massively parallel molecular computer. Recall that nano-scale wires are needed if they are to take advantage of the small in size of molecules.

Lieber has reviewed the work done in his laboratory to synthesize and determine the properties of NWs and nanotubes.[135] He used Au or Fe catalyst nanoclusters to serve as the nuclei for NWs of Si and GeAs with 10 nm diameters and lengths of hundreds of nm. By choosing specific conditions, Lieber was able to control both the length and the diameter of the single crystal semiconductor NW.[136] Silicon NWs doped with B or P were used as building blocks by Lieber to assemble semiconductor nanodevices.[136]   Active bipolar transistors were fabricated by crossing *n*-doped NWs with *p*-type wire base. The doped wires were also used to assemble complementary inverter-like structures.

Heath reported the synthesis of silicon NWs by chemical vapor deposition using SiI_4 as the Si source and Au or Zn nanoparticles as the catalytic seeds at 440°C. [136,150] The wires produced varied in diameter from 14 to 35 nm and were grown on the surface of silicon wafers.   After growth, isolated NWs were mechanically transferred to wafers and Al contact electrodes were put down by standard e-beam lithography and e-beam evaporation such that each end of a wire was connected to a metallic contact. In some cases a gate electrode was positioned at the middle of the wire. Tapping-mode atomic force microscopy (AFM) indicated the wire in this case was 15 nm in diameter.

Heath found that annealing the Zi-Si wires at 550° C produced increased conductance attributed to better electrode/nanowire contacts. Annealing Au-Si wires at 750° C for 30 min increased current about 10^4, an effect attributed to doping of the Si with Au, and lower contact resistance between the wire and Ti/Au electrodes.

Much research has been done to determine the efficacy of SWNTs as NWs in molecular computers.   One problem with SWNTs is their lack of solubility in common organic solvents.   In their synthesized state, individual SWNTs form ropes[151] from which it is difficult to isolate individual tubes. In our laboratory some solubility of the tubes was seen in 1,2-dichlorobenzene.[152] An obvious route to higher solubility is to functionalize SWNTs by attachment of soluble groups through covalent bonding. Margrave and Smalley found that fluorinated SWNTs were soluble in alcohols[153] while Haddon and Smalley were able to dissolve SWNTs by ionic functionalization of the carboxylic acid groups present in purified tubes.[154]

We have found that SWNTs can be functionalized by electrochemical reduction of aryl diazonium salts in their presence.[155] Using this method, about 1 in 20 carbon atoms of the nanotube framework are reacted. We have also found that the SWNTs can be functionalized by direct treatment with aryl diazonium tetrafluoroborate salts in solution, or by in situ generation of the diazonium moiety using an alkyl nitrite reagent.[156]   These functional groups give us handles with which we can direct further, more selective derivatization (Figure 5.2) and we recently reviewed the area of covalent sidewall derivatization of SWNTs.[157]

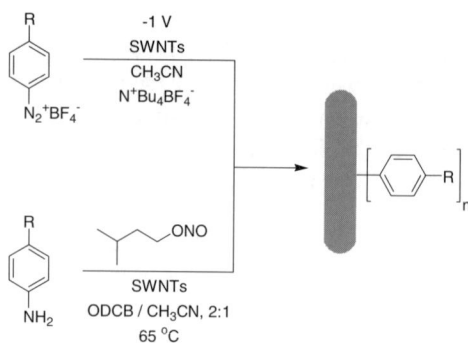

R = halogen, CO_2CH_3, NO_2, *tert*-butyl, COOH, in addition to

and

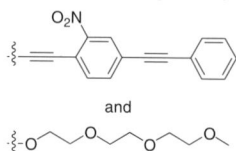

Figure 5.2   Reaction of SWNTs with aryl diazonium compounds. Shown are the electrochemical reactions with pre-formed diazonium salts, and the thermally motivated reaction with in-situ generated diazonium compounds.   Also shown are a number of specific moieties that have been attached via these methods.

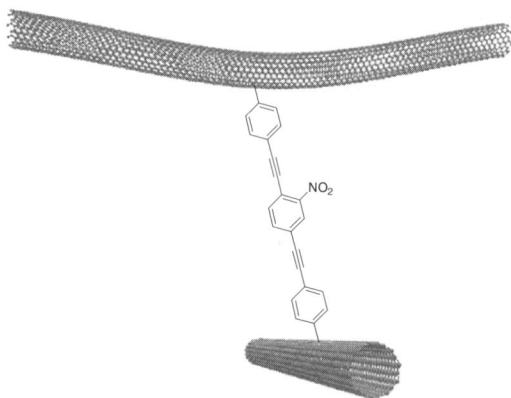

Figure 5.3   An electronically active OPE that has been attached between SWNTs. This is an idealized view. In reality, numerous cross-links were formed in an ill-defined manner.

Using a bifunctional diazonium-tipped oligo(phenylene ethynylene) (OPE) that had shown device-like character in the nanopore, we have been able to cross-link nanotubes as proximal probes across active molecular components (Figure 5.3). From an electronics perspective, the good thing about SWNTs is that all the carbons on the sidewalls are essentially the same; therefore, there is no location for an electron to localize during electronic transport. However, almost all synthetic chemistry is predicated upon the ability to distinguish different carbon atoms in a molecule based upon their subtle electronic and/or steric differences. Thus, from a synthetic perspective, the bad thing about SWNTs is that all the carbons on the sidewalls are essentially the same; therefore, one can not use the plethora of synthetic techniques that favor formation of one type of carbon-carbon bond over another. However, if one did have a crossbar array of nanotubes where the voltage between the two tubes could be selective addressed at discrete locations, then the bridging on the specified cross-junctions might be possible. But thas not yet been demonstrated.

Unfortunately, fluorination and other sidewall functionalization methods can perturb the electronic nature of the SWNT. An approach by Smalley,[151,158] and Stoddart and Heath,[135d] to increasing the solubility without disturbing the electronic nature of the SWNTs was to wrap polymers around the SWNTs but leave individual tubes' electronic properties unaffected. Stoddart and Heath found that the SWNT ropes were not separated into individually wrapped tubes; the entire rope was wrapped. Smalley found that individual tubes were wrapped with polymer; the wrapped tubes did not exhibit the roping behavior. While Smalley was able to demonstrate removal of the polymer from the tubes, it is not clear, however, how easily the SWNTs can be manipulated and subsequently used in electronic circuits. In any case, the placement of SWNTs into controlled configurations has been by a top-down methodology, for the most part. Significant advances will be needed to take advantage of controlled placement at dimensions that exploit a molecule's small size.

Lieber proposed a SWNT-based nonvolatile random access memory device comprising a series of crossed nanotubes wherein one parallel layer of nanotubes is placed on a substrate and another layer of parallel nanotubes, perpendicular to the first set, is suspended above the lower nanotubes by placing them on a periodic array of supports.[135b] The elasticity of the suspended nanotubes provides one energy minima, wherein the contact resistance between the two layers is zero, and the switches (the contacts between the two sets of perpendicular NWs) are "off". When the tubes are transiently charged to produce attractive electrostatic forces, the suspended tubes flex to meet the tubes directly below them, and a contact is made, representing the "on" state. The "on" / "off" state could be read by measuring the resistance at each junction, and could be switched by applying voltage pulses at the correct electrodes. This theory was tested by mechanically placing two sets of nanotube bundles in a crossed mode and measuring the I(V) characteristics when the switch was "off" or "on".

Although they used nanotube bundles with random distributions of metallic and semiconductor properties, the difference in resistance between the two modes was a factor of 10, enough to provide support for their theory.

In another study, Lieber used an STM to determine the atomic structure and electronic properties of intramolecular junctions in SWNTs samples.[135c] Metal-semiconductor junctions were found to exhibit an electronically sharp interface without localized junction states while metal-metal junctions had a more diffuse interface and low-energy states.

A major problem with using SWNTs or NWs is how to guide them in formation of the device structures, i.e. how to put them where you want them. Lieber has studied the directed assembly of NWs using fluid flow devices in conjunction with surface patterning techniques and found that it was possible to deposit layers of NWs with different flow directions for sequential steps.[136a] For surface patterning, Lieber used NH_2- terminated surface strips to attract the NWs; in between the NH_2- terminated strips were either methyl-terminated regions or bare regions, to which the NW had less attraction. Flow control was achieved by placing a poly(dimethylsiloxane) (PDMS) mold, in which channel structures had been cut into the mating surface, on top of the flat substrate. Suspensions of the NWs (GaP, InP, or Si) were then passed through the channels. The linear flow rate was about 6.40 mm/s. In some cases the regularity extended over mm length scales, as determined by scanning electron microscopy (SEM).

While Lieber has shown that it is possible to use the crossed NWs themselves as switches, Stoddart and Heath have synthesized molecular devices that would bridge the gap between the crossed NWs and act as switches in memory and logic devices.[159] The UCLA researchers have synthesized catenanes and rotaxanes that can be switched between states using redox chemistry. For instance Langmuir-Blodgett (LB) films were formed from the catenane, and the monolayers were deposited on polysilicon NWs etched onto a silicon wafer photolithographically. A second set of perpendicular titanium NWs were deposited through a shadow mask, and the I(V) curve was determined. The data, when compared to controls, indicated that the molecules were acting as solid-state molecular switches. As yet, however, there have been no demonstrations of combining the Stoddart switches with NWs.

Carbon nanotubes are known to exhibit either metallic or semiconductor properties. And ~2:1 mixtures are formed under all growth methods known to date. Separation of the two forms is presently impossible on ay usable scale. Avouris and coworkers at IBM have developed a method of engineering both multiwalled nanotubes (MWNTs) and SWNTs using electrical breakdown methods.[160] Shells in MWNT can vary between metallic or semiconductor character. Using electrical current in air to rapidly oxidize the outer shell of MWNTs, each shell can be removed in turn because the outer shell is in contact with the electrodes and the inner shells carry little or no current. Shells are removed until arrival at a shell with the desired properties.

With ropes of SWNTs, Avouris used an electrostatically coupled gate electrode to deplete the semiconducting SWNTs of their carriers. Once depleted, the metallic SWNTs can be oxidized while leaving the semiconductor SWNT untouched. The resulting SWNT, enriched in semiconductors, can be used to form nanotubes-based field-effect transistors (FETs).

The defect tolerant approach to molecular computing using cross bar technology faces several hurdles before it can be implemented. Large numbers of nano-sized wires are used in order to obtain the defect tolerance. How are each of these wires going to be accessed by the outside world? Multiplexing, the combination of two or more information channels into a common transmission medium, will have to be a major component of the solution to this dilemma. The directed assembly of the NWs and attachment to the multiplexers will be quite complicated. Goldstein, et al. has proposed a NanoFabric architecture that would lessen the number of wires;[134] however, the numbers could still be overwhelming. Another hurdle is signal strength degradation as it travels along the NWs. Gain is typically introduced into circuits by the use of transistors. However, placing a transistor at each NW junction is an untenable solution because the size would be limited by the silicon transistor, therefore there is no size advantage to the molecular-based system. Likewise, in the absence of a transistor at each cross point in the crossbar array, molecules with very large "on" : "off" ratios will needed. For instance, if a switch with a 10:1 "on" : "off" ratio was used, then ten switches in the "off" state would appear as an "on" switch. Hence, isolation of the signal via a transistor is essential, but presently the only solution for the transistors' introduction would be for a large solid-state gate below each cross point, again defeating the purpose for the small molecules.

Additionally, if SWNTs are to be used as the crossbars, connection of molecular switches via covalent bonds introduces sp^3-hybridized carbon atom linkages at each junction, disturbing the electronic nature of the SWNT and possibly rendering useless the SWNTs in the first place. Non-covalent bonding of the device molecule to the SWNT will probably not provide the conductance necessary for the circuit to operate. Therefore, continued work is being done to devise and construct crossbar architectures that address these challenges.

5.1.3. *The NanoCell Architecture*

Molecular electronics seeks to build electrical devices to implement computation—logic and memory—using individual or small collections of molecules. These devices have the potential to reduce device size and fabrication costs, by several orders of magnitude, relative to conventional CMOS. However, the construction of a practical molecular computer will require the molecular switches and their related interconnect technologies to behave as large-scale diverse logic, with input/output wires scaled to molecular dimensions. It is unclear whether it is necessary or even possible to control the precise regular placement and interconnection of these diminutive molecular systems. We have therefore developed a NanoCell architecture where we can address the nanoscale via the microscale, and thereby take advantage of the smallness in size of the molecules via our current lithographic tools. The approach also offers enormous defect tolerance and a fabrication simplicity that is unrivaled by other molecular electronic architectures. However, the programming issues become far more difficult, but excellent headway is being made and the programming methods will be described in Chapter 6.

This NanoCell architecture involves an approach where molecular switches are not specifically directed to a precise location and the internal topology is generally disordered. A NanoCell is a two-dimensional (three-dimensional models could also be considered) network of self-assembled metallic particles or islands connected by molecules that show reprogrammable (can be turned "on" or "off") negative differential resistance (NDR) (or other switching and/or memory properties although these are not initially addressed). The NanoCell is surrounded by a small number of lithographically defined access leads at the edges of the NanoCell. Unlike typical chip fabrication, the NanoCell is not constructed as a specific logic gate and the internal topology is, for the most part, disordered. Logic is created in the NanoCell by training it post-fabrication, similar in some respects to a field-programmable gate array (FPGA). Even if this process is only a few percent efficient in the use molecular devices, very high logic densities will be possible. Moreover, the NanoCell has the potential to be reprogrammed throughout a computational process via changes in the "on" and "off" states of the molecules, thereby creating a real-time dynamic reconfigurable hard-wired logic. The CPU of the computer would be comprised of arrays of NanoCells wherein each NanoCell would have the functionality of many transistors working in concert. A regular array of NanoCells is assumed to manage complexity, and ultimately, a few NanoCells, once programmed, should be capable of programming their neighboring NanoCells through bootstrapping heuristics. Alternatively, arrays could be programmed one NanoCell at a time via an underlying CMOS platform.

In order to simplify fabrication we initially consider a NanoCell that ~1 μm^2 possessing 20 input/output leads on the edges that can be contacted by standard lithographic wiring. A simulated self-assembled NanoCell is depicted in Figure 6.1. The black rectangles at the edges are the input/output leads. The entire cell, excluding the outer portions of the contact pads, would be approximately 1 μm^2. A two-dimensional array of the metal nanoparticles (dots) is deposited on an oxide surface (base) with a 90% density in this simulation. A molecular self-assembled monolayer coating each nanoparticle would control the spacing between nanoparticles. Molecular switches would insert into the inert self-assembled monolayer barrier around each nanoparticle via processes that have previously been demonstrated, and thereby inter-link adjacent nanoparticles.[161] Each molecular switch, or group of switches between adjacent nanoparticles, could be set into an "on" state or an "off" state. Based on a nanoparticle diameter of 60 nm and 3 nm spacings by the bridging molecules, each NanoCell will contain approximately 200 to 250 nanoparticles; well within the range of experimental fabrication. Initial experiments on nanoparticle depositions on an oxide surface indicate that they do from a fairly regular grid. We estimate that an average of five molecular switches will have the proper orientation between adjacent particles' facets to join adjacent nanoparticles. Both the oxide surface attachment and the nanoparticle segregation have been experimentally demonstrated. Our preliminary experimental results on two-dimensional gold nanoparticle/conjugated molecule assemblies have shown that we can have current flow over the 1-micron distance. The nanoparticles are kept from coalescing into multi-particle arrays through the use of short alkanethiols self-assembled onto the gold nanoparticles. A series of functional, electrically settable molecular switches will then be introduced. Each molecular switch is terminated on both its ends with molecular alligator clips, such as thiols, and allowed to insert between adjacent nanoparticles via self-assembly with molecule-metal chemical bonding to establish the electrical contacts between the adjacent nanoparticles and between the nanoparticles and nearby input/output leads. We have already demonstrated molecular switch insertion and bridging between nanoparticles, and through these structures re-settable enhanced conductivity and resistivity states have been established using voltage pulses at nearby lithographically defined contact pads.

Once the physical topology of the self-assembly is formed in the NanoCell, it remains static; there is no molecule or nanoparticle dynamic character (other than bond rotations or vibrations) to the highly crosslinked network. The only changeable behavior is in the molecular states: conducting "on" or non-conducting "off", as set by voltage pulses from the periphery of the cell, or as defined by the search algorithms in these simulations.

Several types of room temperature-operable molecular switches have been synthesized and demonstrated in nanopores and atop silicon-chip platforms. The functional molecular switches can be reversibly switched from an "off" state to an "on" state, and/or the reverse, based on stimuli such as voltage pulses. The number of nanoparticles (usually metallic or semiconducting) and the number of the interconnecting molecular switches can vary dramatically based on the chosen size of the NanoCell and on the dimensions of the nanoparticles and molecules chosen.

Within the fabricated NanoCell, the input and output leads could be repetitively interchanged based on the programming needs of the system, thereby demonstrating the pliability of the architecture. Naturally, issues of gain will eventually have to be addressed through either an underlying CMOS layer or clocked circuits programmed into the NanoCell.[162] Even if one CMOS transistor was used for gain at the output from each NanoCell, enormous space savings could be attained since a NanoCell could possess the functionality of numerous transistors working in concert to produce a specified logic function. Furthermore, by capitalizing on the NDR properties of the molecular switches, internal gain elements based upon NDR/nanoparticle/NDR stacks (Goto pairs) could be efficacious.[163]

The functionality of a NanoCell depends largely on the I(V) characteristics and placement of its molecular switches with respect to the nanoparticles. We have demonstrated NDR with a large "on-to-off" ratio from several types of molecular switches based upon nitro-containing OPEs as described in Chapter 4. We will exploit this NDR behavior (rise then decline in the current with increased voltage) in order to build logic devices that exhibit negating functionality such as NAND or XOR responses from these two-terminal devices since two voltage inputs that are high could set the device into an "off" state (right side of the I(V) curve). Switches that do not exhibit the NDR characteristic cannot provide the negating functionality needed for the NanoCell approach.

The object in programming or training a NanoCell is to take a random, fixed NanoCell and turn its switches "on" and "off" until it functions as a target logic device. The NanoCell is then trained post-fabrication by changing the states, "on" or "off", of the molecular switches by imposing voltages at the surrounding input/output leads. Notice how we hope to address, in a broad sense, the internal molecular switches via the 20 surrounding leads. There is a loose analogy to the brain. We have five main input methods to the brain via our five senses. After processing, the information almost always exits from the brain to carry out some mechanical action, whether that is speaking, walking, running, etc. We are unaware of the neuron interconnect pathway in the brain, nevertheless we use it. Likewise, there will be a rich array of nanoscale devices in the NanoCell device that we will access and program (set into either "on" or "off" states) from the few peripheral leads, to carry out the logic function that we so desire.

The simulations described in Chapter 6 provide a proof-of-principle for the overall NanoCell approach. It takes advantage of the diminutive size of molecules within a NanoCell and it appears to configure with an interconnect technology that can be structured to permit the formation of the molecular equivalent of large-scale ordered logic with diverse logic functions. Additionally, it can be selectively connected to lithographically defined input/output, and the size of the solution space lends itself to a high defect- and fault-tolerance. The mode of programmability from disorder is an exciting approach to molecular computing that could prove essential as we move toward molecular-based computing machines.

5.2 Summary

Molecular electronics seeks to build electrical devices to implement computation using individual or small collections of molecules. However, the construction of a practical molecular computer will require the molecular switches and their related interconnect technologies to behave as large-scale diverse logic, with input/output wires scaled to molecular dimensions so that we can take advantage of the ultra-small size of the molecules. The QCA and electrostatics model can solve the heat dissipation problems, but molecular placement and molecular degradation will pose other difficulties. Crossbar-based architectures such as the Teramac or NanoFabric approaches will likewise require precise assembly of molecules with multiplexing obstacles, although programming should be relatively straightforward. The NanoCell approach does not require precise connectivity of the internal molecular switches and the internal topology is generally disordered. It is proposed that using a limited number of input/output lines in the NanoCell, the vast internal array will be programmable. The programming will be challenging, to say the least, but invoking novel search algorithms, we hope to overcome the obstacles.

Chapter 6

Programming the Nanocell

Much of this chapter was adapted from the Rice University Ph.D. Thesis of Summer Husband, a student leading our NanoCell programming effort. Significant contributions from the Thesis work of Christopher Husband are also added. L. Wilson and J. Daniels of our Rice University NanoCell programming team contributed important findings, but to a lesser degree than Husband and Husband.[164,137]

6.1 NanoCell

The NanoCell was inspired by a desire to create a molecular electronics model that takes into account the behavior of molecules. That is, a model that can be realistically fabricated. Unlike the Teramac architecture, each individual molecule within the NanoCell is not addressed by a wire.[133a] Rather, a network of molecules is addressed by relatively small number of leads located at the edges. Additionally the topology of the NanoCell is random. Molecules are allowed to align themselves as they will. Hence many of the very serious obstacles to the implementation of other moletronics designs are avoided, and the size of the molecules is exploited. However, the cost is programmability. As fabricated, the NanoCell serves no useful logical or memory purpose. Rather, it is trained post-fabrication through voltage pulses applied to the input/output (I/O) leads. The design and electrical properties of the NanoCell are addressed in this section, while strategies for training the NanoCell are dealt with in the following section.

266

6.1.1. *Nanocell Design*

The NanoCell is the lowest level logic device in the proposed molecular computer. Its size is approximately $1\mu m^2$. This size was chosen in order to ease lithographic constraints. Along the four edges of the NanoCell are I/O pins. The network within the NanoCell is an array of gold nanoparticles (some metal other than gold could be used) connected with molecules. It can be modeled as a planar graph where the nodes are nanoparticles, and the edges are molecular switches. Realistically, a NanoCell will probably be pseudo-planar with some nanoparticles pushed slightly up and their incident edges overlapping others. However, a planar model is still accurate enough at this point. A picture of a simulated NanoCell is shown in Figure 6.1.[165] The particles are assembled on an insulating surface. The size of the nanoparticles is subject to change, however, currently a 60 nm diameter and 3 nm spacing is used. The spacing between adjacent nanoparticles is controlled by a molecular self-assembled monolayer.[166] This prevents the shorts that would occur if the nanoparticles touched each other. Early experiments on the deposition of nanoparticles onto an oxide surface indicate that they tend to form a fairly regular grid.[166]

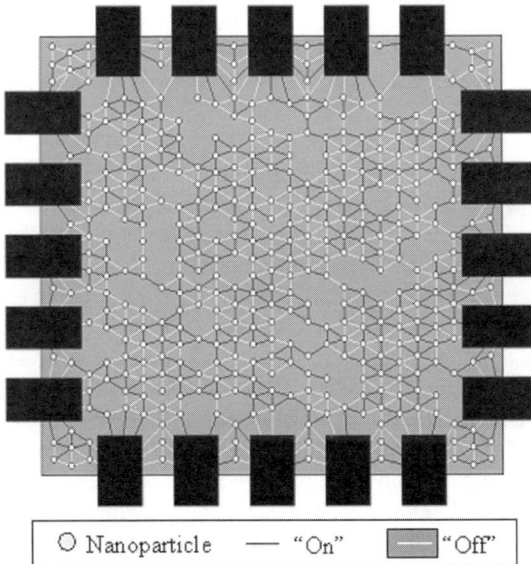

Figure 6.1 This is a simulated NanoCell with five I/O leads on each side. Within the cell is a planar array of metallic nanoparticles and molecules (dark and light lines). The molecules can be in a high conducting state (dark lines) or a low conducting state (whilte lines).

A picture of deposited particles is shown in Figure 6.2. It has been confirmed that molecules can then be deposited between adjacent particles after electroless growth of the nanoparticles.[115a,166] It is estimated that an average of five molecules will be in the correct orientation with respect to adjacent nanoparticle facets to bridge them. These nanoparticles connected by molecular switches have exhibited resettable enhanced conductivity and resistivity states through voltage pulses applied at local contact pads.[82a,b,166] In the next section, properties of the molecular switches are discussed.

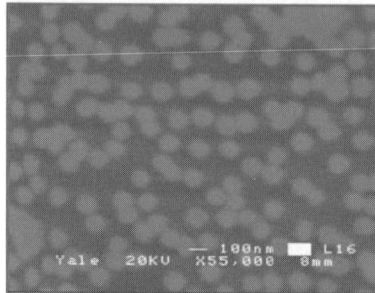

Figure 6.2 The deposition of gold nanoparticles in a physical NanoCell is displayed. Recent depositions have yielded more tightly compacted particles necessary for molecules to interconnect the particles.

6.1.2. *Molecular Switches*

The molecules used in molecular electronics can be characterized by their current as a function of voltage [I(V)] curves. These curves are obtained by testing batches of several thousand molecules together. There are two properties of a molecule that are essential to the programming of the NanoCell. First, they must exhibit two discrete states. The I(V) curve displayed in Figure 6.3 has the two states shown.[82a,b] The higher conducting state is referred to as the "on" state, while the more resistive state is called the "off" state. Suppose a nitroaniline OPE is in the "off" state. It is switched to the "on" state through a voltage pulse. The threshold for this molecule is estimated to be approximately 1.75 V. As long as the applied voltage is below this threshold, the molecule stays in an "off" state. When the applied voltage exceeds 1.75 V, the molecule switches to the "on" state. Now as long as the voltage stays above approximately -1.75 V, the molecule will remain in this state. When the voltage falls below -1.75 V, the molecule switches back to the "off" state. The molecules can be reversibly set to these states. This property offers programmability of an array of molecules. It is theorized that the movement of electrons causes these discrete states; however, a conclusive explanation has not yet been obtained. The specifics of this NanoCell training process are addressed later in this chapter. It should be noted that molecules that exhibit two reversible states have been used as DRAM memory.[82c] The ability to read, write and erase with these molecules is described in Chapter 4.

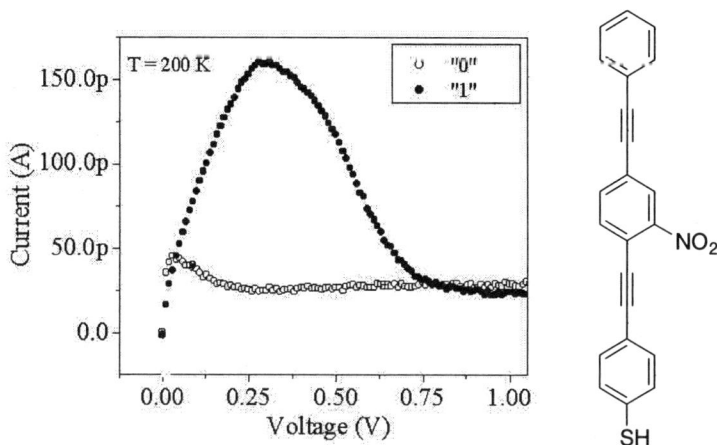

Figure 6.3 Displayed here are "on" and "off" I(V) curves and the chemical structure of the nitroaniline OPE.

The I(V) curves displayed in Figure 6.3 are actually for thousands of molecules in the nanopore cofiguration (see Chapter 4). It is important to determine whether these discrete states occur on the individual molecular scale or are simply a group phenomenon. To address this question, several individual molecules were dispersed in an alkanethiol-based SAM (self-assembled monolayer) and tracked via STM over several hours.[116] Figure 6.4 shows a single molecule turn "on", "off" and back "on" again. Each frame represents three minutes.

Figure 6.4 The switching of a single molecule is monitored over 30 hours by STM. The brighter circles indicate the "on" state, while the darker circles indicate the "off" state.[116]

In addition to resettable states, the NanoCell molecules must show NDR, or negative differential resistance behavior in the high conducting state if the NanoCell is to function as a complex logic device. In the "on" state displayed in Figure 6.3, note that after approximately 0.3 V, the current begins to decrease as the voltage drop increases. This is NDR behavior. It enables negating logic gates such as inverters, NANDs or XORs. Such logic gates are essential in constructing logical statements. In compound propositions, the operators are AND, OR, NOT, implication $(\rightarrow)$ and biconditional $(\leftrightarrow)$. A *functionally complete* set of logical operators is a set of operators such that every compound proposition is logically equivalent to some compound proposition using only operators in the set. Any logical function that generates NOT and AND or OR is functionally complete. Hence, NAND is functionally complete because NOT(X) = NAND(X,1) and AND(X,Y) = NOT(NAND(X,Y)). Similarly, AND and XOR are functionally complete because NOT(X) = XOR(X,1). Hence, NDR is essential if a NanoCell is to perform functions similar to those performed by transistors. NDR has been demonstrated with several different molecules. [82a,b] For over a year and nearly 10^9 switching events, these devices have shown no degradation.

Another property that is helpful is a large "on-to-off" ratio. The molecule shown in Figure 6.3 has about a 5 to 1 "on-to-off" ratio. This is adequate, but NanoCell logic function is greatly enhanced with larger ratios. In simulated NanoCells, the I(V) curve displayed in Figure 6.5 is used. It has NDR and a 1000 to 1 "on-to-off" ratio. This curve has not been reliably obtained experimentally, however, some devices have shown this throught three switching cycles. At low temperatures, similar effects have been observed, and room temperature effects are currently being optimized. [82a,b,166]

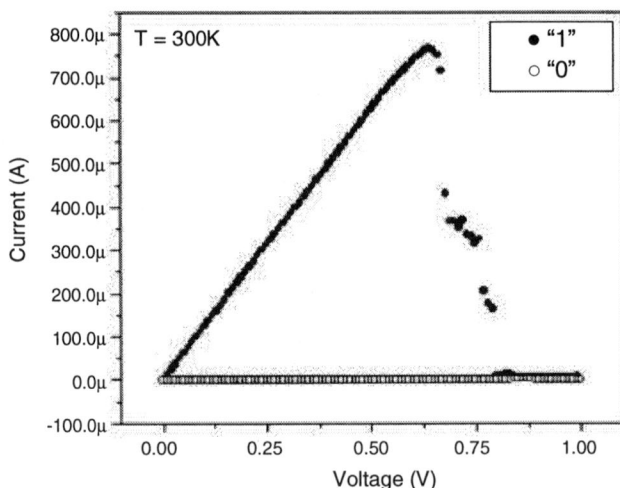

Figure 6.5 This is the I(V) curve used most often in simulating NanoCells. The "on" to "off" ratio is 1000:1. Such behavior has only been repeatedly obtained experimently at very low temperatures (approximately 60 K), but chemists on the NanoCell project expect to see it at room temperature in the near future.

For some logic gates, such as 1-bit adders, molecules with rectifying diode behavior are essential. Such an I(V) curve is displayed in Figure 6.6. This curve is used in some of the simulations presented in this chapter. This precise curve has not been obtained, however, a mononitro OPE was synthesized and its I(V) curve has a similar shape (see Figure 6.7).[82b,167]

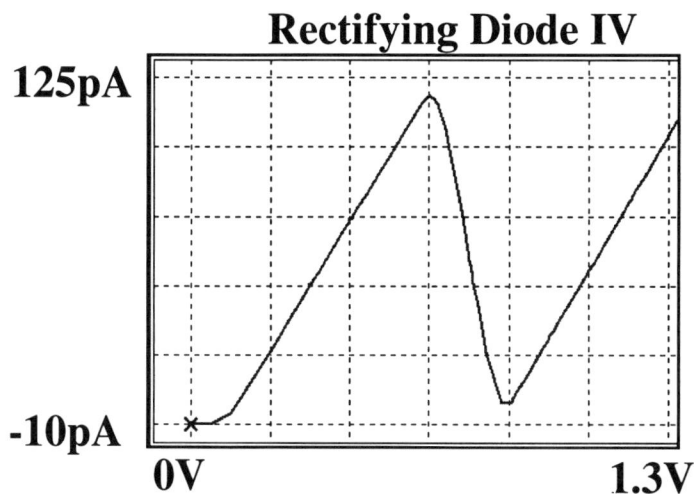

Figure 6.6 This I(V) curve is also used in simulating NanoCells. Although it was not obtained experimentally, similar behavior has been observed (see Figure 6.7).

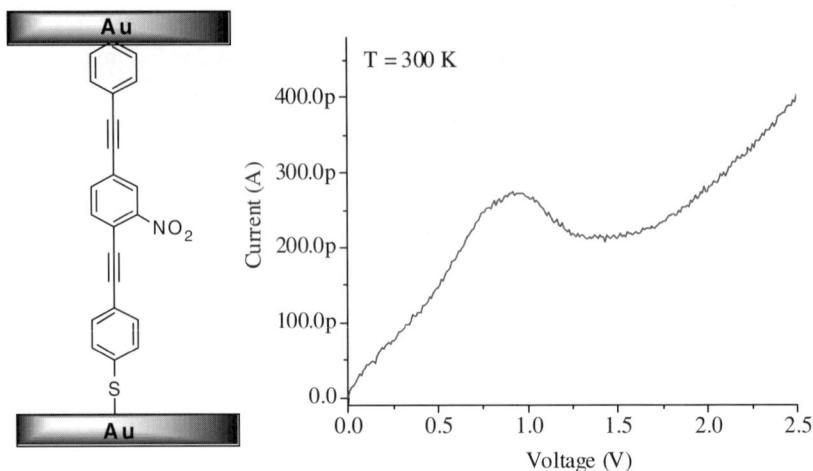

Figure 6.7 This is a molecule that exhibits rectifying diode behavior at room temperature.

The simulated NanoCell is presented in the next section. The precise I(V) curves used in simulations have not always been obtained experimentally. However, in every case, similar behavior has been observed and chemists expect to obtain the curves that are used.

6.1.3. *Simulated Nanocell*

Scientists in the NanoCell group are currently working toward the goal of fabricating a NanoCell. Thus far, chip platforms have been designed and fabricated and gold nanoparticles have been selectively deposited within the NanoCell at densities sufficient for the molecular connections. Pictures of these NanoCells are shown in Figure 6.8.[161] Figure 6.8a depicts a NanoCell before deposition of the gold particles, while Figures 6.8b and 6.8c depict NanoCells post-deposition.

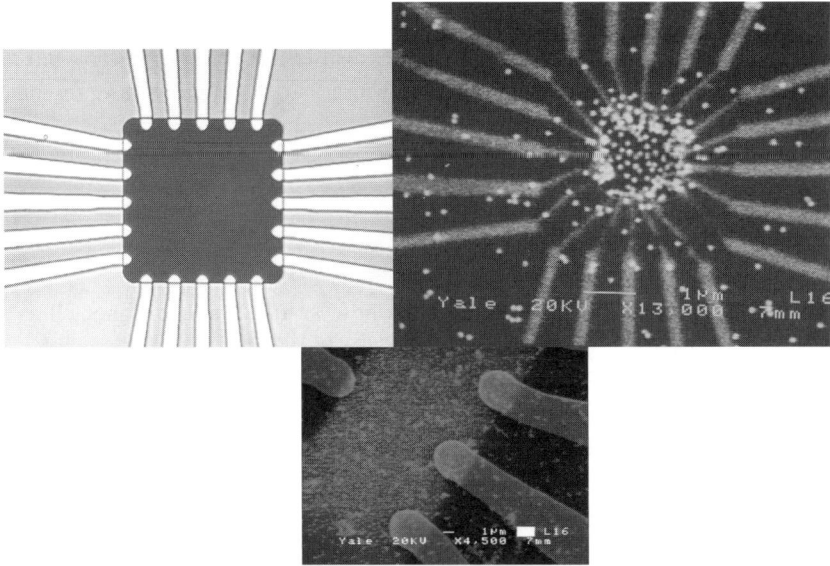

Figure 6.8 Figure a (top left) displays the I/O leads into a NanoCell before deposition. Figures b (top right) and c (bottom center) display NanoCells after deposition of gold particles. These were test NanoCells that are about 10 μm^2 as fabricated by P. Franzon *et al.* and assemblies by D. Allara *et al.*

Our NanoCell computer architecture group works with simulated NanoCells in an attempt to predict the potential functionality of a NanoCell. The Nanocell simulator was written in Microsoft Visual C++. The NanoCell is modeled as a square device (measuring approximately 1 μm^2) but the simulator is written so that a variety of geometries may be easily implemented. The simulator takes a grid size n for the nanoparticles, the number of pins to a side, nanoparticle density, represented by some probability p, and an average m_1 and maximum m_2 number of molecules between adjacent gold particles. As mentioned previously, the interior of a NanoCell is modeled as a planar graph. The nodes (nanoparticles) are laid out on a hexagonal n by n grid with a probability p that any given nanoparticle appears. The edges (molecular switches) are laid out with a Poisson distribution centered around m_1 with a maximum of m_2 edges between adjacent nodes. In initial simulations, the m_2 was set to 1 in order to speed up the training process. Additionally, the molecules were assumed to have symmetric I(V) curves (odd functions), so the edges were not directed. However, realistically the graph should be modeled as a digraph with multiple edges. Later simulations included these characteristics. In fact, some of the logic gates would have been impossible to train without these properties.

After choosing a desired truth table, such as NAND or XOR, I/O pins are set to input and output. Some pins may also be set to high rail, low rail or ground, however, these settings are rarely used in the simulations presented here.

6.1.4. *The Nanocell as a Logic Gate*

Our objective is to determine first how to train a single NanoCell as a logic gate and next to hook several NanoCells together to work in concert. A NanoCell functions as a logic gate based on the signal (current or voltage) through some number of I/O pins designated as output. Each input pin can be either "on" or "off". It is said to be "on" if a high voltage is applied and "off" if a low voltage is applied. The output signal is measured at every possible combination of input pin settings. Hence if there are three input pins, then there are 2^3 different settings. Given a pattern of input voltages then, it is necessary to determine the resulting output current. The electrical engineering circuit solver HSpice is used along with the NanoCell simulator to calculate the output signal of a NanoCell.[168] A NanoCell is translated into a circuit as shown in Figure 6.9, and that circuit is then evaluated with HSpice.

As mentioned previously, the NanoCell was initially modeled as a voltage in – current out device. With this setup, each input pin is connected to a voltage source, while each output pin is connected to a very small resistor to ground (essentially a wire). The current through this resistor is the output signal that determines the functionality of the NanoCell. A demonstration of this circuit is shown in Figure 6.10. When the signal is current out, "on" and "off" thresholds, I_{OL} and I_{OH} must be established. Anything below I_{OL} is considered "off", while anything above I_{OH} is considered "on". Currents between the two thresholds are neither "on" nor "off".

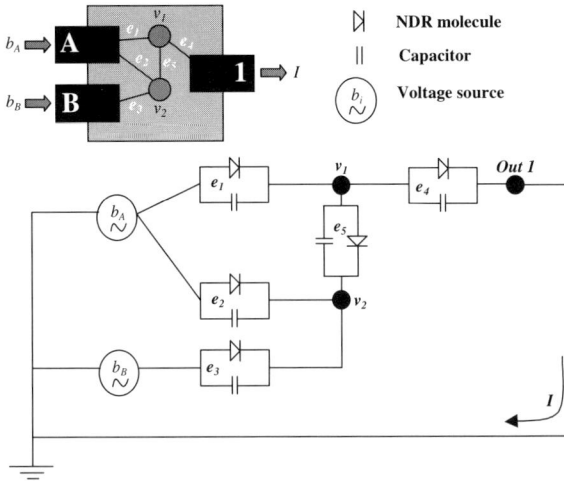

Figure 6.9 This is a circuit with an NDR device along with its Kirchhoff compliant solutions.

In Table 6.1 there is an example of the output of a voltage in – current out NanoCell that can be used both as a NAND and a NOR gate. This NanoCell has two inputs: "A" and "B", and one output, "1". The output current from pin 1 is given at all possible combinations of high and low voltage from pins A and B. If the "off" and "on" current thresholds are set to 275 pA and 350 pA respectively, then the NanoCell functions as a NAND gate. On the other hand if they are set to 450 pA and 500 pA respectively, then the NanoCell functions as a NOR gate. It should be noted that for a NanoCell to work in concert with other NanoCells, a much greater "on-to-off" ratio is essential.

If NanoCells are to work together, then the output of one NanoCell must function as the input of another NanoCell. This implies that uniformity of signal is necessary. If the input signal is voltage, then the output signal must be measured as voltage, as well. Additionally, if the high and low voltage inputs are V_{IL} and V_{IH} respectively, then the output high and low voltages V_{OL} and V_{OH} should be such that

$$V_{OL} = V_{IL}$$
$$V_{OH} = V_{IH}$$
(6.1)

This is called restoring logic. However, the NanoCell as a whole has some resistance, so this is not possible without some outside influence. One such method is to use a bistable latch or Goto pair at each output pin for restoring logic.[163a,169] Consider the very simple circuit displayed in Figure 6.10.

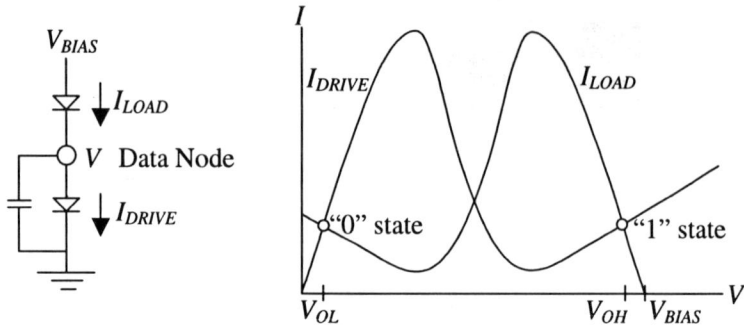

Figure 6.10 This is a bistable latch consisting of metal-NDR-metal-NDR-ground with a capacitor in series with the ground NDR. When V_{BIAS} is high, there are two stable states: "0" and "1". The state is determined by current flowing into the data node.

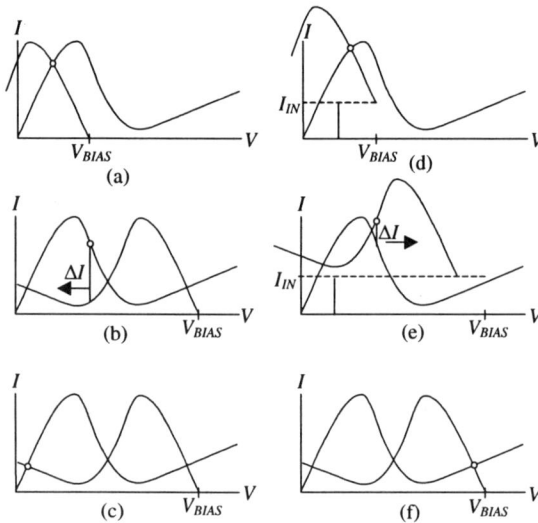

Figure 6.11 This is a bistable latch consisting of metal-NDR-metal-NDR-ground with a capacitor in series with the ground NDR. When V_{BIAS} is high, there are two stable states: "0" and "1". The state is determined by current flowing into the data node.

A current source is connected between two NDRs in series. The first NDR is connected to a voltage source, V_{BIAS}. The second NDR is in parallel with a capacitor and is connected to ground. Think of the current source as the output pin of a NanoCell. Current through the first NDR is denoted I_{LOAD}, and current through the second NDR is I_{DRIVE}. When V_{BIAS} is high, there are three possible voltages at the data node as shown in Figure 6.10. The middle state is unstable, so there are really just two possible choices, V_{OL} and V_{OH} for the voltage at the data node. Clearly, V_{BIAS}, V_{IL} and V_{IH} can be selected so that Eq. 6.1 is satisfied.[168]

The next question is, In which of the two states will the bistable latch reside, and what are the determinants? To set the latch, V_{BIAS} is brought to a low voltage, such that the latch is now in a monostable state as in Figures 6.11a and 6.11d.

Now an input current I_{IN} is supplied as V_{BIAS} is raised back up. The state that the data node latches to, V_{OL} or V_{OH} is determined by the value of I_{IN} as V_{BIAS} increases. Suppose I_{IN} is low as in Figure 6.11a. Then, as V_{BIAS} rises, I_{DRIVE} exceeds I_{LOAD} as in Figure 6.11b. This implies that the current through the capacitor is negative, which means that $\dfrac{dV}{dt}$ is negative, as well. Hence the voltage decreases and latches to V_{OL} as shown in Figure 6.11c. On the other hand, suppose that I_{IN} is relatively high as V_{BIAS} increases (see Figure 6.11d). Then I_{LOAD} exceeds I_{DRIVE} as in Figure 6.11e. Now the current through the capacitor is positive. This implies that $\dfrac{dV}{dt}$ is positive, and the voltage will increase to V_{OH}.[169]

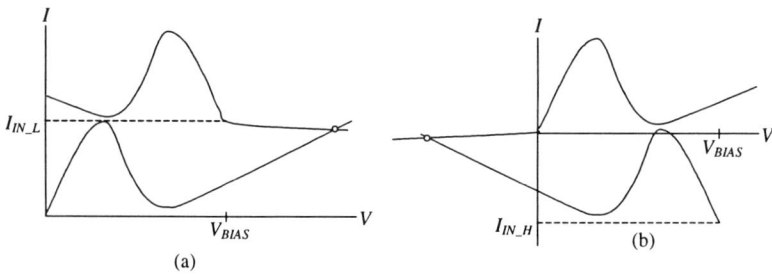

(a)   (b)

Figure 6.12 Figures (a) and (b) demonstrate currents that cause the bistable latch to flip before it is clocked. Current must stay within an acceptable range ($I_{IN\_L} - I_{IN\_H}$) to prevent this phenomenon.

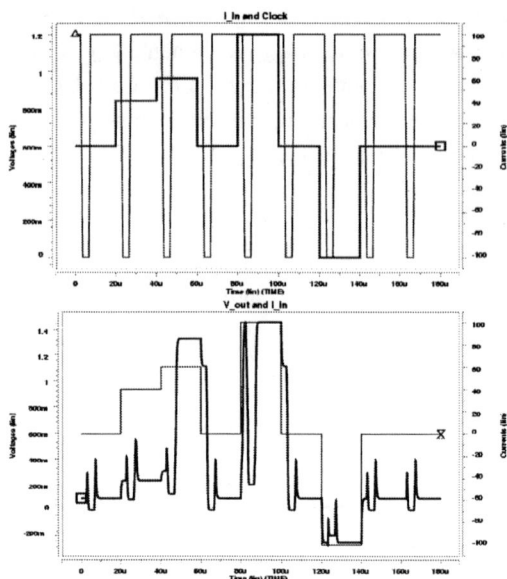

Figure 6.13 In the top graph, clock voltage (V_{BIAS}) and input current are displayed for the bistable latch shown in Figure 6.11. The bottom graph displays the input current and output voltage at the data node. Input current is inserted at the data node while V_{BIAS} is high. Then V_{BIAS} is quickly brought down so that the latch is now in a monostable state (see Figure 6.11a) and then brought back up to a high voltage. Based on the input current, the latch sets either high or low. Note that the fifth input current is too high and causes the latch to flip before it is clocked. Similarly, the seventh input current is too low.

Note that there are also currents that will "break" the latch setup. If I_{IN} is too large, then the latch is forced to a monostable state regardless as in Figure 6.112a. Additionally, if I_{IN} is too small (too negative) the latch is also forced to a monostable state (see Figure 6.12b). Either of these may cause the latch to change states before it has been clocked. This can be a very serious problem when NanoCells are hooked together.[169]

In Figure 6.13 there is a demonstration of the output of the circuit displayed in Figure 6.10. Note that there are currents that latch "on" and "off", as well as currents that "break" the latch. Therefore, if the output pin of a NanoCell is the data node, then using molecular NDRs in the latch we can build a voltage in – voltage out NanoCell with restoring logic. Figure 6.14 demonstrates the circuit for this setup.

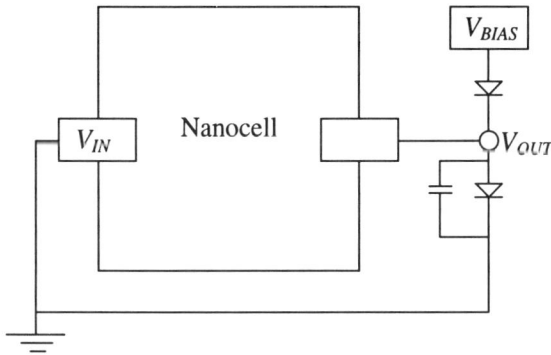

Figure 6.14 This is a NanoCell whose output pin is connected to a bistable latch.

Table 6.1 For the NanoCell outputs shown in the first three columns, thresholds can be set so that the cell functions as a NAND (2nd three columns) or a NOR (3rd set of three columns).

Nanocell Data			NAND "Off" Current Threshold: 275 pA; "On" Current Threshold: 350 pA			NOR "Off" Current Threshold: 450 pA; "On" Current Threshold: 500 pA		
Pin A	Pin B	Pin 1	Pin A	Pin A	Pin A	Pin A	Pin A	Pin A
1 V	1 V	523 pA	0	0	1	0	0	1
1 V	5 V	421 pA	0	1	1	0	1	0
5 V	1 V	413 pA	1	0	1	1	0	1
5 V	5 V	217 pA	1	1	0	1	1	0

6.2 Training a NanoCell

6.2.1. *NanoCell as an Optimization Problem*

The object in training a NanoCell is to take a random NanoCell and turn its switches (edges) "on" and "off" until it functions as some target logic device. The physical position of each nanoparticle and molecular switch is fixed. Existing nodes and edges cannot be deleted and new nodes and edges cannot be added. Hence, the planar graph within the cell is static. In the following sections, two formulations of NanoCell training as an optimization problem are presented.

6.2.1.1. Nanocell Training

Nanocell training occurs by switching the molecules within the cell "on" and "off". Recall that it has been shown that a molecule can be brought from one state to the other by passing a certain voltage pulse across it.[5c] In the molecular based logic device, the NanoCell is treated as a black box. Its topology is unknown. In addition, access to the NanoCell is limited to the I/O pins at the edges.

The NanoCell is trained through a series of voltage pulses applied to the I/O pins. Voltage pulses are applied to the I/O pins, and then the resulting configuration is tested against the desired truth table. Hence, it is a control problem where the controls are the voltage pulses and the states are the switch states along with the various currents and voltages. This is a difficult problem for many reasons. First there are both continuous and discrete state variables. The switch states of the molecules are discrete states, while the voltages and currents are continuous variables. Second, solving the currents and voltages of the system is a nonlinear problem involving a system of ordinary differential equations (ODEs). Additionally, each step in the search space of voltage pulses is, in general, irreversible. For instance, Figure 6.15 depicts a set of switch states that is impossible to reach again once they are altered and the two molecules have the same I(V) curve. When one molecule switches "on", the other must necessarily switch "off."

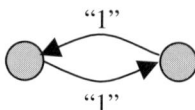

Figure 6.15 Because their voltage drops are the same, when one of these molecules switches "on", the other must switch "off.".

Our initial strategy for training the NanoCell is to model the problem as a *dynamic program*. There are two principal elements of a dynamic programming problem.[170] The first is a discrete time dynamic system whose state transition depends on a control. For the NanoCell training problem, discrete time is a series of observation, action, observation, action, etc. At each observation step we check the functionality of the NanoCell by applying the input voltages corresponding to the desired truth table and observing the resulting output signal. At each action step we apply voltages to the I/O pins to alter the switch states within the NanoCell. Note that while this is a discrete time system, at each discrete action and observation step, a system of continuous time ODE's is solved.

The NanoCell training problem as stated is extremely difficult. We are just now beginning to make progress on this problem using neuro-dynamic programming.[170] These initial results are encouraging, but before attacking this problem, we investigated a simplified version of the NanoCell training problem. The focus of this chapter is this second, simplified problem. The first assumption is that the connections within the NanoCell are known, that is, we know the graph within the NanoCell. Note that in actuality the NanoCell is a black box and knowledge of the interior is limited to measurements taken at the I/O pins. The second assumption is that we have control over the state of each individual switch so that each switch is selectively, reversibly settable. We call training a NanoCell under these conditions *omnipotent training*. We call the first, more realistic problem *mortal training*. In the next section, the problem of omnipotent training is explored in more detail.

6.2.1.2. Omnipotent Training

Before exploring the optimization problems with the assumption of omnipotence and omniscience, it is worthwhile to ask whether such a problem is of practical use or is merely an academic exercise. This molecular electronics project is currently in the proof-of-concept phase. Before determining whether it is possible to train a NanoCell with realistic constraints, we are attempting to verify whether it is theoretically possible. If it becomes clear that it is impossible to train a randomly assembled NanoCell as a 2-bit adder, even with the assumptions of omnipotence and omniscience, then there is no point in trying to train one without these simplifying assumptions. Hence, the optimization problem with the supposition of omnipotence is of practical use.

In the optimization problem with omnipotence, we are given the graph within the NanoCell along with the I(V) curve of the molecule, the target logic device and settings of the I/O pins (i.e., which are set to input and output). The search space is all possible combinations of switch, or edge states. Simply stated, the problem then is:

Given the graph of the NanoCell G, the I(V) curve of the molecule, the target logic device and the settings of the I/O pins,

$$\min \quad f(S) \text{ subject to}$$
$$S \in \{0,1\}^m.$$

Here f involves two steps: (1) evaluate the output currents and (2) compare these currents against the desired logic.

The simplified NanoCell training problem is particularly well suited to genetic algorithms. After presenting the fundamentals of genetic algorithms, a heuristic for solving this optimization problem is presented.

6.2.2. *Genetic Algorithms*

Genetic algorithms (GA) work by taking a population of individuals, represented as strings of 1's and 0's, quantifying their fitness, then recombining them to generate a new population of children. Usually the first generation is randomly created, and then three operators are used to produce each subsequent generation: *selection, crossover* and *mutation*.

First two parents must be selected. They are either selected randomly or more fit individuals are given preference. The random method is obvious. The most simplistic method of giving preference is to use *roulette wheel selection*. Each individual is given a portion of a roulette wheel that is proportional to its fitness. The wheel is spun to select each parent. Hence, the most fit individual is most likely to be selected as a parent, and the least fit individual is the least likely to be chosen. Another method of selection is *tournament selection*. In tournament selection, a subset of n individuals is chosen, and the two most fit of this group are chosen to reproduce next. Tournament selection has the advantage of varying the degree to which the most fit individuals are favored. Favoring them too much can cause the problem of too little population diversity.[171]

Once two parents are selected, they must be recombined to form two new children. *Single point crossover* is the simplest recombination method. If the length of the chromosome (the string of 1's and 0's representing each individual) is m, then some point p between 1 and $m - 1$ is chosen as the crossover point. To create the first child, the first p bits of the first parent are combined with the last $m - p$ bits of the second parent. The second child is created by attaching the first p bits of the second parent to the last $m - p$ bits of the first parent. Alternatively, with *n-point crossover*, n crossover points are selected and the chiidren are produced analogously. This process is demonstrated in Figure 6.16.

In *uniform crossover*, a coin is flipped for each bit of the chromosome. If it is heads, the first child gets this bit from the first parent, and the second child gets this bit from the second parent. If it is tails, then the first child gets this bit from the second parent, and the second child gets this bit from the first parent. Hence, uniform crossover is similar to n-point crossover, except the number of crossover points changes with each new pair of children. Uniform crossover is demonstrated in Figure 6.17. The two parents are both automatically replaced by their children, or of the four individuals, the two most fit are kept and the other two discarded.[171]

Parent 1   0 1|1 1 0 0 1|0|0 1 1|1 0 0   0 1 1 0 0 1 0 0 0 1 0 1 0 0   Child 1
Parent 2   1 1|1 0 0 1 0|0|0 1 0|0 1 1   1 1 1 1 0 0 1 0 0 1 1 0 1 1   Child 2

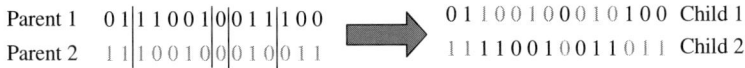

Figure 6.16 In n-point crossover, n points in the vector are randomly chosen, and crossover is performed to produce two new children.

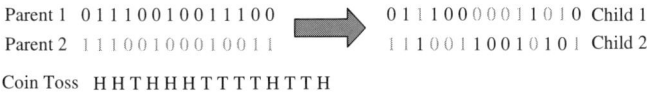

Parent 1   0 1 1 1 0 0 1 0 0 1 1 1 0 0   0 1 1 1 0 0 0 0 0 1 1 0 1 0   Child 1
Parent 2   1 1 1 0 0 1 0 0 0 1 0 0 1 1   1 1 1 0 0 1 1 0 0 1 0 1 0 1   Child 2

Coin Toss   H H T H H H T T T T H T T H

Figure 6.17 In uniform crossover, a coin is flipped for each switch state. For heads, the first child gets the first parent's switch state, and the second child gets the second parent's switch state. For tails, the opposite action is taken.

After crossover, each of the two new children is mutated. Each bit of each new child is flipped with some probability p. Mutation keeps the GA from losing certain chromosomal information. For instance, without mutation if every individual in the current generation has a "0" in some particular bit, then it is impossible for subsequent generations to have a "1" in this bit. Hence, potentially beneficial genetic information is not lost when mutation is used.[171]

After a new generation is produced, the fitness of each new individual is evaluated and the replacement process begins. First, replacement can simply be generational. The older generation is replaced by the newer generation regardless of how the fitness values compare. Another strategy is to choose the two most fit individuals among the two parents and two children. Yet another strategy is to make a slight variation of generational replacement called *elitism*. Each new generation replaces the older generation, but one or two of the most fit individuals of the older generation are always copied into the younger generation.[171]

There are some potential problems with using a genetic algorithm. First a lack of population diversity can cause premature convergence. Crossover becomes ineffective when the chromosomes of the two parents are too similar. Giving too much preference to more highly fit individuals can cause this problem. A balance must be maintained between giving preference based on fitness and preserving population diversity. Another potential problem concerns the fitness function. You may not be measuring what you intend to measure, and will therefore give preference to inferior individuals. These are both problems that arose in training NanoCells, and the solutions are presented later in this chapter.

6.2.3. *Training Process*

The primary factor in determining the functionality of a NanoCell is the I(V) characteristic of the molecule used. Hence, we have found that before beginning to train NanoCells it is essential to first gain an understanding of the particular molecule used.

6.2.3.1. Testing the Molecule

Our first step in working with the molecule is to simply study the I(V) curve of the high conducting state and try to assess which logic gates would make sense. For instance, consider the dinitro I(V) displayed in Figure 6.5. Note that the peak voltage is about 0.6 V, and the valley voltage is about 0.8 V. We look at this curve and see several logical possibilities. First the very simple gates, AND and OR look feasible (these truth tables are shown in Tables 6.2 and 6.3). For both of these gates suppose that the "on" input voltage is such that with one pin "on", the output current is about half of its max and with two pins "on" the output is at its max. Then for two voltage sources attached to a dinitro molecule, output current thresholds I_{OL} and I_{OH} and/or input voltages V_{IL} and V_{IH} can be set so that this single molecule functions as either an AND or an OR. In Figure 6.18 a NanoCell comprised of just three molecules is displayed. This cell can work as either an AND or an OR depending on input voltages. The input voltages and output currents for an AND gate and an OR gate are displayed in Figures 6.19 and 6.20.

Table 6.2 This is the truth table for an AND gate.

AND

In 0	In 1	Out 0
0	0	0
0	1	0
1	0	0
1	1	1

Table 6.3 This is the truth table for an OR gate.

OR

In 0	In 1	Out 0
0	0	0
0	1	1
1	0	1
1	1	1

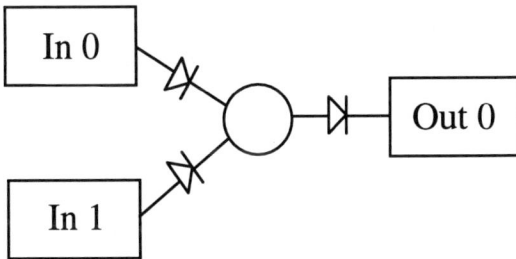

Figure 6.18 These three molecules with two inputs and one output can function as a NAND, AND, OR or XOR.

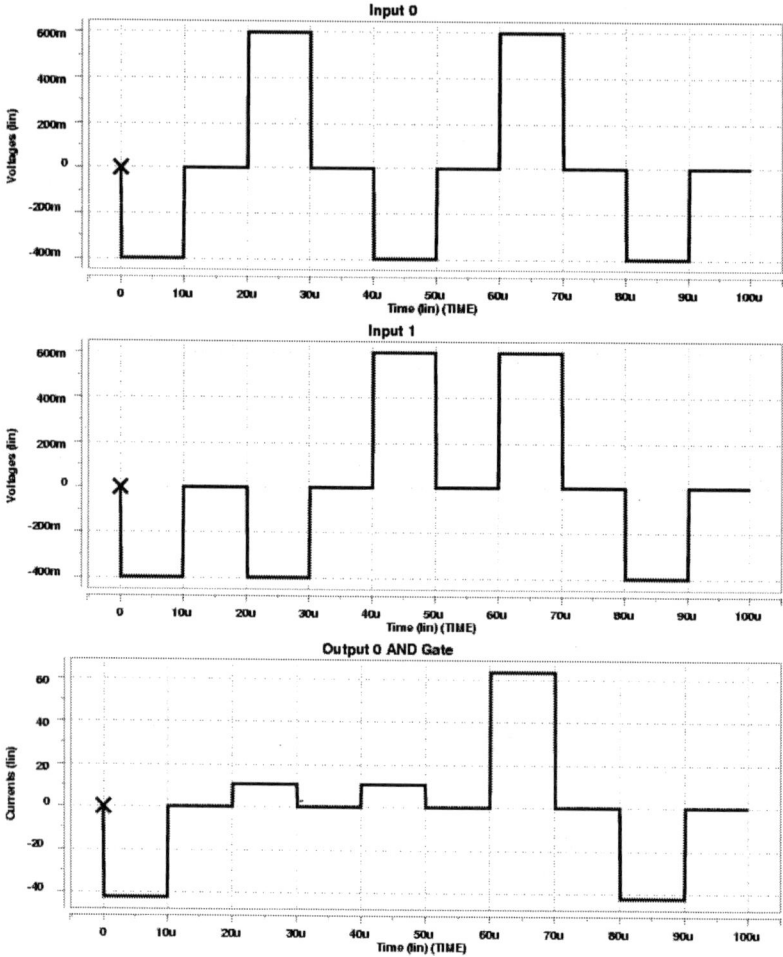

Figure 6.19 The first two graphs are input voltage, and the bottom graph is output current for the 3-molecule AND gate shown in Figure 6.19. Note that current is highest when both inputs are high.

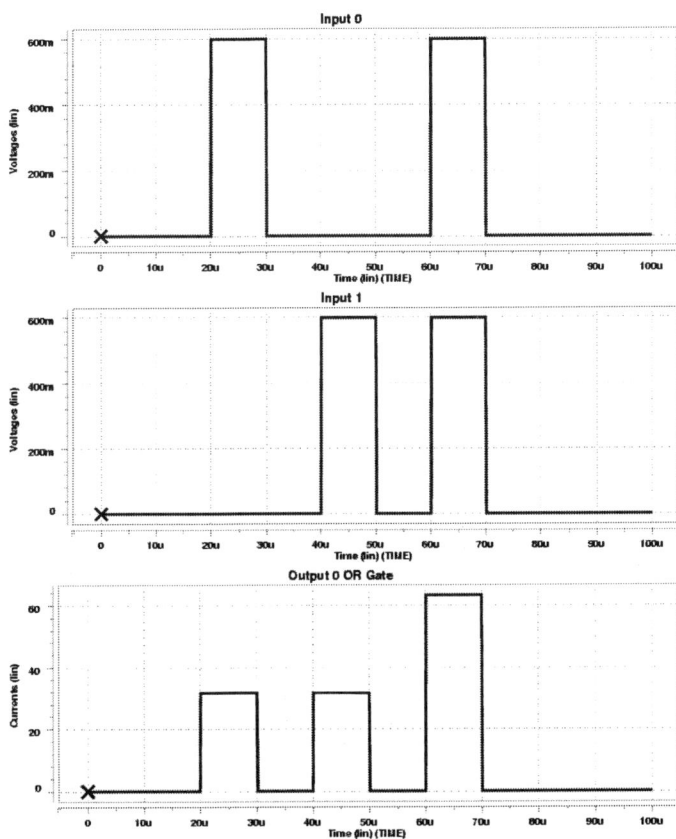

Figure 6.20 The first two graphs are input voltage, and the bottom graph is output current for the 3-molecule OR gate shown in Figure 6.18. Note that the high threshold can be set so the output is high when either one or two inputs are high.

Now consider an inverter whose truth table is shown in Table 6.4. Note that NDR is essential to obtain this logic. Suppose that the "off" input voltage is such that output current is at its peak, and the "on" input voltage is such that output current is at its valley. Then a voltage source hooked to a dinitro (or any NDR molecule) should function as an inverter. In Figure 6.21 a NanoCell with a single molecule functions as an inverter with input voltages and output currents displayed in Figure 6.22.

Table 6.4 This is the truth table for an inverter.

Inverter

In 0	Out 0
0	1
1	0

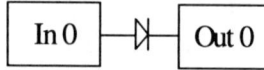

Figure 6.21 This single molecule with one input pin and one output pin functions as an inverter.

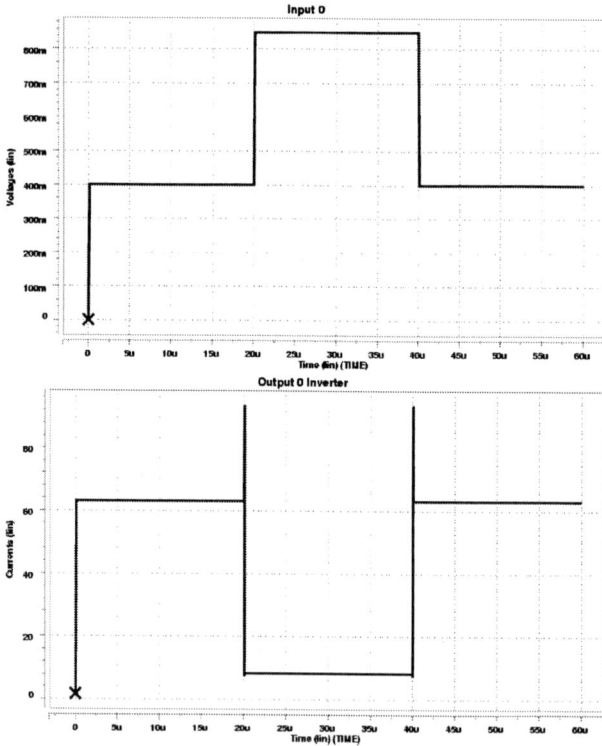

Figure 6.22 The input voltage (top) and output current (bottom) are displayed for a 1-molecule inverter (see Figure 6.21). Note that when the input is high, the output is low, and vice versa. The current spikes are due to simulation issues and would not occur in an actual NanoCell.

Table 6.5 This is the truth table for an XOR gate.

XOR

In 0	In 1	Out 0
0	0	0
0	1	1
1	0	1
1	1	0

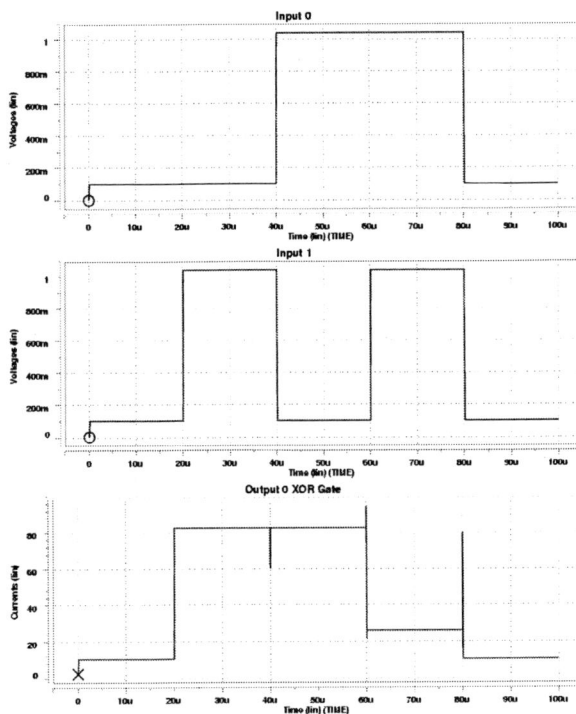

Figure 6.23 The first two graphs are input voltage, and the bottom graph is output current for the 3-molecule XOR gate shown in Figure 6.18. Note that current is high only when exactly one pin is high. The current spikes are due to simulation issues and would not occur in an actual NanoCell.

An XOR also seems feasible with this molecule. Its truth table is shown in Table 6.5. Once again, because this gate requires negation, NDR is vital. Let the "on" voltage be such that output current is near its peak with one input pin "on". Then with two pins "on", the output current should be at its valley. Therefore, it is intuitive that two voltage sources hooked to an NDR should function as an XOR. The NanoCell displayed in Figure 6.18 works as an XOR with output as shown in Figure 6.23.

Now consider a NAND gate whose truth table is displayed in Table 6.6. Again NDR is necessary for this gate. Suppose that the "off" voltage is about ¼ of the peak voltage and the "on" voltage is about ¾ of the peak voltage. Then with both inputs "on", the output current should be pushed into the valley. Output current thresholds can be set so that the output is "on" when zero or exactly one input pin is "on". Once again, the 3-molecule NanoCell displayed in Figure 6.18 functions as a NAND. The corresponding input voltages and output currents are displayed in Figure 6.24. Note that one can obtain an inverter from a NAND gate. If one input is held to high voltage (high rail), then a NAND inverts the second input. Hence the NAND setup works as an inverter if one pin is set to high rail.

What about more complex truth tables? First consider the truth table of a half adder which is shown in Table 6.7. A half adder simply adds two one bit numbers together. Note that output 1 is simply an XOR gate, while output 2 is an AND gate. So this truth table seems achievable with the dinitro molecule, or something similar. Note that the input voltages mentioned above were different for an XOR and for an AND. However, for a half adder, the input voltages must be the same. This is easily resolved by the genetic algorithm in the training process. Lower input voltages can be replaced by longer paths between the input and the output.

Next consider a 1-bit adder, which is very similar to a half adder. The difference is that a 1-bit adder, or full adder, has three inputs, one of which is for a carry-in. Hence, with a full adder every 2-bit number is a possible output.

Table 6.6 This is the truth table for a NAND gate.

NAND

In 0	In 1	Out 0
0	0	1
0	1	1
1	0	1
1	1	0

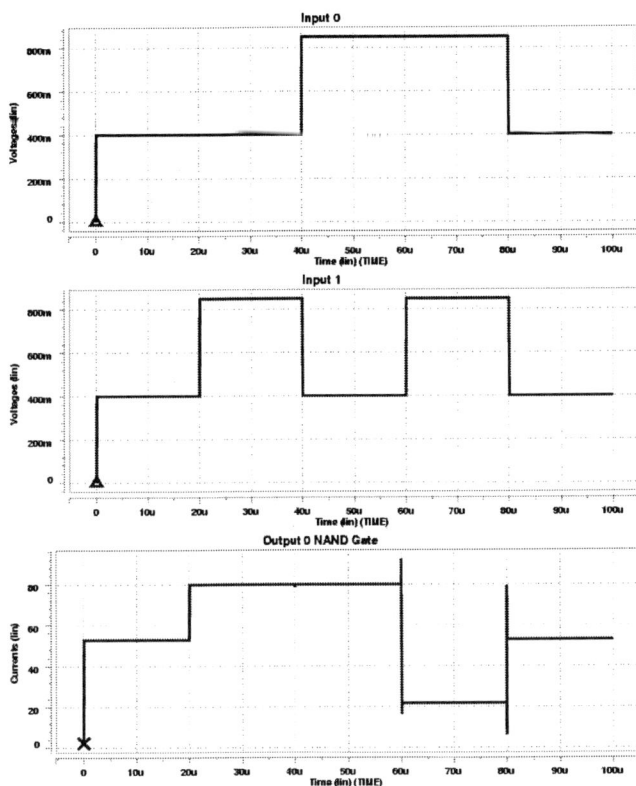

Figure 6.24 The first two graphs are input voltage, and the bottom graph is output current for the 3-molecule NAND gate shown in Figure 6.18. Note that current is low only when both pins are high. The current spikes are due to simulation issues and would not occur in an actual NanoCell.

Table 6.7. This is the truth table for a half adder.

Half Adder

In 0	In 1	Out 0	Out 1
0	0	0	0
0	1	0	1
1	0	0	1
1	1	1	0

Looking at the truth table of 1-bit adder (see Table 6.8), it is not immediately clear whether the dinitro molecule should be sufficient. First consider output 1. It is simply a majority gate on the three inputs. If two are more inputs are "on", then output 1 should be "on". This is synonymous to an AND gate and seems reachable with the dinitro. Now consider output 2. It should be "on" if exactly one or three inputs are "on". Suppose that V_{IH} is the input high voltage. Then V_{IH} should turn output 2 "on", $2 \cdot V_{IH}$ should turn it "off" and $3 \cdot V_{IH}$ should turn it back "on". Initially, we attempted to train NanoCells containing dinitro molecules as full adders and was unsuccessful. Upon more deeply considering the I(V) curve, the system did not appear trainable with this I(V). We decided that we needed a molecule that went from high current to low current and back to high current again. Fortunately, molecules exhibiting this behavior have been synthesized.[82b,167] However, to more easily train adders with good "on-to-off" ratios, we used the idealized version of this curve shown in Figure 6.6. Adder behavior makes sense with this molecule.

Table 6.8 This is the truth table for a 1-bit adder.

1-Bit Adder

In 0	In 1	In 2	Out 0	Out 1
0	0	0	0	0
0	0	1	0	1
0	1	0	0	1
0	1	1	1	0
1	0	0	0	1
1	0	1	1	0
1	1	0	1	0
1	1	1	1	1

Thus far we have discussed the intuition used to train voltage in – current out NanoCells. However, the voltage in – voltage out setup involves bistable latches which are driven by input current. Hence, molecules that work for one setup also work for the other. However, there is an additional step for voltage in – voltage out. The NDR molecules used in the latch must be tested to determine the "on" and "off" input voltages (V_{IH} and V_{IL}) as well as the input currents that break the latch. This can be done by simply graphing the latch NDR, $g(V)$ and $g(V - V_{BIAS})$ as in Figure 6.25. In addition tests on the latch with a simple input current source are performed as in Figure 6.13.

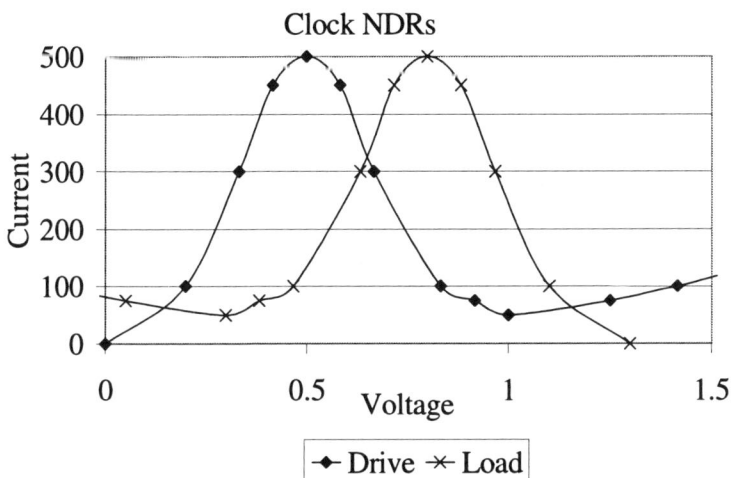

Figure 6.25 This is a clock NDR used in simulating NanoCell training. The resulting low and high voltages, V_{IL} and V_{IH} are shown, as well.

Once the molecules are tested, then full NanoCells are tested to find appropriate input voltages. The voltages that worked for 1 to 4 molecules must be increased to work for a full-sized NanoCell. In addition, for the voltage in – current out setup, appropriate output current thresholds, I_{OL} and I_{OH} must be determined. Running the output of a single NanoCell is usually sufficient to find reasonable settings.

It is our experience that intuition as to which logic gates are possible with a given NanoCell is usually correct. However, it should be noted that if a NanoCell contains different molecules, then different logic gates may become possible. For instance two molecules with different peaks connected in parallel will give the net effect of a single molecule with two peaks as shown in Figure 6.26. This has been observed in experiments by NanoCell collaborators at Motorola.[172] Additionally, intuition should always be questioned. We hypothesized that a molecule with a plateau after the peak would result in superior logic gates. It seems as if they will work for very large voltage ranges. However, in testing the I(V) shown in Figure 6.27, we found that it is actually very difficult to get negation with these. Hence, intuition should always be tested.

NDR Molecule 1

NDR Molecule 2

NDR Molecules 1 &2 in Parallel

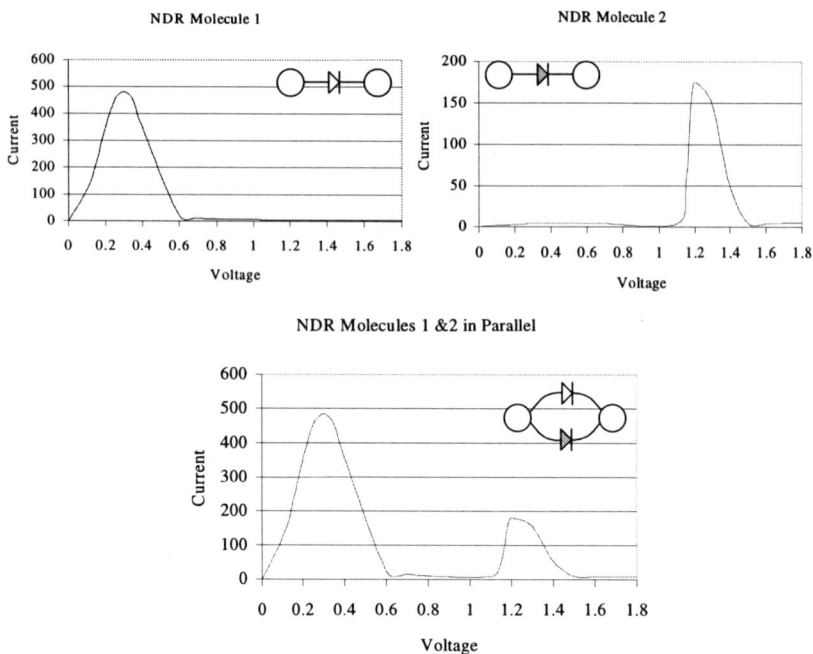

Figure 6.26 These NDR molecules in parallel give two current peaks.

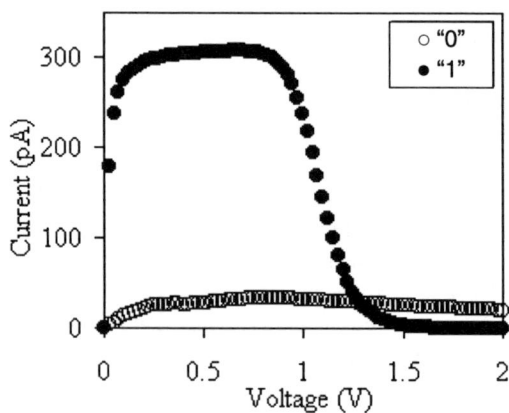

Figure 6.27 This molecule exhibits NDR along with a plateau. It does not work well in simulations.

6.2.3.2. Genetic Algorithm Used in Training Nanocells

A pictorial overview of the NanoCell GA is shown in Figure 6.28. Random configurations of the same NanoCell are generated, each individual is evaluated, then a new population is produced through crossover and mutation.

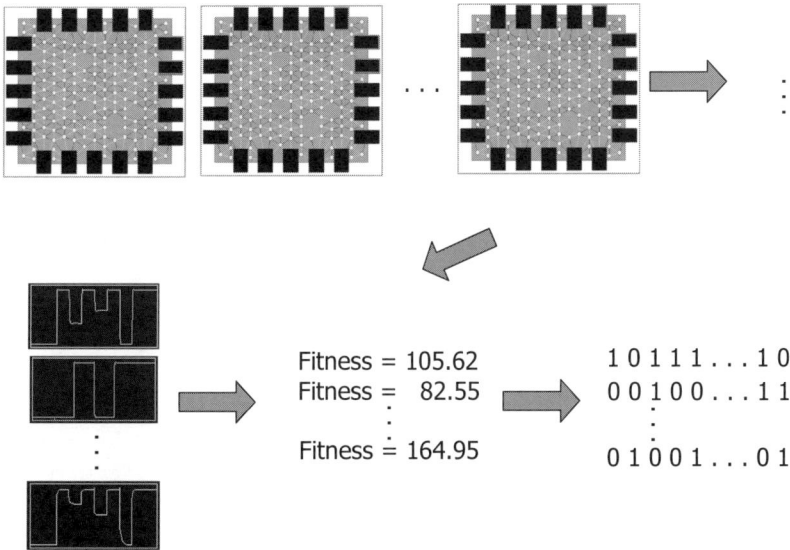

Figure 6.28 This is an illustration of the NanoCell GA. An initial population of random NanoCells is generated where the positions of particles and molecules are fixed, but switch states are randomly altered. These are represented by binary vectors that are sent to the circuit solver HSpice. Output signals are then generated and a fitness is determined for each cell. Based on these fitness scores a new population is generated.

The first step in a GA is to generate an initial population. The NanoCell GA uses a random population of 25 individuals. Tournament selection with a tournament size of 5 is used to choose parents. If two chosen parents have the same fitness, then a new parent is chosen. Both uniform crossover and 5-point crossover are used with comparable success. The mutation probability is $\frac{1}{2l}$ where l is the length of the chromosome, or number of switches. Replacement is generational, so parents are replaced by their children regardless of relative fitness. However, elitism is used so that the two best individuals are always copied into the next generation.

As mentioned previously, a balance must be maintained between population diversity and preference to the highly fit. In this NanoCell GA preference is given by using tournament selection and by saving the two best individuals of each generation. Population diversity is maintained by always replacing parents with children regardless of their fitnesses and by not allowing parents with identical fitnesses to mate. These parameters have worked very well in training NanoCells. In Figure 6.29 there is a graph of generation vs. fitness for an inverter. Note that generational fitness goes to zero even though parents are replaced by their potentially inferior children.

GA Convergence

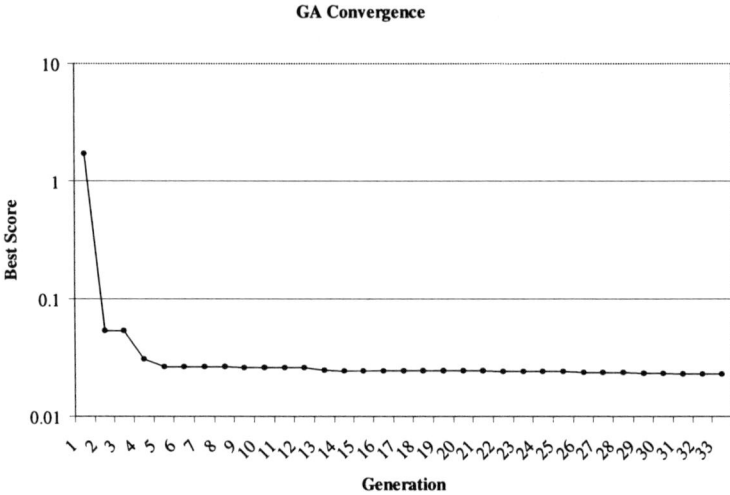

Figure 6.29 Displayed here is generation number vs. best fitness on a logarithmic scale.

We further experimented with training NanoCells using another Monte-Carlo search algorithm, simulated annealing.[173] Simulated annealing produced essentially the same results as the GA. It is our opinion that in this case, the particular base search algorithm is not nearly as important as the manner in which it is adapted to the NanoCell problem.

6.2.4. *Fitness Function*

In adapting the GA to the NanoCell training problem, the fitness function is the most difficult issue. First consider the voltage in – current out setup. Recall that I_{OL} and I_{OH} denote the high and low output current thresholds. Let I_j^i denote the output current for the ith output pin at the jth time step. Then the fitness can be defined

$$F = \sum_{i=1}^{NumOutputs} \left(\sum_{j"on"} \left(I_{OH} - I_j^i \right) + \sum_{j"off"} \left(I_j^i - I_{OL} \right) \right) \tag{6.2}$$

However, suppose the target logic gate is a NAND, and consider the output graphs shown in Figure 6.30, where (a) is the target. The graph shown in (b) is on the correct scale but does not have NAND behavior. On the other hand, the graph shown in (c) has the correct shape but is simply on the wrong scale. Output (c) is definitely the more desirable of the two, but it will get a lower score because its current is too low at almost every measurement. However, output (b) will get a fairly good score because it is only wrong at one measurement. This example is analogous to the illustration of a broken watch. A clock that does not work is still correct twice a day, whereas a clock that is one second slow will never be correct. Hence, it is necessary to measure both the magnitude and the shape of the output current.

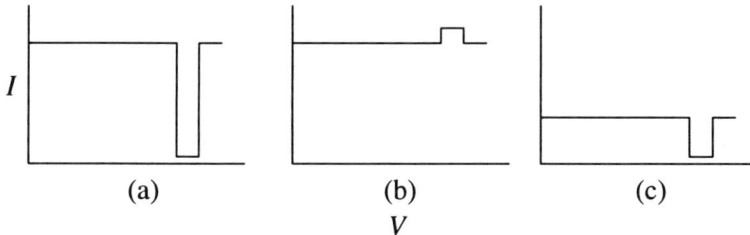

Figure 6.30 If figure (a) is the target output (NAND output), then figure (b) will get a better score than figure (c). However, figure (c) is closer to the desired logic. It is simply on the wrong scale. Scaling the initial score based on the various slopes of the output alleviates this problem.

The fitness function is computed as before in Eq. 6.2, but the score for each individual output is scaled based on the slope of certain portions of the graph before the output scores are summed. The particular portions and the severity of the scaling differ from logic gate to logic gate. We found, for instance, that what works for a NAND does not work for an XOR gate or an AND gate. This makes sense because a potential AND gate should be rewarded if the current for the "1 1" test is greater than the current for the "0 1" or "1 0" test. A NAND gate, on the other hand, should be penalized for such output. Hence, in the NanoCell simulator a current scaling scheme is read in based on the particular truth table.

Scoring a voltage in – voltage out NanoCell with bistable latches is very different. Recall that for a latched output there are only two possible output voltages. Therefore, if there are n different outputs then there are only 2^n possible output graphs and the problem is now a discrete optimization problem. Suppose you are training an XOR and two configurations of the NanoCell have identical output. How can you tell which one is closer in some sense to the desired XOR output? The answer is to consider output current as well as output voltage. Recall that the current through the output pin is what causes the output node to latch high or low. The fitness for a voltage in – voltage out NanoCell is therefore computed by first finding the basic fitness score as in Eq. 6.8 except the output voltage, rather than output current is measured. Next the score is scaled based on output current as in the voltage in – current out scenario. In Figure 6.31 there is an example of the output voltage and current of an XOR gate.

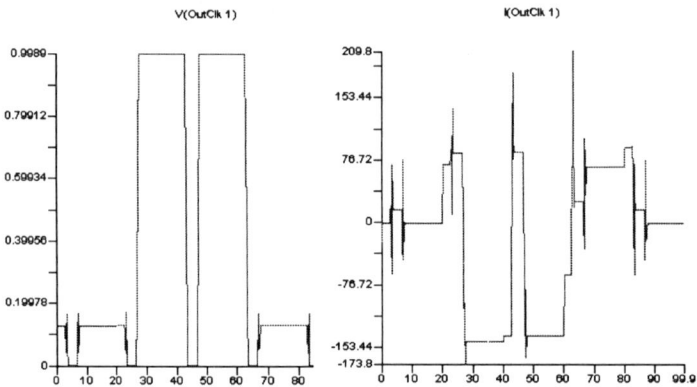

Figure 6.31 On the left, the output voltage of an XOR gate is shown. Note that due to the bistable latch, there are just two possible output voltages. Output current is shown on the right.

6.3 Trained Nanocells

Before we began simulating the training of NanoCells, no one knew whether anything useful could be done with a random array of NDR devices. With the NanoCell simulator, we have shown that in fact NanoCells can be trained as fairly complex logical devices with the simplifying assumption of omnipotent training.

6.3.1. *Voltage In – Current Out*

In this section the results of the simulated training of voltage in – current out NanoCells are presented. Using this design, several complex, negating logic gates were trained, many of them easily. Except for the four NANDs and the 1-bit adder, all of these NanoCells were trained with the version of the simulator that worked with IsSpice.

6.3.1.1. Trained Logic Gates

Inverters, NAND gates, half-adders and 1-bit adders have been found using the NanoCell simulator. For the inverters, NANDS, and half-adders, the simulated I(V) curve displayed in Figure 6.2 was used to characterize the "on" and "off" states of the molecular switches. A simulated I(V) curve with rectifying diode behavior was used for the 1-bit adder.

12 NanoCells were randomly generated, and all 12 were successfully trained as inverters. In Figure 6.33 the output of one of the inverters is shown. The pin labeled "A" is set to input, and the pin labeled "1" is set to output. High input voltage is set at 2 V and low input voltage is set at 0.5 V. The output pin is considered "off" if there is < 7 nA recorded, and "on" if > 700 nA are recorded. Hence, the "on-to-off" ratio is 100 to 1. If the thresholds are allowed to vary with each NanoCell, an "on-to-off" ratio of 1000 to 1 is obtained. The accompanying plots show the voltage as a function of time $(V(t))$ for input A. The corresponding output current through the output pin is also plotted as a function of time. Note that Output "1" is high when Input "A" is low. Likewise, the Output "1" is low when the Input "A" is high, which is the proper truth table sequence for an inverter (see Figure 6.32).

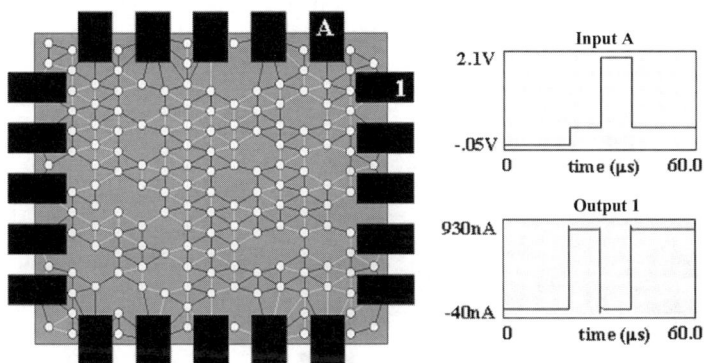

Figure 6.32 This is a NanoCell trained as an inverter. Pin "A" is set to input, and pin "1" is set to output. The input voltage and output current are displayed, as well.

It took an average of four generations to train each inverter. The simulation time depends primarily on the number of molecular switches in the NanoCell. To run a generation of 25 individuals it takes approximately 10 sec if there are 10 switches, 25 sec if there are 100 switches, and 250 sec if there are 1000 switches. Hence four generations took about 160 sec on a 800 MHz desktop PC, virtually all of which was simulation time for IsSpice to operate. In actual physical training time we estimate that this would take on the order of 1 msec since the NanoCell and test electronics can operate at a rate of 100

MHz, thus 100,000 trials can be performed in 1 msec. Encouragingly, three of the inverters were found in the initial random population, before the genetic algorithm began to converge upon a solution. This indicates that the solution space is enormous, which will be helpful when the move is made toward more realistic mortal switching and when considering defect- and fault-tolerant needs wherein multiple solutions are necessary.

In addition to the inverters, input/output settings for NAND gates were discovered. NANDs are particularly attractive since they constitute a functionally complete logic set meaning that any logic function could be created from NAND sets. The settings that were found to yield NAND gates are shown in Figure 6.33. The logic for a NAND requires two distinct inputs. The pins labeled "A" and "B" are the input pins, and the pin labeled "1" is the output. High input voltage is set at 2 V, while low input voltage is set at 0.5 V. The output pin is considered "off" if there is < 6.9 nA recorded, and it is considered "on" if > 345 nA are recorded. Hence, there is a 50 to 1 "on-to-off" ratio. As with the inverters, an even better "on-to-off" ratio (100 to 1 or 1000 to 1) was obtained when the thresholds vary from NanoCell to NanoCell. The accompanying plots show the $V(t)$ for the inputs "A" and "B" and the corresponding output current over time. Note that Output 1 is high when either Inputs "A" or "B" are high, but low when both Inputs "A" and "B" are high, in concert with the accompanying NAND truth table in Table 6.6.

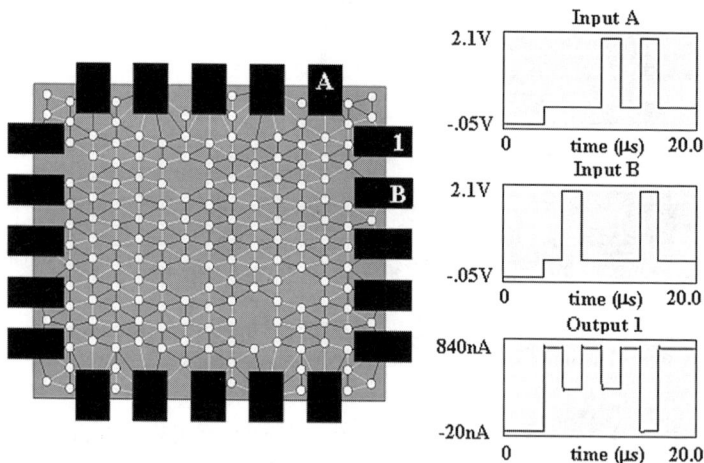

Figure 6.33 This is a NanoCell trained as a NAND gate. Pins "A" and "B" are set to input, and pin "1" is set to output. The input voltages and output current are displayed, as well.

12 NanoCells were randomly generated, and all 12 were successfully trained. However, the pin settings had to be adjusted to get one of the NANDs

to converge since it was too sparse around the original input pins, so the locations of the pins were changed from the upper right corner of the NanoCell to the lower left corner. After this change, the NanoCell was successfully trained as a NAND gate. On average it took about nine generations for each NAND to converge. This took about 6 min to run due to the SPICE simulations. Again, this should take merely milliseconds in actual physical training time (*vide supra*). As with the inverters, two NAND gates were found in the initial random population. Once again, this indicates a vast solution space, which will be extremely helpful when the assumption of omnipotence is dropped.

To test the robustness of the NAND gates, input "B" is held to high voltage while input "A" is swept from "off" to "on" to "off". Next, "A" is set to constant high voltage, and "B" is swept from "off" to "on" to "off". For both of these tests, each NAND gate functions as an inverter with the "on" and "off" thresholds given for the NANDs. As anticipated, this implies that the NAND gates are robust.

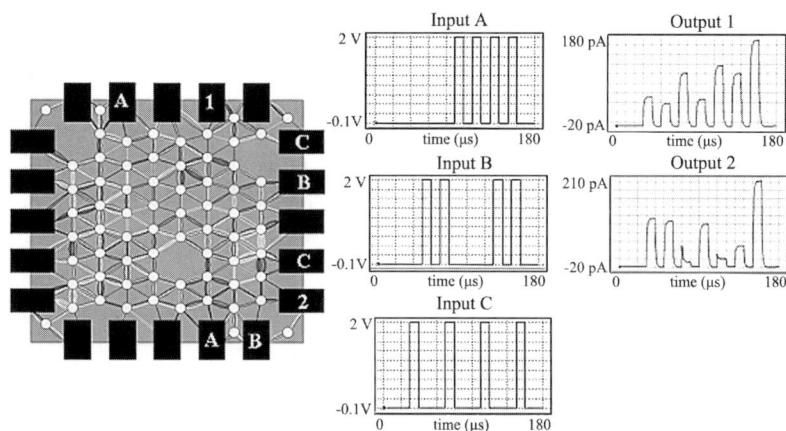

Figure 6.34 This is a NanoCell trained as a 1-bit adder. Pins "A", "B" and "C" are set as inputs, and pins "1" and "2" are set to output. The input voltages and output currents are displayed, as well. This result is particularly significant because two entirely different signals are obtained from output pins in the same cell.

Finally, a 1-bit adder has been trained (Figure 6.34) with a 70-nanoparticle, 1000-molecular switch NanoCell, where the molecules exhibit rectifying diode behavior as displayed in Figure 6.7. In Figure 6.34, the pins labeled "A" are set to the first input, those labeled "B" are set to the second input, and those labeled "C" are set to the third input. The output pins are labeled "1" and "2". High input voltage is set at 1.8 V, while low input voltage is set at 0 V. The output pin is considered "off" if there is < 50 pA recorded. It is considered "on" if > 100 pA are recorded. The accompanying plots show the *V(t)* for the inputs and the corresponding output currents over time. Improved search techniques and the exploitation of different molecules and pin settings should improve upon the 2:1 "on-to-off" ratio exhibited here. This result is significant in that it demonstrates that a NanoCell can be trained as a complex logic device with multiple outputs.

Observations made in the genetic algorithm trials indicated that to train a NanoCell as a NAND, one must simply have enough molecular switches in the "on" state near the input and output pins. Additionally, the number of "on" molecules between input A and the output should be approximately the same as the number between input B and the output. This conclusion indicated that a single NanoCell could be trained as several independent NANDs. We tested this hypothesis on a very large NanoCell – approximately 900 nanoparticles and 9000 molecular switches, with multiple switches between adjacent nanoparticles. The NanoCell is displayed in Figure 6.35 where each two-letter set is the independent NAND input, i.e. "A" and "B", working in concert with the nearby output designated by a numeral, i.e. "1". The molecular switches in each of the four corners are in the "on" state while the molecular switches in the middle of the cell are "off". This establishes a barrier between the four corners. Each of the four corners of this single cell functions as an independent NAND gate with minimally a 15:1 "on-to-off" ratio. This NanoCell might be straightforward to train mortally by initially adding all the molecules into the cell in an "off" state. Next, apply enough voltage to the input pins in each corner to turn "on" most of the molecules in that region. The result should be that each corner functions as an independent NAND.

6.3.1.2. Observations

Several observations were made in training these NanoCells. These observations prove the efficacy of the omnipotence assumption. First, the voltage in – current out model is very robust. The input voltages can be varied by as much as 0.5 V, and the NanoCell retains its logical function. We also learned that to train inverters, XORs and NANDs, it is sufficient to simply have enough molecules in an "on" state. In addition a NanoCell that functions as a NAND will also function as an inverter or an XOR by making very simple changes to the input voltages. For instance, hold one input pin at high voltage and a NAND works as an inverter. Simply lower the input high and low voltages to get an XOR from the same NanoCell (with no change in molecular switch states). Similary, an XOR will function as a NAND or an inverter with similar changes. Recall that three molecules will function as a NAND, XOR, AND or OR. A NAND or XOR can be used as an inverter if one input is held to high voltage. Similar results are found when training full-sized NanoCells.

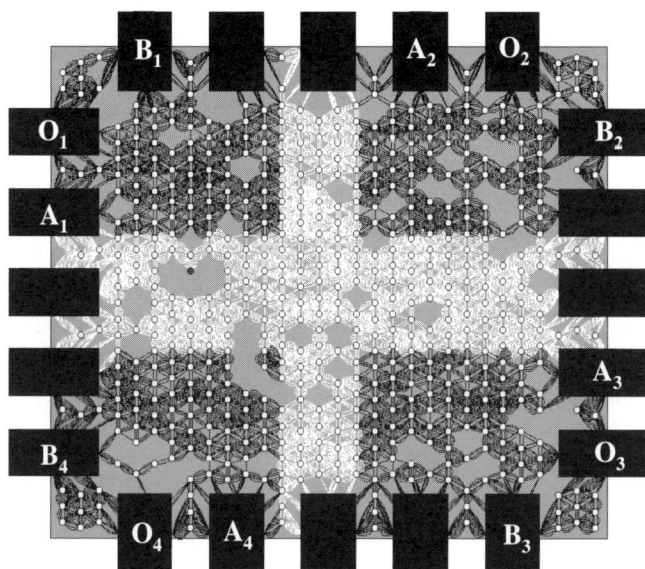

Figure 6.35 This is a NanoCell trained as four independent NAND gates. The molecules in each corner are "on", while others are "off". Pins "A" through "H" are set to input, and pins "1" through "4" are set to output. Pins "A", "B" and "1" form one independent NAND, etc. The "on" to "off" ratio is 15 to 1. If all molecules are "on", there are still four NANDs. The "on" to "off" is just 2 to 1. Note that this NanoCell looks fairly straightforward to train in a mortal fashion.

In fact, training any of the standard 2 input − 1 output logic gates (AND, OR, NAND, XOR, NOR) is very straightforward with the voltage in − current out model. Not all of them were specifically mentioned in the previous section simply because many of them are similar. NAND, NOR and XOR are very similar and AND and OR are very easy to train (they do not even require NDR).

In simulating the omnipotent training of NanoCells, a great deal was discovered about the I(V) characteristics that are desirable in a molecule. First, as mentioned previously, plateaus are not helpful. The current should peak and then come back down (Figure 6.5 vs. Figure 6.27). In addition, basically any NDR molecule will work for a NAND, inverter or any logic gate where the 0 0 . . . 0 input implies an output of 1 1 . . . 1. On the other hand, more is required for logic gates such as XOR, half adder, 1-bit adder or any logic gate where the 0 0 . . . 0 input implies an output of 0 0 . . . 0. Namely, the molecule should conduct very little (the closer to zero, the better) current for negative voltage drops. Initially, we had no success in training such logic gates. However, we later discovered that asymmetric I(V) curves made it possible to get NANDs. We then narrowed this down to the fact that the high conducting I(V) should conduct little to no current for negative voltage drops. Note that for asymmetric I(V) curves, the molecules must be modeled as directed edges because current will flow more in one direction. Therefore, from this point on, the NanoCell graph is modeled as a planar graph with multiple, directed edges.

In addition to these observations, we also found that the NanoCell training problem is vastly more difficult for logic gates with multiple outputs. One strategy for dealing with this problem is to use a separate set of inputs in a separate portion of the NanoCell for each unique output. This strategy was utilized for the 1-bit adder displayed in Figure 6.34. The disadvantage is that this requires much more power to drive the NanoCell, and many more pins are needed for input.

6.3.2. *Voltage In – Voltage Out*

In this section, the results of voltage in – voltage out training are presented. Recall that with this model, the output pins are latched (see Figure 6.14). This model offers the advantage of uniformity of signal and restoring logic but at the cost of robustness of the logic gates. Note that in this section the NanoCells are smaller with one I/O pin to each side. We came up with a new idea for a NanoCell computer architecture that seems easier to train mortally. Instead of attempting to train very complex logic such as a 2-bit adder in a single NanoCell, limit the NanoCell to one I/O pin per side. Then let a half adder be the most complex NanoCell. Then two half adders along with an XOR make a 2-bit adder. Hence, a 2-bit adder could be constructed in a smaller space with three 4-pin NanoCells than with one 20-pin NanoCell. A picture of this architecture is shown in Figure 6.36.[174] These 13 cells make up a 4-bit adder that requires 7 clock cycles, represented by contrasting shades in the figure. Note that the cells with just arrows are simply NanoCells that copy an input to the subsequent clock cycle (some copy two inputs). These are called delays or cornerturns and are easily trainable. A memory design is also shown, as well as a picture of four of these smaller NanoCells wired together.

Recall that with the voltage in – voltage out model, the output pins are connected to bistable latches. A potential architecture that includes this feature is displayed in Figure 6.37.[174]

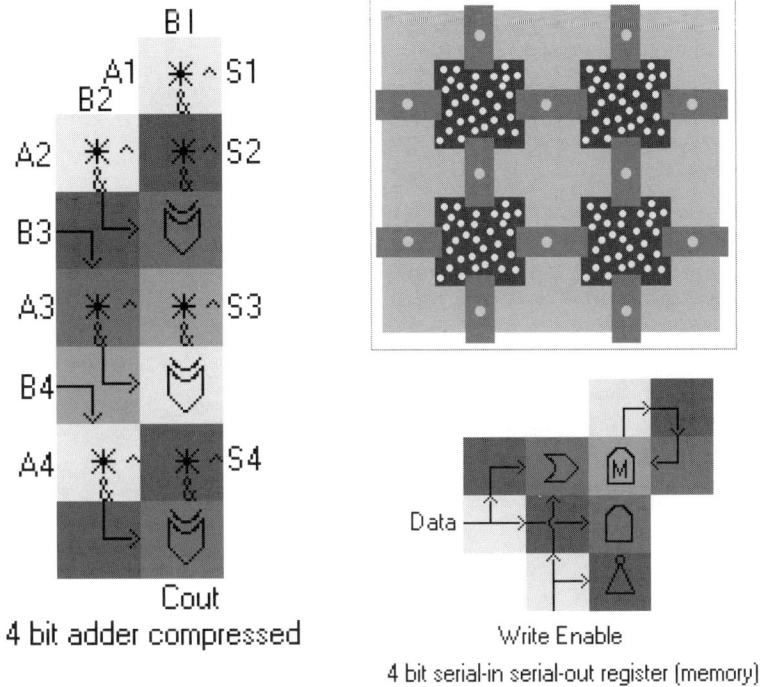

Figure 6.36 This is a molecular electronics architecture using 4-pin NanoCells. The picture on the left depicts a four bit adder comprised of corner turns (L,), half adders (*) and XORs (the little shield symbols). Each shade represens a different clock cycle. The figure on the right illustrates four NanoCells wired together.

Note that with these smaller NanoCells every molecule is within a few molecules of a pin, so each edge should be reachable with a voltage pulse from a pin. Therefore, for this section the NanoCells are smaller than those shown in the previous section.

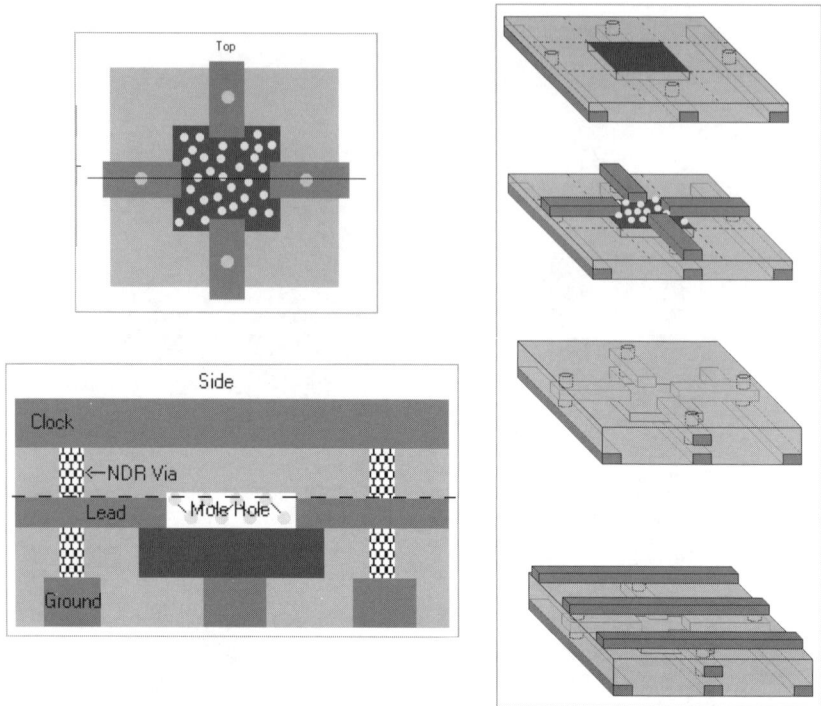

Figure 6.37 This is an architecture that includes bistable latches. The side view demonstrates the bistable latch outside the NanoCell. The figures on the right illustrate a potential process for constructing such a NanoCell.

6.3.2.1. Trained Logic Gates

Several voltage in – voltage out logic gates have been trained. The I(V) curve of the NDR molecule used in most of these cells is shown in Figure 6.25. If this precise NDR was not used, then one with a similar I(V) was used. Note that the latched voltages are shown in the figure.

The first voltage in – voltage out NanoCells were trained as XORs to show that negation is feasible with this model. One of these XORs in a bigger NanoCell is shown in Figure 6.38. The input voltages, output voltage and input current are shown. The dotted line in each graph is the clock voltage. Note that the output voltages exactly equal the input voltages of 0.12 V and 1 V. After demonstrating the feasibility of the design, we began training the individual NanoCells required for the circuit shown in Figure 6.36. First–we trained cornerturns as they converge easily (see Figure 6.39). Then we trained XORs in 4-pin NanoCells. An example of a trained XOR is shown in Figure 6.40. Next we trained an AND gate, which is one of the outputs of a half adder.

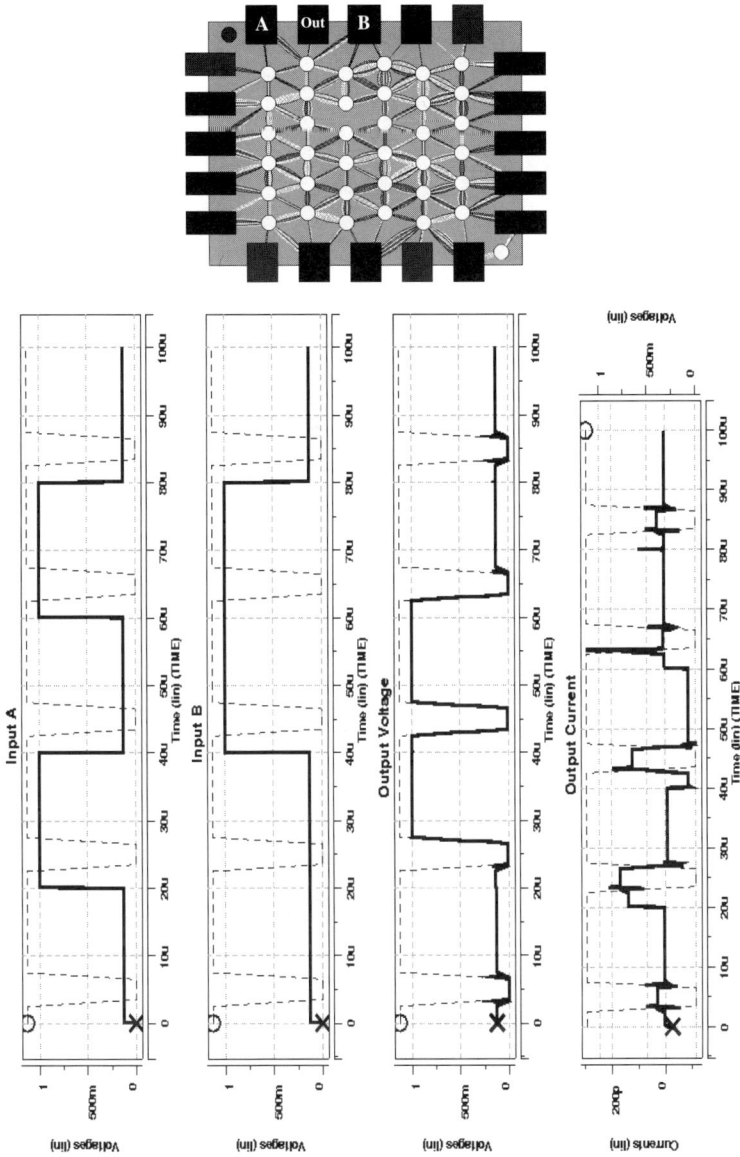

Figure 6.38 This is a 20 pin NanoCell trained as an XOR gate. Pins "A" and "B" are set to input, and pin "Out" is set to output. Input voltages and output current and voltage are shown. The dotted line is the clock voltage for the bistable latch.

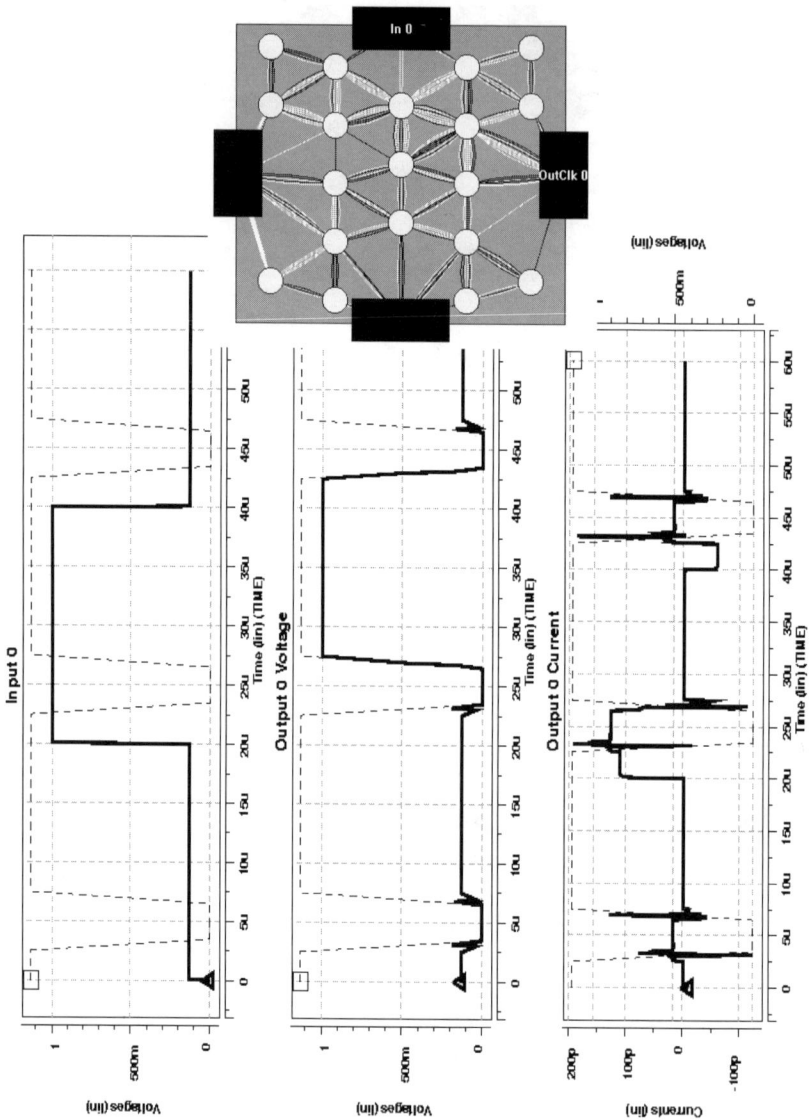

Figure 6.39 This is a 4 pin NanoCell trained as a corner turn. Pin "In 0" is set to input, and pin "OutClk 0" is set to output. Input voltage and output current and voltage are shown. The dotted line is the clock voltage for the bistable latch.

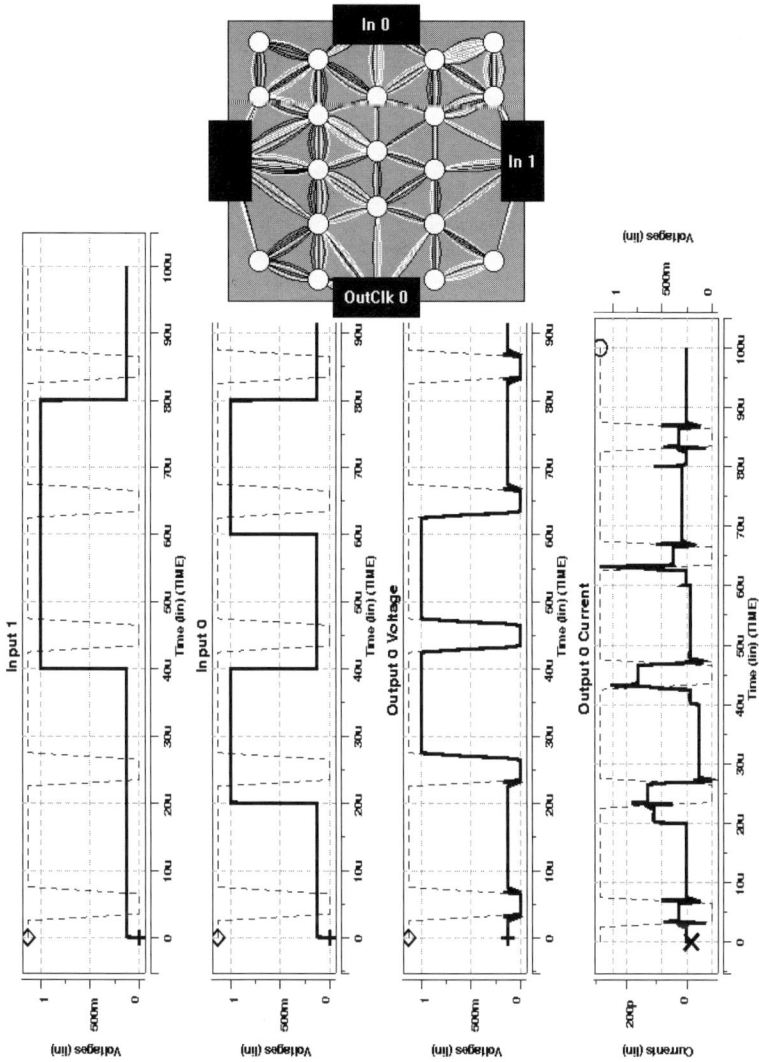

Figure 6.40 This is a 4 pin NanoCell trained as an XOR gate. Pins "In 0" and "In 1" are set to input, and pin "OutClk 0" is set to output. Input voltages and output current and voltage are shown. The dotted line is the clock voltage for the bistable latch.

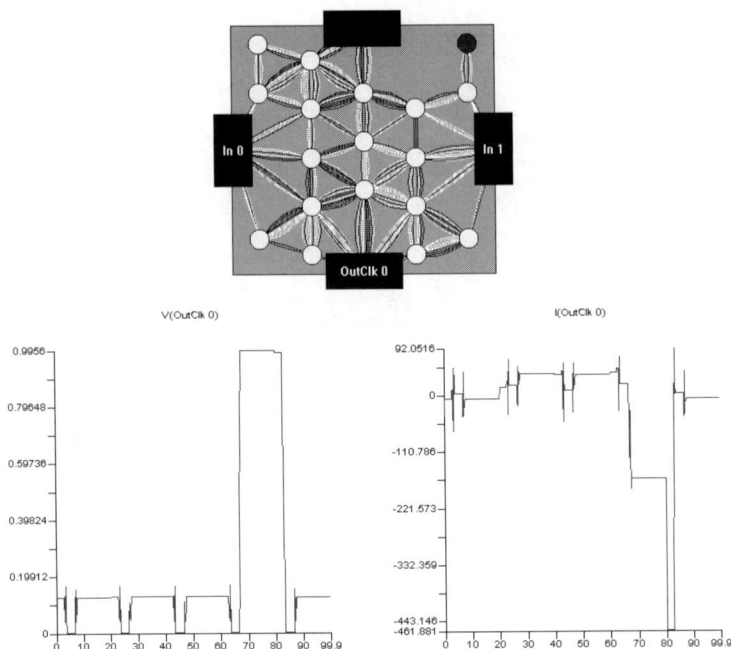

Figure 6.41 This is a 4 pin NanoCell trained as an AND gate. Pins "In 0" and "In 1" are set to input, and pin "OutClk 0" is set to output. Output current and voltage are shown.

A picture of one of these NanoCells is shown in Figure 6.41. After determining that each output of a half adder could be trained separately, we began training half adders. Approximately ten NanoCells have been successfully trained as half adders to date. A picture of one such NanoCell is shown in Figure 6.42. Training these cells is a very significant accomplishment in that only one set of inputs were used, and a fairly complex device was obtained from a relatively small array of molecules. After successfully training the pieces of the 4-bit adder shown in Figure 6.36, we began to try to hook NanoCells together. The details of this work are given in a later section, but due to the obstacles encountered, we began to train simpler voltage in – voltage out logic gates. OR gates were trained, and one such NanoCell is shown in Figure 6.43. Seeking to train the simplest negating gate, we trained several inverters (see Figure 6.44).

For each of the logic gates mentioned in this section, several were successfully trained. With the exception of half adders, none of the NanoCells failed to train as the target logic device. The half adders were more difficult to train, and approximately 2/3 of the attempts at training ended in success. This figure could improve with superior molecular I(V) characteristics or with alterations to the search algorithm that tailor it more to a two output problem.

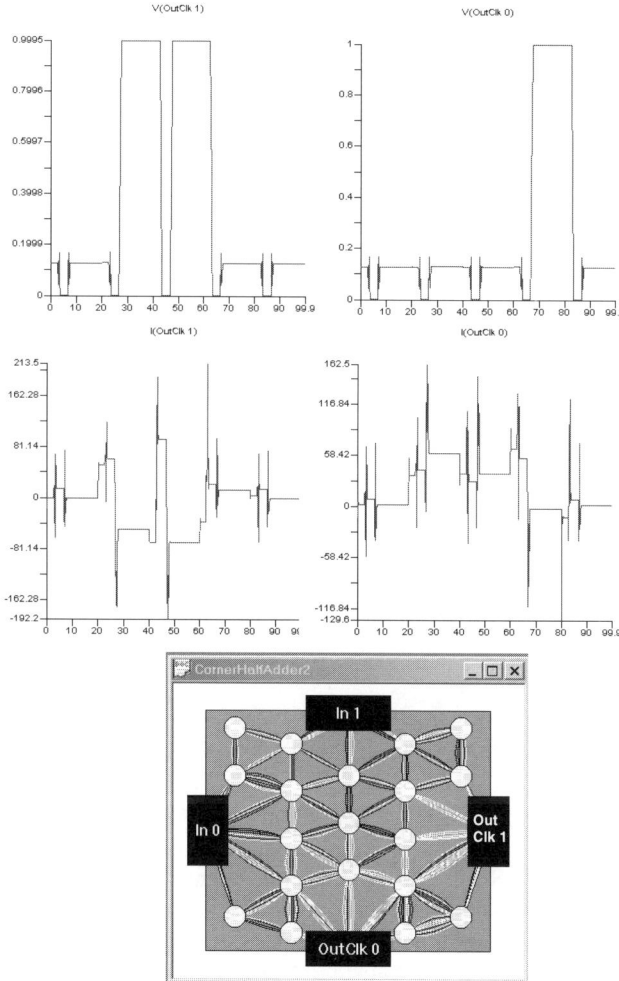

Figure 6.42 This is a 4 pin NanoCell trained as a half adder. Pins "In 0" and "In 1" are set to input, and pins "OutClk 0" and "OutClk 1 are set to output. Output currents and voltages are shown.

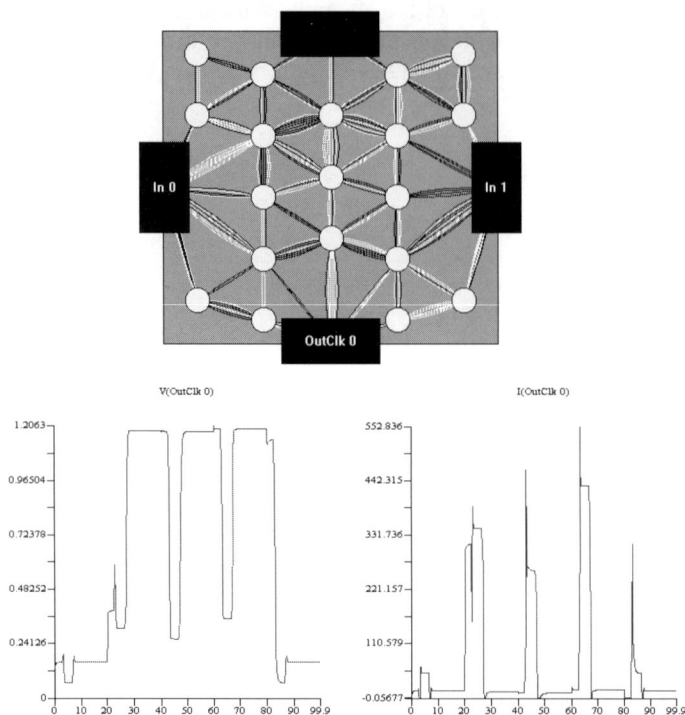

Figure 6.43 This is a 4 pin NanoCell trained as an OR gate. Pins "In 0" and "In 1" are set to input, and pin "OutClk 0" is set to output. Output current and voltage are shown.

6.3.2.2. Observations

The first observation in training voltage in – voltage out NanoCells is that while bistable latches offer the benefit of restoring logic, they come at the cost of a very sensitive logic gate. Small changes in input voltage or capacitances can break the gate. After much study, we have narrowed the problem down to one primary cause. The latch is set high or low by the input current while V_{BIAS} is rising. Note that for any of the trained NanoCells, the current is always changing drastically at this point. On the other hand, for the bistable latch tests shown in Figure 6.13, the input current is always constant. Therefore, the bistable latches are much more robust when the input current is constant. We are currently investigating ways to alleviate this problem by either altering the

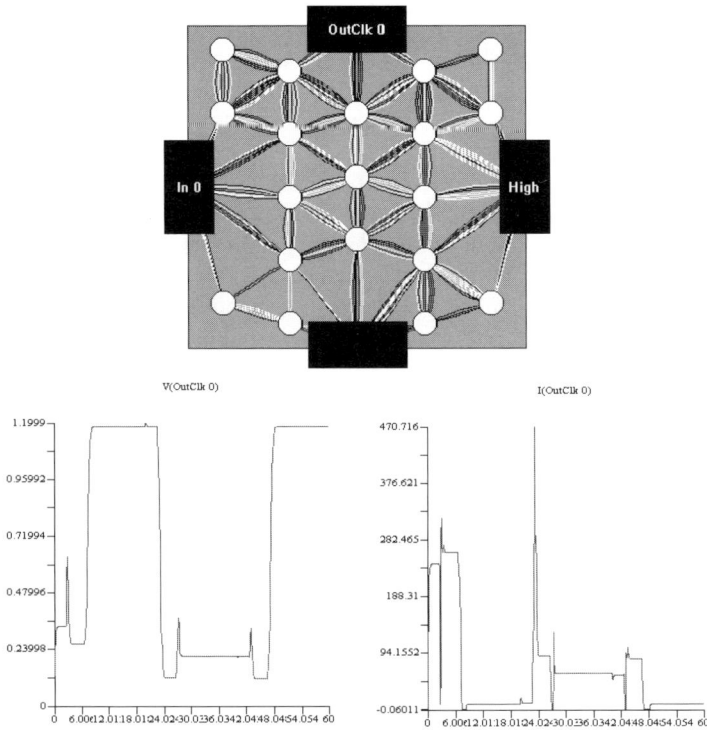

Figure 6.44 This is a NanoCell trained as an inverter. Pins "In 0" and "High" are set to input and high rail respectively, and pin "OutClk 0" is set to output. Output current and voltage are shown.

bistable latch model or employing an alternative means to achieve restoring logic.

Another potential problem with bistable latches is current drain through input pins. This can make the bistable latches on output pins less effective. This problem is easily solved by allowing no molecules to be directed into an input pin. Hence, it is not possible for current to flow in this direction.

6.3.3 *Hooking NanoCells Together*

The point of using a voltage in – voltage out model with bistable latches for restoring logic is to eventually hook multiple NanoCells together to work in unison. This has been a very difficult problem and we are still trying to gain a better understanding of the obstacles involved. A good deal of progress has been made, though.

First as shown in the previous sections, NanoCells have been trained that exhibit both uniformity of signal and restoring logic. While complex gates such as half adders were trained, we were unsuccessful in hooking half adders together. Also, there was some success in hooking together less complex gates. First, a half adder was successfully connected to a corner turn gate. The NanoCells along with their input and output graphs are displayed in Figures 6.45 and 6.46. Note that the output of pin 12 serves as the input of pin 14, then this signal is copied to pin 13 on the subsequent clock cycle. In addition, three OR gates were successfully hooked together. The NanoCell diagram along with input and output graphs are shown in Figures 6.47 and 6.48.

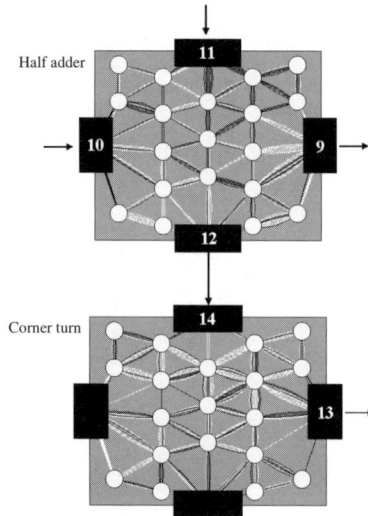

Figure 6.45 Depicted here is a NanoCell trained as a half adder connected to a NanoCell trained as a corner turn. The architecture that uses these cells is displayed in Figure 6.36. The input and output signals are shown in Figure 6.46.

Although multiple complex logic gates were not successfully connected together, it is significant that multiple gates were connected and that a negating gate was hooked to another NanoCell. In attempting to hook NanoCells together, we found that NanoCells whose output current is approximately the same shape as their output voltage tend to be easier to connect to other NanoCells. In the next section, strategies for dealing with this interconnect problem are presented, and proofs involving both necessary and sufficient conditions for training NanoCells are outlined.

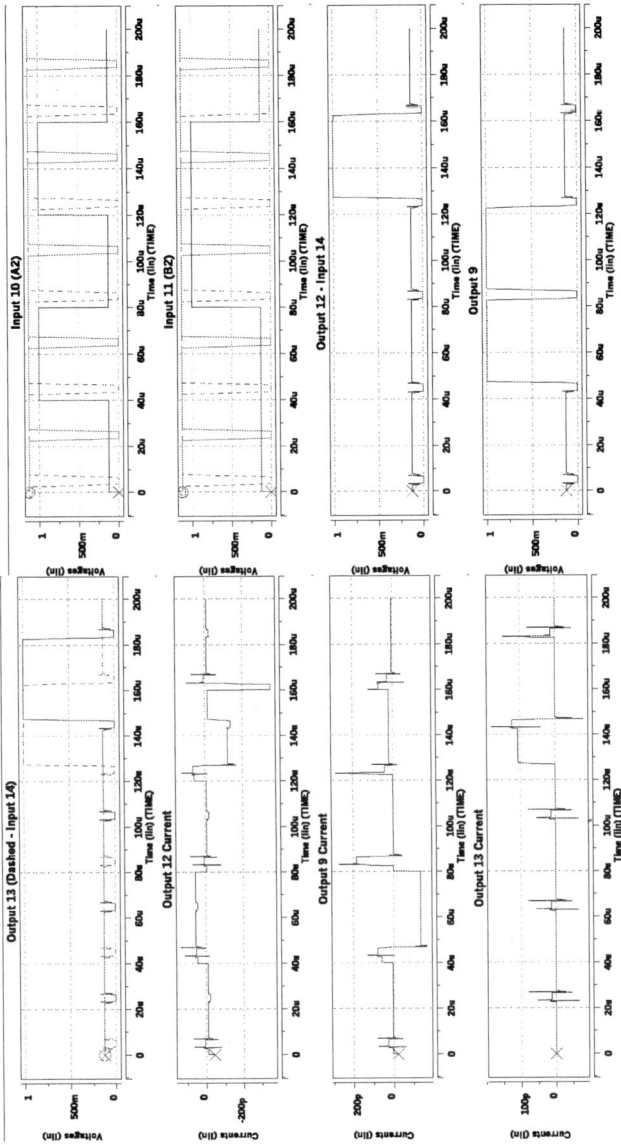

Figure 6.46 Input voltages and output currents and voltages are displayed here for the half adder and corner turn shown in Figure 6.45.

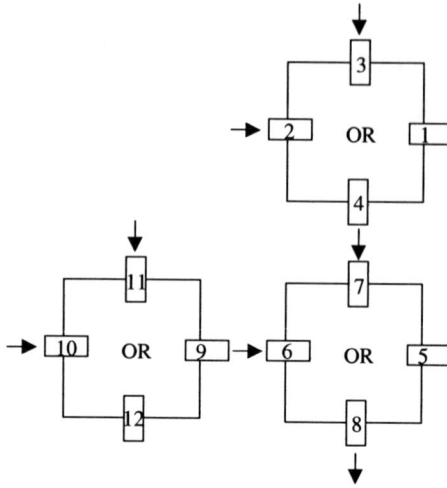

Figure 6.47 This figure displays three OR gates hooked together. The input and output traces are displayed in Figure 6.48.

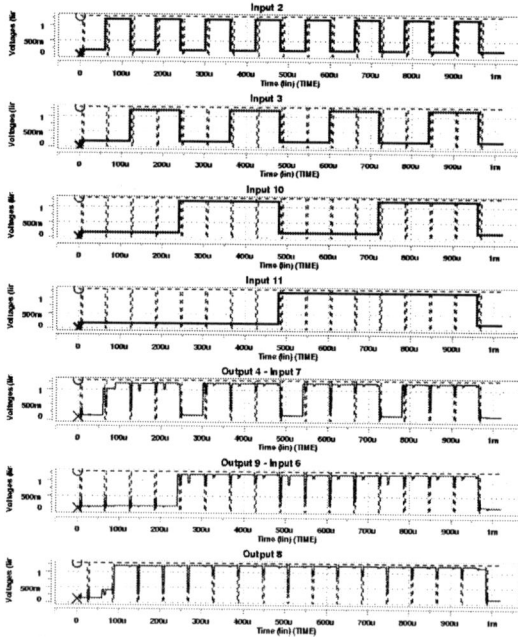

Figure 6.48 These are input and voltages and output voltages for the three NanoCell OR gates shown in Figure 6.47. Note that outputs of two of the NanoCells serve as inputs for the third Nanocell.

6.4 NanoCell Proofs

In this section, some proofs involving NanoCells are presented. After a proof concerning biconnected components in NanoCells, there are several results on computing the probability that a random NanoCell will function as a particular logic gate.

6.4.1. *Biconnected Components in a NanoCell*

In initial tests, some NanoCells would not run in SPICE. After examining these cells, a common characteristic was found. As in Figure 6.49, there was some molecular switch, or edge, not on a path between two nonfloating I/O pins. No current can flow through such a switch, so SPICE does not know how to handle them. Hence, it is necessary to find and eliminate all switches that are not on a path between two nonfloating, or active, pins before evaluating the NanoCell in SPICE. The following algorithm is used to delete all such irrelevant switches. Recall that a *block* in a graph is a maximal 2-connected subgraph. If there are only two nonfloating pins, then the problem has already been solved. The following theorem suggests an algorithm for finding all edges on a path between the two "active" pins.

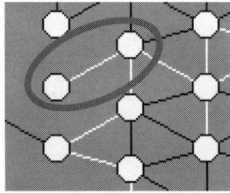

Figure 6.49 This is an edge that will conduct no current because it is not on a path between two voltage sources. Such edges are deleted before a NanoCell is evaluated in HSpice.

Theorem 6.1: Let G be a graph and let a and b be nodes in G. Let $G + ab$ be the graph obtained from G by adding the edge ab. A node $v \in V \setminus \{a,b\}$ lies on an ab path in $G \Leftrightarrow v$ is in the block which contains a and b in $G + ab$.

Proof: Follows directly from proof of Theorem 6.3.

Based on this theorem, we formulated and proved Theorem 6.3, and implemented the following algorithm.

1. Add a cycle among the nonfloating pins to create the graph G'.
2. Find the block containing all nonfloating, or active, pins.
3. Delete all edges not contained in this block.

Some definitions are necessary before presenting the next theorem. Let G be a graph and let x,U be such that $x \in V(G)$ and $U \subseteq V(G)$. An x,U fan is a set of x,U paths such that any two of them intersect only at the node, x. Note that if $x \in U$, then one of the paths can be the path of length 0 from x to x.[175] The following theorem is useful in proving that the algorithm above works.

Theorem 6.2: A graph is k-connected if and only if it has at least $k + 1$ vertices and, for every choice of x,U with $|U| \geq k$, it has an x,U fan of size k.[175]

Theorem 6.3 proves that this algorithm works.

Theorem 6.3: Let G be a graph and let $A = \{a_1, \ldots, a_t\}$ be a subset of the nodes with $t \geq 2$. Let G' be the graph obtained from G by adding the edges $e_i = a_i a_{i+1}$ for $i = 1, \ldots, t - 1$ and the edge $e_t = a_t a_1$. A node $v \in V \setminus A$ lies on some $a_i a_j$ ($i \neq j$) path in $G \Leftrightarrow v$ is in the block of G' that contains $a_1, \ldots, a_t$.

Proof: First note that since $a_1, \ldots, a_t$ lie on a cycle in G', they are in the same block B in G'. To prove the forward direction, let $v \in V \setminus A$ be a node that lies on an $a_i a_j$ path in G. Let a_{i_1} be the first node in A as you move from v to a_i along the path. Similarly, let a_{j_1} be the first node in A as you move from v to a_j along the path. Clearly now, there are two disjoint paths from v to the cycle through the nodes in A. Hence, v lies on a cycle with each node in A and is contained in the block B as demonstrated in Figure 6.50.

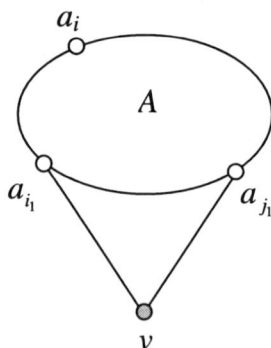

Figure 6.50 As shown here, v lies on a cycle with every node in A. Therefore, v is contained in the 2-connected component, B.

To prove the other direction, suppose that $v \in B$ in the graph G'. Then consider the 2-connected subgraph B. Let $U = \{a_1, a_2 \ldots a_t\}$. By *Theorem* 6.2, there exists a v,U fan consisting of paths P_1 and P_2 that intersect only at v. Let a_i be the first node in A from v along P_1, and let a_j be the first node in A from v along P_2. This is illustrated in Figure 6.51. Hence there exists an a_i-v-a_j path in G.

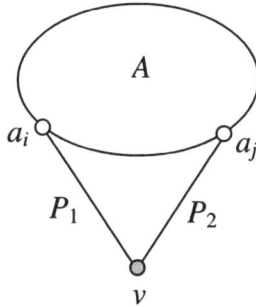

Figure 6.51 As the figure demonstrates, v must lie on some path between nodes in A.

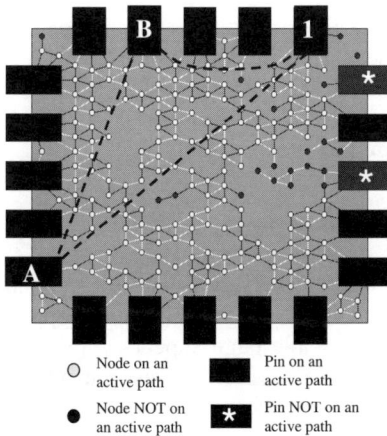

Figure 6.52 If the pins labeled "A", "B" and "1" are voltage sources, then the light nodes and black pins lie on some path between active pins.

Before a NanoCell is evaluated in SPICE, this algorithm is implemented to determine which nodes are contained in the block B. Nodes that are not contained in this block are deleted along with all of their incident edges. This process is demonstrated in Figure 6.52. In this NanoCell, there are 3 active pins: Input A, Input B, and Output 1. To find nodes that are not on a path between two of these active pins, first a cycle is added between A, B, and 1. This is indicated by the dotted lines. Next the block containing A, B, and 1 is found, and all edges with one or more endpoints not in this block are deleted. In the figure, nodes on some active path are white, and nodes on no active path are black. Likewise, pins on no active path are indicated by a "*", and pins on some active path are not.

6.4.2. *Observing Sufficient Conditions*

Previously in this chapter it was noted that to train voltage in – current out inverters and NANDs the NanoCell just needs enough molecules in the "on" state. In working with NanoCells composed of just a few molecules we found that for inverters, a path of molecules in an "on" state between the input and output is both necessary and sufficient. If the molecules are not symmetric, then they must be oriented so that current flows from the input to the output. The conditions for NANDs are similar. First, obviously it is necessary that there is a path from each input to the output. In training NanoCells and working with a few molecules, we observed that it is sufficient to have paths from each input to the output, some of which should intersect.

After observing these conditions, we decided to try to prove that such conditions are satisfied in a random NanoCell. Working with both large NanoCells (approximately 200 nanoparticles, 1000 molecular switches) and smaller NanoCells (about 30 nanoparticles and 250 molecular switches), the results are presented in the following sections.

6.4.3. *Isolated Nanoparticles*

As mentioned previously, a NanoCell can be modeled as a planar graph on n nodes where each possible node appears in the graph with probability q and each possible edge appears with probability p. In an actual NanoCell, there are multiple, directed edges, but one can initially assume that the graph is not directed. In fact, when considering isolated nanoparticles it does not matter whether the edges are directed. It is shown in a subsequent section that for many problems it is equivalent to assume single edges with some altered probability p'. Hence for this section, each graph has single undirected edges. In addition, let each node appear in the graph ($q = 1$) and let each edge appear with some probability p.

Isolated nanoparticles are not a concern in the NanoCell. This is merely a good place to start proving theorems about a NanoCell. First, a more precise NanoCell model is necessary. Suppose that a NanoCell has n gold nanoparticles (nodes) and t terminals. Note that in Figure 6.53 almost every nanoparticle is adjacent to 6 particles or terminals and node has more than 6 neighbors. Additionally, the only nodes with less than 6 neighbors are those on the outside near the terminals. Take this model but add edges where necessary between nodes and terminals so that $d(v) = 6$ for each node (not terminals) v.

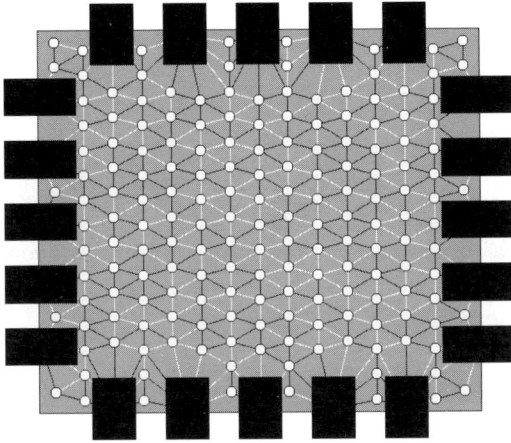

Figure 6.53 This is the NanoCell used for the isolated nanoparticle results.

Suppose we want to compute the expected number of isolated nodes. Let $X(G)$ be the number of isolated nodes in the graph G with n nodes and possible edges as shown in Figure 6.54 (with additional edges added to pins so that $d(v) = 6$ for each nanoparticle, but not necessarily for each pin) each with probability p. Then

$$X = X_1 + X_2 + \cdots + X_n \text{ where}$$
$X_i = 1$ if the ith node of G is isolated and 0 otherwise.

Note that since the degree of each node is equal and the expectation is linear

$E(X_i) = E(X_1)$ for $i = 1 \ldots n$
$E(X) = nE(X_1)$.

Therefore,
$E(X) = n(1 - p)^6$.[176]

Suppose now that we want to compute the probability that a random NanoCell graph has at least t isolated nanoparticles where $t > 0$. Consider the following inequality.

Theorem 6.4: (Markov's Inequality) If $X \geq 0$ and $t > 0$, then

$$P(X \geq t) \leq \frac{E(X)}{t} \text{ .[176]}$$

Note that since $P(X = 0) = 1 - P(X \geq 1)$,

$$P(X = 0) \geq 1 - E(X)$$

$$P(X = 0) \geq 1 - n(1 - p)^6$$

To get an upper bound on $P(X = 0)$, consider the following,

Theorem 6.5: If $E(X) \neq 0$,

$$P(X = 0) \leq \frac{E(X^2)}{E(X)^2} - 1 \text{ .[176]}$$

$E(X^2)$ is computed below.

$$E(X^2) = \sum_{i=1}^{n} E(X_i^2) + \sum_{i \neq j} E(X_i X_j) \text{ here } X_i X_j \text{ is an ordered pair}$$

$$E(X^2) = E(X) + \sum_{i \neq j, v_i v_j \notin E(G)} E(X_i X_j) + \sum_{i \neq j, v_i v_j \in E(G)} E(X_i X_j)$$

$$E(X^2) = E(X) + \sum_{i \neq j, v_i v_j \notin E(G)} (1-p)^6 (1-p)^6 + \sum_{i \neq j, v_i v_j \in E(G)} (1-p)^6 (1-p)^5$$

$$E(X^2) = n(1-p)^6 + (n(n-1) - 2m)(1-p)^{12} + 2m(1-p)^{11}$$

Now use this value to compute an upper bound on $P(X = 0)$.

$$P(X = 0) \leq \frac{n(1-p)^6 + (n(n-1) - 2m)(1-p)^{12} + 2m(1-p)^{11}}{n^2(1-p)^{12}} - 1$$

Simplify to obtain the following.

$$P(X = 0) \leq \frac{1}{n}\left(\frac{1}{(1-p)^6} - 1\right) + \frac{2m}{n^2}\left(\frac{p}{1-p}\right)$$

Now we can state the following theorem.

Theorem 6.6: Given a NanoCell with n nanoparticles and m edges where each nanoparticle has degree 6 and each edge has probability p, let X represent the number of isolated nanoparticles. Then,

$$1 - n(1-p)^6 \le P(X = 0) \le \frac{1}{n}\left(\frac{1}{(1-p)^6} - 1\right) + \frac{2m}{n^2}\left(\frac{p}{1-p}\right)$$

Note that in this model the endpoints of some edges are terminals, or I/O pins, and these are not counted as nodes in the graph. In Figure 6.54 the expected number of isolated nodes in a 177 node, 626 edge NanoCell is graphed as a function of edge probability p. In Figure 6.55 both upper and lower bounds are displayed for $P(X = 0)$.

Expected Number of Isolated Nanoparticles

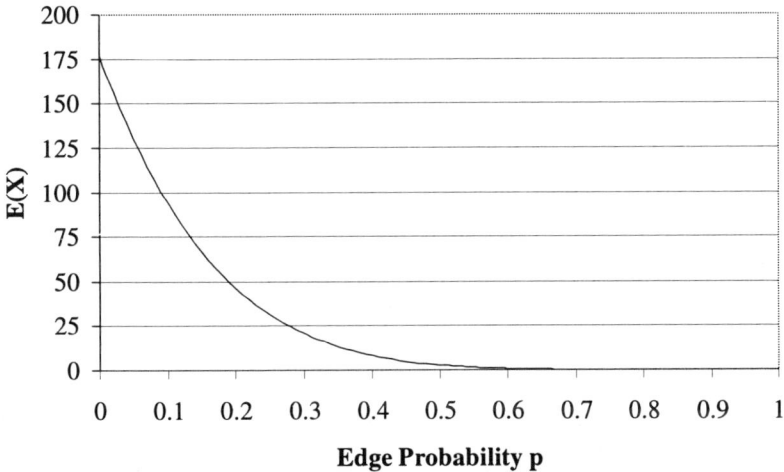

Figure 6.54 This is the expected number of isolated nanoparticles in a 175-nanoparticle Nanocell.

Bounds on P(X=0)

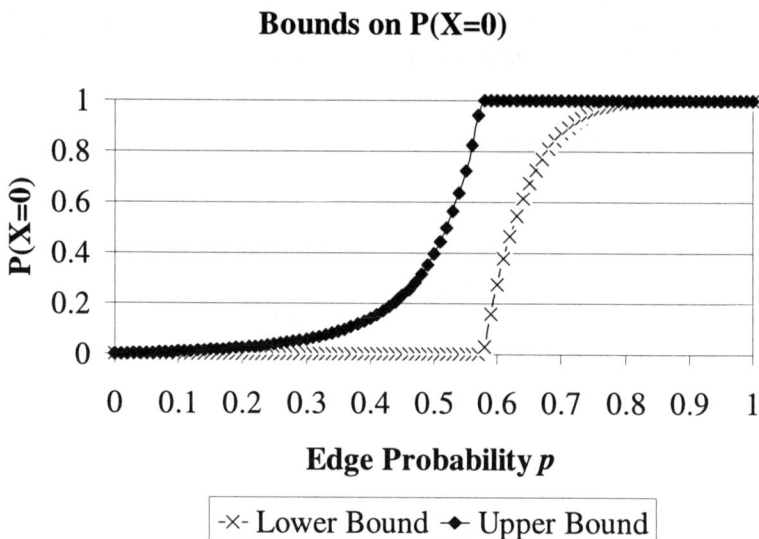

Figure 6.55 Displayed here are bounds on P(X = O), or the probability that there are no isolated nanoparticles in the NanoCell.

6.4.4. *Necessary and Sufficient Conditions for Inverters and NANDs*

As mentioned previously, it is necessary in any logic gate to have a path of "on" molecules from the inputs to the output. In the following sections, the specific conditions for inverters and NANDs are discussed.

6.4.4.1. Inverters

In some cases, as in an inverter in a small NanoCell, a path of "on" molecules is necessary and sufficient. To get an inverter in a large NanoCell, the shortest path must not be too long. So a relatively short path of "on" molecules is necessary and sufficient.

Consider the NanoCell displayed in Figure 6.56 with the left pin set to input and the right pin set to output. We can find 5 edge disjoint paths from the input to the output, 4 of length 5 and one with length 4 as shown in Figure 6.57. The shortest path is length 4, and there is only one such path. The input and output are each degree five, so clearly this is an optimal packing of internally disjoint input-output paths. If one of these paths exists in the NanoCell, then the NanoCell can be used as an inverter. Hence, the existence of one such path is a sufficient condition for the NanoCell to function as an inverter.

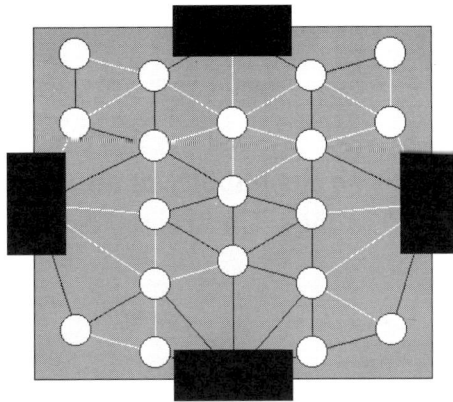

Figure 6.56 For this 4 pin NanoCell, bounds and exact probabilities are computed for inverters and NANDs.

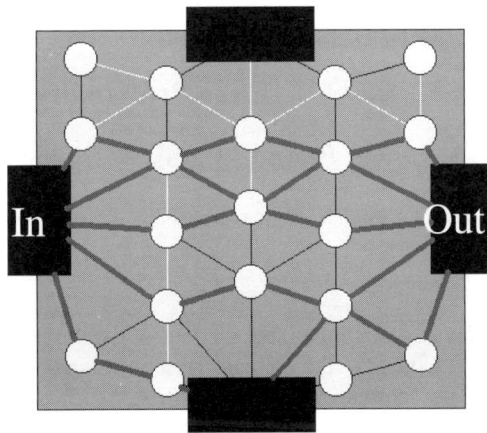

Figure 6.57 In this NanoCell there are five edge disjoint paths of length four or five.

For each edge, let p be the probability of edge is "on". Assume first that there are single, undirected edges. Let $P_{I/O}$ represent the probability that there exists some path from the input to the output. Let $A_1, A_2, \ldots A_5$ be the 5 paths, where A_1 is the path of length 4. Then the probability that path A_i exists is p^4 if $i = 1$ and p^5 otherwise. Hence the probability the A_i does not exist is $1 - p^4$ or $1 - p^5$. The probability that no such path exists is

$\left(1-p^4\right)\left(1-p^5\right)^4$. Therefore the probability that some such path exists is $1-\left(1-p^4\right)\left(1-p^5\right)^4$. This bound is displayed in Figure 6.58.

Bounds on Probability of an Inverter with Single Undirected Edges

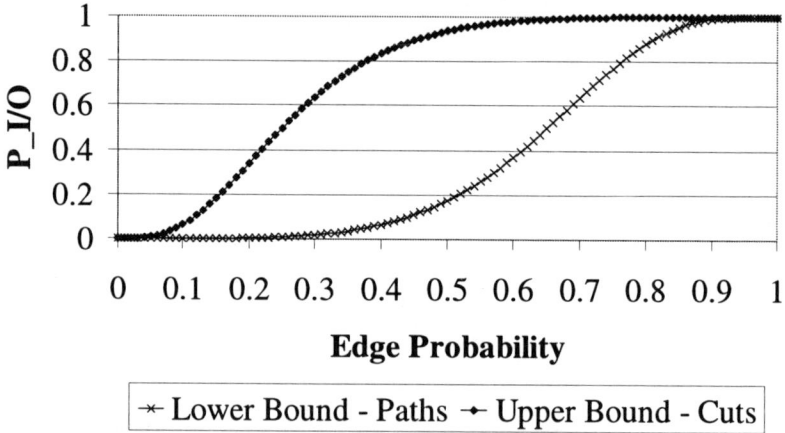

Figure 6.58   These are bounds on the probability of obtaining an inverter from the NanoCell shown in Figure 6.8.

Disjoint edge cuts can be used in a similar way to obtain an upper bound on $P_{I/O}$. Consider the 4 disjoint input-output edge cuts displayed in Figure 6.59. The shortest path has length 4, so clearly there are at most 4 disjoint edge cuts. Let B_1, B_2, B_3, B_4 represent these four edge sets where $|B_1|=|B_2|=5$ and $|B_3|=|B_4|=9$. If no edge in some cut B_i exists, then there is no path from the input to the output. Therefore, the existence of at least one edge in each of these four cuts is a necessary condition. The probability that no edge exists in B_i is either $(1-p)^5$ or $(1-p)^9$. The probability that some edge in B_i exists is then $1-(1-p)^5$ or $1-(1-p)^9$. Hence the probability that some edge exists in each B_i is $\left(1-(1-p)^5\right)^2\left(1-(1-p)^9\right)^2$. Now we have the following result.

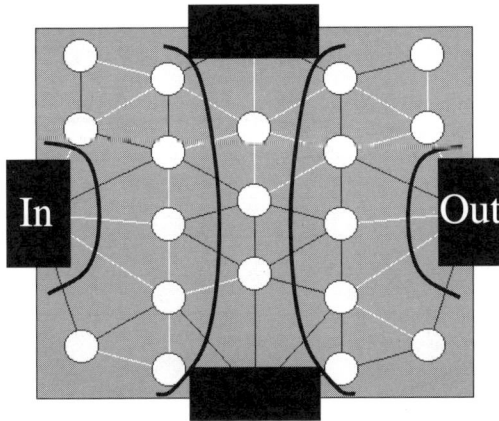

Figure 6.59 There are four disjoint edge cuts in this NanoCell. These are used to determine a lower bound on the probability that this NanoCell will function as an inverter.

Theorem 6.7: The probability that the NanoCell displayed in Figure 6.5r with single undirected edges functions as an inverter is bounded as follows:

$$1-\left(1-p^{4}\right)\left(1-p^{5}\right)^{4} \le P_{I/O} \le \left(1-\left(1-p\right)^{5}\right)^{2}\left(1-\left(1-p\right)^{9}\right)^{2}.$$

These bounds are displayed in Figure 6.59.

These bounds are easily converted for NanoCells with multiple edges. Suppose that there are 6 edges between any pair of adjacent nanoparticles. Then the probability that at least one edge between two given adjacent nodes in "on" is $1-\left(1-p\right)^{6}$. Hence, in the above bound replace p with $1-\left(1-p\right)^{6}$ to get the following result.

Theorem 6.8: The probability that the NanoCell displayed in Figure 6.56 with each edge replaced with six undirected edges functions as an inverter is bounded as follows:

$$1-\left(1-s^{4}\right)\left(1-s^{5}\right)^{4} \le P_{I/O} \le \left(1-\left(1-s\right)^{5}\right)^{2}\left(1-\left(1-s\right)^{9}\right)^{2} \quad \text{where}$$

$s = 1-\left(1-p\right)^{6}.$

These bounds are displayed in Figure 6.60.

Bounds on Probability of an Inverter with
Multiple Undirected Edges

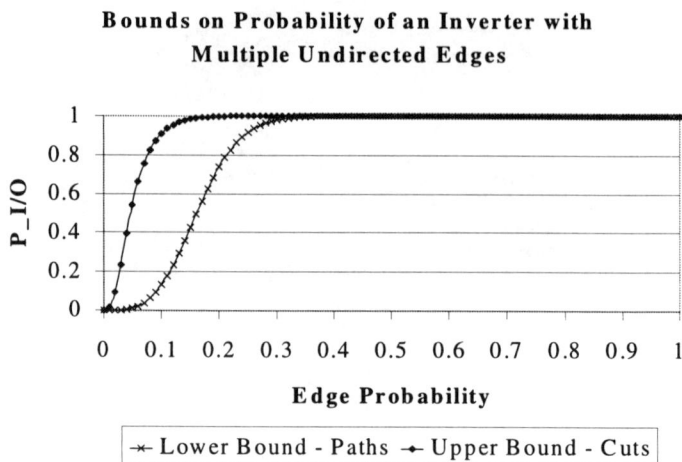

Figure 6.60 Shown here are bounds on the probability of obtaining an inverter from the NanoCell shown in Figure 6.56 with six edges between adjacent nanoparticles.

The molecules in a NanoCell will have direction. Current will flow more freely in one direction than the other, so we should compute this bound for multiple, directed edges. Suppose that there is a 50/50 chance that a molecule will be oriented in either direction. For a directed path to exist from the input to the output, the edge must be directed from left to right. Similarly, in each edge cut there must be an edge directed from left to right. Therefore, replace p with $0.5p$ in the theorem above to obtain Theorem 6.9.

Theorem 6.9: The probability that the NanoCell displayed in Figure 6.56 with each edge replaced by six directed edges functions as an inverter is bounded as follows:

$$1-\left(1-s^4\right)\left(1-s^5\right)^4 \leq P_{I/O} \leq \left(1-(1-s)^5\right)^2\left(1-(1-s)^9\right)^2$$

where $s = 1 - (1 - 0.5p)^6$.

These bounds are diplayed in Figure 6.61.

Bounds on Probability of an Inverter with Multiple Directed Edges

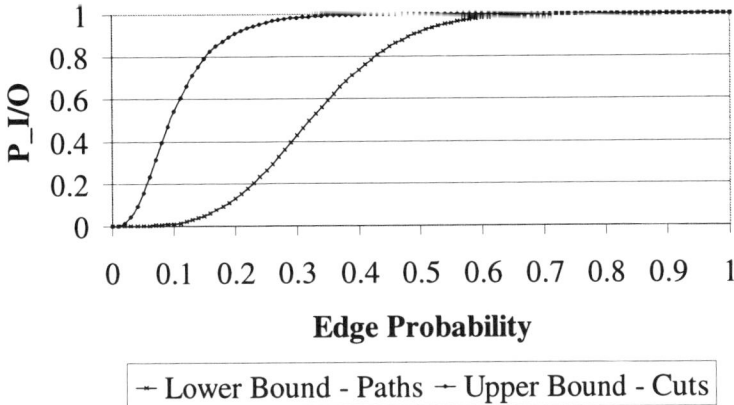

Figure 6.61 Shown here are bounds on the probability of obtaining an inverter from the NanoCell shown in Figure 6.56 with six edges between adjacent nanoparticles. Three edges are oriented one way, and the other three edges are oriented in the opposite direction.

Two-terminal reliability theory can be used to compute $P_{I/O}$ exactly for small NanoCells. Given a graph G, with terminals $s, t_1, t_2, \ldots t_k$, and each edge assigned some probability, the *reliability* of G, $R(G, s, t_1, t_2, \ldots t_k)$, is defined to be the probability that there exists a path in G from s to each t_i. Let $S \subseteq E$ be the set of functioning edges in G. Define $\Psi(S) = 1$ if G is operating, or has a path from s to each t_i and $\Psi(S) = 0$ otherwise. Let p_e be the probability that e is "on", then,

$$R(G, s, t_1, t_2, \ldots t_k) = \sum_{S \subseteq E} \left(\Psi(S) \prod_{e \in S} p_e \prod_{e \notin S} (1 - p_e) \right).^{177}$$

Using the result above, one can obtain the following theorem.

Theorem 6.10: (The Factoring Theorem) In a graph G with each edge assigned some probability p_e,

$$R(G, s, t_1, t_2, \ldots t_k) = p_e R(G \cdot e, s, t_1, t_2, \ldots t_k) + (1 - p_e) R(G - e, s, t_1, t_2, \ldots t_k)$$

This theorem can be used recursively to compute $R(G, s, t_1, t_2, \ldots t_k)$ for relatively small graphs. The performance is improved by including reliability preserving reductions.

First consider the *series reduction* illustrated in Figure 6.62. The reliability of a graph is preserved when a non-terminal node v of degree 2 and its incident edges, e and f, are replaced with an edge between the neighbors of v. The probability of the new edge should be $p_e p_f$.[177]

Figure 6.62 This is an illustration of the series reduction that is used in the Factor algorithm.

In a *parallel reduction*, two parallel edges e and f are replaced with a new edge of probability $p_e + p_f - p_e p_f$. As mentioned previously, this operation is reliability preserving. The process is demonstrated in Figure 6.63.[177]

Figure 6.63 This is an illustration of the parallel reduction that is used in the Factor algorithm.

There are also reductions that change the reliability but the change is some known ratio. The first of these is a *degree 2 reduction*. In this case, a terminal node of degree 2 whose neighbors are both terminals is deleted along with its incident edges e and f. An edge of probability $\dfrac{p_e p_f}{1 - (1 - p_e)(1 - p_f)}$ is added between the two neighbors. The resulting graph G' satisfies

$$R(G', s, t_1, t_2, \ldots t_k) = \frac{R(G, s, t_1, t_2, \ldots t_k)}{1 - (1 - p_e)(1 - p_f)}.^{[177]}$$

The next reduction is the *degree 1 reduction*, in which a terminal node of degree 1 and its incident edge e are deleted. The other endpoint of e is designated as a terminal vertex, and the resulting graph G' satisfies

$$R(G', s, t_1, t_2, \ldots t_k) = \frac{R(G, s, t_1, t_2, \ldots t_k)}{p_e}.^{[177]}$$

An algorithm using the Factor Theorem along with these reductions is given below. [177] Note that initially, $G_{mult} = 1$.

Factor(G)

if ($\exists$ no $t_i s$ path for some i)

ıeturn 0

while ($\exists$ reductions)

apply series reductions

apply parallel reductions

apply degree 1 reductions (update G_{mult})

apply degree 2 reductions (update G_{mult})

if ($|E(G)| = 1$)

return $G_{mult} p_e$

else

select an edge $e \in E(G)$

$G^{contract} = G \cdot e$

$G^{delete} = G - e$

$G_{mult}^{contract} = p_e$

$G_{mult}^{delete} = 1 - p_e$

return $G_{mult} \cdot (\text{Factor}(G^{contract}) + \text{Factor}(G^{delete}))$

We used this algorithm to compute the reliability of the NanoCell displayed in Figure 6.56 where the pins designated as input and output are the terminals. For a graph with single undirected edges, *Theorem* 4 was used to compute the exact probability that the NanoCell functions as an inverter. For multiple undirected edges, we can compute the exact probability of an inverter by reducing the graph to a simple graph with adjusted edge probabilities. Replace the 6 edges each of probability p with a single edge of probability $s = 1 - (1-p)^6$. The Factor Theorem does not work for directed graphs because edge contractions are invalid when all edges between the adjacent nanoparticles are oriented in one direction. However, we can use it to compute another lower bound that is tighter for some values of p. To use the Factor Theorem, we need an edge in each direction between any pair of adjacent nanoparticles. Compute this probability by considering the number of edges between a pair of nodes that are "on." If there are no or exactly one edge, then there clearly cannot be an edge oriented in either direction. The probability of exactly i "on" edges is

$$\binom{6}{i} p^i (1-p)^{6-i}.$$

To get the probability that there is an edge in each direction given that exactly q edges are "on," multiply this by the number of instances where there are edges in each direction divided by the total number of possible edge orientations. Sum these from $i = 2...6$ to get

$$s = \sum_{i=2}^{6} \frac{\binom{6}{i} p^i (1-p)^{6-i} (2^i - 2)}{2^i}.$$

Now replace each edge probability p in Figure 6.56 with s given above to compute a lower bound on the probability that the graph with multiple directed edges contains a path of "on" molecules from the input to the output. Note that this is actually the exact probability that there exists a directed path from the input to the output and from the output to the input. Hence this is the exact probability that the NanoCell functions as an inverter with either pin set to input or output. The exact probabilities and bounds for all three cases are displayed in Figures 6.64, 6.65 and 6.66.

Probability that a Nanocell with Single Undirected Edges is an Inverter

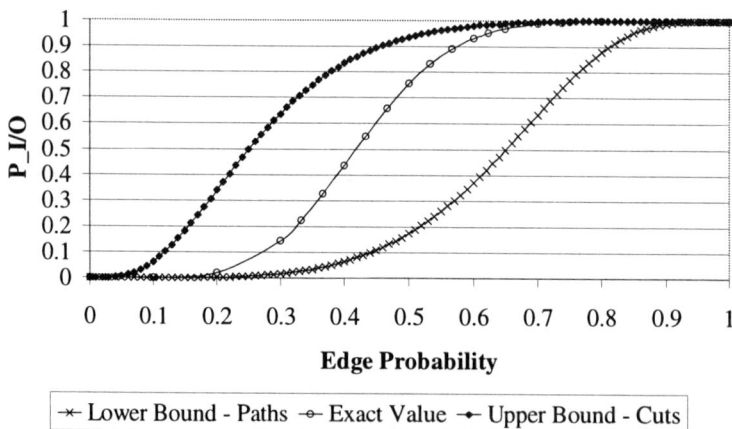

Figure 6.64 This is the exact probability of obtaining an inverter from the NanoCell shown in Figure 6.56. The edges are assumed to be single and undirected. The upper and lower bounds are displayed, as well.

Probability that a Nanocell with Multiple Undirected Edges is an Inverter

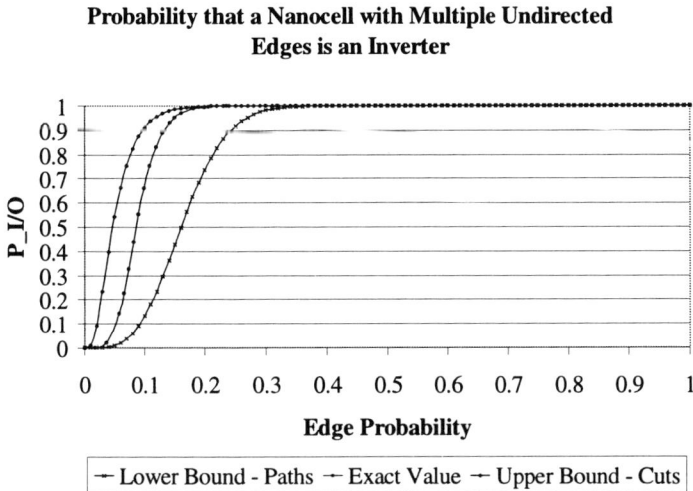

Figure 6.65 This is the exact probability of obtaining an inverter from the NanoCell shown in Figure 6.56. It is assumed that there are six undirected edges between adjacent nanoparticles. The upper and lower bounds are displayed, as well.

Probability that a Nanocell with Multiple Directed Edges is an Inverter

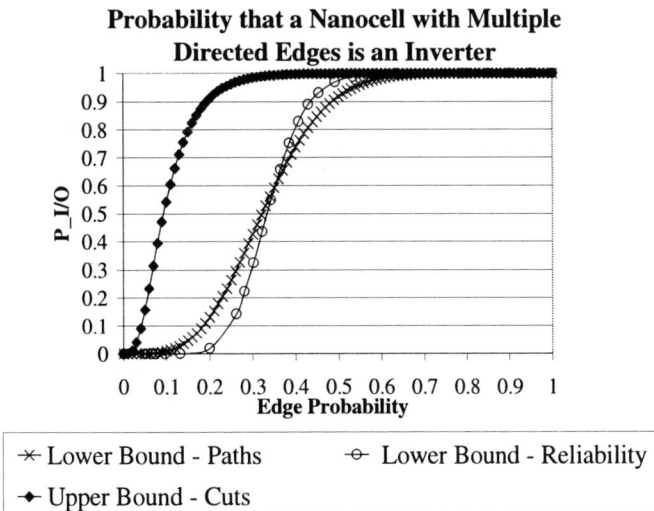

Figure 6.66 These are bounds on the probability of obtaining an inverter from the NanoCell shown in Figure 6.56. It is assumed that there are six directed edges between adjacent nanoparticles, and that there is at least one edge oriented in each direction.

Clearly the algorithm displayed above is exponential in the number of edges, m, and is therefore not feasible for the larger NanoCells displayed earlier in the chapter. Fabricated NanoCells may be this size, so it is important to get an understanding of the probability that some such NanoCell functions as an inverter. We decided to find computational results for these NanoCells. Figure 6.67 displays a NanoCell with approximately 175 nanoparticles where the edges are distributed with a Poisson distribution centered around 5 with at most 10 edges between neighbors. In the figure, every possible node is shown; however, the NanoCell is modeled with each node appearing with some probability q. We ran tests on random NanoCells, varying the node probability from 0, 0.1, . . . 1 and varying the center of the edge distribution from 0.25, 0.5, . . . 9. One hundred NanoCells were evaluated at each combination of node probability and edge center. A breadth-first-search was conducted to determine whether there was a path of length 10 or less from the input to the output (assuming undirected edges initially, because it is possible to get inverters with symmetric molecules). The results of these simulations are displayed in Figure 6.68. As you can see, there is a very distinct line where NanoCells function less than 50% of the time beneath the line and more than 90% of the time above it. These tests were also run for the directed case. The results are shown in Figure 6.69.

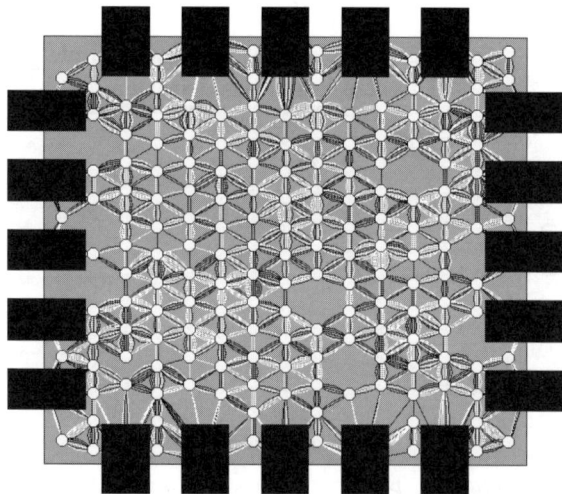

Figure 6.67 This is the NanoCell used to obtain computational results on inverters and NANDs.

Inverter Prob. (15 by 15 Nanocell)

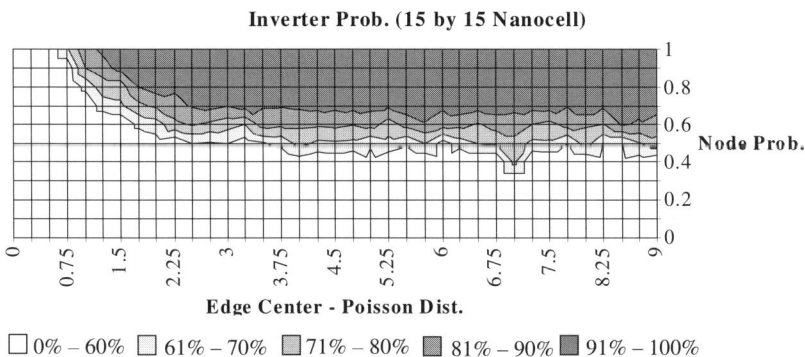

Edge Center - Poisson Dist.

□ 0% – 60% □ 61% – 70% ▨ 71% – 80% ▦ 81% – 90% ▩ 91% – 100%

Figure 6.68 This is the probability of obtaining an inverter based on random switch states generated for the NanoCell displayed in Figure 6.67. The node probability is varied from 0 to 1, and the center of the edge distribution is varied from 0 to 9. One hundred random NanoCells are generated at each combination and tested to determine inverter functionality. For these results, the edges are assumed to be undirected.

Inverter Prob. (15 by 15 Nanocell, directed edges)

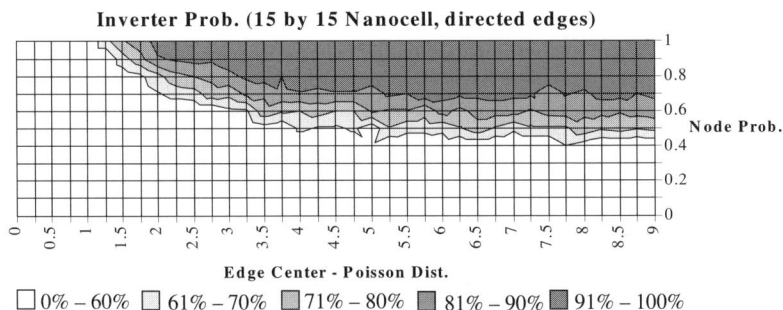

Edge Center - Poisson Dist.

□ 0% – 60% □ 61% – 70% ▨ 71% – 80% ▦ 81% – 90% ▩ 91% – 100%

Figure 6.69 This is the probability of obtaining an inverter based on random switch states generated for the NanoCell displayed in Figure 6.67. The node probability is varied from 0 to 1, and the center of the edge distribution is varied from 0 to 9. One hundred random NanoCells are generated at each combination and tested to determine inverter functionality. For these results, the edges are assumed to be directed. There is a 50/50 chance that a given edge is oriented in either direction.

6.4.4.2. NANDs

For a NAND, we observed in the training process that it is sufficient to have paths of "on" molecules from each input to the output that intersect and are not too long. Clearly, for a NAND or any other logic gate, it is necessary to have a path of "on" molecules from the inputs to the output. We computed the exact probability of obtaining such paths for the NanoCell displayed in Figure 6.56 using the algorithm described in the previous section. The right pin is set to output, and the left and top pins are set to input. The results of this calculation at various edge probabilities are shown in Figure 6.70. Results are shown for single undirected edges, multiple undirected edges and multiple directed edges.

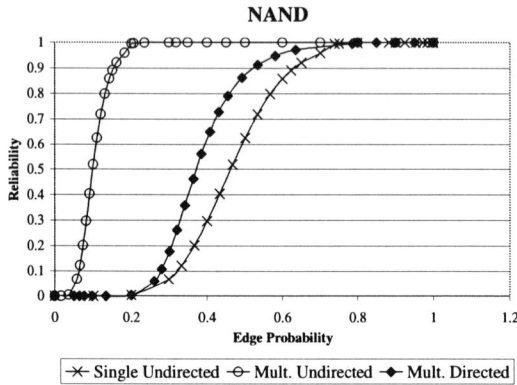

Figure 6.70 These are upper bounds on the probability of obtaining a NAND from the NanoCell shown in Figure 6.56. The top pin and left pins are set to input, while the right pin is set to output. Results for three cases are shown: single undirected edges, multiple undirected edges and multiple directed edges.

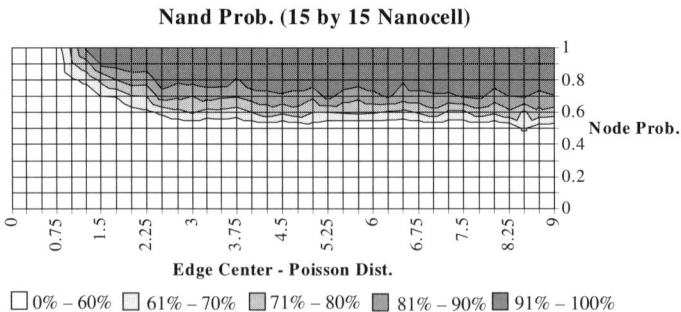

Figure 6.71 This is the probability of obtaining a NAND based on random switch states generated for the NanoCell displayed in Figure 6.67. The node probability is varied from 0 to 1, and the center of the edge distribution is varied from 0 to 9. One hundred random NanoCells are generated at each combination and tested to determine NAND functionality. For these results, the edges are assumed to be undirected.

After obtaining the NAND results for small NanoCells, the computational NAND results for the larger NanoCell shown in Figure 6.67 were determined. Once again, tests on random NanoCells were carried out, varying the nanoparticle probability from 0, 0.1, . . . 1 and varying the center of the edge distribution from 0.25, 0.5, . . . 9. At each combination of node probability and edge center, 100 NanoCells were evaluated. A breadth-first-search was conducted to determine whether there was a path of length 10 or less from each of the two inputs to the output. The results of these simulations for undirected edges are displayed in Figure 6.71. Results for the directed case are shown in Figure 6.72. Once again, there is a very distinct line where NanoCells function less than 50% of the time beneath the line and more than 90% of the time above it.

Nand Prob. (15 by 15 Nanocell, directed edges)

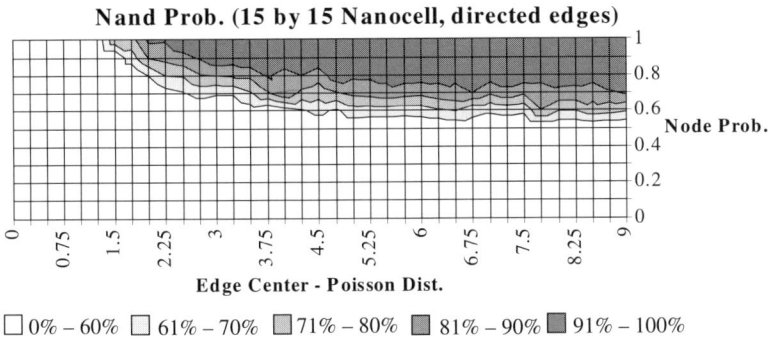

Figure 6.72 This is the probability of obtaining a NAND based on random switch states generated for the NanoCell displayed in Figure 6.67. The node probability is varied from 0 to 1, and the center of the edge distribution is varied from 0 to 9. One hundred random NanoCells are generated at each combination and tested to determine NAND functionality. For these results, the edges are assumed to be directed. There is a 50/50 chance that a given edge is oriented in either direction.

Based on these results, it is clear that a random NanoCell that is dense enough will very likely function as a NAND or an inverter.

6.4.5. *Defect- and Fault-Tolerance*

Explicit evidence of the size of NAND gate solution spaces was found using SPICE to *exhaustively evaluate every possible combination of switch states for 50 NanoCells*. Small cells of 5-16 switches were used because the number of switch state combinations is 2^n, where n is the number of switches. The scores of every possible set of switch states for a cell with 14 molecular switches (16,384 possible states) is displayed in Figure 6.73. The *x*-axis represents every combination of switch states and the *y*-axis, the corresponding score. The combinations were evaluated in Gray code order; hence, two adjacent combinations differ by exactly one bit.[171] For the NanoCell results displayed in Figure 6.73, 13% of the possible switch states functioned as a NAND. In the 50 test NanoCells, 3-19% of the switch states functioned as NANDs. This implies that it will probably not be difficult to mortally train a NAND. It also indicates that a NanoCell trained as a NAND is defect tolerant because the performance of the logic gate does not depend on a single set of switch states.

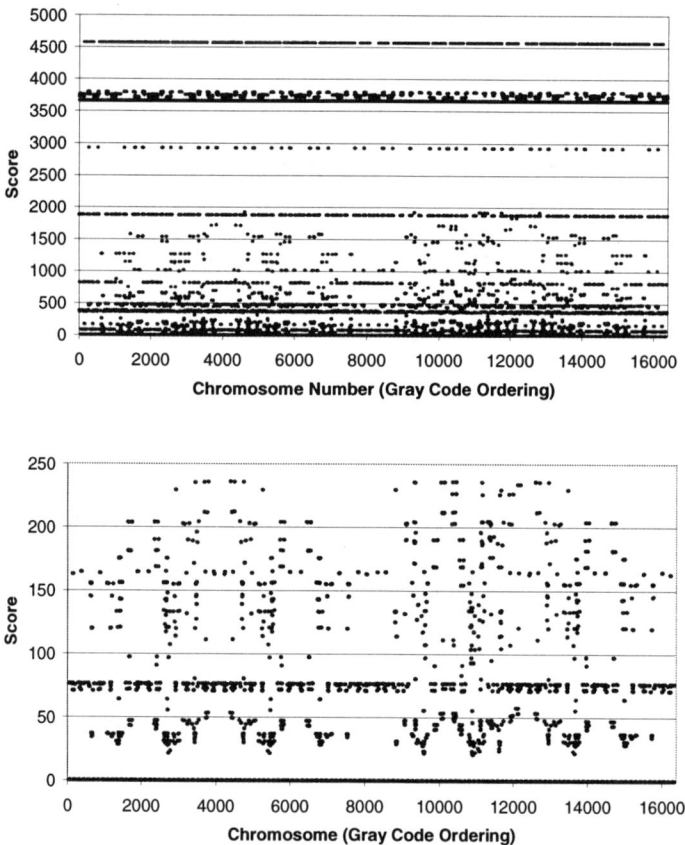

Figure 6.73 Displayed is the solution space for configurations of a 14-switch NanoCell that function as a NAND gate. The x-axis represents each combination of switch states. The y-axis represents the corresponding scores, where a score of zero indicates that the NanoCell functions as a NAND with these switch states. The bottom graph magnifies the range from 0 to 250. The space of solutions is clearly dense.

Furthermore, as a check of defect tolerance, large, multiple-switch NanoCells were tested for defect tolerance through the SPICE interface. With all switches in the "on" position, the cell showed NAND logic with "on-to-off" output current thresholds of approximately 20:1. Switches were then chosen at random and set to the "off" position and the cell was evaluated periodically (data not shown). The average NanoCell tested had 1826 switches, and >60% of these switches could be turned to the "off" non-conducting state before the cell lost NAND functionality with the minimum output "on-to-off" ratio set-point of 10:1. This indicates a high tolerance for numerous faults in the NanoCell architecture.

6.4.6. *Training More Robust Nanocells*

In training NanoCells we discovered that occasionally a certain NanoCell would only function as the desired logic gate if the input voltages were applied in a certain order. Suppose we want to train the NanoCell whose output is shown in Figure 6.74 as an inverter. Table 6.9 displays the desired output and actual output for each truth test. Only the fourth test fails. The output should be "off" but is actually "on". Note that if the truths had been applied as "off", "on", "off", then the NanoCell would have tested as an inverter. This is the way NanoCells are currently trained. Each possible truth is tested once, with the exception of the case where every input is "off". This case is tested both first and last. However, for some NanoCells this may not be sufficient. Note that in Table 6.9, each output transition is tested. That is, each truth is tested when the output was previously "off" and when it was previously "on". This provides a method for training more robust NanoCells.

Now the question is how to test every output transition for any given truth table. A naïve solution would be to test every input transition as shown in Figure 6.75 for a 2-input 1-output NanoCell.

A total of $2^n(2^n+1)$ truth tests are applied using this method, where n is the number of distinct input pins and 2^n is the number of distinct truth tests. Note that applying input tests in this systematic way implies that a few of the input transitions are duplicated. Also, the output transitions are the important factor. It is only necessary to test every possible input transition if each input maps to a different output. This does not happen in any of the logic gates that we tested or plan to test in the future. Therefore, it is worthwhile to devise a method of efficiently testing every output transition.

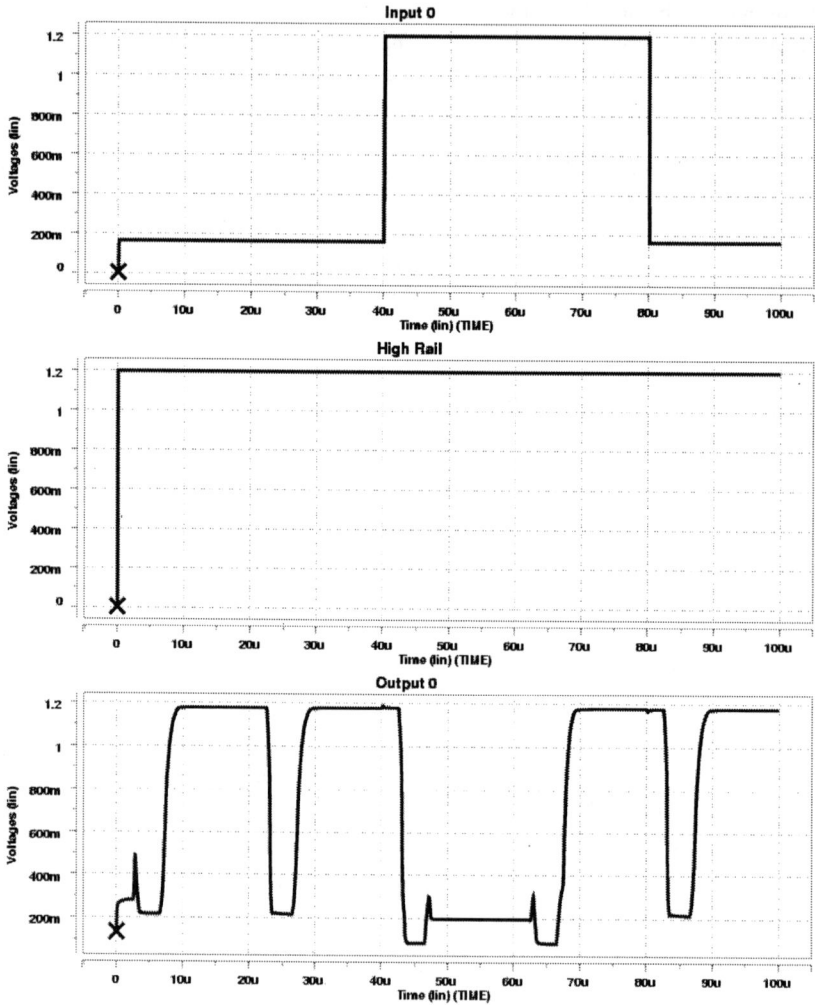

Figure 6.74 Input and output voltages are displayed for a broken inverter. The input signal is inverted only if the output was previously high.

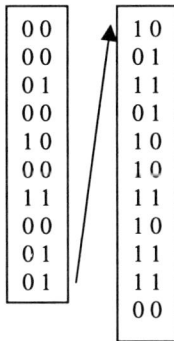

0 0		1 0
0 0		0 1
0 1		1 1
0 0		0 1
1 0		1 0
0 0		1 0
1 1		1 1
0 0		1 0
0 1		1 1
0 1		1 1
		0 0

Figure 6.75   In this figure, every possible input transition is tested for a 2-input truth table.

Table 6.9   This is the actual output and desired output of the broken inverter displayed in Figure 6.75

Input	Desired Output	Actual Output
0	1	1
0	1	1
1	0	0
1	**0**	**1**
0	1	1

It was determined that this can be formulated as a graph theory problem on a directed, bipartite graph. Consider the truth table for a NAND displayed in Table 6.6. Let the input truth tests, "0 0", "0 1", "1 0" and "1 1" be the nodes in one partite set, I. Let the outputs, "0" and "1" be the nodes in the other partite set, O. We want to apply each truth test when the output was previously "off", so draw an edge from "0" to each of the input nodes. Likewise, add an edge from "1" to each input node. For a NAND gate, the truths "0 0", "0 1" and "1 0" map to the output "1", while the truth "1 1" maps to the output "0". So add an edge from nodes "0 0", "0 1" and "1 0" to output node "1", and add an edge from "1 1" to "0". This graph, G, is displayed in Figure 6.76.   To test every possible output transition, we must find a supergraph of G, G' such that G' has a directed Eulerian circuit. Therefore, in G' it must be the case that $d^+(v) = d^-(v)$ for each node v. Each input node has only one out-edge and two in-edges, so add another copy of each of the out-edges as shown in Figure 6.77. Now $d^+(v) = d^-(v)$ for each input node v. However, $d^+("0") - d^-("0") = 2$ and $d^-("1") - d^+("1") = 2$. Add two more parallel edges for each edge on the path from "1" to "1 1" to "0" as in Figure

6.79. Now $d^+(v) = d^-(v)$ for each node v, and the graph G' has a directed Eulerian circuit and corresponding truth tests as shown in Table 6.10.

Table 6.10 In the table above, every output transition is tested for a NAND gate.

Input	Output
0 0	1
0 1	1
1 0	1
1 1	0
0 1	1
1 1	0
1 0	1
1 1	0
1 1	0
0 0	1
0 0	1

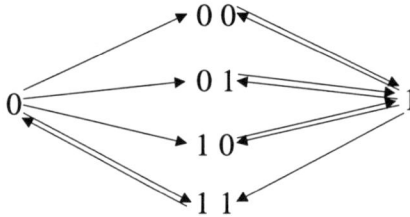

Figure 6.76 This is the graph G with the edges that are required to test every output transition.

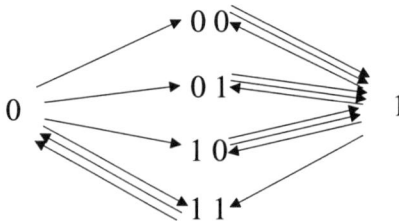

Figure 6.77 This is the graph $\hat{G}$, a supergraph of G where $d^+(v) = d^-(v)$ for every input node v.

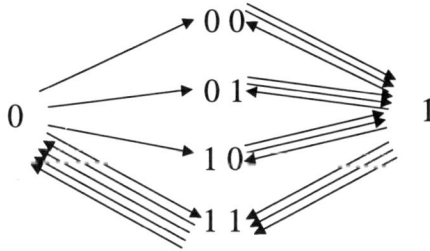

Figure 6.78 This is the graph G', a supergraph of G where $d^+(v) = d^-(v)$ for each node v.

Clearly, the algorithm given above works for NANDs, but now we need to formulate a general algorithm that works for any given truth table. Consider the algorithm below.

1. Given a truth table with n distinct inputs and the p distinct outputs corresponding to the 2^n truths, construct the bipartite graph $G = (I, O, E)$ as follows.
   a. Let each combination of on/off inputs be a node in the partite set I, and let each unique output be a node in the partite set O.
   b. Add a directed edge from each node in O to each node in I.
   c. Add a directed edge from each node in I to its corresponding output in O.

2. Construct the Eulerian supergraph G' as follows when $p = |O|$
   a. In the graph G, $d^-(v) = p$ and $d^+(v) = 1$ for each input node v. Repeat the single out edge $p - 1$ times to get $d^-(v) = d^+(v) = p$ for each input node v. Call this graph $\hat{G}$.
   b. While there exists output node u_1 such that $d^-(u_1) > d^+(u_1)$, add a path of length two along existing edges from u_1 to some input node v to its corresponding output u_2 where $d^-(u_2) < d^+(u_2)$.

The proof that the algorithm above results in an Eulerian supergraph is given below.

Theorem 6.11: Given a bipartite graph G corresponding to a truth table with 2^n truth tests and p distinct outputs, the supergraph G' constructed in the algorithm above is Eulerian.

Proof: We must show that $d^+(v) = d^-(v)$ for each node $v \in V(G')$.

Clearly this is true for each input node v. Let $\hat{G}$ be the supergraph of G at the completion of step 2a, so $d^+(v) = d^-(v)$ for each input node v. To show that

$d^+(u) = d^-(u)$ for each output node in G', we first must show that if there exists some output node u_1 such that $d^-(u_1) > d^+(u_1)$ in $\hat{G}$, then there exists output node u_2 such that $d^-(u_2) < d^+(u_2)$ in $\hat{G}$. We know that $\sum_{v \in V(\hat{G})} d^+(v) = \sum_{v \in V(\hat{G})} d^-(v)$ and $d^+(v) = d^-(v)$ for each input node v, so $\sum_{v \in O(\hat{G})} d^+(v) = \sum_{v \in O(\hat{G})} d^-(v)$. This implies that if there exists some output node u_1 such that $d^-(u_1) > d^+(u_1)$ in $\hat{G}$, then there exists output node u_2 such that $d^-(u_2) < d^+(u_2)$ in $\hat{G}$. Now we must show that there exists a path of length two from u_1 to u_2. Let v be an input node that maps to output u_2. There is an edge from u_1 to each input node, and there is an edge from v to u_2. Therefore, G' is Eulerian.

Clearly if an Eulerian circuit in G' is given, then applying the truth tests in the order of the input nodes in the circuit (with the first input node tested first and last) will result in the testing of every output transition. The vast majority of NanoCell training time is spent simulating the circuit in HSpice, so it is important to determine how many truth tests are necessary to test every output transition. Once again, let 2^n be the number of truth tests, let p be the number of distinct outputs and let t_i be the number of input nodes that map to the ith output node. Note that $2^n \geq p$ and $t_i \geq 1$ for each i. Let $m(H)$ denote the number of edges in the graph H. In G' we can count the number of edges incident with output nodes to determine the number of truth tests. In the graph G,

$$d^-(v) = p \text{ and } d^+(v) = 1 \text{ for each } v \in I(G)$$

$$d^-(u_i) = t_i \text{ and } d^+(u_i) = 2^n \text{ for each } u_i \in O(G)$$

$$m(G) = \sum_{i=1}^{p} (t_i + 2^n).$$

In the graph $\hat{G}$,

$$d^-(v) = p \text{ and } d^+(v) = p \text{ for each } v \in I(\hat{G})$$

$$d^-(u_i) = pt_i \text{ and } d^+(u_i) = 2^n \text{ for each } u_i \in O(\hat{G})$$

$$m(\hat{G}) = \sum_{i=1}^{p} (pt_i + 2^n).$$

In the graph G',

$$m(G') = \sum_{i=1}^{P} \left(pt_i + 2^n + \left| pt_i - 2^n \right| \right).$$

Every other node in this circuit is an input node, and the first input node is repeated as the last truth test. Therefore, the number of truth tests is given by

$$\frac{1}{2} \sum_{i=1}^{P} \left(pt_i + 2^n + \left| pt_i - 2^n \right| \right) + 1.$$

Table 6.11 lists the number of truth tests for some common logic gates testing every output transition and testing every input transition with the algorithm given at the beginning of the section. Based on Table 6.11, output transitions offer a clear advantage over input transitions.

Table 6.11 The number of input transitions vs. the number of output transitions are displayed for various truth tables.

Logic Gate	Input Transitions $2^n(2^n + 1)$	Output Transitions $\frac{1}{2} \sum_{i=1}^{P} \left(pt_i + 2^n + \left| pt_i - 2^n \right| \right) + 1$
Inverter	6	5
AND	20	11
OR	20	11
XOR	20	9
NAND	20	11
Half Adder	20	15
1-Bit Adder	72	41
2-Bit Adder	1056	321

Now we need an algorithm for obtaining an Eulerian circuit from G'. There is a well known algorithm for finding an Eulerian circuit in a directed graph.[175] Recall that an *in-tree* of a directed graph is a tree rooted at some node v such that for every node u in the tree there exists a directed path from u to v. Consider the following algorithm.

1. Given a directed graph G, construct an in-tree T rooted at some node $v \in V(G)$ and containing every node in G.
2. Begin the Eulerian circuit at v and only choose an edge in T when there is no other option.

Note that in G' each input node has exactly one out neighbor (see Figure 6.78). Therefore in the algorithm for finding an Eulerian circuit, the only decision is where to go from each output node. Given this structure, it seems as if there should be an even simpler algorithm for finding a directed Eulerian circuit. Consider the following algorithm.

1. Given a graph G' corresponding to some truth table, start at some input node v_1 upong reaching any input node, choose any one of its outgoing edges.
2. Upon reaching any output node arbitrarily choose one of its outgoing edges choosing an edge to v_1 only if there is no other option.

The proof that this algorithm works is given below.

Theorem 6.12:   The algorithm above constructs a directed Eulerian circuit C from the Eulerian graph G'.

Proof: First note that the algorithm is finite because it marks edges at each iteration and concludes when there are no more unmarked edges. Let v_1 be the start input node. We can construct an in-tree rooted at v_1 by letting the first level be all the output nodes (recall that there is an edge directed from each output to each input). Construct the second level by adding the edge from each input other than v_1 to its corresponding output. The tree T now covers every node in G'. For the algorithm given for arbitrary directed graphs, T can now serve as the in-tree. However, the edges on the second level (from input nodes other than v_1 to their corresponding outputs) are unnecessary because each input node has only one unique out neighbor. Therefore, it is equivalent to simply take the edge from any output to v_1 only after traversing all other possible edges from the output.

Note that the algorithm runs in $O(m)$.

The output transitions algorithm is not something that we currently use in NanoCell training because of the significant increase it would cause in training time. However, in the future we may use the algorithm to evaluate the robustness of trained NanoCells. Nanocells that do not perform the desired logic for every output transition would then be trained further.

In summary, graph theory is useful in the training process and in determining the probability of obtaining certain logic gates. Based on the results shown in this section, NANDs and inverters should be easy to train in a mortal fashion. One simply needs enough "on" molecules. This is very encouraging as the transition is made from omnipotent training to mortal training.

6.5 Future Work

In this section, various areas of future research are discussed. First improvements in training individual NanoCells are covered, and then strategies for hooking NanoCells together are addressed. This is followed by a section on dropping the assumption of omnipotence for more realistic mortal training. The section concludes with the subject of proofs concerning trainability.

Our NanoCell computer architecture group has made a great deal of progress in the area of omnipotently training individual NanoCells. However, there are still some interesting issues to explore. First of all, we should experiment with alternative I(V) curves. It may be possible to get more complex logic gates using different NDR-based I(V) curves. For instance, a 2-bit adder has proved impossible to train using just one type of NDR molecule. However, it may be possible to train this logic gate using different curves. Collaborators at North Carolina State University are currently exploring various I(V) curves in an attempt to get transistor behavior out of a NanoCell. In addition to experimenting with the I(V) curve of a single molecule, we should try training NanoCells that contain a mixture of molecules. The molecules could be NDR devices with different peaks and valleys as well as simple resistors. Note that a mixture of molecules with varying "on" and "off" voltages would definitely be useful when mortally training a NanoCell, as it would reduce the size of groups that switch simultaneously. This would allow for more selective switching.

Another potentially advantageous addition to training a NanoCell is "connectability". In other words, train a NanoCell so that it is easily connected to other NanoCells. One strategy for doing this is to pick some target such as the overall resistance of a NanoCell. When two NanoCells are hooked together, one looks simply like a resistor to the other. Therefore, if a NanoCell were trained with some resistor hooked to each output and trained for some target resistance, then that NanoCell should be easily wired to another NanoCell with similar overall resistance.

The robustness of NanoCells can be further enhanced by the training process. First, the output transitions algorithm presented in the preceding algorithm could be utilized. This will ensure that every output transition functions properly. The downside is the dramatic increase in simulation time. Nearly all of the NanoCell training time is devoted to simulating the electrical properties of the cell. Therefore, increasing the number of tests drastically increases the training time. In addition to testing output transitions, NanoCell robustness can be increased by testing for noise margin.[178] The noise margin of a circuit is tested by applying a range of voltages to the input pins. The question is, For what range of input voltages are the outputs valid? The goal is to obtain a very large range of input voltages for which the outputs voltages are correct.

Although, the problem of omnipotently training a NanoCell has been thoroughly explored in this chapter, there are still some potential improvements that should be explored. In the next section the issue of hooking NanoCells together is addressed.

The issue of hooking NanoCells has proved difficult so far. As mentioned previously, the bistable latches can be fairly unstable. They tend to work for only very small ranges of input voltages and current drain can cause problems. Fortunately we have devised a few strategies for overcoming these difficulties. The first potential solution is to substitute bistable latches with transistors to restore logic and isolate NanoCells. This tactic would almost certainly work, however we would lose a great deal of the size advantage. Therefore, hooking NanoCells to transistors is currently considered our final option when all else fails. Fortunately, there are other approaches that show promise.

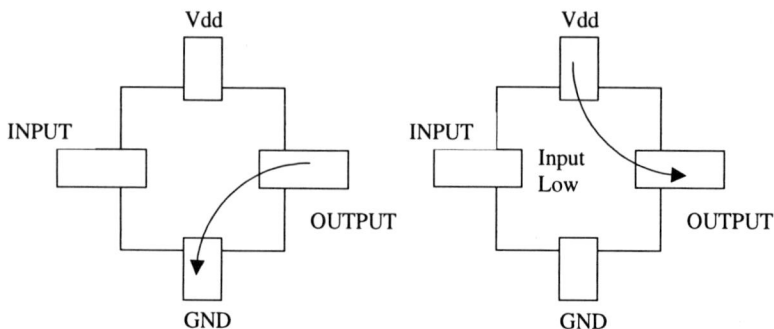

Figure 6.79 This is an illustration of an inverter that functions similar to a transistor. The power for the cell comes from the high rail instead of the input pin.

Another such method is to attempt to train a NanoCell to function as a transistor. Consider the NanoCell inverter shown in Figure 6.79. As the figure demonstrates, current flows either from high rail to ground or from high rail to the output. No current is drawn from the input pin. The advantage here is that current drain on the input is no longer a problem and thus hooking NanoCells together is straightforward. The drawback with this option is that it is not currently clear what kind of molecular I(V) curves would result in this behavior. As was mentioned in the NanoCell training section, NanoCells train successfully as logic gates that make sense based on their molecular I(V) curves. For instance, the I(V) curve shown in Figure 6.5 looks like it should give rise to an XOR. Enough voltage will push the current over the hump. However, it does not look like it would work for a 1-bit adder. For one of the outputs of a 1-bit adder, if one input is "on" then the output is "on", but if two inputs are "on" then output is "off", and if three inputs are "on" then the output is "on" again. On the other hand, the I(V) curve shown in Figure 6.7 seems reasonable. A small voltage results in a high current, a medium voltage drop gives rise to a low current, and a high voltage produces a high current again. The point of this explanation is to demonstrate that the molecular I(V) curve should make sense. None of the I(V) curves displayed in this chapter seem to give rise to the behavior shown in Figure 6.79. Past experience has shown then that attempting to train such a device would be unsuccessful. This does not mean that it is impossible to obtain such a device using molecules. It just implies that different I(V) curves are necessary.

The most promising method for hooking NanoCells together remains the bistable latch option. It just needs to be made more stable. Fortunately, there is another way to implement bistable latches. As presented in previous sections, latches are set to high or low based on input current as the clock is rising (see Figure 6.10). However, setting the latch as the clock is changing can lead to unstable circuits. A better approach is to find a way to set a latch while the clock is constant. In their paper, our collaborators, Nackashi and Franzon, demonstrate such an approach.[179] The clock voltage first sets the latch to low voltage then rises to a point (and stays constant for a few moments) where relatively little additional current will set the latch to a high state. This process is demonstrated in Figure 6.80. Initial tests indicate that this approach is very effective. Tests on a single molecule show that a wide range of input voltages result in an inverter. Conversely, only a very narrow range worked with the first bistable latch setup. We are currently in the process of training NanoCells with these latches.

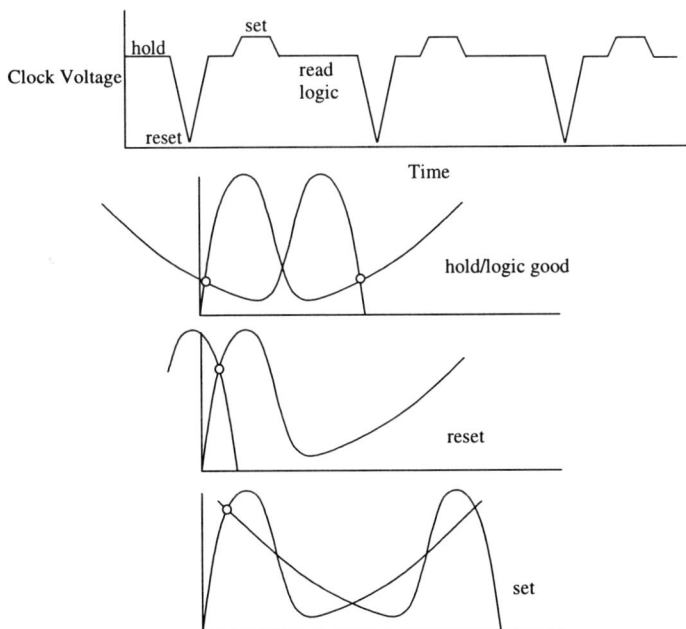

Figure 6.80 This is an alternative latching scheme with three stages: hold, reset and set. First the clock is at the hold state, then it comes down to its reset voltage. In the set state, it takes relatively little current to set the latch to a high state. Once the latch is set, the clock moves to the read or hold state.

The NanoCell training to date has been conducted under the assumption of omnipotence. However, NanoCells will be trained in a mortal fashion, so it is essential that mortal techniques be investigated in the near future. This is an issue that we are exploring.

One potential strategy is to first use the intermediate supposition of omniscience. That is, assume knowledge of the graph within the NanoCell but limit control of molecular states to voltage pulses applied to the I/O pins. Our collaborator, Pat Lincoln, suggested using the capacitance of gold nanoparticles along with omniscience to train a NanoCell. For instance, consider a path of nanoparticles and molecules of length 5. Suppose that the goal is to alternate between "on" and "off" and all molecules are initially "off". Note that the capacitance of the nodes implies that one a line of switches will turn "on" linearly and not simultaneously. Bring the voltage on the first node up long enough to turn the first three molecules "on". Next bring the voltage on the same node down long enough to turn the first two molecules "off" before bringing it back up to turn the first switch back "on". Turn on the last switch by increasing the voltage on the last node and then decreasing it quickly. This process is demonstrated in Figure 6.81.

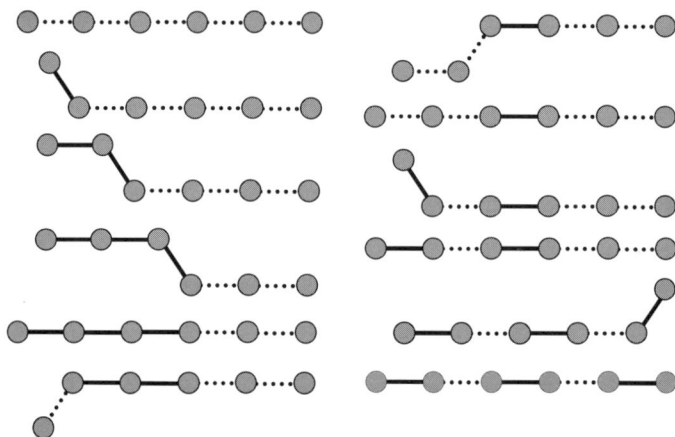

Figure 6.81 This is a demonstration of a strategy for using capacitance of nanoparticles to set a linear group of molecules from "0 0 0 0 0" to "1 0 1 0 1".

Clearly, a NanoCell is not composed of simple paths of molecules, and it is therefore unlikely that there exists a series of voltage pulses that uniquely turns on each individual molecule. Rather, molecules will certainly be switched in groups. The size of these groups can be greatly reduced by using a variety of molecules with varying "on" and "off" voltages. However, determining these groups from knowledge of the graph within the NanoCell will most certainly be a difficult task. A more effective strategy may be to simply treat the NanoCell as a black box, and use appropriate techniques to train it.

One mortal training strategy that we are currently exploring is neuro-dynamic programming.[180] This technique is basically a biased random walk. During training, a NanoCell "learns" which voltage pulses tend to be the most beneficial in each given set of switch states based on previous experience. Throughout the process, the NanoCell is treated as a black box. Therefore, this is a method that could be applied to a physical NanoCell.

In previous sections, NanoCell training proofs were presented. In this section, some possible methods for improving these results are explored. First, the results are for voltage in – current out NanoCells, and NanoCells will actually be voltage out. Therefore, the results should be adapted to this case. This will be extremely simple if the voltage in – current out observations hold for voltage in – voltage out with the new bistable latches.

Another potential improvement is to prove the observed sufficient conditions for NANDs. Namely, compute the probability that there are intersecting paths of sufficiently short length from each input to the output.

Computing the all-terminal reliability of a NanoCell is another open NanoCell problem. It would be interesting to compute the probability that there exists a path between any two I/O pins. Computational results would be beneficial in evaluating the quality of any bounds. Independent edge cuts or Gomory-Hu cut trees may prove useful in determining a necessary condition for larger NanoCells.[181]

The necessary and sufficient conditions given for inverters can be more closely adapted to realistic NanoCells. First node failures should be factored in. Second, the edge distribution should follow a Poisson distribution rather than a constant probability for each edge.

Observations should be made for some of the more complex logic gates such as half adders or 1-bit adders. Once necessary and sufficient conditions are well defined, proofs should be generated. Note that one necessary condition will always be a path of "on" molecules from the inputs to the output. It should also be noted that NanoCell proofs may become much more complex when a variety of molecules are within each cell.

6.6 Conclusions

Over the past two years, our NanoCell programming team has made huge advances in training cells. No one had any idea whether a randomly assembled array of high and low conducting NDR devices could be trained to serve some useful logical purpose. We have shown that indeed, such arrays of molecules do function as logical devices in simulations. It is important to note that even logic gates involving multiple outputs have proven trainable in a NanoCell. In addition, we have developed a process for training NanoCells by first studying a small number of molecules, then applying search techniques such as genetic algorithms. We can also predict whether a logic gate is trainable with any given set of molecular "on" and "off" I(V) curves. The strategy of observing necessary and sufficient conditions, then proving bounds seems to be valuable, as well.

The advances made by the computer architecture group have enabled us to offer valuable advice to the rest of the team. We are able to suggest the most easily trainable NanoCells. We have also recommended that the NanoCell be reduced in size from that shown in Figure 6.1 to that shown in Figure 6.39.

There are still many substantial obstacles to overcome. Large circuits of simulated NanoCells working in concert must be obtained. This achievement should be realized shortly, however. In addition, results seen in simulation should be validated by tests on physical NanoCells. Simulations should be adapted to reflect what is seen in these experimental cells. After obtaining working physical NanoCells, the issues of size and power must be addressed. What kind of size advantage is possible with a NanoCell? What kind of power is required to run a chip of many thousands of the cells? Here bistable latches may prove a liability as they draw constant current.

Even taking these issues into account, the NanoCell team has made tremendous advances over the past few years. We are confident that all of these questions will be adequately addressed and look forward to further research in this area.

Bibliography

1. Christensen, C. M. "The Innovator's Dilemma," Harper Collins, New York, **2000**.
2. Rivette, K. G.; Kline, D. "Rembrandts in the Attic," Harvard Business School Press: Boston, **2000**.
3. Brumfiel, G. "Bell Labs Launches Inquiry Into Allegations of Data Duplication," *Nature* **2002**, *417*, 367.
4. Simon, H. A. "Administrative Behavior," Macmillan: New York, **1959**.
5. For some recent reviews on the topic, see: (a) Tour, J. M. *Acc. Chem. Res.* **2000**, 33, 791-804. (b) Tour, J. M.; Rawlett, A. M; Kozaki, M.; Yao, Y.; Jagessar, . C.; Dirk, S. M.; Price, D. W.; Reed, M. A.; Zhou, C.-W.; Chen, J.; Wang, W.; Campbell, I. *Chem. Eur. J.* **2001**, 7, 5118. (c) Reed, M. A.; Tour, J. M. *Scientific American*, June **2000**, 86. (d) Kwok, K. S.; Ellenbogen, J. C. *Materials Today*, February **2002**, 28. (e) *Atomic and Molecular Wires*; Joachim, C.; Roth, S., eds.; NATO Applied Sciences, Vol. 341, Kluwer: Boston, **1997**. (f) Molecular Electronics: Science and Technology; Aviram, A.; Ratner, M., eds.; *Annal. New York Acad. Sci.,* Vol. 852, **1998**. (g) Advances in Chemistry, Series 240, Molecular and Biomolecular Electronics; Birge, R. R. ed.; American Chemical Society, **1994**. (h) Joachim, C.; Gimzewski, J. K.; Aviram, A. *Nature* **2000**, 408, 541-548. (i) Goldhaber-Gordon, D.; Montemerlo, M. S.; Love, J. C.; Opiteck, G. J.; Ellenbogen, J. C. *Proc. IEEE* **1997**, 85, 521-540.
6. Overton, R. *Wired* **2000**, 8 (7), 242.
7. Service, R. F. *Science* **2001**, 294, 2442.
8. Petty, M. C., Bryce, M. R., Bloor, D., *Introduction to Molecular Electronics,* Oxford University Press, New York, **1995**.
9. Adleman, L. M. *Sci. Am.* **1998**, 279, 54.
10. (a) Preskill, J. *Proc. R. Soc. Lond. A* **1998**, 454, 385. (b) Preskill, J., *Proc. R. Soc. Lond. A* **1998**, 454, 469. (c) Platzman, P. M., Dykman, M. I. *Science* **1999**, 284, 1967. (d) Kane, B. *Nature* **1998**, 393, 133.
11. (a) Anderson, M. K. *Wired* **2001**, 9 (9), 157. (b) Anderson, M. K. *Wired* **2001**, 9 (9), 152.
12. Wolf, S. A.; Awschalom, D.D.; Buhrman, R. A.; Daughton, J. M.; von Molnar, S.; Roukes, M. L.; Chtchelkanova, A. Y.; Treger, D. M. *Science* **2001**, 294, 1488.

13. Raymo, F. M., Giordani, S. *Org. Lett.* **2001**, 3, 1833.
14. Moore, G. E. *Electronics* **1965**, 38.
15. Hand, A. *Semiconductor International* **2001**, 24 (6), 15.
16. Selzer, R. A.; Friml, W. R.; Gagnon, J. B.; Macklin, R. H.; Rauch, F. L.; Siegert, H.; Simon, K. US Patent 6,295,332, 25 September **2001**.
17. Packan, P. *Science* **1999**, 285, 2079.
18. Whitney, D. E. *Res. Eng. Des.* **1996**, 8, 125.
19. Hand, A. *Semiconductor International* **2001**, 24 (8), 62.
20. Golshan, M., Schmitt, S. *Ultrapure Water* **2001**, 18, 34.
21. Hoggan, E. N.; Carbonell, R. G.; DeSimone, J. M.; Cramer, G. L.; Stewart, G. M. *CleanTech98 Proceedings* **1998**, 137-146; Witter Publishing, Flemington, NJ.
22. *Handbook of Conducting Polymers,* Skotheim, T. A., Ed.; Dekker: New York, **1986**.
23. (a) Tour, J. M.; Jones, L., II; Pearson, D. L.; Lamba, J. S.; Burgin, T.; Whitesides, G. W.; Allara, D. L.; Parikh, A. N.; Atre, S. *J. Am. Chem. Soc.* **1995**, 117, 9529. (b) Dhirani, A.; Zehner, R. W.; Hsung, R. P.; Guyot-Sionnest, P., Sita, L. R. *J. Am. Chem. Soc.* **1996**, 118, 3319.
24. Bumm, L. A.; Arnold, J. J.; Cygan, M. T.; Dunbar, T. D.; Burgin, T. P.; Jones, L., II, Allara, D. L.; Tour, J. M.; Weiss, P. S. *Science* **1996**, 271, 1705.
25. Schumm, J. S.; Pearson, D. L.; Tour, J. M. *Angew. Chem. Int. Ed. Engl.* **1994**, 33, 1360.
26. Pearson, D. L.; Schumm, J. S.; Tour, J. M. *Macromolecules* **1994**, 27, 2348.
27. Tour, J. M. *Chem. Rev.* **1996**, 96, 537.
28. (a) Wegner, G. In *Thermoplastic Elastomers, A Comprehensive Review,* Legge, N. R.; Holden, G.; Schroeder, H. E., Eds.; Hanser: New York, **1987**, 405. (b) Young, J. K.; Nelson, J. C.; Moore, J. S. *J. Am. Chem. Soc.* **1994**, 116, 10841. (c) Nelson, J. C.; Young, J. K.; Moore, J. S. *J. Org. Chem.* **1996**, 61, 8160.
29. Suffert, J.; Ziessel, R. *Tetrahedron Lett.* **1991**, 32, 757.
30. Oligomer lengths were simulated using standard molecular modeling procedures. All calculations were performed on a Power Macintosh 8100/80 AV using *Personal CAChe* version 3.7 for both structure drawing and minimization. The *CAChe* mechanics application implements a standard MM2 force field. All energy calculations were minimized over a large number of iterations to convergence at local minima nearest in energy to the starting compounds' energies.
31. Ulman, A. *An Introduction to Ultrathin Organic Films,* Academic: Boston, **1991**.
32. Samuel, I. D. W.; Ledoux, I.; Delporte, C.; Zyss, J.; Pearson, L.; Tour, J. M. *Chem. Mater.* **1996**, 8, 819.

33. For sulfur quench without acylation, see: Jones, E.; Moodie, I. M. *Org. Synth.* **1970**, 50, 104. Note that we also used of thienylthiols as alligator clips, however, dimerization reactions via conjugate addition on thiol tautomers ensued. For analogous dimerizations, see: Ponticello, G. S.; Habecker, C. N.; Varga, S. L.; Pitzenbeger, S. M. *J. Org. Chem.* **1989**, 54, 3223.

34. Wu, J.; Chi, C,; Wang, X.; Li, J.; Zhao, X.; Wang, F. *Syn. Commun.* **2000**, 30, 4293-4298.

35. (a) Uhlenbroek, J. H.; Bijloo, J. D. *Rec. Trav. Chim.* **1960**, 79, 1181. (b) Reinecke, M. G.; Adickes, H. W.; Pyrn, C. *J. Org. Chem.* **1971**, 36, 2690.

36. (a) Pham, C. V.; Mark, H. B.; Zimmer, H. *Synth. Commun.* **1986**, 16, 689. (b) Tamao, K.; Kodama, S.; Nakajima, I.; Kumada, M.; Minato, A.; Suzuki, K. *Tetrahedron* **1982**, 38, 3347. (c) Campaigne, E.; Yokley, O. E. *J. Org. Chem.* **1963**, 28, 914.

37. (a) Bharathi, P.; Patel U.; Kawaguchi, T.; Pesak, D. J.; Moore, J. S. *Macromolecules* **1995**, 28, 5955. (b) *Polymer-Supported Reactions in Organic Synthesis*, Hodge, P.; Sherrington, D. C., Eds.; Wiley: New York, **1980**.

38. Negishi, E.; King, A. O.; Tour, J. M. *Org. Synth.* **1985**, 64, 44.

39. Doyle, M. P.; Bryker, W. J. *J. Org. Chem.* **1979**, 44, 1572.

40. Lamba, J. S. S.; Tour, J. M. *J. Am. Chem. Soc.* **1994**, 116, 11723.

41. (a) Sonogashira, K; Tohda, Y.; Hagihara, N. *Tetrahedron Lett.* **1975**, 4467. (b) Stephens, R. D.; Castro, C. E. *J. Org. Chem.* **1963**, 28, 3313. (c) Sonogashira, K. ; Trost, B. M., Ed.; Pergamon: Oxford, 1991; Vol. 3, pp 521-549.

42. Moore, J. S.; Weinstein, E. J.; Wu, Z. Y. *Tetrahedron Lett.* ***1991, 32, 2465.***

43. (a) Crowley, J. I.; Rapoport, H. *Acc. Chem. Res.* **1976**, 9, 135. (b) Leznoff, C. C. *Chem. Soc. Rev.* **1974**, 3, 65. (c) Leznoff, C. C. *Acc. Chem. Res.* **1978**, 11, 327. (d) Dewitt, S. H.; Czarnik, A. W.; Ellman, J. A. *Acc. Chem. Res.* **1996**, 29, 114. (e) Giralt, E.; Rizo, J.; Pedroso, E. *Tetrahedron* **1984**, 40, 4141.

44. Yan, B.; Kumaravel, G.; Anjaria, H.; Wu, A.; Petter, R. C.; Jewell, C. F., Jr.; Wareing, J. R. *J. Org. Chem.* **1995**, 60, 5736.

45. Grindley, T. B.; Johnson, D. F.; Katritzky, A. R.; Keogh, H. J.; Thirkettle, C.; Topsom, R. D. *J. Chem. Soc., Perkin Trans. 2* **1974**, 3, 282.

46. An analogous equation utilizing resin weight differences has been derived by Moore. See reference 28c.

47. (a) Bengton, G.; Keyaniyan, S.; Demeijere, A. *Chem. Ber.* **1986**, 119, 3607. (b) Hurst, D. T.; McInnes, A. G. *Can. J. Chem.* **1965**, 43, 2004. (c) Gaffney, B. L.; Jones, R. A. *Tetrahedron Lett.* **1982**, 23, 2257.

48. Giralt, E.; Rizo, J.; Pedroso, E. *Tetrahedron* **1984**, 40, 4141.

49. McLafferty, F. W.; Turecek, F. *Interpretation of Mass Spectra*, 4th ed.; University Science: Mill Valley, California, **1993**.

50. Briscoe, G. B.; Humphries, S. *Talanta* **1969**, 16, 1403.
51. Zervais, L.; Photaki, I.; Ghelis, N. *J. Am. Chem. Soc.* **1963**, 85, 1337.
52. Incomplete carbon combustion in high MW materials with high unsaturated carbon content is a commonly observed property. See reference 40 and (a) Wallaow, T. I.; Novak, B. M. *J. Am. Chem. Soc.* **1991**, 113, 7411. (b) Stephens, E. B.; Tour, J. M. *Macromolecules* **1993**, 26, 2420.
53. (a) Takahashi, S.: Kuroyama, Y.; Sonogashira, K.; Hagihara, N.; *Synthesis* **1980**, 627-630. (b) Blum, J.; Baidossi, W.; Badrieh, Y.; Hoffmann, R. E.; Schumann, H.; *J. Org. Chem.* **1995**, 60, 4738.
54. Pearson, L.; Tour, J. M.; *J. Org. Chem.* **1997**, 62, 1376.
55. Adams, R. D.; Barnard, T.; Rawlett, A. ; Tour, J. M. *Eur. J. Inorg. Chem.* **1998**, 429.
56. Seminario, J. M.; Zacarias, A. G.; Tour, J. M. *J. Am. Chem. Soc.* **1998**, 120, 3970.
57. Satyamurthy, N.; Barrio, J. R.; Bida, G. T.; Phelps, M. E. *Tetrahedron Lett.* **1990**, 31, 4409.
58. Jones II., L. ; Schumm, J. S. ; Tour, J. M. *J. Org. Chem.* **1997**, 62, 1388.
59. Allara, D. L.; Dunbar, T. D.; Weiss, P. S.; Bumm, L. A.; Cygan, M. T.; Tour, J. M.; Reinerth, W. A.; Yao, Y.; Kozaki, M.; Jones II., L. *N.Y. Acad. Sci., Molecular Electronics: Science and Technology*; Aviram, A.; Ratner, M., Eds.; Ann. N.Y. Acad. Sci., 1998, Vol. 852, pp. 349.
60. (a) Brown, R.; Jones, W. E.; Pinder, A. R.; *J. Chem. Soc.* **1951**, 2123. (b) Bordwell, F. G.; Hewett, W. A. *J. Org. Chem.* **1957**, 22, 980-981.
61. Tour, J. M.; Kozaki, M.; Seminario, J. M. *J. Am. Chem. Soc.* **1998**, 120, 8486.
62. (a) Gribble, G. W.; Leese, R. M. *Synthesis* **1977**, 172. (b) Gribble, G. W.; Kelly, W. J.; Emery, S. E.; *Synthesis* **1978**, 763. (c) Gribble, G. W.; Nutaitis, C. F.; *Tetrahedron Lett.* **1985**, 6023.
63. Reed, M. A.; *Proceedings of the IEEE: Special Issue on Nanoelectronics* **1999**, 87, 652-658.
64. (a) Yao, Y.; Tour, J. M. *Macromolecules* **1999**, 32, 2455. (b) Lamba, J. J. S.; Tour, J. M. *J. Am. Chem. Soc.* **1994**, 116, 11723.
65. Olah, G. A.; Arvanaghi, M.; Ohannesian, L. *Synthesis* **1986**, 770.
66. Moroni, M. ; Moigne, J. Le ; Pham, T. A. ; Bigot, J.-Y. *Macromolecules*, **1997**, 30, 1964.
67. Cacchi, S. ; Fabrizi, G. ; Moro, L. *J. Org. Chem.* **1997**, 62, 5327 and references therein.
68. Appel, R. ; Kleinstück, R.; Ziehn, K.-D. *Angew. Chem., Int. Ed. Engl.* **1971**, 10, 132.
69. (a) Collier, C. P.; Wong, E. W.; Belohradský, M.; Raymo, F. M.; Stoddart, J. F.; Kuekes, P. J.; Williams, R. S.; Heath, J. R. *Science* **1999**, 285, 391. (b) Collier, C. P.; Mattersteig, G.; Wong, E. W.; Luo, Y.; Beverly, K.; Sampaio, J.; Raymo, F. M.; Stoddart, J. F.; Heath, J. R. *Science* **2000**, 289,

1172. Heath et al, are currently evaluating the efficacy of **91** in their assembled system. (c) Metzger, R. M.; Chen, B.; Hopfner, U.; Lakshmikantham, M. V.; Vuillaume, D.; Kawai, T.; Wu, X.; Tachibana, H.; Hughes, T. V.; Sakurai, H.; Baldwin, J. W.; Hosch, C.; Cava, M. P.; Brehmer, L.; Ashwell, G. J. *J. Am. Chem. Soc.* **1997**, 119, 10455. (d) Metzger, R. M. *Acc. Chem. Res.* **1999**, 32, 950.

70. B. C. Ranu, D. C. Sarkar, R. Chakraborty, *Synth. Comm.* **1992**, 22, 1095.

71. Corey, E. J.; Székely, I.; Shiner, C. S. *Tetrahedron Lett.* **1977**, 3529.

72. Kilic, E.; Tuzun, C. *Org. Prep. Proced., Int.* **1990**, 22 (4), 485.

73. Wulfman, D. S.; Cooper, C. F. *Synthesis* **1978**, 924.

74. Kaczmarek, L.; Nowak, B.; Zukowski, J.; Borowicz, P.; Sepiol, J.; Grabowska, A. *J. Mol. Struct.* **1991**, 248, 189.

75. (a) Romero, F. M.; Ziessel, R. *Tetrahedron Lett.* **1995**, 36, 6471. (b) Benin, V.; Kaszynski, P.; Pink, M.; Young, Jr. V. G. *J. Org. Chem.* **2000**, 65, 6388.

76. Littler, B. J.; Ciringh, Y.; Lindsay, J. S. *J. Org. Chem.* **1999**, 64, 2864.

77. For several background procedures that were used or modified for these studies, see: (a) Lee, C. H.; Lindsey, J. S. *Tetrahedron* **1994**, 50, 11427. (b) Wang, Q. M.; Bruce, D. W. *Synlett.* **1995**, 1267. (c) Wagner, R. W.; Johnson, T. E.; Li, F.; Lindsey, J. S. *J. Org. Chem.* **1995**, 60, 5266. (d) Nierengarten, J. F.; Schall, C.; Nicoud, J. F. *Angew. Chem. Int. Ed. Engl.* **1998**, 37, 1934. (e) Alder, A. D.; Longo, F. R.; Finarelli, J. D.; Goldmacher, J.; Assour, J.; Korsakoff, L. *J. Org. Chem.* **1967**, 32, 476. (f) Khoung, R. G.; Jaquinod, L.; Smith, K. M. *Chem. Comm.* **1997**, 1057.

78. Jagessar, R. C. ; Tour, J. M. *Org. Lett.* **2000**, 2, 111.

79. Austin, W. B.; Bilow, N.; Kellegham, W. J.; Lau, K. S.Y. *J. Org. Chem.* **1981**, 46, 2280.

80. Tao, Y.-T.; Hietpas, G. D.; Allara, D. L. *J. Am. Chem. Soc.* **1996**, 118, 6724.

81. Xu, Z. F., Moore, J. S. *Angew. Chem.* **1993**, 105, 1394-1396; *Angew Chem. Int. Ed. Engl.* **1993**, 32, 1354.

82. (a) Chen, J.; Reed, M. A.; Rawlett, A. M.; Tour, J. M. *Science* **1999**, 286, 1550. (b) Chen, J.; Wang, W.; Reed, M. A.; Rawlett, A. M.; Price, D. W.; Tour, J. M. *Appl. Phys. Lett.* **2000**, 77, 1224. (c) Reed, M. A.; Chen, J. Rawlett, A. M.; Price, D. W.; Tour, J. M. *App. Phys. Lett.* **2001**, 3735.

83. (a) Balabin, I. A.; Onuchic, J. *Science* **2000**, 290, 114. (b) Gupta, N.; Linschitz, H. *J. Am. Chem. Soc.* **1997**, 119, 6384. (c) Rabenstein, B.; Ullmann, G. M. Knapp, E. *Biochemistry*, **2000**, 39, 10487.

84. Illescas, B. M. ; Martin, N. *J. Org. Chem.* **2000**, 65, 5986.

85. (a) Vondrak, T.; Cramer, C. J.; Zhu, X.-Y. *J. Phys. Chem.* **1999**, 103, 8915. (b) Creager, S.; Yu, C. J.; Bamdad, C.; Oconnor, S.; Maclean, T.; Lam, E.; Chong, Y.; Olsen, G. T.; Luo, J. Y.; Gozin, M.; Kayyem, J. F. *J. Am. Chem. Soc.* **1999**, 121, 1059.

86. Zollinger, H. *Diazo Chemistry I*; VCH Verlaggesellschaft mbH: Weinheim, **1994,** Vol. 273.
87. Kosynkin, D. V.; Tour, J. M. *Organic Lett.* **2001,** 3, 993.
88. (a) Sun, S.-S.; Lees, A. J. *Inorg. Chem.* **1999,** 38, 4181. (b) Sun, S.-S.; Lees, A. J. *J. Am. Chem. Soc.* **2000,** 122, 8956.
89. Chanteau, S. H.; Tour, J. M. *Tetrahedron Lett.* **2001,** 42, 3057.
90. Dirk, S. M.; Mickelson, E. T.; Henderson, J. C.; Tour, J. M. *Organic Lett.* **2000,** 2, 3405.
91. Shieh, W.; Carlson, J. A. *J. Org. Chem.* **1992,** 57, 379.
92. Price, D. W.; Dirk, S. M.; Rawlett, A. M.; Chen, J.; Wang, W.; Reed, M. A.; Zacarias, A.; Seminario, J. M.; Tour, J. M. *Mat. Res. Soc. Symp. Proc.* **2000,** submitted.
93. Palmgren, A.; Thorarensen, A.; Bäckvall, J.E. *J. Org. Chem.* **1998,** 63, 3764.
94. Bäckvall, J.E.; Byström, S.E.; Nordberg, R.E. *J.Org.Chem.* **1984,** 49, 4619.
95. Jacob, III, P.; Callery, P.S.; Shulgin, A.T.; Castagnoli, Jr., N. J. *Org. Chem.* **1976,** 41, 3627.
96. (a) Ziessel, R.; Suffert, J.; Youinou, M.-T. *J. Org. Chem.* **1996,** 61, 6535. (b) Champness, N. R.; Khlobystov, A. N.; Majuga, A. G.; Schröder, M.; Zyk, N. V. *Tetrahedron Lett.* **1999,** 40, 5413. (c) Whiteford, J. A.; Lu, C. V.; Stang, P. J. *J. Am. Chem. Soc.* **1997,** 119, 2524.
97. Rosen, G. M.; Tsai, P.; Barth, E. D.; Dorey, G.; Casara, P.; Spedding, M.; Halpern, H. J. *J. Org. Chem.* **2000,** 65, 4460.
98. Shuartsbery, M. S.; Moroz, A. A.; Kiseleva, O. D. *Bull Acad. Sci. USSR Div. Chem. Sci.* **1981,** 30, 608.
99. (a) Seneci, P. *Solid-Phase Synthesis and Combinatorial Technologies*; Wiley: New York, **2000.** (b) Jung, G. Ed.; *Combinatorial Chemistry-Synthesis, Analysis, Screening;* Wiley-VCH: Weinheim, **1999.**
100. (a) Natzi, A.; Ostrech, J. M.; Houghten, R. A. *Chem. Rev.* **1997,** 97, 449. (b) Fruchtel, J. S.; Jung, G. *Angew. Chem. Int. Ed.* **1996,** 35, 17.
101. (a) Jandeleit, B.; Schaefer, D. J.; Powers, T. S.; Turner, H. W.; Weinberg, W. H. *Angew. Chem. Int. Ed.* **1999,** 38, 2494. (b) Senkan, S. *Angew. Chem. Int. Ed.* **2001,** 40, 312.
102. (a) **9b,** Purchased from Aldrich Chemical Co. (b) **9c,** Akula, M. R.; Kabalka, G. W. *J. Labelled Cpd. Radiopharm.* **1999,** 42, 959. (c) **9d,** Kruger, G.; Keck, J.; Null, K.; Pieper, H. *Arzneim-Forsch./Drug Res.* **1984,** 34, 1612. (d) **9e,** Alabaster, C. T.; Bell, A. S.; Campbell, S. F.; Ellis, P.; Henderson, C. G. *J. Med. Chem.* **1988,** 31, 2048.
103. (a) Malenfant, P. R. L.; Fréchet, J. M. *J. Chem. Commun.* **1998,** 2657. (b) Briehn, C. A.; Kirschbaum, T.; Bäuerl, P. *J. Org. Chem.* **2000,** 65, 352. (c) Briehn, C. A.; Schiedel, M.-S.; Bonsen, E. M.; Schuhmann, W.; Bäuerl, P. *Angew. Chem. Int. Ed.* **2001,** 40, 4680.

104. (a) Bräse, S.; Enders, D.; Köbberling, J.; Avemaria, F. *Angew. Chem. Int. Ed.* **1998**, 37, 3413. (b) Bräse, S.; Schroen, M. *Angew. Chem. Int. Ed.* **1999**, 38, 1071. (c) Dahmen, S.; Bräse, S. *Org. Lett.* **2000**, 2, 3563. (d) Bräse, S.; Dahmen, S.; Pfefferkorn, M. *J. J. Comb. Chem.* **2000**, 2, 710.
105. Anderson, A. *Chem. Eur. J.* **2001**, 7, 4706.
106. Bräse, S.; Dahmen, S. *Chem. Eur. J.* **2000**, 6, 1899.
107. Huang, S.; Tour, J. M. *J. Org. Chem.* **1999**, 64, 8898.
108. IRORI, 9640 Towne Contre Drive, San Diego, CA 92121, phone (858) 259-2421.
109. In this preliminary study, we used sets of colored glass beads to label the MacroKans. For larger libraries, we use the full combinatorial system with radio frequency (RF) tags from IRORI (research in progress).
110. Korzeniowski, S. H.; Leopold, A.; Beadle, J. R.; Ahern, M. F.; Sheppard, W. A. *J. Org. Chem.* **1981**, 46, 2153.
111. Stewart, M. S.; Tour, J. M., submitted for review.
112. (a) Gryko, D. T.; Zhao, F.; Yasseri, A. A.; Roth, K. M.; Bocian, D. F.; Kuhr, W. G.; Lindsey, J. S. *J. Org. Chem.* **2000**, 65, 7356. (b) Clausen, C.; Gryko, D. T.; Yasseri, A. A.; Diers, J. R.; Bocian, D. F.; Kuhr, W. G.; Lindsey, J. S.. *J. Org. Chem.* **2000**, 65, 7371.
113. Cai, L.; Yao, Y.; Yang, J.; Price, D. W. Jr., Tour, J. M. *Chem Mater.* **2002**, 14, 2905.
114. (a) Joachim, C.; Gimzewski, J. K. *Proceedings of the IEEE*, **1999**, 86 (1), 184. (b) Klein, D. L.; Roth, R.; Lim, A. K. L.; Alivisatos, A. P.; McEuen, P. L. *Nature* **1997**, 389, 699. (c) Park, J.; Pasupathy, A. N.; Goldsmith, J. I.; Chang, C.; Yaish, Y.; Petta, J. R.; Rinkoski, M.; Sethna, J. P.; Abruna, H. D.; McEuen, P. L.; Ralph, D. C. *Nature* **2002**, 417, 722. (d) Liang, W.; Shores, M. P.; Bockrath, M.; Long, J. R.; Park, H. *Nature* **2002**, 417, 725.
115. (a) Dunbar, T. D.; Cygan, M. T.; Bumm, L. A.; McCarty, G. S.; Burgin, T. P.; Reinerth, W. A.; Jones, L. II; Jackiw, J. J.; Tour, J. M.; Weiss, P. S.; Allara, D. L. *J. Phys. Chem. B* **2000**, 104, 4880. (b) Cygan, M. T.; Dunbar, T. D.; Arnold, J. J.; Bumm, L. A.; Shedlock, N. F.; Burgin, T. P.; Jones, L., II, Allara, D. L.; Tour, J. M.; Weiss, P. S. *J. Am. Chem. Soc.* **1998**, 120, 2721.
116. Donhauser, Z. J.; Mantooth, B. A.; Kelly, K. F.; Bumm, L. A.; Monnell, J. D.; Stapleton, J. J.; Price, D. W. Jr.; Rawlett, A. M.; Allara, D. L.; Tour, J. M.; Weiss, P. S *Science* **2001**, 292, 2303.
117. (a) Fan, F.-R. F., Yang, J.; Cai, L.; Price, D. W.; Dirk, S. M.; Kosynkin, D.; Yao, Y.; Rawlett, A. M.; Tour, J. M.; Bard, A. J. *J. Am. Chem. Soc.* **2002**, 124, 5550. (b) Fan, F.-R. F., Yang, J.; Price, D. W.; Dirk, S. M.; Kosynkin, D.; Tour, J. M.; Bard, A. J. *J. Am. Chem. Soc.* **2001**, 123, 2454.
118. Allara, D. L. personal communication.

119. Chen, J.; Reed, M. A.; Asplund, C. L.; Cassell, A. M.; Myrick, M. L.; Rawlett, A. M.; Tour, J. M.; Van Patten, P. G. *Appl. Phys. Lett.* **1999**, 75, 624.

120. Reed, M. A.; Zhou, C.; Muller, C. J.; Burgin, T. P.; Tour, J. M. *Science* **1997**, 278, 252.

121. Zhou, C.; Deshpande, M. R.; Reed, M. A.; Jones, L., II; Tour, J. M. *Appl. Phys. Lett.* **1997**, 71, 611.

122. Seminario, J. M.; Tour, J. M. *Molecular Electronics: Science and Technology*; Aviram, A.; Ratner, M., Eds.; *Ann. N.Y. Acad. Sci.* **1998**, Vol. 852, pp. 68.

123. Seminario, J. M.; Zacarias, A. G.; Tour, J. M. *J. Am. Chem. Soc.* **1999**, 121, 411.

124. Zhou, C. "Atomic and Molecular Wires", Ph.D. thesis, Yale University **1999**.

125. Reed, M. A.; Price, D. W. Tour, J. M., unpublished data.

126. (a) Seminario, J. M.; Zacarias, A. G.; Tour, J. M. *J. Am. Chem. Soc.* **2000**, 122, 3015. (b) Seminario, J. M. ; Zacarias, A. G. ; Derosa, P. A. *J. Phys. Chem. A.* **2001**, 105, 791.

127. (a) Reinerth, W. A.; Tour, J. M. *J. Org. Chem.* **1998**, 63, 2397. (b) Reinerth, W. A.; Burgin, T. P.; Dunbar, T. D.; Bumm, L. A.; Arnold, J. J.; Jackiw, J. J.; Zhou, C.-w.; Deshpande, M. R.; Allara, D. L.; Weiss, P. S.; Reed, M. A.; Tour, J. M. *Polym. Mater., Sci. Engin. (Am. Chem. Soc., Div. Polym. Mater.),* **1998**, 78, 178.

128. (a) Chen, J.; Calvet, L. C.; Reed, M. A.; Carr, D. W.; Grubisha, D. S.; Bennett, D. W. *Chem. Phys. Lett.* **1999**, 313, 741. (b) Chen, J. ; Wang, W.; Klemic, J. ; Reed, M. A. ; Axelrod, B. W. ; Kaschak, D. M. ; Rawlett, A. M. ; Price, D. W. ; Dirk, S. M. ; Tour, J. M. ; Grubisha, D. S. ; Bennett, D. W. *Ann. N.Y. Acad. Sci., Molecular Electronics II* **2002**, 960, 69.

129. Schön, J. H.; Meng, H.; Bao, Z. *Science* **2001**, 294, 2138. (b) Schön, J. H.; Meng, H.; Bao *Nature* **2001**, 413, 713 with correction **2001**, 414, 470.

130. (a) Rao, P. D.; Dhanalekshmi, P.; Littler, B. J.; Lindsey, J. S. *J. Org. Chem.* **2000**, 65, 7323. (b) Gryko, D. T.; Clausen, C.; Roth, K. M.; Dontha, N.; Bocian, D. F.; Kuhr, W. G.; Lindsey, J. S. *J. Org. Chem.* **2000**, 65, 7345. (c) Clausen, C.; Gryko, D. T.; Dabke, R. B.; Dontha, N.; Bocian, D. F.; Kuhr, W. G.; Lindsey, J. S. *J. Org. Chem.* **2000**, 65, 7363. (d) Li, J.; Gryko, D.; Dabke, R. B.; Diers, J. R.; Bocian, D. F. Kuhr, W. G.; Lindsey, J. S. *J. Org. Chem.* **2000**, 65, 7379.

131. Adapted in part from a book chapter, Tour, J. M.; James, D. K. "Molecular Electronic Computing," *CRC Encyclopedia of Nanoscience, Engineering, & Technology*, in press.

132. (a) Lent, C. S.; Tougaw, P. D. *Proceedings of the IEEE,* **1997**, 85(4), 541. (b) Douglas, P.; Lent, C. S.; Tougaw, P. D.; Porod, W. *J. Appl. Phys.* **1994**,

75, 1818. (c) Tour, J. M.; Kozaki, M.; Seminario, J. M. *J. Am. Chem. Soc.* **1998**, *120*, 8486.

133. (a) Heath, J. R., Kuekes, P. J. Snider, G. R., Williams, R. S. *Science* **1998** 280, 1716. (b) Heath, J. R. *Pure Appl. Chem.* **2000**, 72, 11.

134. Goldman, S. C.; Budiu, M. *Proc. 28th Annual Int. Symp. On Comp. Arch.* **2001**, June.

135. (a) Hu, J.; Odom, T. W.; Lieber, C. M. *Acc. Chem. Res.* **1999**, 32, 435. (b) Rueckes, T.; Kim, K.; Joselevich, E.; Tseng, G. Y.; Cheung, C.-L.; Lieber, C. M. *Science* **2000**, 289, 94. (c) Ouyang, M.; Huang, J.-L.; Cheung, C.-L.; Lieber, C. M. *Science* **2000**, 291, 97. (d) Star, A.; Stoddart, J. F.; Steuerman, D.; Diehl, M.; Boukai, A.; Wong, E. W.; Yang, X.; Chung, S.-W.; Choi, H.; Heath, C. M. *Angew. Chem. Int. Ed. Engl.* **2001**, 40, 1721. (e) Bozovic, D.; Bockrath, M.; Hafner, J. H.; Lieber, C. M.; Park, H.; Tinkham, M. *App. Phys. Lett.* **2001**, 78, 3693.

136. (a) Huang, Y., Duan, X., Wei, Q., Lieber, C. M. *Science* **2001**, 291, 630. (b) Gudiksen, M. S., Wang, J., Lieber, C. M. *J. Phys. Chem. B* **2001**, 105, 4062. (c) Cui, Y., Lieber, C. M. *Science* **2001**, 291, 851. (d) Chung, S.-W., Yu, J.-Y, Heath, J. R. *App. Phys. Lett.* **2000**, 76, 2068.

137. Tour, J. M., Van Zandt, W. L., Husband, C. P., Husband, S. M., Wilson, J., Franzon, P., Nackashi, D., *IEEE Trans. Nanotech.* **2002**, in press.

138. McEuen, P. L.; *Science* **1997**, 278, 1729.

139. Stewart, D. R.; Sprinzak, D.; Marcus, C. M.; Duruoz, C. I.; Harris, J. S. Jr. *Science* **1997**, 278, 1784,.

140. Rajeshwar, K., de Tacconi, N. R., Chenthamarakshan, C. R. *Chem. Mat.* **2001**, 13, 2765.

141. Leifeld, O.; Hartmann, R.; Muller, E.; Kaxiras, E.; Kern, K.; Grutzmacher, D. *Nanotechnology*, **1999**, 19, 122.

142. Sear, R. P.; Chung, S.-W.; Markovich, G.; Gelbart, W. M.; Heath, J. R. *Phys. Rev. E* **1999**, 59, 6255.

143. (a) Markovich, G.; Collier, C. P.; Henrichs, S. E.; Remacle, F.; Levine, R. D.; Heath, J. R. *Acc. Chem. Res.* **1999.**, 32, 415. (b) Weitz, I. S.; Sample, J. L.; Ries, R.; Spain, E. M.; Heath, J. R. *J. Phys. Chem. B* **2000**, 104, 4288.

144. Snider, G. L.; Orlov, A. O.; Amlani, I.; Zuo, X.; Bernstein, G. H.; Lent, C. S.; Merz, J. L.; Porod, W. *J. Appl. Phys.* **1999**, 85, 4283.

145. Snider, G. L.; Orlov, A. O.; Amlani, I.; Bernstein, G. H.; Lent, C. S.; Merz, J. L.; Porod, W. *Jpn. J. Appl. Phys. Part I: Reg. Papers and Short Notes* **1999**, 38, 7227.

146. Toth, G., Lent, C. S. *J. Appl. Phys.* **1999**, 85, 2977.

147. (a) Amlani, I.; Orlov, A. O.; Snider, G. L.; Lent, C. S.; Bernstein, G. H. *Appl. Phys. Lett.* **1998**, 72, 2179. (b) Amlani, I.; Orlov, A. O.; Snider, G. L.; Lent, C. S.; Porod, W.; Bernstein, G. H. *Superlattices and*

Microstructures **1999**, 25, 273. (c) Bernstein, G. H.; Amlani, I.; Orlov, A. O.; Lent, C. S.; Snider, G. L. *Nanotechnology* **1999**, 10, 166.

148. Amlani, I.; Orlov, A. O.; Toth, G.; Bernstein, G. H.; Lent, C. S.; Snider, G. L. *Science* **1999**, 284, 289.

149. (a) Orlov, A. O.; Amlani, I.; Toth, G.; Lent, C. S.; Bernstein, G. H.; Snider, G. L. *Appl. Phys. Lett.* **1999**, 74, 2875. (b) Lent, C. S. *Science*, **2000**, 288, 1597. (c) Bandyopadhyay, S., dEbate response: what can replace the transistor paradigm?, *Science* **2000**, 288, 29 June.

150. Yu, J.-Y., Chung, S.-W., Heath, J. R. *J. Phys. Chem. B.* **2000**, 104, 11864.

151. Ausman, K. D.; O'Connell, J. J.; Boul, P.; Ericson, L. M.; Casavant, M. J.; Walters, D. W.; Huffman, C.; Saini, R.; Wang, Y.; Haroz, E.; Billups, E. W.; Smalley, R. E. *Proceedings Of XVth International Winterschool on electronic properties of novel materials Euroconference Kirchberg*, Tirol, Austria, **2000**.

152. Bahr, J. L.; Mickelson, E. T.; Bronikowski, J. J.; Smalley, R. E.; Tour, J. M. *Chem. Commun.* **2001**, 2001, 193.

153. Mickelson, E. T.; Chiang, I. W.; Zimmerman, J. L.; Boul, P. J.; Lozano, J.; Liu, J.; Smalley, R. E.; Hauge, R. H.; Margrave, J. L. *J. Phys. Chem. B.* **1999**, 103, 4318.

154. Chen, J.; Rao, A. M.; Lyuksyutov, S.; Itkis, M. E.; Hamon, M. A.; Hu, H.; Cohn, R. W.; Eklund, P. C.; Colbert, D. T.; Smalley, R. E.; Haddon, R. C. *J. Phys. Chem. B.* **2001**, 105, 2525.

155. Bahr, J. L.; Yang, J.; Kosynkin, D. V.; Bronikowski, M. J.; Smalley, R. E.; Tour, J. M. *J. Am. Chem. Soc.* **2001**, 123, 6536.

156. Bahr, J. L., Tour, J. M. *Chem. Mater.* **2001**, 13, 3823.

157. Bahr, J. L.; Tour, J. M. *J. Mater. Chem.* **2002**, *12*, 1952-1958.

158. O'Connell, M. J.; Boul, P.; Ericson, L. M.; Huffman, C.; Wang, Y.; Haroz, E.; Kuper, C.; Tour, J.; Ausman, K. D.; Smalley, R. E. *Chem. Phys. Lett.* **2001**, 342, 265.

159. Pease, A. R.; Jeppesen, J. O.; Stoddart, J. F.; Luo, Y.; Collier, C. P.; Heath, J. R. *Acc. Chem. Res.* **2001**, 34, 433.

160. Collins, P. G., Arnold, M. S., Avouris, P. *Science* **2001**, 292, 706.

161. Allara, D. L.; Reed, M. A.; Tour, J. M. **2001**, unpublished data.

162. Nackashi, D. P.; Franzon, P. D. *Proc. SPIE* **2001**, *4236*, 80.

163. (a) Goto, E. *Proc. IRE* **1959**, *47*, 1304. (b) Ellenbogen, J. C.; Love, J. C. *Proc. IEEE* **2000**, *88*, 386.

164. This chapter was taken in large part from the thesis of Summer M. Husband, Rice University, **2002**. Work that will appear in the Thesis of Christopher Husband, Rice University, expected **2003**, will more specifically address the mortal training concepts.

165. Allara, D. L.; Franzon, P. D.; Reed, M. A.; Tour, J. M. unpublished data, **2001**.

166. Allara, D. L.; Reed, M. A.; Tour, J. M. presented at DARPA research conference Santa Fe, New Mexico, **2001**.
167. Chen, J., Wang, W., Reed, M. A., Rawlett, A. M., Tour, J. M. *Mater. Res. Soc. Symp. Proc.* **2001** (Molecular Electronics), 5.
168. Vladimirescu, A., *The Spice Book*, **1994**, New York, NY; Wiley.
169. Mathews, R. H., et. al., *Proc. IEEE*, **1999** (87), 596.
170. Bertsekas, D. P.; Tsitsiklis, J. N. *Neuro-Dynamic Programming*, **1996**, Belmont, Mass., Athena Scientific.
171. Goldberg, D. E., *Genetic Algorithms in Search, Optimization, and Machine Learning*, **1989**, Reading, MA; Addison Wesley.
172. Amlani, I., Rawlett, A. M., Nagahara, L. A., Tsui, R. K. *Appl. Phys. Lett.* **2002**, 80 (15), 2761.
173. Pham, D. T., Karaboga, D., *Intelligent Optimisation Techniques: Genetic Algorithms, Tabu Search, Simulated Annealing and Neural Networks* **2000**; London; Springer.
174. Husband, C. P., *Thesis Proposal: Mortally Training Nanocells*, Rice University, **2002**.
175. West, D. B., *Introduction to Graph Theory* **1996**, Upper Sadle River, NJ; Prentice Hall.
176. Palmer, E. M. *Graphical Evolution: An Introduction to the Theory of Random Graphs* **1985**, New York, NY; Wiley Intersciences.
177. Harms, D. H., Kraetzl, M., Colbourn, C. J., Devitt, J. S., *Network Reliability: Experiments with a Symbolic Algebra Environment* **1995**, New York, NY; CRC Press.
178. Dally, W. J., Poulton, J. W., *Digital Systems Engineering* **1998**, New York, NY; Cambridge University Press.
179. Nackashi, D. P., Franzon, P. D., *Proc. SPIE* **2001**, 4236, 80.
180. Sutton, R. S., Barto, A. G., *Reinforcement Learning* **1998**, Cambridge, MA; MIT Press.
181. Cook, W. J.; Cunningham, W. H., Pulleybank, W. R., Schrijver, A., *Combinatorial Optimization* **1998**, New York, NY; Wiley-Interscience.

Index

About the Author

James M. Tour, a synthetic organic chemist, received his Bachelor of Science degree in chemistry from Syracuse University, his Ph.D. in synthetic organic and organometallic chemistry from Purdue University, and postdoctoral training in synthetic organic chemistry at the University of Wisconsin and Stanford University. After spending 11 years on the faculty of the Department of Chemistry and Biochemistry at the University of South Carolina, he joined the Center for Nanoscale Science and Technology at Rice University in 1999 where he is presently the Chao Professor of Chemistry, Professor of Computer Science, and Professor of Mechanical Engineering and Materials Science. Tour's scientific research areas include molecular electronics, chemical self-assembly, conjugated oligomers, electroactive polymers, combinatorial routes to precise oligomers, polymeric sensors, flame retarding polymer additives, carbon nanotube modification and composite formation, synthesis of molecular motors and nanotrucks, use of the NanoKids concept for K-12 education in nanoscale science, and methods for retarding chemical terrorist attacks. Tour is a co-founder of Molecular Electronics Corp. He has served as a visiting scholar at Harvard University, on the *Chemical Reviews* Editorial Advisory Board, the National Defense Science Study Group, the Governor's Mathematics and Science Advisory Board, in addition to numerous other professional committees and panels. Tour has won several national awards including the National Science Foundation Presidential Young Investigator Award in Polymer Chemistry and the Office of Naval Research Young Investigator Award in Polymer Chemistry. Tour's outside interests include enjoying the wife of his youth and their four children, and studying the ancient saying and writings of selected wise individuals, specifically Moses through John. For more, see: http://www.jmtour.com